Lecture Notes on Data Engineering and Communications Technologies

Volume 23

Series editor

Fatos Xhafa, Technical University of Catalonia, Barcelona, Spain
e-mail: fatos@cs.upc.edu

The aim of the book series is to present cutting edge engineering approaches to data technologies and communications. It publishes latest advances on the engineering task of building and deploying distributed, scalable and reliable data infrastructures and communication systems.

The series has a prominent applied focus on data technologies and communications with aim to promote the bridging from fundamental research on data science and networking to data engineering and communications that lead to industry products, business knowledge and standardisation.

More information about this series at http://www.springer.com/series/15362

Fatos Xhafa · Leonard Barolli
Michal Greguš
Editors

Advances in Intelligent Networking and Collaborative Systems

The 10th International Conference on Intelligent Networking and Collaborative Systems (INCoS-2018)

 Springer

Editors
Fatos Xhafa
Universitat Politècnica de Catalunya
Barcelona, Spain

Michal Greguš
Faculty of Management
Comenius University in Bratislava
Bratislava, Slovakia

Leonard Barolli
Department of Information
    and Communication Engineering, Faculty
    of Information Engineering
Fukuoka Institute of Technology
Fukuoka-shi, Fukuoka-ken, Japan

ISSN 2367-4512 ISSN 2367-4520 (electronic)
Lecture Notes on Data Engineering and Communications Technologies
ISBN 978-3-319-98556-5 ISBN 978-3-319-98557-2 (eBook)
https://doi.org/10.1007/978-3-319-98557-2

Library of Congress Control Number: 2018950535

This Springer imprint is published by the registered company Springer Nature Switzerland AG
The registered company address is: Gewerbestrasse 11, 6330 Cham, Switzerland

Welcome Message from the INCoS-2018 Organizing Committee

Welcome to the 10th International Conference on Intelligent Networking and Collaborative Systems (INCoS-2018), which is held from 5th to 7th September 2018 at Comenius University in Bratislava, Slovakia.

INCoS is a multi-disciplinary conference that covers the latest advances in intelligent social networks and collaborative systems, intelligent networking systems, mobile collaborative systems, secure intelligent cloud systems, etc. Additionally, the conference addresses security, authentication, privacy, data trust, and user trustworthiness behavior, which have become crosscutting features of intelligent collaborative systems. With the fast development of the Internet, we are experiencing a shift from the traditional sharing of information and applications as the main purpose of the networking systems to an emergent paradigm, which locates people at the very center of networks and exploits the value of people's connections, relations, and collaboration. Social networks are playing a major role as one of the drivers in the dynamics and structure of intelligent networking and collaborative systems.

Virtual campuses, virtual communities, and organizations strongly leverage intelligent networking and collaborative systems by a great variety of formal and informal electronic relations, such as business-to-business, peer-to-peer, and many types of online collaborative learning interactions, including the virtual campuses and e-Learning and MOOCs systems. Altogether, this has resulted in entangled systems that need to be managed efficiently and in an autonomous way. In addition, the conjunction of the latest and powerful technologies based on cloud, mobile and wireless infrastructures is currently bringing new dimensions of collaborative and networking applications a great deal by facing new issues and challenges. INCoS-2018 paid a special attention to cloud computing services, storage, security, and privacy, data mining, machine learning and collective intelligence, cooperative communication and cognitive systems, management of virtual organization and enterprises, big data analytics, e-learning, virtual campuses and MOOCs, among others.

In all, the principal aim of this conference is to stimulate research that will lead to the creation of responsive environments for networking and, at longer term, the development of adaptive, secure, mobile, and intuitive intelligent systems for collaborative work and learning. INCoS-2018 addresses a large number of themes and focuses on the following research tracks:

- Data mining, machine learning and collective intelligence
- Fuzzy systems and knowledge management
- Grid and P2P distributed infrastructure for intelligent networking and collaborative systems
- Nature's inspired parallel collaborative systems
- Security, organization, management and autonomic computing for intelligent networking and collaborative systems
- Software engineering, semantics and ontologies for intelligent networking and collaborative systems
- Wireless and sensor systems for intelligent networking and collaborative systems
- Service-based systems for enterprise activities planning and management
- Next-generation secure network protocols and components
- Big data analytics for learning, networking and collaborative systems
- Cloud computing: services, storage, security and privacy
- Intelligent collaborative systems for work and learning, virtual organization and campuses
- Social networking and collaborative systems
- Intelligent and collaborative systems for c-health

As in all previous editions, INCoS-2018 counted on with the support and collaboration of a large and internationally recognized TPC covering all main themes of the conference. This conference edition comprises 28 regular papers. Additionally, five workshops were organized in conjunction with the conference, in which 21 workshop papers were accepted.

The successful organization of the conference is achieved thanks to the great collaboration and hard work of many people and conference supporters. First and foremost, we would like to thank all the authors for their continued support to the conference by submitting their research work to the conference and for their presentations and discussions during the conference days. We would like to thank TPC members and external reviewers for their work by carefully evaluating the submissions and providing constructive feedback to authors. We would like to thank the track chairs for their work on setting up the tracks and the respective TPCs and also for actively promoting the conference and their tracks. We would like to appreciate the work of PC Co-chairs and Workshops Co-chairs for the successful organization of workshops in conjunction with main conference.

We would like to acknowledge the excellent work and support by the International Advisory Committee. Our gratitude and acknowledgment for the conference keynote speakers, Peter Hellinckx, University of Antwerp, Belgium;

Juraj Laifr, SAP Education, Slovakia; and Peter Linhardt, FineSoft Ltd., Slovakia, for their interesting and inspiring keynote speeches.

We greatly appreciate the support by Web Administrators Donald Elmazi, Yi Liu, Miralda Cuka, and Kosuke Ozera, Fukuoka Institute of Technology (FIT), Japan.

We would like to give special thanks to the members of the local organizing committee from Comenius University in Bratislava, Slovakia, for excellent arrangements for the conference.

We are very grateful to Springer as well as several academic institutions for their endorsement and assistance.

Finally, we hope that you will find these proceedings to be a valuable resource in your professional, research, and educational activities!

General Co-chairs

Michal Greguš   Comenius University in Bratislava, Slovakia
Leonard Barolli   Fukuoka Institute of Technology, Japan

Program Co-chairs

Peter Balco   Comenius University in Bratislava, Slovakia
Flora Amato   University of Naples, Italy

Message from the INCoS-2018 Workshops Chairs

Welcome to the workshops of the 10th International Conference on Intelligent Networking and Collaborative Systems (INCoS-2018), which is held from 5th to 7th September 2018 at Comenius University in Bratislava, Slovakia.

In this edition of the conference, there are held five workshops, which complemented the INCoS main themes with specific themes and research issues and challenges, as follows.

1. 10th International Workshop on Information Network Design (WIND-2018)
2. 8th International Workshop on Adaptive Learning via Interactive, Collaborative and Emotional approaches (ALICE-2018)
3. 6th International Workshop on Frontiers in Intelligent Networking and Collaborative Systems (FINCoS-2018)
4. 4th International Workshop on Theory, Algorithms and Applications of Big Data Science (BDS-2018)
5. 1st International Workshop on Machine Learning in Intelligent and Collaborative Systems (MaLICS-2018)

We would like to thank the workshop organizers for their great efforts and hard work in proposing the workshop, selecting the papers, the interesting programs and for the arrangements of the workshop during the conference days. We are grateful to the INCoS-2018 Conference Chairs for inviting us to be the workshops co-chairs of the conference.

We hope you will enjoy the workshops programs!

INCoS-2018 Workshops Co-chairs

Natalia Kryvinska   Comenius University in Bratislava, Slovakia
Hiroyoshi Miwa   Kwansei Gakuin University, Japan
Omar Hussain   UNSW Canberra, Australia

INCoS-2018 Organizing Committee

Honorary Chair

Makoto Takizawa   Hosei University, Japan

General Co-chairs

Michal Greguš   Comenius University in Bratislava, Slovakia
Leonard Barolli   Fukuoka Institute of Technology, Japan

Program Co-chairs

Peter Balco   Comenius University in Bratislava, Slovakia
Flora Amato   University of Naples, Italy

Workshops Co-chairs

Natalia Kryvinska   Comenius University in Bratislava, Slovakia
Hiroyoshi Miwa   Kwansei Gakuin University, Japan
Omar Hussain   UNSW Canberra, Australia

International Advisory Committee

Joseph Tan   McMaster University, Canada
Vincenzo Loia   University of Salerno, Italy
Masato Tsuru   Kyushu Institute of Technology, Japan
Christine Strauss   University of Vienna, Austria
Albert Zomaya   University of Sydney, Australia
Amélia Ferreira da Silva   CEOS.PP, Instituto Politécnico do Porto,
                                     Portugal
Ana Azevedo   CEOS.PP, Instituto Politécnico do Porto,
                                     Portugal
Manuel Moreira da Silva   CEOS.PP, Instituto Politécnico do Porto,
                                     Portugal

International Liaison Co-chairs

Pavel Kromer   Technical University of Ostrava, Czech Republic
Santi Caballé   Open University of Catalonia, Spain
Kin Fun Li   University of Victoria, Canada

Award Co-chairs

Marek Ogiela   AGH University of Science and Technology,
                                    Poland
Fang-Yie Leu   Tunghai University, Taiwan

Web Administrators

Donald Elmazi   Fukuoka Institute of Technology, Japan
Yi Liu   Fukuoka Institute of Technology, Japan
Miralda Cuka   Fukuoka Institute of Technology, Japan
Kosuke Ozera   Fukuoka Institute of Technology, Japan

Local Arrangement Co-chairs

Peter Balco   Comenius University in Bratislava, Slovakia
Martina Drahošová   Comenius University in Bratislava, Slovakia

Steering Committee

Fatos Xhafa   Technical University of Catalonia, Spain
Leonard Barolli   Fukuoka Institute of Technology, Japan

Track 1: Data Mining, Machine Learning and Collective Intelligence

Track Co-chairs

Carson K. Leung   University of Manitoba, Canada
Thomas Lenhard   Comenius University in Bratislava, Slovakia

TPC Members

Alfredo Cuzzocrea   University of Trieste, Italy

Fan Jiang   University of Northern British Columbia, Canada
Wookey Lee   Inha University, Korea
Oluwafemi A. Sarumi   Federal University of Technology, Akure,
                                     Nigeria
Syed K. Tanbeer   University of Manitoba, Canada
Tomas Vinar   Comenius University in Bratislava, Slovakia
Kin Fun Li   University of Victoria, Canada

Track 2: Fuzzy Systems and Knowledge Management

Track Co-chairs

Marek Ogiela   AGH University of Science and Technology,
                                     Poland
Morteza Saberi   UNSW Canberra, Australia
Chang Choi   Chosun University, Korea

TPC Members

Inseop Na   Chosun University, Korea
Christian Esopsito   University of Napoli "Federico II", Italy
Xin Su   Hohai University, China
Makoto Ikeda   Fukuoka Institute of Technology, Japan
Evjola Spaho   Polytechnic University of Tirana, Albania
Lidia Ogiela   Pedagogic University of Cracow, Poland

Track 3: Grid and P2P Distributed Infrastructure
for Intelligent Networking and Collaborative Systems

Track Co-chairs

Aneta Poniszewska-Maranda   Lodz University of Technology, Poland
Michal Gregus   Comenius University in Bratislava, Slovakia

TPC Members

Remy Dupas   Université de Bordeaux, France
Jose Luis Vazquez-Poletti   Universidad Complutense de Madrid, Spain
Dominikus Heckmann   Ostbayerische Technische Hochschule
                                     Amberg-Weiden, Germany

Serhat Ozekes   Uskudar University, Turkey
Witold Maranda   Lodz University of Technology, Poland
Mohammad Shojafar   University of Padua, Italy

Track 4: Nature's Inspired Parallel Collaborative Systems

Track Co-chairs

Mohammad Shojafar   University of Rome, Italy
Zahra Pooranian   University of Padua, Italy

TPC Members

Jemal Abawajy   Deakin University, Australia
Shahaboddin Shamshirband   Norwegian University of Science
                                       and Technology, Norway
Ali DehghanTanha   Sheffield University, UK
Tooska Dargahi   Salford University, Manchester, UK
Hossein Soleimani   University of Padua, Italy
Mehdi Sookhak   Arizona State University, USA
Tohid Alizadeh   Nazarbayev University, Kazakhstan
Mohammad M. Tajiki   Tarbiat Modares University, Iran
Ibrahim Al-Shourbaji   University of Jazan, Saudi Arabia

Track 5: Security, Organization, Management and Autonomic Computing for Intelligent Networking and Collaborative Systems

Track Co-chairs

Jungwoo Ryoo   Pennsylvania State University, USA
Simon Tjoa   St. Pölten University of Applied Sciences,
                                       Austria

TPC Members

Nikolaj Goranin   Vilnius Gediminas Technical University,
                                       Lithuania
Kenneth Karlsson   Lapland University of Applied Sciences, Finland

Peter Kieseberg	SBA Research, Austria
Hyoungshick Kim	Sungkyunkwan University, Korea
Hae Young Lee	DuDu IT, Korea
Moussa Ouedraogo	Wavestone, Luxembourg
Sebastian Schrittwieser	St. Pölten University of Applied Sciences, Austria
Syed Rizvi	Pennsylvania State University, USA

Track 6: Software Engineering, Semantics and Ontologies for Intelligent Networking and Collaborative Systems

Track Co-chairs

Kai Jander	University of Hamburg, Germany
Alexander Pokahr	Helmut Schmidt University/University of the Bundeswehr, Hamburg, Germany

TPC Members

Santi Caballé	Universitat Oberta de Catalunya, Spain
Ghassan Beydoun	University of Technology Sydney, Australia
Massimo Ficco	Università degli Studi della Campania Luigi Vanvitelli, Italy
Dirk Bade	University of Hamburg, Germany
Julian Kalinowski	University of Hamburg, Germany
Francisco García Peñalvo	Universidad de Salamanca, Spain
Antoni Pérez Navarro	Universitat Oberta de Catalunya, Spain
Nestor Mora Nuñez	Universidad de Cádiz, Spain
David Bañeres	Universitat Oberta de Catalunya, Spain
Martin Gaedke	Technische Universität Chemnitz, Germany
Akihiro Fujihara	Chiba Institute of Technology, Japan
Jordi Conesa Caralt	Universitat Oberta de Catalunya, Spain

Track 7: Wireless and Sensor Systems for Intelligent Networking and Collaborative Systems

Track Co-chairs

Do van Thanh	Telenor & Norwegian University of Science & Technology, Norway
Salem Lepaja	AAB College in Prishtina, Kosovo

TPC Members

Boning Feng	Oslo Metropolitan University, Norway
Van Thuan Do	Wolffia AS, Norway
Niels Jacot	Wolffia AS, Finland
Nor Shahniza Kamal Bashad	Universiti Teknologi MARA, Malaysia
Lule Ahmedi	University of Prishtina, Kosovo
Arianit Maraj	Telecom of Kosovo and AAB College, Kosovo
Nysret Musliu	Vienna University of Technology, Austria
Astrit Ademaj	TTTech Computertechnik AG, Austria

Track 8: Service-Based Systems for Enterprise Activities Planning and Management

Track Co-chairs

Corinna Engelhardt-Nowitzki	University of Applied Sciences, Austria
Natalia Kryvinska	Comenius University in Bratislava, Slovakia

TPC Members

Maria Bohdalova	Comenius University in Bratislava, Slovakia
Ivan Demydov	Lviv Polytechnic National University, Ukraine
Jozef Juhar	Technical University of Košice, Slovakia
Nor Shahniza Kamal Bashah	Universiti Teknologi MARA, Malaysia
Eric Pardede	La Trobe University, Australia
Francesco Moscato	University of Campania, Italy
Tomoya Enokido	Rissho University, Japan

Track 9: Next-Generation Secure Network Protocols and Components

Track Co-chairs

Xu An Wang	Engineering University of CAPF, China
Mingwu Zhang	Hubei University of Technology, China

TPC Members

Ximeng Liu	Singapore Management University, Singapore
Hui Tian	Huaqiao University, China
Xuefeng Liu	Xidian University, China
Yinbin Miao	Xidian University, China
Huaqun Wang	Nanjing University of Posts and Telecommunications, China
Jie Chen	East China Normal University, China
Qi Jiang	Xidian University, China
Yong Yu	Shaanxi Normal University, China
Debiao He	Wuhan University, China
Jia Yu	Qingdao University of Science and Technology, China
Jinbo Xiong	Fujian Normal University, China
Wei Ren	China University of Geosciences, China
Fushan Wei	Information Engineering University, China
Zhiquan Liu	Jinan University, China
An Wang	Beijing Institute of Technology, China
Lei Zhang	East China Normal University, China

Track 10: Big Data Analytics for Learning, Networking and Collaborative Systems

Track Co-chairs

Santi Caballe Llobet	Universitat Oberta de Catalunya Spain
Francesco Orciuoli	University of Salerno, Italy

TPC Members

Jordi Conesa	Open University of Catalonia, Spain
Michalis Feidakis	Piraeus University of Applied Sciences, Greece
Nicola Capuano	University of Salerno, Italy
Nestor Mora	Universitat Oberta de Catalunya, Spain
Jorge Moneo	University of San Jorge, Spain
David Gañán	Universitat Oberta de Catalunya, Spain
Arun Kumar	Vellore Institute of Technology, India
Antonio Sarasa	Universidad Complutense de Madrid, Spain
Giuseppe Fenza	University of Salerno, Italy
Juan Gomez-Romero	Universidad de Granada, Spain
Giuseppe D'Aniello	University of Salerno, Italy

Fernando Bobillo	University of Zaragoza, Spain
Diego Dermeval	Federal University of Alagoas, Brasil
Angelo Gaeta	University of Salerno, Italy
Carmen De Maio	University of Salerno, Italy
Maria Cristina Gallo	University of Salerno, Italy

Track 11: Cloud Computing: Services, Storage, Security and Privacy

Track Co-chairs

| Javid Taheri | Karlstad University, Sweden |
| Shuiguang Deng | Zhejiang University, China |

TPC Members

Daning Hu	University of Zurich, Switzerland
Xuyun(Sean) Zhang	University of Auckland, New Zealand
Qinglin Zhao	Macau University of Science and Technology, Macau, China
Shangguang Wang	Beijing University of Posts and Telecommunications, China
Yutao Ma	Wuhan University, China
Bernabé Dorronsoro Díaz	University of Cadiz, Spain
Hamid Arabnejad	Dublin City University (DCU), Ireland
Luis Veiga	Universidade de Lisboa/INESC-ID Lisboa, Portugal
Bahman Javadi	Western Sydney University, Australia
MohamadReza Hoseiny	University of Sydney, Australia
Ejaz Ahmed	University of Malaya, Malaysia
Assad Abbas	COMSATS Institute of Information Technology, Pakistan
Nikolaos Tziritas	Chinese Academy of Sciences, China
Osman Khalid	COMSATS Institute of Information Technology, Pakistan
Kashif Bilal	Qatar University, Qatar

Track 12: Intelligent Collaborative Systems for Work and Learning, Virtual Organization and Campuses

Track Co-chairs

Nikolay Kazantsev   National Research University Higher School
                                 of Economics, Russia
Monika Davidekova   Comenius University in Bratislava, Slovakia

TPC Members

Ondrej Krehel   IANS Boston, USA
Elena Subertova   University of Economics in Bratislava, Slovakia
Milos Bodis   University of Economics in Bratislava, Slovakia
Eva Litavcova   University of Presov, Slovakia
Miron Pavlus   University of Presov, Slovakia
Martin Noga   University of Zilina, Slovakia

Track 13: Social Networking and Collaborative Systems

Track Co-chairs

Nicola Capuano   University of Salerno, Italy
Dusan Soltes   Comenius University in Bratislava, Slovakia

TPC Members

Dagmar Cagáňová   Slovak University of Technology in Bratislava,
                                 Slovakia
Iveta Šimberová   Brno University of Technology, Czech Republic
Stefan Janovjak   Comenius University in Bratislava, Slovakia
Zdeňka Konečná   Brno University of Technology, Czech Republic
Daniel Olejár   Comenius University in Bratislava, Slovakia
Andrej Ferko   Comenius University in Bratislava, Slovakia
Francisco Chiclana   De Montfort University, UK
Jordi Conesa   Open University of Catalonia, Spain
Thanasis Daradoumis   University of the Aegean, Greece
Cherie Ding   Ryerson University, Canada
Angelo Gaeta   University of Salerno, Italy
Agathe Merceron   Beuth University of Applied Sciences Berlin,
                                 Germany

Krassen Stefanov	Sofia University "St. Kliment Ohridski", Bulgaria
Marco Temperini	Sapienza University of Rome, Italy
Alberto Trombetta	University of Insubria, Italy
Jian Yu	Auckland University of Technology, New Zealand

Track 14: Intelligent and Collaborative Systems for e-Health

Track Co-chairs

| Massimo Esposito | Institute for High Performance Computing and Networking, National Research Council of Italy, Italy |
| Giovanni Luca Masala | University of Plymouth, UK |

TPC Members

Stavros Shiaeles	University of Plymouth, UK
Liz Stuart	University of Plymouth, UK
Enrico Grosso	Università di Sassari, Italy
Marco Palomino	University of Plymouth, UK
Bogdan Ghita	University of Plymouth, UK
Marco Pota	Institute for High Performance Computing and Networking, National Research Council of Italy, Italy
Aniello Minutolo	Institute for High Performance Computing and Networking, National Research Council of Italy, Italy
Raffaele Guarasci	Institute for High Performance Computing and Networking, National Research Council of Italy, Italy
Fiammetta Marulli	Institute for High Performance Computing and Networking, National Research Council of Italy, Italy
Emanuele Damiano	Institute for High Performance Computing and Networking, National Research Council of Italy, Italy

Track 15: Low Power Wan: Ad Hoc Networks, Energy Harvesting

Track Co-chairs

Miroslav Voznak	VSB-Technical University of Ostrava, Czech Republic
Lukas Vojtech	Czech Technical University in Prague, Czech Republic

TPC Members

Dinh-Thuan Do	Ton Duc Thang University, Vietnam
Floriano De Rango	University of Calabria, Italy
Homero Toral-Cruz	University of Quintana Roo, Mexico
Remigiusz Baran	Kielce University of Technology, Poland
Mindaugas Kurmis	Klaipeda State University of Applied Sciences, Lithuania
Radek Martinek	VSB-Technical University of Ostrava, Czech Republic
Mauro Tropea	University of Calabria, Italy
Gokhan Ilk	Ankara University, Turkey

INCoS-2018 Reviewers

Alizadeh Tohid
Amato Flora
Barolli Admir
Barolli Leonard
Caballé Santi
Capuano Nicola
Chen Xiaofeng
Cui Baojiang
Daradoumis Thanasis
Elmazi Donald
Enokido Tomoya
Esposito Christian
Fenza Giuseppe
Ficco Massimo
Fiore Ugo
Fujihara Akihiro
Fun Li Kin

Gañán David
Hsing-Chung Chen
Hussain Farookh
Hussain Omar
Ikeda Makoto
Joshua Hae-Duck
Kolici Vladi
Köppen Mario
Koyama Akio
Kromer Pavel
Kryvinska Natalia
Kulla Elis
Leu Fang-Yie
Li Yiu
Loia Vincenzo
Ma Kun
Mangione Giuseppina Rita

Matsuo Keita
Messina Fabrizio
Miguel Jorge
Miwa Hiroyoshi
Natwichai, Juggapong
Nadeem Javaid
Nalepa Jakub
Nishino Hiroaki
Nowakowa Jana
Ogiela Lidia
Ogiela Marek
Palmieri Francesco
Pardede Eric
Poniszewska-Maranda Aneta
Rahayu Wenny
Rawat Danda
Shibata Masahiro

Shibata Yoshitaka
Snasel Vaclav
Spaho Evjola
Suganuma Takuo
Sugita Kaoru
Takizawa Makoto
Terzo Olivier
Tsukamoto Kazuya
Tsuru Masato
Uchida Masato
Uchida Noriki
Wang Xu An
Woungang Isaac
Younas Mohammad
Zhang Mingwu
Zomaya Albert

Welcome Message from WIND-2018 Workshop Organizers

Welcome to the 10th International Workshop on Information Network Design (WIND-2018), which is held in conjunction with the 9th International Conference on Intelligent Networking and Collaborative Systems (INCoS-2018), which is held from 5th to 7th September 2018 at Comenius University in Bratislava, Slovakia.

Nowadays, the Internet is playing a role in social and economical infrastructure and is expected to support not only comfortable communication and information dissemination but also any kind of intelligent and collaborative activities in a dependable manner. However, the explosive growth of its usage with diversifying the communication technologies and the service applications makes it difficult to manage efficient sharing of the Internet. In addition, an inconsistency between Internet technologies and the human society forces a complex and unpredictable tension among end users, applications, and Internet Service Providers (ISPs).

It is thought, therefore, that the Internet is approaching a turning point and there might be the need for rethinking and redesigning the entire system composed of the human society, nature, and the Internet. To solve the problems across multiple layers on a large-scale and complex system and to design the entire system of systems toward future information networks for human/social orchestration, a new tide of multi-perspective and multi-disciplinary research is essential. It will involve not only the network engineering (network routing, mobile and wireless networks, network measurement and management, high-speed networks, etc.) and the net-worked applications (robotics, distributed computing, human–computer interactions, Kansei information processing, etc.), but the network science (providing new tools to understand and control the huge-scale complex systems based on theories, e.g., graph theory, game theory, information theory, learning theory, statistical physics) and the social science (enabling safe, secure, and human-centric application principles and business models).

The information network design workshop aims at exploring ongoing efforts in the theory and application on a wide variety of research fields related to the design of information networks and resource sharing in the networks. The workshop provides an opportunity for academic/industry researchers and professionals to share, exchange, and review recent advances on information network design research. Original contribution describing recent modeling, analysis, and experiment on network design research with particular, but not exclusive, regard to:

- Large-scale and/or complex networks
- Cross-layered networks
- Overlay and/or P2P networks
- Sensor and/or mobile ad hoc networks
- Delay/disruption tolerant networks
- Social networks
- Applications on networks
- Fundamental theories for network design

We would like to thank the organizing committee of INCoS-2018 for giving us the opportunity to organize the workshop. We also like to thank our program committee members and referees and, of course, all authors of the workshop for submitting their research works and for their participation.

We wish all participants and contributors to spend an event with high research impact, interesting discussions, and exchange of research ideas, to pave future research cooperations.

WIND-2018 Workshop Co-chairs

Masaki Aida	Tokyo Metropolitan University, Japan
Mario Koeppen	Kyushu Institute of Technology, Japan
Hiroyoshi Miwa	Kwansei Gakuin University, Japan
Masato Tsuru	Kyushu Institute of Technology, Japan
Masato Uchida	Waseda University, Japan

WIND-2018 Organizing Committee

WIND-2018 Workshop Co-chairs

Masaki Aida	Tokyo Metropolitan University, Japan
Mario Koeppen	Kyushu Institute of Technology, Japan
Hiroyoshi Miwa	Kwansei Gakuin University, Japan
Masato Tsuru	Kyushu Institute of Technology, Japan
Masato Uchida	Waseda University, Japan

Program Committee

Masaki Aida	Tokyo Metropolitan University, Japan
Yoshiaki Hori	Saga University, Japan
Hideaki Iiduka	Meiji University, Japan
Mario Koeppen	Kyushu Institute of Technology, Japan
Kenichi Kourai	Kyushu Institute of Technology, Japan
Hiroyoshi Miwa	Kwansei Gakuin University, Japan
Kei Ohnishi	Kyushu Institute of Technology, Japan
Masahiro Sasabe	Nara Institute of Science and Technology, Japan
Kazuya Tsukamoto	Kyushu Institute of Technology, Japan
Masato Tsuru	Kyushu Institute of Technology, Japan
Masato Uchida	Chiba Institute of Technology, Japan
Neng-Fa Zhou	City University of New York, USA

Welcome Message from ALICE-2018 Workshop Organizers

Welcome to the 8th International Workshop on Adaptive Learning via Interactive, Collaborative and Emotional approaches (ALICE-2018), which is held in conjunction with the 10th International Conference on Intelligent Networking and Collaborative Systems (INCoS-2018), which is held from 5th to 7th September 2018 at Comenius University in Bratislava, Slovakia. The 8th International Workshop ALICE-2018 aims at providing a forum for innovations in adaptive technologies for e-learning especially designed to improve the engagement of students in learning experiences. Works that combine adaptive techniques with approaches based on gamification, affective computing, social learning, storytelling, interactive video, and new forms of assessment are welcome as well as works describing approaches based on educational data mining, learning, and academic analytics aimed at improving learners' motivation and teachers' experience also taking into account the aspects related to multimedia and security. Special emphasis is given to approaches based on P2P, parallel, grid, cloud, and Internet computing as well as on computational intelligence and knowledge-based technologies.

The workshop covers topics of interest, within the scope of adaptive e-learning:

- Gamification and serious games for education
- Educational use of digital storytelling
- Motivation, metacognition, and affective aspects of learning
- Collaborative and social learning
- Collaborative learning for massive open online courses (MOOCs)
- New approaches to formative assessment
- Learning, academic and business analytics
- Multimedia systems for e-learning
- Conversational agents and learning analytics for collaborative learning
- Security aspects of e-learning
- Cloud, cluster, and P2P computing for e-Learning
- Knowledge-based models and technologies for e-learning
- Intelligent tutoring systems

We would like to thank all authors of the workshop for submitting their research works and their participation. We would like to express our appreciation to the reviewers for their timely review and constructive feedback to authors.

We are looking forward to meet you again in the forthcoming editions of the workshop.

ALICE-2018 Workshop Organizers

Santi Caballé   Open University of Catalonia, Spain
Nicola Capuano   University of Salerno, Italy

ALICE-2018 Organizing Committee

Workshop Organizers

Santi Caballé (Co-chair)   Open University of Catalonia, Spain
Nicola Capuano (Co-chair)   University of Salerno, Italy

Program Committee

Jordi Conesa   Universitat Oberta de Catalunya, Spain
Thanasis Daradoumis   University of the Aegean, Greece
Giuliana Dettori   Italian National Research Council, Italy
Sara de Freitas   Coventry University, UK
Angelo Gaeta   University of Salerno, Italy
David Gañán   Universitat Oberta de Catalunya, Spain
Isabel Guitart   Universitat Oberta de Catalunya, Spain
Jorge Miguel   Universitat Oberta de Catalunya, Spain
Giuseppina Rita Mangione   Institute of Educational Documentation,
                                   Innovation and Research, Italy
Néstor Mora   Universitat Oberta de Catalunya, Spain
Anna Pierri   University of Salerno, Italy
Modesta Pousada   Universitat Oberta de Catalunya, Spain
Antonio Sarasa   Universidad Complutense de Madrid, Spain
Marco Temperini   Sapienza University of Rome, Italy

Welcome Message from FINCoS-2018 Workshop Organizer

Welcome to the 6th International Workshop on Frontiers in Intelligent Networking and Collaborative Systems (FINCoS-2018), which is held in conjunction with the 10th International Conference on Intelligent Networking and Collaborative Systems (INCoS-2018), which is held from 5th to 7th September 2018 at Comenius University in Bratislava, Slovakia.

The FINCoS-2018 covers the latest advances in the inter-disciplinary fields of intelligent networking, social networking, collaborative systems, cloud-based systems, and business intelligence, which lead to gain competitive advantages in business and academia scenarios. The ultimate aim is to stimulate research that will lead to the creation of responsive environments for networking and, at longer term, the development of adaptive, secure, mobile, and intuitive intelligent systems for collaborative work and learning.

Industry and academic researchers, professionals, and practitioners are invited to exchange their experiences and present their ideas in this field. Specifically, the scope of FINCoS-2018 comprises research work and findings on intelligent networking, cloud and fog distributed infrastructures, security and privacy and data analysis. We would like to thank all authors of the workshop for submitting their research works and their participation. We would like to express our appreciation to the reviewers for their timely review and constructive feedback to authors.

We are looking forward to meet you again in the forthcoming editions of the workshop.

FINCoS-2018 Workshop Organizer

Leonard Barolli   Fukuoka Institute of Technology, Japan

FINCoS-2018 Organizing Committee

Workshop Organizer

Leonard Barolli   Fukuoka Institute of Technology, Japan

Program Committee

Santi Caballé   Open University of Catalonia
Xiaofeng Chen   Xidian University, China
Makoto Ikeda   Fukuoka Institute of Technology, Japan
Kin Fun Li   University of Victoria, Canada
Shengli Liu   Shanghai Jiaotong University, China
Janusz Kacpryzk   Polish Academy of Science, Poland
Hiroaki Nishino   University of Oita, Japan
Makoto Takizawa   Hosei University, Japan
David Taniar   Monash University, Australia
Xu An Wang   CAPF Engineering University, P.R. China

Welcome Message from BDS-2018 Workshop Organizers

Welcome to the 3rd International Workshop on Theory, Algorithms and Applications of Big Data Science (BDS-2018), which is held in conjunction with the 10th International Conference on Intelligent Networking and Collaborative Systems (INCoS-2018), which is held from 5th to 7th September 2018 at Comenius University in Bratislava, Slovakia.

Diverse multi-disciplinary approaches are being continuously developed and advanced to address the challenges that big data research raises. In particular, the current academic and professional environments are working to produce algorithms, theoretical advance in big data science, to enable the full utilization of its potential, and better applications.

The proposed workshop focuses on the dissemination of original contributions to discuss and explore theoretical concepts, principles, tools, techniques, and deployment models in the context of big data. Via the contribution of both academics and industry practitioners, the current approaches for the acquisition, interpretation, and assessment of relevant information will be addressed to advance the state-of-the-art big data technology.

The workshop covers the following topics:

- Contributions should focus on (but not limited to) the following topics
- Statistical and dynamical properties of big data
- Applications of machine learning for information extraction
- Hadoop and big data
- Data and text mining techniques for big data
- Novel algorithms in classification, regression, clustering, and analysis
- Distributed systems and cloud computing for big data
- Big data applications
- Theory, applications and mining of networks associated with big data
- Large-scale network data analysis
- Data reduction, feature selection, and transformation algorithms
- Data visualization
- Distributed data analysis platforms

- Scalable solutions for pattern recognition
- Stream and real-time processing of big data
- Information quality within big data
- Threat detection in big data

We would like to thank the organizing committee of INCoS-2018 for giving us the opportunity to organize the workshop and the local arrangement chairs for facilitating the workshop organization.

We are looking forward to meet you again in the forthcoming editions of the workshop.

Workshop Organisers

Marcello Trovati	Edge Hill University, UK
Mark Liptrott	Edge Hill University, UK
Jeffrey Ray	Edge Hill University, UK

BDS-2018 Organizing Committee

Workshop Organisers

Marcello Trovati	Edge Hill University, UK
Mark Liptrott	Edge Hill University, UK
Jeffrey Ray	Edge Hill University, UK

Workshop PC Members

Georgios Kontonatsios	Edge Hill University, UK
Richard Conniss	University of Derby, UK
Ovidiu Bagdasar	University of Derby, UK
Peter Larcombe	University of Derby, UK
Stelios Sotiriadis	University of Toronto, Canada
Jer Hayes	IBM Research, Dublin Lab, Ireland
Xiaolong Xu	Nanjing University of Post and Telecommunications, China
Nan Hu	Nanjing University of Post and Telecommunications, China
Tao Lin	Nanjing University of Post and Telecommunications, China

Welcome Message from MaLICS-2018 International Workshop Organizer

Welcome to the 1st International Workshop Machine Learning in Intelligent and Collaborative Systems (MaLICS-2018), which is held in conjunction with the 10th International Conference on Intelligent Networking and Collaborative Systems (INCoS-2018), which is held from 5th to 7th September 2018 at Comenius University in Bratislava, Slovakia.

The era of big data is here and now. The amount of data produced every day grows tremendously in most real-life domains, including medical imaging, genomics, text categorization, computational biology. Hence, data-driven machine learning-powered approaches are consistently gaining research attention, and they are applied in multiple fields, with intelligent and collaborative systems not being an exception. In this workshop, we strive to present current advances on novel ideas and practical aspects concerning intelligent and collaborative systems, which benefit from machine learning, and deep learning in particular. Also, we hope to identify and highlight challenges, which are being faced by research and industrial communities in the field.

This workshop covers the latest advances in machine- and deep learning-powered intelligent systems that lead to gain competitive advantages in business and academia scenarios. The ultimate aim is to stimulate research that will lead to the creation of robust intelligent and collaborative systems applicable in a variety of fields (ranging from medical image analysis to smart delivery systems).

The workshop covers the topics of:

- Deep learning and neural networks: applications, techniques, and tools
- Soft computing techniques for design of intelligent and collaborative systems
- Computer vision and image processing in intelligent and collaborative systems
- Bio-inspired algorithms for intelligent and collaborative systems
- Hybrid algorithms for intelligent and collaborative systems
- Heuristic and meta-heuristic algorithms in intelligent and collaborative systems
- Machine learning in intelligent and collaborative systems
- Advanced data analysis in intelligent and collaborative systems

- Approaches, techniques, and challenges in parallelising machine learning-powered intelligent systems
- Practical applications of intelligent and collaborative systems
- Automated design and auto-tuning of deep learning and machine learning systems
- Smart delivery systems (incl. autonomous vehicles)
- Medical image analysis in intelligent decision-support systems
- Learning systems: approaches, techniques, and tools
- Multi- and hyperspectral imaging, analysis and processing

We are looking forward to meet you again in the forthcoming editions of the workshop.

MaLICS-2018 Workshop Organizer

Jakub Nalepa   Silesian University of Technology & Future
                                Processing, Poland

INCoS-2018 Keynote Talks

IoT and CPS Two Worlds Growing Together: The Embedded Software Point of View

Peter Hellinckx

University of Antwerp, Antwerp, Belgium

Abstract. IoT and CPS are growing together. Autonomous vehicles are probably the best example. The car as a CPS will communicate or even be part of an IoT environment enriching environment knowledge and optimizing the CPS control. However, new innovations come with new challenges. In this talk, I will focus on different challenges in the distributed embedded software context. More specifically, I will talk about WCET and schedulability analysis, (real-time) resource-oriented real-time code placement/movement, and testing of emergent behavior.

SAP Leonardo, Your Digital Innovation System

Juraj Laifr

SAP Education, Bratislava, Slovakia

Abstract. Never before have there been so many promising breakthrough technologies available—and so many businesses ready to capitalize on them. SAP Leonardo delivers new capabilities in future-forward technologies, which add tremendous value to the company's digital journey. SAP Leonardo is a holistic digital innovation system that seamlessly integrates future-facing technologies and capabilities into the SAP cloud platform, using our design thinking services. This powerful portfolio enables people to rapidly innovate, scale new models, and continually redefine the business.

Centralized e-Health National Wide Concept to Reduce the Complexity of Integration

Peter Linhardt

Partner FineSoft Ltd., Kosice, Slovakia

Abstract. Implementation of e-solution in the field of health care is the must in order to achieve the efficiency, quality of care, and implementation of new methods for care provisioning. All subsystems and points of care have to communicate with EHR silos, central, and local applications, and it underlines of the importance of proper protocols, data format, and integration issues, as well the importance of proper project management for testing and delivery in real-time conditions of cloud and infrastructure. All this effort is essential in order to ensure the wide portfolio of benefits for all group of healthcare stakeholders and for delivery of the future application in e-commerce and m-commerce environment.

Contents

**The 8th International Workshop on Adaptive Learning via
Interactive, Collaborative and Emotional approaches (ALICE-2018)**

**The 6th International Workshop on Frontiers in Intelligent
Networking and Collaborative Systems (FINCoS-2018)**

The 4th International Workshop on Theory, Algorithms and Applications of Big Data Science (BDS-2018)

The 1st International International Workshop Machine Learning in Intelligent and Collaborative Systems (MaLICS-2018)

10th International Conference on International Conference on Intelligent Networking and Collaborative Systems (INCoS-2018)

Findings from a Success Factor Analysis for SaaS Usage

Dietmar Nedbal[(⊠)] and Mark Stieninger

University of Applied Sciences Upper Austria, Wehrgrabengasse 1-3,
4400 Steyr, Austria
{dietmar.nedbal,mark.stieninger}@fh-steyr.at

Abstract. Research in the area of SaaS revealed that several different aspects are critical to the usage of such cloud services. The drawback is that usually all of the identified success factors seem to be equally important. Therefore, this paper provides insights from a survey on the ranking of 18 relevant success factors according to their perceived actual performance and priority among 274 individuals. The results show that the priority of several factors like the image of the adopter, a regulatory framework, observability, trialability, and energy efficiency are perceived as sub relevant. The three factors with the highest priority are security, trustfulness, and ease of use of the service.

1 Introduction

Cloud computing is a model to provide convenient, flexible, on-demand access to a shared pool of configurable IT resources provided as services [1]. Thus, the term cloud services refers to any IT services that are provisioned from a cloud computing provider and accessed by a customer. According to the NIST definition of cloud computing [19], which is widely used in research and practice [20], the service model of cloud computing is divided into three basic categories: IaaS (Infrastructure as a Service), PaaS (Platform as a Service), and SaaS (Software as a Service). With the proliferation of cloud computing services (short: cloud services), many different success factors that are critical to a successful usage of cloud computing were identified [2–4]. We understand success factors in the context of this research as specific areas, which are critical to the successful usage from the customer perspective [3]. The difficulty is that usually all of the identified factors seem to be equally important and where to start taking actions for improvement from a management perspective is not always obvious.

Based on citation/co-citation analysis, Wang et al. found that 68% of important scientific papers regarding cloud computing published between 2004 and 2014 were non-empirical (conceptual, literature review or modeling) papers [5]. This implies that although several empirical papers in the field of cloud computing can be found in IS research [6–12], there is still a lack of empirical studies carried out in different settings (e.g. cross-cultural, cross-continent).

Motivated by the difficulty of addressing the most important factors and the lack of empirical studies, the contribution of this paper is to undertake a survey-based evaluation and ranking of common success factors for cloud services from current literature

© Springer Nature Switzerland AG 2019
F. Xhafa et al. (Eds.): INCoS 2018, LNDECT 23, pp. 3–15, 2019.
https://doi.org/10.1007/978-3-319-98557-2_1

among individuals. Specifically, aiming at the use by end users (not administrators, or platform/software developers), the focus is laid on cloud services provided as SaaS.

The remainder of the paper is structured as follows. Section 2 presents the methodology comprising the identification of success factors, the method for data collection, a description of the sample, and the method of success factor analysis. In Sect. 3, we present the results of the empirical study. This is followed by a discussion (Sect. 4) and conclusion in Sect. 5.

2   Methodology

We considered three different methods for the evaluation: Analytic Hierarchy Process (AHP), a ranking-based Delphi study, and success factor analysis. AHP was not used due to the excess of resulting questions (pairwise comparison of alternatives). Delphi studies are also widely used and accepted for gathering data from respondents within their domain of expertise. A panel could have been gathered and identified as well, but consensus-building by using several iterations was not focus of this research. The success factor analysis was finally chosen, as it allows arranging several factors according to their perceived importance.

2.1   Success Factor Analysis

The success factor analysis is an established method in IS research for the evaluation of a companies' success [13], which has already been adapted for example to the field of knowledge management [14, 15] or in the context of Enterprise 2.0 [16]. The objective of the success factor analysis in this context is the measurement of the contribution of certain factors to the success of cloud services provided as SaaS, by identifying strengths as well as weaknesses and utilising the analysis to deduce actions to extend strengths and to dismantle weaknesses [14]. The application of the success factor analysis was in line with previous research [13, 14] comprising the following steps:

- identification of success factors
- determination of survey participants
- formulation of questions for the survey
- perform data collection
- evaluate the data and show the empirical results
- discuss the results
- report the findings

Strengths and weaknesses are measured by the two dimensions "performance" and "priority". Performance is seen as the perceived actual contribution of a success factor to the usage of the cloud service. The second dimension "priority" measures the importance of the success factor for the usage of the service. Calculating the average values of the two dimensions for each factor allows identifying gaps and clustering the factors. The gaps analysis shows the difference between priority and performance per success factor. Lowering the gaps by starting with factors having a high perceived difference is an advisable starting point for improvement. The method also allows clustering the factors into four areas or quadrants. These quadrants provide different recommendations for improvement [17]:

- Factors in quadrant I "Improve" have a relatively low performance and high priority.
- Quadrant II contains factors that are sub relevant with both under averaged priority and performance.
- Factors in quadrant III "Well Done" have a relatively high performance as well as priority.
- Quadrant IV "Exceeding Performance" indicates that the factors have a relatively high performance but are under averaged priority.

The main focus has to be laid on quadrant I containing high priority factors that need to be examined in relation to the measures and whose performance has to be improved.

2.2 Success Factors for Cloud Services

Due to the vast diversity of available cloud services from a customer's point of view [18] there is a need to clarify the services that the study is aiming at. The following common services in particular were targeted:

- Cloud storage and/or backup services that provide disk space in data centers, which can be provisioned on-demand.
- Cloud office solutions that deliver tools for word processing, spreadsheet and other office programs.
- Cloud e-mail solutions that provide mailing functionalities without the need to operate an own e-mail server.

The success factor analysis has to be based on a validated list of success factors [14]. Therefore, we needed to identify them for our context. Drawing on previous literature, several different factors that are critical to a successful usage of cloud computing were identified. Garrison et al. identify trust, technical capabilities and individual managerial skills as relevant factors for successfully deploying cloud computing [2]. Walther et al. provide a comprehensive list of success factors in the context of software as a service via literature research [3]. For this research, we use the success factors identified in previous work which were empirically elaborated on the basis of theoretical models using qualitative interviews among organizations [4]. This resulted in 18 success factors relevant to individuals as shown in Table 1 with a short description.

2.3 Data Collection and Sample Description

One way to explore the notion of cloud services is to think about "Everything as a Service" (XaaS). This allows the exploration of the resource "human capacity" in addition to many others discussed in recent literature (e.g. software, infrastructure, platform, process). Online labor markets such as Amazon's Mechanical Turk (www. mturk.com) or CrowdFlower (www.crowdflower.com) are referred to as Human as a

Table 1. Relevant success factors for usage of cloud services.

ID	Description/statement	Short
1	The image or status of the adopter using cloud services [4]	Image
2	Cloud services are better than the solutions they supersede. This includes load relieving of the network infrastructure, enhancement of service availability, removal of hardware maintenance and the flexibility to provide the right amount of resources in general [3, 4]	Relative advantage
3	Security and safety aspects when using cloud services. This includes measures for privacy management and measures against data loss or denial of service [3, 4]	Security and safety
4	The technical infrastructure required to support cloud services is available [4]	Facilitating conditions
5	Costs for implementation and operation of a cloud service solution in comparison with conventional solutions [3, 4]	Costs
6	Usefulness of cloud services (e.g. fulfill my individual needs and enhance the fulfillment of tasks) [3, 4]	Perceived usefulness
7	Required effort associated with the use of a cloud service (i.e. the attractiveness of the cloud service in relation to the effort to use and an in-house system) [3, 4]	Effort expectation
8	Availability of appropriate cloud services (e.g. up-time of the service as well as suitability for a certain task) [3, 4]	Availability
9	Freedom of decision to use a cloud service [4]	Voluntariness of use
10	The interoperability of cloud solutions with existing systems and experiences/needs [3, 4]	Compatibility
11	The improvement in task performance when using cloud services. This includes assurance of performance while scaling resources rapidly on a large scale [3, 4]	Task performance
12	Additional technological characteristics of the cloud services like the technological maturity, or simplification of administration [2, 4]	Technological characteristics
13	Trustfulness and reliability of cloud services [2, 4]	Trust
14	Ease of use of cloud services (i.e. it is user-friendly and easy to understand with no additional skills required) [2, 4]	Ease of use
15	Legal (e.g. governmental) or contractual (e.g. service level agreements) regulations imposed by the cloud service provider [3, 4]	Regulatory framework
16	Monitoring and measuring the impact of using cloud services [4]	Observability
17	The opportunity to test cloud services for a limited time [4]	Trialability
18	The energy consumption when using cloud services [4]	Energy efficiency

Service (HaaS) [21] in the light of cloud computing. While the cloud workers (or "click workers") are paid for their responses, Paolacci et al. [22] suggest that sample errors, such as coverage error, and risks, such as dishonest responses, are low or moderate compared to traditional recruiting methods for laboratory, traditional web study and web studies through purpose built websites. Furthermore, the efficacy of using HaaS for behavioral research has been explored in the domains of political science [23], linguistics [24], psychology [25], economics [26] and computer science [27, 28]. Although they are no experts in cloud computing, individuals using cloud services are deemed to be qualified to answer certain questions on this topic as they are using them on a regular basis.

Therefore, an online survey was designed and subjects were recruited using Amazon's Mechanical Turk. The survey was available for participation for four weeks in April 2014. In order to prevent repeated submissions by an individual participant, the unique identifiers assigned to each user by Amazon's Mechanical Turk ("Worker ID"), were verified to be unique prior to the data analysis. The survey was only available to participants with an approval rate of at least 97% and who previously completed at least 500 approved tasks.

At the beginning of the survey the focus of the survey was clarified to the participants. As the survey was executed in English language, the participants were asked to indicate their level of English proficiency in order to avoid misunderstandings due to language deficiencies. Furthermore, the participants were asked to provide some demographic data such as age, sex, and nationality. Then, the terms cloud computing and cloud services were explained together with some SaaS examples. The participants specifically had to state, whether they are familiar with each of the following cloud services: cloud storage and/or backup services, cloud office solutions, and cloud e-mail solutions. To be included in the sample, the participants needed to be familiar with at least one of the cloud services.

The main part of the survey included the identified relevant success factors together with a statement for clarification of the factor itself as stated in Table 1. A standardized questionnaire with a five point Likert scale was used to indicate the current performance (excellent, good, average, poor, very poor) and priority (very high, high, medium, low, very low) of each factor. We also monitored the overall time needed to fill out the survey as additional indicator for the seriousness of the answer. Additionally, the participants were required to provide a code to the Mechanical Turk system that was displayed within the text on the last page of the survey. Only completed surveys with a correct code were included in the sample.

Overall, the sample includes valid responses from 274 individuals, with more men (62%, n = 171) than women (38%, n = 103) participating. 68% (n = 185) of the participants can be categorized as digital natives. For this categorization we followed the definition of Palfrey and Gasser [29], where digital natives are born after the year 1980. The geographical distribution shows that the majority of them were located in North America (42%, n = 116), Asia (28%, n = 77) and Europe (23%, n = 64). The participation from other continents (Africa, Australia, South America) was low (combined around 6%). Consequently, the sample can be considered heterogeneous. While participation in online labor markets, such as Amazon Mechanical Turk are popular in Asia, this study was able to generate a sample with a good mix in respect to sex, age and location.

3   Empirical Results

This section reports the results from the cluster analysis using four quadrants and the gaps analysis. In addition, results from the sample divided into the factors age, sex and continent are presented.

Figure 1 shows the evaluation of the relevant success factors from the sample as cluster analysis. The four clusters are divided by the average priority (2.21) in the horizontal and the average performance (2.39) in the vertical. The resulting quadrants show factors below or above the average in terms of priority and performance. The 18 indicators examined are assigned based on their measured differences to one of the four fields. The results show that mostly all factors are either in quadrant II or III, emphasizing the overall good acceptance of cloud services provided as SaaS. Nearly all

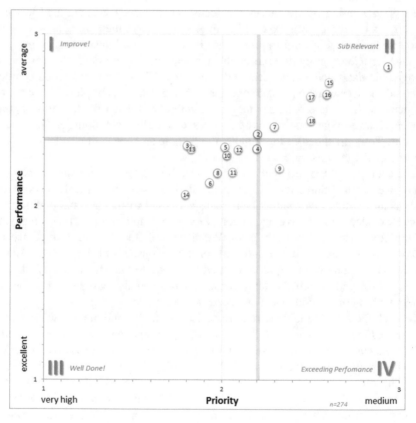

Fig. 1. Performance and priority of the factors arranged in four quadrants.

factors are rated either with both a high priority and good performance (quadrant III "well done"), or with a medium priority and average performance (quadrant II "sub relevant). Factors with a best-rated performance and high priority include ease of use (ID = 14), perceived usefulness (ID = 6), and technological availability (ID = 8). The least important factor was the image (ID = 1) of the adopter. Interestingly, not a single factor ended up in quadrant I ("improve"). The factor closest to be improved turned out to be "security and safety" (ID = 3).

Figure 2 shows the factors by their difference in performance and priority. The three factors with the highest gap in performance and priority are security and safety (ID = 3), trust (ID = 13), and costs (ID = 5).

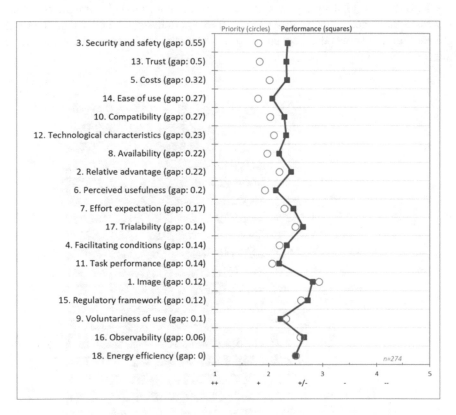

Fig. 2. Gap analysis of the success factors showing the difference in performance and priority.

Wang et al. suggest that several other factors like gender, education, as well as organizational, psychological and social factors have an impact on "digital fluency" [32]. We therefore also had a look at the sample divided into the factors age, sex and continents. We used the first factor "age" to distinguish between digital natives and digital immigrants. Commonly two characteristics, age and accessibility, are used for defining digital natives [30]. We did not use accessibility as our sample solely consisted of subjects that had access to cloud technologies. Specifically, we split the sample into subjects born before the year 1980 (referred to as "digital immigrants") and born in 1980 or later ("digital natives"). The results show that digital natives tend to evaluate the current performance slightly better. Trust (ID = 13) shows the highest difference with 62.70% of digital natives ranking the factor either excellent (1) or good (2) and only 37.08% of digital immigrants sharing this opinion. Digital immigrants actually only rated the factor compatibility (ID = 10) to a certain degree better than the digital natives. Figure 3 shows the frequencies of all factors given by respondents divided by their age.

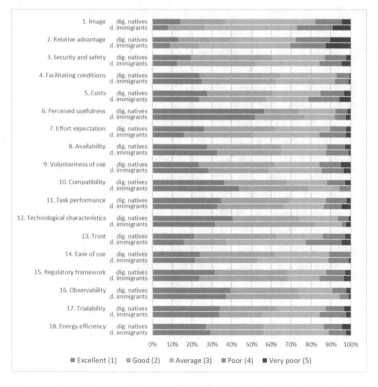

Fig. 3. Differences in performance by age: digital natives (n = 185) and immigrants (n = 89).

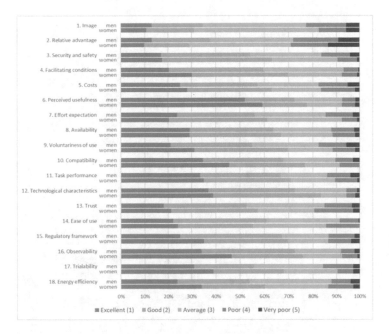

Fig. 4. Differences in performance by gender: women (n = 103) and men (n = 171).

In terms of gender, results show that women evaluated the current performance slightly better than men did. Women rated a total of 15 factors to some extent better than men (combined result of excellent and good). Men did only evaluate the factors relative advantage (ID = 2), ease of use (ID = 14), and image (ID = 1) better than women. Figure 4 shows the results in performance with the sample divided into male and female subjects.

Looking at the sample divided by continents (Fig. 5), the results show some differences between Europe, North America and Asia. In a combined result of the ratings "excellent" and "good", North Americans gave 10 factors better performance ratings than the other continents. Europeans see five factors ahead and Asian subjects only three. The highest difference is seen in relative advantage (ID = 2) with 52.56% Asians and only 21.88% Europeans ranking the factor either excellent or good. The second highest difference is seen in the improvement in task performance (ID = 11), with 78.45% of the North American subjects seeing a notable improvement when using cloud services but only 51.28% Asians sharing this opinion.

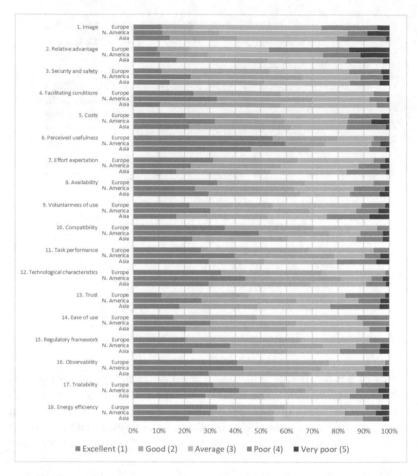

Fig. 5. Differences in performance by continent: Europe (n = 64), North America (n = 116), and Asia (n = 77).

4 Discussion

The results illustrate that although the priority of the factors investigated is with an average of 2.21 rated as "high", there are some factors that are perceived as sub relevant or even exceeding their performance. However, with nearly all investigated critical success factors being either "well done" or "sub relevant", the survey draws a rather positive picture of cloud services provided as SaaS. The growth and revenue numbers of the IT industry in the cloud sector also reflect this result. With a saturation of such services, it is getting harder to set yourself apart from other SaaS providers offering similar services. Thus, vendors could explicitly focus on high priority factors like security, trust, or ease of use to improve or specialize their services. This specialization also led to several additional service models like security as a service, backup as a service and terms like XaaS.

The factors with top priority are also in line with current findings from literature. SaaS involves the use of software via internet technology and therefore poses security requirements on the application, physical & virtual level [31]. Security has been identified as crucial for the widespread use of cloud services in general [1]. Furthermore, security is connected with our notion of trust, i.e. reliability and trustfulness of cloud services. Usually, trust cannot be directly enforced. A system that is correct and secure also increases the level of trust for this system [31].

The results also indicate some similarities and differences in the rating of cloud success factors between groups divided by age, gender and continent. Results show that digital immigrants tended to be more critical in the rating (i.e. lower overall performance) of the factors. It has to be noted that we used an exact cut-off year of birth and treated digital natives and digital immigrants as mutually exclusive groups which has been criticized in recent literature [30]. We also had a look at gender differences and differences according to the geographic location in our sample, which yielded some tendencies like those that female subjects tend to rate the performance of the factors better than their male counterparts. Combinations such as female digital natives or male immigrants from Europe, etc. would have been interesting as well, but the sample size did not allow us such fine-grained subgroups.

5 Limitations and Conclusion

Cloud computing has shaped the IT industry for more than a decade and has received a lot of attention in IS research. This paper provides insights from a survey on the ranking of 18 relevant success factors according to their actual performance and their priority among individuals. The empirical investigation shows that safety and security, trust, ease of use, perceived usefulness and availability are of highest priority within a group of the success factors investigated. The highest differences between priority and performance were found in the factors trust and security.

Although the sample showed a mix in respect to sex, age and location, there are a number of factors that limit or may bias the results. With 274 valid responses, the sample size is considered moderate for the purpose of this research. A larger sample with rigorous selection would yield results that are more accurate. The selection itself is another limitation, as we had no influence on the selection of the individuals. We also need to note that the result may be biased towards a positive attitude concerning SaaS of the subjects. Although 18 success factors have been identified, the list might not be up to date since a lot of progress has been made in this topic. Furthermore, the success factors are not independent. As stated, they are at least partly overlapping which might result in a bias. While sex, age and continent were analysed, several other factors that might have a great impact to service adaption have not been discussed (e.g. education, or organizational, psychological and social factors). The paper uses success factor analysis as a practical and easy-to-use instrument that allows drawing attention-grabbing results primarily for managers, but is not targeted to be used for theory building. Treating ordinal Likert scales as interval scales has long been controversial and is another limitation of the success factor analysis. Therefore, additional analysis of the data divided into age, sex and continents only included frequencies (e.g. standard deviation is seen inappropriate and was omitted).

Nevertheless, we conclude that the paper provides interesting insights into common success factors for the cloud service model SaaS. It could narrow the gap between theory and practice by the conduction of a cross-continent empirical study using success factor analysis as method for evaluation.

References

1. Rong, C., Nguyen, S.T., Jaatun, M.G.: Beyond lightning. A survey on security challenges in cloud computing. Comput. Electr. Eng. **39**, 47–54 (2013)
2. Garrison, G., Kim, S., Wakefield, R.L.: Success factors for deploying cloud computing. Commun. ACM **55**, 62 (2012)
3. Walther, S., Plank, A., Eymann, T., Singh, N., Phadke, G.: Success factors and value propositions of software as a service providers – a literature review and classification. In: AMCIS 2012 Proceedings (2012)
4. Stieninger, M., Nedbal, D.: Diffusion and acceptance of cloud computing in SMEs: towards a valence model of relevant factors. In: Proceedings of the 47th Hawaii International Conference on System Sciences, pp. 3307–3316 (2014)
5. Wang, N., Liang, H., Jia, Y., Ge, S., Xue, Y., Wang, Z.: Cloud computing research in the IS discipline: a citation/co-citation analysis. Decis. Support Syst. **86**, 35–47 (2016)
6. Borgman, H.P., Bahli, B., Heier, H., Schewski, F.: Cloudrise: exploring cloud computing adoption and governance with the TOE framework. In: 46th Hawaii International Conference on System Sciences (HICSS), pp. 4425–4435 (2013)
7. Benlian, A., Hess, T.: Opportunities and risks of software-as-a-service: findings from a survey of IT executives. Decis. Support Syst. **52**, 232–246 (2011)
8. Li, Y., Chang, K.-c.: A study on user acceptance of cloud computing: a multi-theoretical perspective. In: Proceedings of the AMCIS (2012)
9. Mital, M., Chang, V., Choudhary, P., Pani, A., Sun, Z.: Adoption of cloud based Internet of Things in India: a multiple theory perspective. Int. J. Inf. Manag.
10. Park, S.C., Ryoo, S.Y.: An empirical investigation of end-users' switching toward cloud computing: a two factor theory perspective. Comput. Hum. Behav. **29**, 160–170 (2013)
11. Wu, W.-W.: Developing an explorative model for SaaS adoption. Expert Syst. Appl. **38**, 15057–15064 (2011)
12. Tjikongo, R., Uys, W.: The viability of cloud computing adoption in SMME's in Namibia. In: IST-Africa 2013 Conference Proceedings, pp. 1–11 (2013)
13. Heinrich, L.J.: Informationsmanagement. Planung, Überwachung und Steuerung der Informationsinfrastruktur. Oldenbourg Verlag, München; Oldenbourg, München (1999)
14. Lehner, F.: Measuring KM success and KM service quality with KnowMetrix: first experiences from a case study in a software company. In: Karagiannis, D., Jin, Z. (eds.) Knowledge Science, Engineering and Management, pp. 335–346. Springer, Heidelberg (2009)
15. Lehner, F., Amende, N., Wildner, S., Haas, N.: KnowMetrix - Erfahrungen mit der Erfolgsbewertung im Wissensmanagement in einem mittelständischen Unternehmen. In: Hinkelmann, K., Wache, H. (eds.) WM2009: 5th Conference on Professional Knowledge Management, 25–27 March 2009, Solothurn, Switzerland, pp. 470–479 (2009)

16. Nedbal, D., Auinger, A., Hochmeier, A., Holzinger, A.: A systematic success factor analysis in the context of enterprise 2.0: results of an exploratory analysis comprising digital immigrants and digital natives. In: Huemer, C., Lops, P. (eds.) E-Commerce and Web Technologies. 13th International Conference, EC-Web 2012, vol. 123, pp. 163–175. Springer, Heidelberg (2012)

17. Auinger, A., Nedbal, D., Hochmeier, A.: An enterprise 2.0 project management approach to facilitate participation, transparency, and communication. Int. J. Inf. Syst. Proj. Manag. **1**, 43–60 (2013)

18. Garg, S.K., Versteeg, S., Buyya, R.: A framework for ranking of cloud computing services. Future Gener. Comput. Syst. **29**, 1012–1023 (2013)

19. Mell, P., Grance, T.: The NIST definition of cloud computing (2011)

20. Stieninger, M., Nedbal, D.: Characteristics of cloud computing in the business context: a systematic literature review. Glob. J. Flex. Syst. Manag. **15**, 59–68 (2014)

21. Irani, L.: The Cultural Work of Microwork. New Media Soc. **17**, 720–739 (2015)

22. Paolacci, G., Chandler, J., Ipeirotis, P.G.: Running experiments on Amazon Mechanical Turk. Judgment Decis. Making **5**, 411–419 (2010)

23. Berinsky, A.J., Huber, G.A., Lenz, G.S.: Evaluating online labor markets for experimental research: Amazon.com's Mechanical Turk. Polit. Anal. **20**, 351–368 (2012)

24. Sprouse, J.: A validation of Amazon Mechanical Turk for the collection of acceptability judgments in linguistic theory. Behav. Res. **43**, 155–167 (2011)

25. Buhrmester, M., Kwang, T., Gosling, S.D.: Amazon's Mechanical Turk: a new source of inexpensive, yet high-quality, data? Perspect. Psychol. Sci. **6**, 3–5 (2011)

26. Horton, J.J., Rand, D.G., Zeckhauser, R.J.: The online laboratory: conducting experiments in a real labor market. Exp. Econ. **14**, 399–425 (2011)

27. Ratinov, L., Roth, D., Downey, D., Anderson, M.: Local and global algorithms for disambiguation to wikipedia. In: Proceedings of the 49th Annual Meeting of the Association for Computational Linguistics: Human Language Technologies, vol. 1, pp. 1375–1384. Association for Computational Linguistics, Stroudsburg (2011)

28. Milne, D., Witten, I.H.: Learning to link with wikipedia. In: Proceedings of the 17th ACM Conference on Information and Knowledge Management, pp. 509–518. ACM, New York (2008)

29. Palfrey, J.G., Gasser, U.: Born Digital. Understanding the First Generation of Digital Natives. Basic Books, New York (2008)

30. Wang, E., Myers, M.D., Sundaram, D.: Digital natives and digital immigrants: towards a model of digital fluency. In: ECIS 2012 Proceedings (2012)

31. Zissis, D., Lekkas, D.: Addressing cloud computing security issues. Future Gener. Comput. Syst. **28**, 583–592 (2012)

32. Wang, Q., Myers, M.D., Sundaram, D.: Digital natives and digital immigrants. Bus. Inf. Syst. Eng. **5**, 409–419 (2013)

Distributed Computation for Protein Structure Analysis

Nobuyuki Tsuchimura[✉] and Adnan Sljoka

Department of Informatics, School of Science and Technology,
Kwansei Gakuin University, Sanda, Hyogo, Japan
tutimura@kwansei.ac.jp, adnanslj@gmail.com

Abstract. To understand how proteins function it is crucial to understand the connection between their structure, flexibility and dynamics.
In the field of bioinformatics and computational biology, there is a strong
interest to develop software and computational tools that analyzes various properties of protein structures. The software we are using for protein
flexibility and dynamics analysis generally assumes a single task, single
thread environment. To more efficiently elucidate the function of proteins, we need to perform large-scale calculations on many structures.
To improve computational speed of such large scale analysis, we decided
to perform parallel distributed computation with the conventional software. We designed a simple protocol dedicated to this software over http
and achieved a speedup of 550 times with 600 CPU cores. With such
speed ups, we are able to perform faster high-throughput computations
on large number of protein structures.

1 Introduction

We are interested in analyzing protein functions using both available experimental structural data from the protein data bank (PDB) and computational
models of protein structures. Proteins are a constituent element of living organ-
ism. They are the working engines of almost all events in the cells and play
crucial roles in immunity, catalysis, metabolism, structure, they transport and
store other molecules such as oxygen etc and are often targets in therapeutics.
Proteins are composed of hundreds of amino acids attached by peptide bods
which form complicated 3-dimensional structures. Understanding flexibility and
rigidity and how various components move in proteins 3-dimensional structure
is a critical component to understand how protein functions.

Experimental biochemistry and biophysics-based technologies for analyzing
protein structures and their dynamics such as X-ray crystallography, NMR,
Cryo-EM, mass spectrometry etc are vital to investigate how proteins function. However, experiments are often costly, require expensive equipment, are
generally difficulty to design and interpret and can take substantial amount of
time to perform. Proteins are highly dynamic sampling various conformations
of a highly-dimensional and complicated energy landscapes [9]. Moreover, protein motions can occur on multiple time scales, further complicating analysis

© Springer Nature Switzerland AG 2019
F. Xhafa et al. (Eds.): INCoS 2018, LNDECT 23, pp. 16–23, 2019.
https://doi.org/10.1007/978-3-319-98557-2_2

of protein motions. As such, there is an increasing demand for computational modelling and simulations of protein motions.

Various computational techniques (i.e. MD simulations, graph theoretical methods, normal mode analysis etc) based on biophysics, mathematical modelling, graph theory, statistics and simulations have been designed for analyzing protein structures and their motions which have been verified with experimental data. Molecular dynamics (MD) simulations remains the most prevalent way to computationally simulate protein motions. However, MD simulations rely on very complicated force fields and are limited by immense computational resources which cannot simulate motions on longer (ms-s) time scales which are often associated with functionally relevant motions that occur in enzyme catalysis, large amplitude domain-domain motions and other biological phenomena.

In our work we routinely calculate flexibility and motions of proteins based on established biophysics techniques and mathematical models in rigidity theory [2,8]. One advantage of such approach is we can obtain very fast computational predictions of protein flexibility and are not confined by time scale issues that are encountered in MD simulations.

Our aim is to confirm and validate the correctness of our mathematical models and computational predictions with experimental data (in collaborations with experimentalists) and at the same time determine and predict novel protein function insights that are difficult or not capable to obtain with experiments. To obtain meaningful and statistically significant results we often need to analyze a large number of protein structures and models and repeat such analysis with different parameters. This naturally requires many computations and calculations.

Recent studies have focused on analysis of flexibility of specific motions and parts of proteins in particular antibody proteins and also probing allosteric communication with rigidity theory methods. In particular we used rigidity theory-based FIRST method for analyzing protein flexibility [3] and Rigidity Transmission Allostery (RTA) method and software [5] to probe allosteric communication in proteins [1,5,6]. Allostery refers to an effect of binding at one site of a protein that affects another, often significantly distant functional site on the protein, enabling regulation of the protein function. Allostery is an important part of how proteins function, yet the underlying mechanism of allosteric transmission is largely not well understood. RTA analysis is based on degree of freedom transmissions and conformational coupling between distant sites (residues, atoms etc) on a protein structure which models to predict allostery.

In both rigidity and allostery analysis we often need to analyse many structures and to test properties and communication between many pairs of different sites. This can increase the required amount of calculations and computations. Fortunately this problem is suitable for distributed computation as we can deal with a series of independent and identical tasks. We built a distributed computing framework dedicated to the software and reduced the computing time of 6 days with one core to 15 min with 150 computers each with 4 cores. It means that the calculation speed has been increased by 550 times with 600 cores.

In Sect. 2, we describe the importance of analyzing protein function and rigidity theoretical model for flexibility predictions of protein structures. In Sect. 3, we describe the method of distributing particular tasks of the software and our design choice. In Sects. 4 and 5 we provide a few practical techniques and offer some conclusions.

2   Protein Analysis

Rigidity theory is the study of rigidity and flexibility of structural frameworks, specified by geometric constraints on points and rigid bodies [8]. Advancements in mathematical rigidity theory have opened up possibilities and new methods for computational predictions of protein flexibility and their dynamics. Starting with a 3-dimensional protein structure (PDB), rigidity-based method FIRST [2] and several other similar methods generates a mechanical linkage and a corresponding multigraph as a model of a protein, built up of atoms (vertices) and edges (covalent bonds, hydrogen bonds, hydrophobics etc). Hydrogen bonds are given energy strengths and hydrogen bond cutoff energy value is selected, where weak bonds are ignored. The pebble game algorithm is applied on the multigraph, checking the combinatorial counts in the Molecular Theorem which decomposes the protein into flexible and rigid regions. To model and calculate allosteric communication, which is distant communication between distant remote sites in the protein structure, special extensions of pebble game algorithm and other mathematical rigidity theory advances and algorithms are applied [4–6]. Such analysis, methods and software has been applied to obtain deep understanding of protein flexibility, antibody dynamics and allosteric communication in enzymes and receptors [1,3,4]. For allostery, calculations of flexibility, more specifically degrees of freedom of remote pairs of sites is calculated and the dependence of how conformational degrees of freedom of one site affect degrees of freedom of distinct remote sites. In the next sections we consider parallelized computation as a way to speed up calculations of flexibility and allostery of many sites.

3   Distributing Method

Parallel computing libraries such as PVM (Parallel Virtual Machine [12]) and MPI (Message Passing Interface [13]) are available. If we develop rigidity protein analysis programs from scratch, we can perform parallel computation with such a computation library. However, we already have a program that runs in a single thread environment and to maintain the same behaviour and outcome we did not modify the program.

3.1   Design

Referring to distributed computing model such as SETI@home [14] and GIMPS [10], we consider how to distribute and consolidate the input and output of the target program on the network. Our computer environment is 151 units of Fujitsu Celsius J530 (CPU Xeon E3-1226v3 (3.30 GHz, 4 core), RAM 8 GB, SSD 500 GB) connected by Gigabit Ethernet, running on Linux 4.4.x, OpenJDK 1.8.0, Perl 5.12.3. For clarity we will assume that only one server is used and the server manages instances of problems and aggregates the corresponding results. We have a total of 600 clients (150 units each having four cores). Each client repeats the following operations: connects to the server, receives instances of the prepared problems, performs the indicated calculations, and sends the results back to the server.

3.2   Feasibility

Our aim is to repeat similar calculations for one protein with varying parameters. We distribute a file representing a protein structure (about 1 MB) only once to a client, then we can describe an instance of problem by sending several additional parameters. The result is also an enumeration of numerical values. Data transfer of about 1 KB is enough for one input/output. Adjusting the calculation time at the client around 1 min requires a communication speed of 1 KB/60 s per one client. When 600 clients are concurrently connected to the server, the communication bandwidth required for the server side is 10 KB/s = 0.08 Mbps, 10 TCP/IP connections per second, so there is sufficient margin in the band of Gigabit Ethernet. (Actually, we spent more bandwidth on distributing protein files. See Sect. 4.3.)

For now, we are holding a sufficient number of reliable clients. If we accept participation of volunteer-based clients like SETI@home or BONIC [11], we should also consider redundancy to detect incorrect results as well as adjust communication frequency.

4   Practical Techniques and Result

After the basic mechanism of distributed computation is established and operation has started, a new bottleneck appears every time the number of clients is increased to 10, 100, or 600, or each time the number of instances is increased, and each time we make an adjustment. Here, we discuss the bottlenecks, how we dealt with them and the results of speedup.

4.1   Disconnect Nonresponsive Clients

As in our previous distributed computation [7] we found that it is important to disconnect a client which is nonresponsive for certain number of seconds. Otherwise, the server will not complete the aggregation process while waiting for the results, and the memory usage will increase. This cause the server to not respond to responsive clients. To deal with a nonresponsive client, we simply reassigned tasks to another client. This allows us to complete the aggregation process (i.e. collection of results of clients) and the release of memory stabilizes the whole operation. We decided to reassign a task to another client when its processing time exceeds three times the average processing time of the same protein.

4.2   Waiting Time Without Task

When all the instances are processed, it is necessary to adjust the frequency with which the clients connect to the server while waiting for the next instance to be entered. If the frequency is too high, sending "no task" reply increases the load on the server. If the frequency is too low, it takes time for the task to reach all clients when a new instance is entered. The number of clients is 10 to 600, and we change it dynamically as necessary. Therefore, assuming that the number of clients is n, and the frequency at which the server is accessed per second is k times, we adjusted the waiting time of the clients to satisfy the following condition:

$$k = 5 \times n^{0.36}.$$

In this case, the access frequency is as shown in Table 1. If there are 10 clients, it takes 0.9 s to reach every client, and if there are 600 clients, it takes 12 s. When the number of clients is small, the load on the server is low.

Table 1. The access frequency seen from the server and the client (times per second).

n	10	100	600
Server	11.5	26.2	50.0
Client	1/0.9	1/3.8	1/12.0

4.3   Task Assignment

We found that 600 clients is sufficiently large for analysis of one protein. Distribution of protein files from the server occurred every 10 min and adjustment was necessary to avoid concentration.

When a new protein was ready to be processed by all clients at the same time, a request for transferring next protein file occurs in the server at the same time, and task distribution of the server was stalled. Therefore, we devised to gradually change the assignment of proteins processed by clients.

Figure 1 illustrates the change in assignment of clients as a function of time. The horizontal axis represents time and the vertical axis is the number of allocated clients. Case 1 shows where all client tasks are assigned to one protein. When the task of the protein is finished, then clients are assigned to the next protein. The bold lines represent substituting of protein and client's request to transfer to the next protein file. In Case 1, the frequency of transfer requests is high.

In Case 2, clients are divided into 3 groups, and 3 proteins are assigned to each group. The frequency of transfer requests is less than in Case 1. Generally, because the amount of computation for each instance (protein) is different, it is expected that file transfer request to the server will not be performed at the same time, but in reality, it may occur when we are handling similar protein structures. To avoid such case, we consider a realistic refinement as next method.

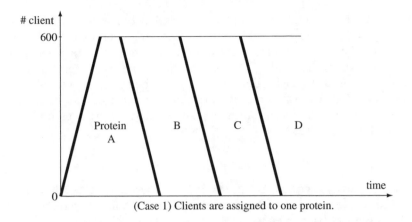

(Case 1) Clients are assigned to one protein.

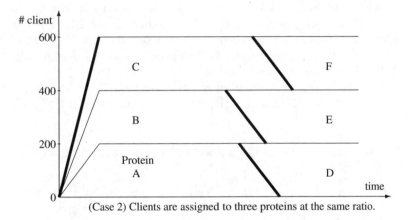

(Case 2) Clients are assigned to three proteins at the same ratio.

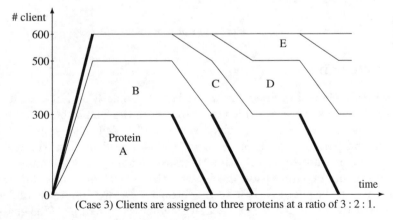

(Case 3) Clients are assigned to three proteins at a ratio of 3 : 2 : 1.

Fig. 1. Task assignment.

Case 3 is the method we adopted. We assign different number of clients for each protein. This figure shows the case of assignment at a ratio of $3 : 2 : 1$. Calculation of instances with a large number of clients finishes quickly. For example, when the task of Protein A ends, then we assign the clients so that the ratio of B, C, D become $3 : 2 : 1$. Although the transfer request is slightly more than Case 2, since it does not concentrate, the maximum load on the server can be reduced. Actually, we adopted 10 proteins and a rate of $1/6 : 1/7 : 1/8 : \dots : 1/15$.

4.4 Result

After some tuning described in this section, we succeeded in finishing a specific instance in 15 min as shown in Fig. 2, which takes 137 h in a single core. It means that the speed has been increased 550 times by 600 CPU cores.

Since this protein is relatively large (i.e. it contains a large number of residues), and we can generate many tasks, 600 cores are effectively utilized. With a smaller protein, similar speedup can not be achieved. If there are many small proteins, similar speedup can be achieved on average by the allocation method described in Sect. 4.3.

The size of the program we created is 1200 lines (60 KB) in Java language for server application and 200 lines (5 KB) for client application in Perl language.

```
[2016/12/20 02:14:22]
Writting result finished successfully! :norovirus/SAGA_A.pdb
#task : 9857
average time: 50.09 sec/task
total time: 493694 sec (8228.24 min = 137 hour)
real  time: 885 sec (14.76 min)
speed up: 557.38 times
```

Fig. 2. Log of calculation of a norovirus protein.

5   Conclusion

In this paper, we have described our method of distributed computation. In order to analyze protein structures, which is a problem in the field of bioinformatics, it became necessary to execute the software repeatedly while changing the protein structures and parameters. In order to guarantee the identity of the execution result, we decided not to modify this software, but instead we focused on creating a distributed computation mechanism that transfers inputs and outputs of clients. Given the characteristic of this problem, by dividing tasks and saving input/output data transfer, we can achieve significant speedups. This enables us to perform more efficient analysis of protein function.

Acknowledgements. The authors would like to acknowledge Naoki Katoh for letting us start collaborative research through CREST project. The authors were supported by JST CREST Grant Number JPMJCR1402 (Japan).

References

1. Deng, B., Zhu, S., Macklin, A.M., Xu, J., Lento, C., Sljoka, A., et al.: Suppressing allostery in epitope mapping experiments using millisecond hydrogen/deuterium exchange mass spectrometry. MAbs **9**(8), 1327–36 (2017). https://doi.org/10.1080/19420862.2017.1379641
2. Jacobs, D.J., Rader, A.J., Kuhn, L.A., Thorpe, M.F.: Protein flexibility predictions using graph theory. Proteins **44**(2), 150–65 (2001). https://doi.org/10.1002/prot.1081
3. Jeliazkov, J.R., Sljoka, A., Kuroda, D., Tsuchimura, N., Katoh, N., Tsumoto, K., Gray, J.J.: Repertoire analysis of antibody CDR-H3 loops suggests affinity maturation does not typically result in rigidification. Front. Immunol. **9**, 413 (2018). https://doi.org/10.3389/fimmu.2018.00413
4. Kim, T.H., Mehrabi, P., Ren, Z., Sljoka. A., Ing, C., Bezginov, A., et al.: The role of dimer asymmetry and protomer dynamics in enzyme catalysis. Science **355**(6322) (2017). https://doi.org/10.1126/science.aag2355
5. Sljoka, A.: Algorithms in rigidity theory with applications to protein flexibility and mechanical linkages. Ph.D. thesis, Graduate Programme in Mathematics and Statistics, York University, Toronto, Canada (2012)
6. Sljoka, A., Wilson, D.: Probing protein ensemble rigidity and hydrogen-deuterium exchange. Phys. Biol. **10**(5), 056013 (2013). https://doi.org/10.1088/1478-3975/10/5/056013
7. Tsuchimura, N.: Computational results for Gaussian moat problem. IEICE Trans. Fundam. **E88-A**, 1267–1273(2005)
8. Whiteley. W: Counting out to the fexibility of molecules. Phys. Biol. **2**, S116–S126 (2005). https://doi.org/10.1088/1478-3975/2/4/S06
9. Wildman, H.K., Kern, K.: Dynamic personalities of proteins. Nature **450**, 964–972 (2007). https://doi.org/10.1038/nature06522
10. Woltman, G., Kurowski, S.: GIMPS. https://www.mersenne.org/
11. The BOINC project: Berkeley Open Infrastructure for Network Computing. http://boinc.berkeley.edu/
12. The PVM Project Team: PVM. https://www.csm.ornl.gov/pvm/
13. Software in the Public Interest, Inc.: Open MPI. https://www.open-mpi.org/
14. University of California Berkeley: SETI@home. http://setiathome.ssl.berkeley.edu/

Solutions for Secure Collaboration, Selection of Methodology, Implementation and Case Studies for Their Use

Juraj Zelenay[1(✉)], Peter Balco[2], Michal Greguš[2], and Ján Luha[3]

[1] MPI Consulting s.r.o., Bratislava, Slovakia
juraj.zelenay@mpicons.sk
[2] Faculty of Management, Comenius University, Odbojárov 10, P.O. BOX 95,
Bratislava, Slovakia
{peter.balco,michal.gregus}@fm.uniba.sk
[3] Faculty of Medicine, Comenius University, Bratislava, Slovakia
jan.luha@lf.uniba.sk

Abstract. Many of us never start thinking about secure information communication and collaboration until it is too late, after a security incident has been identified. The problem is that awareness of the issue remains quite low, coupled with a lack of continuing education in the field. Research on the subject and the level of security awareness was undertaken at Comenius University in Bratislava by the Faculty of Management at the end of 2017 and beginning of 2018, with a wide range of organisations participating. A number of interesting results came out of the analysis, all of which need to be followed up. Our paper discusses the outcome of experiments related to the study's findings, while pointing out possible solutions in the form of process changes and the use of cloud-based tools and technology in order to eliminate risks and potential threats.

1 Introduction

According to global research by PriceWaterhouseCoopers, cybercrime accounts for 13% of all economic crime in Slovakia and is trending higher [1, 4]. In addition to crime, there is even economic espionage aimed at sensitive data, patents, formulae, new technologies, corporate strategic plans and more - in a word "corporate secrets". Data obtained in this way are easy to sell. There are known cases where competitors recruit hackers, highly-skilled thieves that steal strategic data for big money through a combination of social engineering, system failures and unethical behaviour. The consequences of such data loss are fatal for companies. Therefore, the topic appears today to be extremely timely and likewise always in flux because both sides are constantly learning new techniques for "hacking" into systems and counter-strategies to combat it. It seems likely that the situation is never going to stabilise and there will always be a need to think up new mechanisms to defend against novel types of attacks.

Today's conventional SMEs fail to prioritise IT and security, with management at most of these companies having no extensive understanding of the details. They follow the principal of "mainly it works and doesn't cost a lot". The potential risks are never

© Springer Nature Switzerland AG 2019
F. Xhafa et al. (Eds.): INCoS 2018, LNDECT 23, pp. 24–33, 2019.
https://doi.org/10.1007/978-3-319-98557-2_3

really weighed. While data is often the most precious asset a company or individual has today, many times it is the outcome of several years of work and has great value. Secure communication between remote participants and the exposure to risks from failure to comply with the principles of it are both underestimated in today's world [2].

Damage that may be caused by such breaches is sometimes quite significant and corrective action requires considerable resources and effort. Many companies have faced considerable problems when unauthorised people seize their sensitive information, contracts, patents, designs and plans or they are disclosed at the wrong time. Compromised data can cause a loss of competitive advantage in the market and ruin a company's reputation. The risks resulting from the loss or leak of data are almost fatal for many businesses and their commercial activities. Certainly, confidential information leaked to the public or a document falling into the wrong hands can cause many problems, at minimum stress, but also a lot of explaining, loss of goodwill, a financial impact from breach of contract regarding confidentiality, lost business and more. Many times sensitive data is leaked due to user error while either operating a program or (what happens more frequently) due to a violation of secure communication rules when sending send sensitive data through their own free personal e-mail account or to the incorrect e-mail address. A special risk category is ignorance, where users either "don't know that they don't know" or are unaware about the sensitivity of the data and that they are not allowed to send it except through a secure channel. Not to mention the problems caused when a computer is stolen, as no thief would have difficulty removing an unencrypted hard disk. Data mined from a stolen laptop or a large company employee's unsecured notebook can yield quite interesting results.

Many security-conscious managers at SMEs and people in liberal professions (such as physicians, architects, designers, consultants, tax advisers, attorneys, auditors and technical inspectors) are aware that "using anything" and the status quo of "not doing anything" is not correct, and that the various free software and services available may not necessarily be the right choice for them, but they would be better off "looking around" for more suitable options. Many manufacturers today offer secure communications and collaboration tools, but are they suitable for "my specific type of business"? How can I navigate through a myriad range of safe storage services and tools and what criteria should I be using to select them? How can a SME implement a "package" of secure communication and data storage tools without spending a lot of time and money? It is not for nothing that people are talking about data as the 21st century's "gold". Whoever has data about us and the artificial intelligence algorithms to exploit it will be able to control us. Therefore, it is essential to make it impossible for anyone who wants to obtain our data illegally, or at the very least to make it as extremely difficult as possible.

For a layman, it is quite difficult to navigate what the market offers. Unfortunately, no company would ever boast about data leaks and intrusions to their systems, so the statistics are not entirely accurate. Successful penetrations by professional hackers may not be detected immediately and, although we are told that nothing had happened, there are still consequences (like the release of leaked data). Only afterward is it possible to trace what exactly happened when the error occurred, where the weak link causing the leak is and what preventive action can be taken in future [3].

2   Research Objectives and Data Processing Methodology

Our research, which is still ongoing, seeks to gain insight into the level of maturity at SMEs toward developing and putting in place secure collaboration solutions. Concurrently, we wish to develop guidance in this segment for companies which are using cloud technologies coming out of Europe for secure intercommunication both within the organisation and by project teams and business partners. Another objective is to identify each solution's pros and cons, either eliminating or reducing the impact of the exposure to possible risks in this area.

The outcome will be a set of recommendations for secure communications and how not to jeopardise corporate secrets and sensitive data. These recommendations are not just for SMEs, but for anyone aware of how serious the issue has become. When the word "company" is mentioned, it means any institution concerned about addressing this problem. Subsequently, options for either avoiding or preventing attacks caused by so-called "social engineering" are verified, with research aimed at better understanding and navigation of the issue.

Our research included selecting the methodology for collecting data through questionnaires both in paper form and submitted electronically through www.survio. com. In addition, questions were defined exactly, with respondents offered the option of writing out full answers to questions. This approach was used in order to obtain the most objective answers from respondents about this topic. The questionnaires concentrated on determining the level of maturity in seeking solutions and contained 24 questions that were classified as general, process-related and technology-related. The Faculty of Management at Comenius University in Bratislava, where over 1,000 students currently study alongside people employed at companies, took the opportunity to distribute the questionnaire among employers. They were first collected at the end of last year and another collection of questionnaires is scheduled for this year. In collecting the information, we examined several hypotheses relative to the lack of awareness among the respondents observed.

3   Solutions for Secure Communication

When ordinary users gradually become aware of the need to migrate to a better solution, it is necessary to help them select the right services and provide support for operational use. Basically, average users need three services to be secure:

- Email;
- Data storage;
- Communication-chat tools.

But choosing from the myriad of services available on the market when sites claim theirs to be better than others is a daunting task. Before answering this question, it is worth looking at how SMEs see security at their own organisations.

The next section discusses the methodology behind their choices, how to compare different services and on which parameters to mainly focus (Table 1).

Table 1. SME perspective on security in their organisations

Question	Result	Value in %
Has your organisation had a security incident?	Yes	35
Has your existing security solution been replaced by something more sophisticated?	Yes	60
Have you tested other, more sophisticated solutions?	Yes	91.9
If a new solution would be developed for you, would you want to implement it?	Yes	92.0
Are you willing to pay for a more sophisticated security solution?	Yes	59.4
Would you migrate to cloud services for secure communication?	Yes	73.1

3.1 Selecting Suitable Services

The first step necessarily involves clarifying whether "you're going to the cloud" or insisting on "data about me on my notebook or USB flash drive are secured the best". When these aspects are evaluated and the path of the cloud is chosen, there are some questions about sorting out the options offered, where answers are expected. These are afterward checked and tested; hence a longer time should be spent making a decision. Of course, you have no desire for either a "vendor lock-in", or such "digging for a long-term supplier" with no option of switching, or for a change that would be too expensive. It is worth taking time to decide upon a system and not taking what first appears, although it may take longer to find something that "fits to a T". The selected service will have the company or your data available on all the devices you have and from which you have granted access. It is important to be aware that data in the cloud are safeguarded 24/7 by professionals certainly more prepared for security challenges that someone not regularly addressing the issue.

3.2 Selection Criteria

Ease of use - services that are cumbersome to control, where it is necessary to constantly look for something in the manuals and in today's world the approach is not necessarily effective. Control has to be easy and transparent from the very start. The simpler the better, although not at the expense of security.

   Level of support - first-level support can resolve most of the reported problems, while "response time" by support staff is important, too. Long response times tend to discourage users, while communication in Slovak or Czech is an advantage. If these languages are not available, it is advisable to have a cloud broker to provide the necessary support.

   Ease of administration - managing approaches, devices and security policies - everything should be easily manageable, or else the administrator would always have to be asked to make the necessary settings or changes.

   References and reputation - it is advisable to check references for the providers and the services to be brought from the provider, as ultimately the provider will have to be entrusted with valuable data. Therefore, it is appropriate to exercise some caution. If there are doubts about a company's reputation and that of its owners and statutory representatives, then it is worth being careful.

Free trial period - usually 14–30 days and afterwards deciding, something very useful when there are several choices available.

Charges for service or "the prices are soaring" - here everything has to be considered, whether the service is worth the money, what your mode does not need to address, exposure to risk when a service is not ordered, if what is used today will continue to be used tomorrow and whether to decide for another similar service. Financially evaluating the risks can be sometimes very challenging, for example, loss of reputation, violation of regulation, etc. There are occasions when someone's data or data about someone (found in "free" versions) will be sold by the provider to a third party. The outcome is spam, unsolicited advertising, trivial information no one would ever want to disclose and often with no idea that someone has been "paid" for it. With the advent of regulation GDPR, this should be no longer a major problem.

Security architecture - here three types of storage data are distinguished. Their advantages and vulnerabilities are examined below:

- Local, on-premises storage of encrypted data
- Cloud storage with server-side and in-transit encryption
- Cloud storage with client-side, end-to-end encryption.

There is an option of gaining access following the loss of a password - so-called "zero knowledge authentication", where passwords are stored on servers and cannot be misused. But this feature is appreciated mainly by those concerned about someone having been given their passwords to enter without their knowledge. Common concerns are expressed mainly by people in countries that do not fully respect human rights and freedoms and threats exist to divulge sensitive secrets (e.g., journalists, opposition politicians).

Encryption used - whether end-to-end encryption sufficiently protects against attacks.

This includes remote deletion of data to cover cases where if a device is loss or stolen, data will need to be promptly deleted from the device. Many services today offer this option and an administrator can quickly manage it once the incident has been reported and the identity of the person reporting it has been established (in companies).

DRM technology - this new technology enables still more stringent protection of the contents of documents in MS-Office (Word, Excel, PowerPoint), prevents printing and copying of documents and other safeguards. It is suitable for work areas with especially sensitive documents. This technology makes it extremely difficult for unauthorised persons to access data. Naturally, photographing mobile screens cannot be avoided, but an "office mode" can be created for sensitive data which do not enable mobile devices and cameras to be used with them (Fig. 1).

How secure is data after login - Does the provider have access to the data? This question is quite critical because in most countries, if law enforcement agencies (and also secret services and various underworld organisations) wish to get access to your data, the provider will be obliged upon a court order to make your data accessible together with the acquired metadata (such as access times and IP addresses). Similarly, if the provider's administrators decide to have a look at your data, they are able do it.

		Attacker			
		Provider (or its employees)		Hacker	
		Plausibility (Motivation)	Success probability	Plausibility (Motivation)	Success probability
Approach	**1. Local, on-premises storage of encrypted data**	Low	Low	Moderate	**Significant**
	2. Cloud storage without client-side encryption	Moderate	**Large**	**Large**	Moderate
	3. Cloud storage with client-side, end-to-end encryption	Low	Low	Moderate	Moderate

Fig. 1. Vulnerabilities and a comparison of data leakage possibilities by security method [10].

A good cloud service is designed so no one has access except you (Zero Access to User Data). There is no reason for a provider to know what data you stored, as you have paid for storage and not them to snoop (even if it were well meant). Are there any loopholes or "back doors" to decrypting your data? What about encryption libraries? Is only trusted open-source used with no back doors? Nobody really knows how to change the situation regarding the assessment of stored data content and somebody may one day declare your data to be harmful and will provide them to different authorities to cause you trouble (you could be prosecuted or persecuted, and it is quite enough when fully enough for CPS just to immediately revoke access to your data for breach of contract). Therefore, this is a fundamental condition for selection and not to be taken lightly. These clauses in the contract need to be read closely and with an understanding of them.

Legal jurisdiction over data - considered a very important issue, especially when "something happens" and the laws your provider would then have to follow. If data is stored in Europe, where the regulations are stricter than elsewhere, you are safe. Switzerland offers the greatest protection in this area, so we always recommend locating data there.

All user data is protected by the Swiss Federal Data Protection Act (DPA) and the Swiss Federal Data Protection Ordinance (DPO) which offers some of the strongest privacy protection in the world for both individuals and corporations. As ProtonMail is outside of US and EU jurisdiction, only a court order from the Cantonal Court of Geneva or the Swiss Federal Supreme Court can compel us to release the extremely limited user information we have.

It is best to use a so-called "battle card", comparing among the different relative products/services. Of course, it is always necessary to test whether all the claims are true.

3.3   Selecting Suitable Services

Well-known data storage services such as Tresorit, Dropbox, Box, OneDrive, GogleDrive were compared by us.

Our focus was on various attributes such as data storage terms in Europe, end-to-end encryption and policy restrictions regarding IP addresses from which the service can be accessed. Next, there was a comparison of whether Mac and Linux desktop applications also exist for these services since these users are frequently browsing. It is necessary for a team working together not to be limited by the operating system that has to be used. Another important criterion was whether the service has DRM - only Tresorit and OneDrive had it. Something very important is two-factor authentication. The entire comparison of individual services by attribute is displayed in the image below.

The comparison involved both physical testing of the services and also using www.slant.co, where individual services are compared to each other in the event that the results from the earlier comparison were very close.

3.4   Results from Comparison

Tresorit was seen by us as having the most adequately secure data storage. It is followed by ProtonMail, a secure email service based in Switzerland whose founders saw guaranteed privacy as their primary mission. Like Tresorit, ProtonMail is designed so data is only managed and protected, but no one has access to them. In third place is Threema, a service for secure chat communication. Their communication has never been breached (Table 2).

4   Practical Case Studies

Since 2014, data and two-way communications with our tax adviser, cooperating consultants, web and graphic designers, customers and others has been stored by us on TresorIT for Business, [11] a cloud service provided by MPI Consulting spol. s r.o. Its easy, user-friendly service allows communication virtually anywhere in the outside world with no compromising of security. Anybody cooperating with us through TresorIT is granted clear privileges particularly for each safe. After any editing, additions and deletions, a notification is sent to use of the changes that have been made.

– What is most appreciated is the functionality for sending secure links to files, where attachments are not emailed, but instead only the link to them in secure data storage is sent. This ensures that sensitive data will not be sent as an insecure attachment to the email and keeps them still under our control.
– Another advantage is that previous versions are also stored and they can be recovered when necessary. Two-factor authentication is also set up, lowering the risk of penetration by unauthorised people.

Sensitive data are accessed on pre-approved devices (in our case, the OSX operating system, iOS and Windows) practically anywhere in the world (of course, the

Table 2. Comparison of cloud services

Criteria	Tresorit	Dropbox	Box	OneDrive	GoogleDrive
Encryption and Security					
Encryption at rest and in Transit	y	y	y	y	y
End-to-End encrypted Storage	y				
End-to-End encrypted sharing	y				
Zero-knowledge authentication	y				
2-step verification	y	y	y	y	y
HIPAA-compliance	y	y	y	y	y
Encryption at rest and in Transit	y	y	y	y	y
Server location	**EU**	**US + EU**	**US + EU**	**US**	**US**
Storage and file management					
Sync any folder	y				
Desktop sync app	y	y	y	y	y
Linux sync app	y	y			
Network drive	y	y	y	y	
Selective synchronization	y	y		y	y
Mobility					
Edit files on mobile devices	y	y	y	y	y
Automatic camera upload	y	y	y	y	y
Remote wipe of mobile devices	y	y	y		y
Passcode lock on mobile devices	y				
Cross device support	y	y	y	y	y
Mobile app rating	4.3	4.4	4.2	4.4	4.3
Deployment					
Support center	y	y	y	y	y
Live chat support	y	y	y		
Deployment support for SMEs	y			y	
On premises deployment	y				
Custom branding	y	y	y		y

system sends a warning of access from a new country, device or browser). Over the entire time this technology has been used, there has never been a security incident and no secure data has ever been stolen or misused. It is very easy to invite someone to collaborate and assign him or her appropriate privileges. In the unlikely event that someone leaves a project or company, it is not difficult to withdraw assigned privileges almost immediately. At the same time, data that do not belong together are kept logically separated. A major advantage is that "you always have your company with you" because the relevant applications can be called up on mobile devices or through the Internet anytime they are needed. The data is always up to date and there is no need to wait for nightly synchronisation. It is very practical, comfortable and spectacular when talking with customer and business partners. Necessary documents can be viewed online and presented with the words "Is this what you had in mind?" Such data storage forces data to be constantly saved to be "in order". Nothing ever needs to be transferred to a USB flash drive, where it would be at risk of damage, loss or theft, so there is no exposure when transmitting work between the office and home.

- Business and private data are kept strictly separated
- Storage capacity is sufficient
- There is a strong administrative section 100% separate from your own data that allows almost immediate blocking of access if the device is lost, so no data would be misused.

4.1   Customer Projects Based on Tresorit and Types of Protection

Solutions have been developed by us for several SME customers for secure data storage and to create for them a method of secure date exchange both inside the company and between the company and its business partners. Such technology has to be provided in order to monitor staff activity whenever there were suspicions and to promptly block staff members from sensitive data if they were to be suddenly dismissed.

4.1.1   Basic Protection
Basic protection means access privileges to a so-called "vault". Only invited users (those who have been granted access) have access to the vault and the data inside it. Depending on the privilege level granted, users may invite others to share data kept in the vault. Local synchronisation can be deactivated and all files opened only in memory, with no traces remaining on the drive even after they are closed.

4.1.2   Enhanced DRM Protection
For highly important MS-Office based documents whose processing should be "for your eyes only", even more security should be deployed with DRM (digital rights management). Documents secured in this way cannot be opened without authorisation and no unauthorised persons are ever permitted to open them. This higher form of protection involves the ability to block some functions to users, such as printing a document or copying part of a text. If an appropriate user is "denied" access to the vault for any reason, the document cannot be opened anytime after the revocation of privileges, even though it remains synchronised in the user's local disk. All that can be done is to delete it. This functionality should be deployed in "office mode" when working with especially sensitive data and where the prevention of leaks by any technical means is required.

4.2   Practical Recommendations and Implementation of Solutions

Implementation involves installing Tresorit in a device, signing into the service (selecting a login and password) and setting security policies that can be later adapted to specific customer needs. There is an option of setting these policies individually, either by users themselves or through a cloud services broker.

5 Conclusion

From a future perspective, it is probable that the value of your data (whether private or corporate) is going to rise. Increasingly more consideration is being given to protecting them because the golden rule of data protection is that "once leaked, it can never be put back". Today we are witnessing the use of unfair practices to obtain confidential information and data about adversaries and competitors that otherwise would have not been naturally acquired. Information security has become a critical issue and will be increasingly more so. Technologies, management approaches and distribution of information are changing, and so are protection methods. Our research will be continuing, more information is going to be analysed and new data protection approaches and solutions are going to be offered.

References

1. PriceWaterhouseCoopers 2016 Global Economic Crime Survey, Slovakia (2017)
2. ISACA: IT Control Objectives for Cloud Computing, Controls and Assurance in the Cloud (2011). ISBN 978-1-60420-182, ISACA J., 50–51 (2016)
3. Controls and Assurance in the Cloud (2011). ISBN 978-1-60420-182, ISACA J. (2016)
4. PricewaterhouseCoopers Slovakia: Industry 4.0: Budovanie Digitálneho podniku. Building Digital Businesses (2017)
5. Tobias Hoellwarth: Cloud Migration, pp. 80–83 (2012). ISBN 978-3-8266-9224
6. Sosinsky, B.: Cloud Computing Bible. Wiley, Indianapolis (2010)
7. Antonopoulos, N., Gillam, L.: Cloud Computing: Principles, Systems and Applications. Springer, London (2012)
8. Krutz, R.L., Vines, R.D.: Cloud Security: A Comprehensive Guide to Secure Cloud Computing. Wiley, Indianapolis (2010)
9. Wang, L., Ranjan, R., Chen, J., Benatallah, B.: Cloud Computing: Methodology, Systems, and Applications. CRC Press, Boca Raton (2011). ISBN: 9781439856413
10. www.tresorit.com
11. www.slant.com
12. www.protonmail.ch
13. www.threema.ch

Cuckoo Optimization Algorithm Based Job Scheduling Using Cloud and Fog Computing in Smart Grid

Saqib Nazir[1], Sundas Shafiq[1], Zafar Iqbal[2], Muhammad Zeeshan[1],
Subhan Tariq[1], and Nadeem Javaid[1(✉)]

[1] COMSATS University, Islamabad 44000, Pakistan
nadeemjavaidqau@gmail.com
[2] PMAS Agriculture University, Rawalpindi Islamabad 46000, Pakistan
http://www.njavaid.com

Abstract. The integration of Smart Grid (SG) with cloud and fog computing has improved the energy management system. The conversion of traditional grid system to SG with cloud environment results in enormous amount of data at the data centers. Rapid increase in the automated environment has increased the demand of cloud computing. Cloud computing provides services at the low cost and with better efficiency. Although problems still exists in cloud computing such as Response Time (RT), Processing Time (PT) and resource management. More users are being attracted towards cloud computing which is resulting in more energy consumption. Fog computing is emerged as an extension of cloud computing and have added more services to the cloud computing like security, latency and load traffic minimization. In this paper a Cuckoo Optimization Algorithm (COA) based load balancing technique is proposed for better management of resources. The COA is used to assign suitable tasks to Virtual Machines (VMs). The algorithm detects under and over utilized VMs and switch off the under-utilized VMs. This process turn down many VMs which puts a big impact on energy consumption. The simulation is done in Cloud Sim environment, it shows that proposed technique has better response time at low cost than other existing load balancing algorithms like Round Robin (RR) and Throttled.

Keywords: Cloud Computing · Fog Computing · Smart Grid
Cuckoo Optimization Algorithm · Round Robin · Throttled

1 Introduction

The demand of electricity has increased enormously from last couple of decades. For example, surveys shows that electricity consumption, only in US has increased to 2.5% annually over the last 20 to 30 years. Traditional grid system fails to distribute the large amount of electricity to the users. In addition traditional energy distribution system does not provides the facility to end users

© Springer Nature Switzerland AG 2019
F. Xhafa et al. (Eds.): INCoS 2018, LNDECT 23, pp. 34–46, 2019.
https://doi.org/10.1007/978-3-319-98557-2_4

to control or monitor their energy requirements. Therefor the idea of SG system come into existence for better and efficient distribution of energy. SG has a two way communication [1], it gives freedom to end users to manage their energy consumption limits. Unlike traditional grid system where user do not have opportunity to manage their energy needs. The idea in SG is to make a live connection between the electricity provider and the users. In this way providers and users will know how much energy is required. To make this connection between the users and providers, integration of Cloud and Fog computing environment with SG is proposed.

Cloud computing is the process of providing computing services to the users over the Internet, these services includes storage and databases, online servers, software and many more [2]. The users have to pay for these services to the cloud providers. Cloud computing is one hot topic these days, companies are moving their non-cloud based products to cloud computing. Cloud can be used to access a large amount of data more quickly, this is where the idea of fog computing come into existence. Since many companies are using cloud computing services now a days, this has increased the number of customers of cloud. So there is a lot of user's requests traffic at cloud, it becomes almost difficult for the cloud to cope up with user requirements. Fog or 'Fogging' is used for this purpose. Fog simply shifts the cloud services [3] to the edge of the network. Fog works as a middle layer between the cloud and users. Simply speaking fog have added more reliable services to the cloud. Fog provides much quicker services to the users than cloud. Moreover it has better security parameters than cloud which is a big issue in cloud computing.

Talking about the integration of cloud and fog with SG, users make many requests at a time. These requests can be generated for electricity requirements or other services. Using cloud and fog with SG a network between the end users and the cloud system is made. User communicates with the fog and it proceeds these requests to the cloud [4]. In proposed case user makes requests for electricity to the fog and fog then response back with the required services. These services are provided by the nearest available source. In proposed scenario there are 3 different levels at the top level there exists a centralized cloud. In second level fogs are installed, each fog have multiple VMs installed in it. Energy consumption area is at the lowest level. There are group of buildings, these group of buildings are divided into different clusters. Each cluster is directly connected with fog in that region. Request are generated from these clusters of buildings to fogs. Data on the fog is stored temporarily, fog send it to the cloud for permanent storage.

The basic purpose is to provide users with their required services in least amount of time, the time taken by the fog to response back to request is called the Response Time (RT) [5]. This is possible only when there is balance between the user's requests and the fogs. Fog takes different amount of time to process a request, the time taken by a fog to process a request is called the Processing Time (PT). When a fog is over utilized it will take a lot of time to process the request, on the other hand an under-utilized fog will perform these operations in less time. In this research we have used COA for the optimization of over and underutilized fogs.

The rest of the paper is organized as follows. In Sect. 2 motivation and related work is presented in Sect. 3. The proposed system model and COA is described in Sect. 4. Finally we discuss the simulation results and conclusion in Sects. 5 and 6.

2   Motivation

A cloud-fog based model is presented in [1]. An integration of cloud and fog is done for the effective resource distribution in smart buildings. The purpose of this work is to provide services to the end users with minimum RT. Different load balancing algorithms like Throttled and RR are used for job scheduling in Micro Grids (MGs) [2]. In this research our focus is in the combination of cloud and fog for optimization problems. Cloud Computing is becoming an essential part of our lives. Many companies are using this technology now a days like Amazon, Google, and Microsoft. Companies are moving their non-cloud based application towards cloud computing [6]. The increase in number of cloud computing customers is causing more energy consumption and more user request traffic in cloud data centers. So it is essential to manage the energy system and requests coming from users [7].

In this paper COA is used for load balancing purpose. Multiple requests comes from users at a time, management of these requests is not an easy task. In this scenario there are fogs connected to the clouds and the VMs. User put a request to the fog via VM, sometimes some fogs are over-utilized means they are dealing with too many requests. The over-utilized fogs cannot satisfy the need of request allocated to it. Some fogs on the other hand are under-utilized. These fogs are not getting as many requests as over-utilized fogs but they are consuming the same amount of energy as the over-utilized fogs. So our goal is to first of all detect these over-utilized and under-utilized fogs. If possible migrate some of VMs from over-utilized fogs to other fogs and reduce their utilization, or move all the VMs of under-utilized fogs to over-utilized fogs and switch them into sleep mode. In our case all the hosts or fogs that are not under-utilized are considered as over-utilized. By switching the under-utilized fogs into the sleep mode and migrating all of their VMs to over-utilized fogs will give more opportunity to over-utilized fogs to deal better with the request being generated by the users. Putting some fogs into sleep mode will make a big impact on the consumption of electricity.

3   Related Work

In [1], the authors proposed an integration of cloud and fog based environment for effective resource distribution in residential areas. Different load balancing algorithms like RR, Throttled are used for with service broker policies for resource allocation. A new service broker policy is introduced with the name of dynamic service proximity for the selection of most suitable fog to perform a task. Later

Table 1. Related work

Algorithm	Objectives	Achievements	Limitations
Dynamic Service Proximity [1]	Load Balancing and Energy Management	Selection of nearest and best Fog to do a job	Maximum Cost
Practical Swarm Optimization (PSO) VM Load Balancer [2]	Efficient Resource Utilization	PSO achieved better load balancing technique than other	Maximum RT
Shortest Job First [3]	Resource Allocation	Better Response Time	Ignored Optimization of PT
PSO with Simulated Annealing (PSOSA) [4]	Efficient Resource Allocation	Reduced RT, PT, Cost	Maximum VM Cost
COA for Job Scheduling [5]	Load Balancing and Job Scheduling	Scheduled Cloud Services	Unnoticed Cost of VM
Hybrid of COA and PSO [6]	For Clustering and Load Balancing	Global Optimum Solution	Insensitive to Centroids initialization
COA [7]	Switching the VMs to required locations	Reduced Energy Consumption	RT is not Measured
COA with Group Technology (GT) [8]	VMs and Cloud Computing servers consolidation	Efficient resource distribution by consolidation VMs and Cloud servers	Unnoticed VM cost

on results of this policy are compared with other existing techniques like RR and Throttled (Table 1).

A cloud-fog based SG model is presented in [2] for efficient utilization of resources. The focus of authors in this paper is to manage the information flow in SG infrastructure based on cloud-fog. PSO is used for load balancing purposes. The results of PSO are compared with RR and throttled. PSO however shows better results than other two techniques.

For better resource allocation in Smart Buildings, authors in [3] have proposed a cloud to fog to consumer based framework. A three layer network having: cloud, fog and consumer layer is proposed. Authors presented Shortest Job First algorithm for optimization of RT, PT, request per hour and cost in term of MGs and VMs. Although proposed model outperformed other techniques on the bases of RT, and cost of VMs.

In [4], authors proposed a new algorithm PSO with Simulated Annealing (PSOSA) for load balancing. The idea is to balance the load of users request on cloud and reduce power generation. In system model clusters of buildings are connected to fogs and fog is connected to cloud to provide the required services to the users. Later on simulation results of PSOSA are compared with RR, Throttled and Cuckoo Search (CS) to check the performance of proposed model.

COA is used for task scheduling optimization in cloud computing in [5]. Authors focused on minimizing the overall execution time and cost time. These parameters have been improved using load balancing algorithm. Authors proposed a hybrid algorithm consisting of improved COA and PSO for load balancing and clustering in [6].

In [7], the authors describes COA for switching parameters. This paper concentrates on switching and migrating resources to the required users.

A discrete COA with group technology to solve the problem of consolidation in cloud computing environment is discussed in [8]. This problem consists of cloud computing server consolidation with VM for the resource allocation.

4   Proposed Model

In this paper a cloud-fog based model is presented using COA. Fog computing works as intermediate between user request and cloud. Fog is relatively new term than cloud computing. Unlike cloud, fog is a user level server with better RT than cloud. Fog also provides better security parameters than cloud where security is a big issue [8]. Fog although added more services to the cloud by reducing the load on cloud computing. In proposed model, a three level architecture is presented as shown in Fig. 1. At the top level there exists a centralized cloud which communicates with the fog at the second level and at the lowest level is the energy consumption area like group of buildings. All the devices at each level communicates with each other to process the requests generated by the users. Each building has a controller attached to it, these controllers communicate with fog and then fog communicates with cloud. The fog is directly connected with the buildings. There are different clusters of buildings. Each clusters have 10 to 20 buildings, further these buildings have 40 to 60 apartments. The requests are generated from the apartments (end users) to fog, then fog communicate with available VM. VM response back and provide the required actions.

If somehow VM is not available fog communicates with cloud to find another available VM. In this paper the world is divided into six continents, each continent has two fogs. These fogs are then connected with centralized cloud. Fog respond to the requests generated by the end users.

Our goal is to generate results using COA and compare these results with other two existing algorithms Throttled and RR. Main focus is to evaluate the efficiency of proposed algorithm as compare to other two techniques.

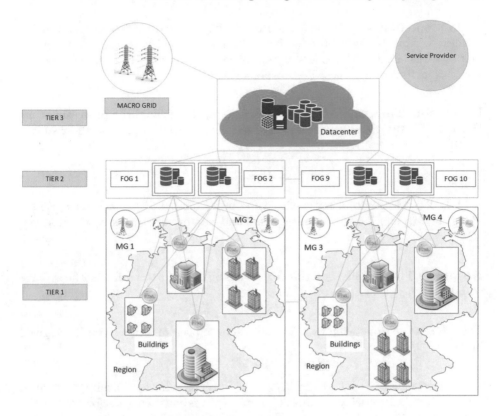

Fig. 1. Proposed system model

4.1   Problem Formulation

In proposed scenario we have considered a three tier model. Bottom layer consists of smart buildings connected with fog which is placed at the middle layer and fog is further connected with the cloud at the top level. The cloud Data Centers (DC) are present at top layer with n number of Servers (S).

$$DC = \{dc_1, dc_2, dc_3 \cdots dc_n\} \tag{1}$$

$$S = \{s_1, s_2, s_3 \cdots s_n\} \tag{2}$$

In this paper, system performance will be evaluated by analyzing three parameters using COA, average PT, average RT and the total cost. The fog layer consists of number of VMs and MGs. VMs processes the different number of tasks T at a time, the mathematical representation of the set of VMs and tasks T is shown below

$$VM = \{vm_1, vm_2, vm_3 \cdots vm_n\} \tag{3}$$

$$T = \{t_1, t_2, t_3, \cdots, t_n\} \tag{4}$$

In proposed model all the VMs are working in parallel. The tasks are being generated from the users randomly. VMs have to deal with the heterogeneity of the tasks. Each VM process a request generated from user on the bases of processing time of a request and response time.

Processing Time of VM. Next we are going to calculate the processing time (PT). PT is the time to process a request and provide user with the required results.

$$PT_{total} = \sum_{i=1}^{n} \sum_{j=1}^{m} (PT_{ij} * R_{ij}) \tag{5}$$

PT denotes the total processing time of VM assigned to each task. Where PT_{ij} and R_{ij} shows the processing time and number of request received at n^{th} VM.

Response Time of VM. In this paper, the objective is to minimize the average response time.

$$minimize(RT_{avg}) \tag{6}$$

Total RT is the time taken by a VM to response back to a task. The RT is given below

$$RT_i = FT_i - AT_i + T_{delay} \tag{7}$$

RTi denotes total RT of the VM and FT represents the finishing time for a task at VM i. The delay is caused when a task is not being performed on the desired time. So when the we are able to minimize the delay it will automatically minimizes the total RT. There are two type of time delays being considered in this paper, transfer time and the latency rate. DT is calculated as follows:

$$DT = T_{transfer} + T_{latency} \tag{8}$$

where latency delay is the delay in the desired outcome of a user request from the VM. Transfer time is the time taken by a VM to transfer a request from user to the server.

$$T_{transfer} = DT/Bandwidth \tag{9}$$

here bandwidth is represented by:

$$Bandwidth = Total_{Bandwidth}/m \tag{10}$$

m shows total number of requests pending to be processed.

Cost. Next parameter is the total cost evaluation. We calculated the VM cost, data transfer cost and the MG cost in proposed model. In proposed model we compute the VM cost on the bases of the Multiple Instructions per Second (MIPS) and the request size.

$$Total_{cost} = VM_{cost} + MG_{cost} + DC_{cost} \tag{11}$$

The total cost consists of cost at data centers, VM cost and MG cost. Where data transfer cost is evaluated on the bases of total data transmitted. The total number of MGs installed in the system gives the MG cost. Different fogs have different amount of VMs installed in it so the cost of VMs is also changing. One VM cost in one hour is 0.1\$. So the VMcost with is as follows:

$$VM_{cost} = 1 * VM_{cost}/hour * totalVMs \qquad (12)$$

Formulation of MG cost is represented below:

$$MG_{cost} = MG^i_{cost} * MG^i \qquad (13)$$

Here MG cost shows the total cost which is equal to the cost at i^{th} VM.

4.2   Cuckoo Optimization Algorithm

Cuckoo is a lazy bird, which lays eggs in the other bird's nests. The eggs laid by the cuckoo are the similar to the existing eggs of the nests. These nests are called the host's nests monitored by the host birds. The similarity between eggs increases the growth probability of cuckoo [9], later on these eggs turn into the mature cuckoo. This is how the process goes on, life of cuckoo depends on the probability of growth of these eggs. Sometimes these eggs may be detected by the host and removed or destroyed from the nests. The increasing number of survived eggs in an area is used by the cuckoo for optimization. So group of cuckoos from the other locations migrate toward this area.

The mature cuckoo randomly lays eggs at different locations, the process continues until nests with the maximum number of survived eggs is found. Cuckoos from other places gather around it. COA is based on the life of cuckoo bird. COA is fast and efficient algorithm so it is suitable for optimization problems [10]. COA is best for solving discrete problems so we have to change the nature of continual processes to discrete. First of all some host nests are selected randomly. The characteristic of these host nests are given in array, which is called the habitat for cuckoos.

At this point profit of hosts is evaluated by the COA. Egg Laying Radius (ELR) [9] is set for the cuckoo to lay eggs. Each cuckoo lay eggs based on ELR randomly. When cuckoo is done laying eggs, profit value of this nest is calculated. COA also uses the K-mean clustering method to cluster the population of cuckoos. When the clustering is done, it will generate the different number of groups of cuckoos. COA now calculates the fitness of all groups. At the end, group with best fitness is selected for the optimization. The goal point is set for the other cuckoos to gather around.

COA consists of different parts. At first, detection of under and over-utilized fogs by applying the COA. When over-utilized hosts are detected, the algorithm tends to move some of their VMs to some other hosts. For under-utilized hosts a simple policy is defined in our scheme, all the hosts which are not the over-utilized host are considered as under-utilized hosts. The idea is to transfer the VMs of this host to other hosts and turn them into sleep mode. If they are not switched into the sleep mode they will remain active, they are not dealing with as many request as over-utilized hosts are but they are using the same amount of energy as the over-utilized host [11]. When hosts are detected next step is to migrate their VMs.

Algorithm 1. COA Algorithm

1: Input: List of VMs, List of Tasks
2: Calculate population of host nests n,
3: Evaluate fitness f,
4: Calculate Number of eggs,
5: **while** k>maximum generation **do**
6:     K=k+1
7:     Randomly select cuckoo
8:     **if** (f(x) >f(y)) **then**
9:         Replace k with new solution
10:     **end if**
11:     Abandon the worst nests
12:     Evaluate the fitness again
13:     Find current best index
14: **end while**
15: Return optimal VM
16: Show results

Detection of Over Utilized Fogs. An over-utilized fog which is receiving too many request at a time cannot serve them all. As a result this will maximize the RT of a fog. To tackle this problem and to find the over-utilized fogs we have employed the COA. COA is relatively new and fast scheme for optimization. Although it does need to change the nature of continual processes to the discrete process. By K-mean clustering [6] different number of groups are generated, the algorithm then calculates the mean profit of these groups. The groups with the maximum profit values are selected. At this point over-utilized fogs are determined.

Detection of Under Utilized Fogs. When a fog does not receive many requests and work load is low then it is under utilized fog. In this paper we mark all the fogs under utilized if they are not over-utilized. In case of energy consumption under utilized hosts use the same amount of energy as the over-utilized fogs. To overcome this problem simply transfer all the VMs of these fogs to the other fogs and turn them off. This process is performed for all the fogs and it will turn off as many fogs as possible. As a result, number of active fogs and energy consumption will be decreased.

Selection Policy. When a fog becomes over utilized, we should migrate its VMs to other fogs by keeping in mind that it would not create other fogs over utilized. The migration process should be performed in such a way that it will keep the balance between the fogs. In case migration process makes other fogs over utilize there is no need of migration.

4.3 Throttled

Throttled is a load balancing technique, where a single task is assigned to a VMs at a time. As long as task is being performed by the VM, status of this VM remains busy. It means a VM cannot perform multiple tasks at the same time.

4.4 Round Robin

RR is simple algorithm for resource allocation to all the fogs by providing them with equal amount of time slice. This time slice is defined by the user. In this paper, RR is used for the load balancing purpose at each VM.

5 Simulation Results

In this paper we have used (Cloud Analyst) tool for simulation purpose. We have compared the results of our proposed load balancing algorithm with other techniques like Throttled and RR. Basically our model consists of six regions with two fogs per region. Each fog is dealing with the number of clusters. These clusters have number of buildings in it. Request are being generated by the apartments for electricity needs and fogs are dealing with these requests via VMs. The simulations are performed to compare the average RT, average PT and costs at each load balancing algorithm.

5.1 Average Response Time

Since we have used COA in this research, the results of COA are compared with other algorithms like RR and Throttled to check its efficiency. RT is the time taken by any fog to response to the request generated by the user [12]. At the on peak hours it could be very low and at the off peak hour's response of fog is very fast. So in this section we have compared the average RT of each cluster according to all the techniques. Average RT at each cluster is shown in Fig. 2. The results shows that most of the time average RT of COA is relatively equal to the average RT of both RR and Throttled. Sometimes it may be lower to average RT of RR or Throttled.

Figure 2 shows the graphical comparison of each optimization algorithm according to the average RT. The idea is to choose a closest fog to get the better RT.

The important thing is that average RT of COA is higher than RR. As shown in the table cluster four (C4) have average RT of 52.89 at COA and average RT of 52.76 at RR. This shows the performance of COA is better than RR.

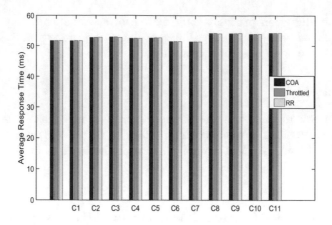

Fig. 2. Average response time

5.2   Average Processing Time

PT is the time taken by a fog to process a request generated by the user. The comparison shows that COA is taking less time to service any request allocated to it. The PT taken by fog 1 at COA is less than Throttled and RR. Due to the less network delay fog 1 have maximum average PT. Similarly fog 9 have more requests allocated to it so the PT is maximum.

Figure 3 shows the graphical comparison of average PT at each fog between COA, Throttled and RR.

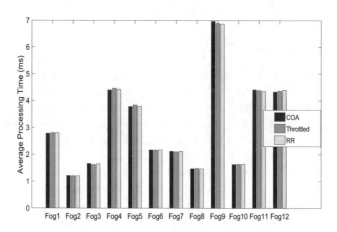

Fig. 3. Average processing time

5.3   Cost Optimization

Each fog is servicing multiple requests in an hour. Our goal is to minimize the cost in on peak hours. There are twelve different fogs in all the regions each fog have VM cost, MG cost and data transfer cost. In this section the total costs of all these parameters are compared according to all three algorithms. Figure 4 shows the total cost of each fog using COA, RR and Throttled. The comparison shows that the COA has minimum total cost at most of the fogs as compare to other techniques.

Figure 4 illustrate that maximum peaks are observed at fog 4, 9 and 12. This shows that these fogs are currently dealing with many requests.

The discussion shows that using COA better performance can be obtained. Although at some points RR and Throttled shows better results than COA. But most of the time COA is cost efficient and gives better response time.

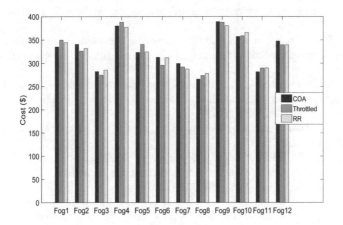

Fig. 4. Total cost

6   Conclusion

Data traffic at cloud servers is being increased day by day. Users make requests to servers to fulfill their energy needs. Since traditional grids are being converted into SG, there exists a live connection between the consumers and the energy providers. Due to the increase in the number of users and their request a mechanism is required to balance the load of these request. In this paper a load balancing algorithm is proposed with the integration of cloud and fog computing to balance the user's requests in residential areas. We used COA to balance the load of user's requests. This implementation provides the required features for load balancing, job scheduling, connectivity and the energy management. The simulations are performed in Cloud Analyst. The results of proposed load balancing algorithm are compared with existing techniques. However results shows that performance of COA are better than existing optimization algorithms.

References

1. Fatima, I., Javaid, N., Iqbal, M.N., Shafi, I., Anjum, A., Memon, U.: Integration of cloud and fog based environment for effective resource distribution in smart buildings. In: 14th IEEE International Wireless Communications and Mobile Computing Conference (IWCMC-2018) (2018)
2. Zahoor, S., Javaid, N., Khan, A., Muhammad, F.j., Zahid, M., Guizani, M.: A cloud-fog-based smart grid model for efficient resource utilization. In: 14th IEEE International Wireless Communications and Mobile Computing Conference (IWCMC-2018) (2018)
3. Javaid, S., Javaid, N., Tayyaba, S.K., Sattar, N.A., Ruqia, B., Zahid, M.: Resource allocation using Fog-2-Cloud based environment for smart buildings. In: 14th IEEE International Wireless Communications and Mobile Computing Conference (IWCMC-2018) (2018)
4. Yasmeen, A., Javaid, N., Iftkhar, H., Rehman, O., Malik, M.F.: Efficient resource provisioning for smart buildings utilizing fog and cloud based environment. In: 14th IEEE International Wireless Communications and Mobile Computing Conference (IWCMC-2018) (2018)
5. Abbasi, M.J., Mohri, M.: Scheduling tasks in the cloud computing environment with the effect of Cuckoo optimization algorithm. Int. J. Comput. Sci. Eng. **3**, 1–9 (2016)
6. Bouyer, A., Hatamlou, A.: An efficient hybrid clustering method based on improved cuckoo optimization and modified particle swarm optimization algorithms. Appl. Soft Comput. **67**, 172–182 (2018)
7. Mareli, M., Twala, B.: An adaptive Cuckoo search algorithm for optimisation. Appl. Comput. Inf. (2017)
8. Tavana, M., Shahdi-Pashaki, S., Teymourian, E., Santos-Arteaga, F.J., Komaki, M.: A discrete Cuckoo optimization algorithm for consolidation in cloud computing. Comput. Ind. Eng. **115**, 495–511 (2018)
9. Tusiy, S.I., Shawkat, N., Ahmed, M.A., Panday, B., Sakib, N.: Comparative analysis on improved Cuckoo search algorithm and artificial Bee colony algorithm on continuous optimization problems. Int. J. Adv. Res. Artif. Intell. **4**, 14–19 (2015)
10. Chen, S.-L., Chen, Y.-Y., Kuo, S.-H.: CLB: a novel load balancing architecture and algorithm for cloud services. Comput. Electr. Eng. **58**, 154–160 (2017)
11. Moghaddam, M.H.Y., Leon-Garcia, A., Moghaddassian, M.: On the performance of distributed and cloud-based demand response in smart grid. IEEE Trans. Smart Grid (2017)
12. Capizzi, G., et al.: Advanced and adaptive dispatch for smart grids by means of predictive models. IEEE Trans. Smart Grid (2017)
13. Faruque, A., Abdullah, M., Vatanparvar, K.: Energy management-as-a-service over fog computing platform. IEEE Internet Things J. **3**(2), 161–169 (2016)
14. Kumar, N., Vasilakos, A.V., Rodrigues, J.J.P.C.: A multi-tenant cloud-based DC nano grid for self-sustained smart buildings in smart cities. IEEE Commun. Mag. **55**(3), 14–21 (2017)
15. Reka, S.S., Ramesh, V.: Demand side management scheme in smart grid with cloud computing approach using stochastic dynamic programming. Perspect. Sci. **8**, 169–171 (2016)

Detection of Defects on SiC Substrate by SEM and Classification Using Deep Learning

Shota Monno, Yoshifumi Kamada, Hiroyoshi Miwa[✉], Koji Ashida, and Tadaaki Kaneko

Graduate School of Science and Technology, Kwansei Gakuin University, 2-1 Gakuen, Sanda-shi, Hyogo, Japan
{shota.m,441k,miwa,dsv00027,kaneko}@kwansei.ac.jp

Abstract. In recent years, next generation power semiconductor devices using semiconductors with large band gap such as SiC (Silicon Carbide) attract attention. It is very important to detect crystal defects, surface processing defects including polishing, defects contained in the SiC substrate, defects included in the epitaxial growth film, defects caused by the device forming process, and so on. This is because elucidating the cause of the detected defect and investigating the influence on device quality and reliability lead to development of a better manufacturing method. Recently, observation with a low energy scanning electron microscope (LE-SEM) which is more accurate than C-DIC and PL has been put to practical use. As a result, crystal information of just below the outermost surface can also be obtained. However, since image processing techniques targeting SEM images of SiC substrates have not existed so far, it has not been possible to efficiently and automatically extract defects from enormous amounts of data. In this paper, we propose a method for detecting defects on SiC substrate by SEM and classifying them using deep learning.

1 Introduction

In recent years, next generation power semiconductor devices using semiconductors with large band gap such as SiC (Silicon Carbide) attract attention. For example, we can realize a device having a high power and high temperature resistance by using SiC as compared with the conventionally used Si device [1].

It is very important to detect crystal defects, surface processing defects including polishing, defects contained in the SiC substrate, defects included in the epitaxial growth film, defects caused by the device forming process, and so on. This is because elucidating the cause of the detected defect and investigating the influence on device quality and reliability lead to development of a better manufacturing method.

© Springer Nature Switzerland AG 2019
F. Xhafa et al. (Eds.): INCoS 2018, LNDECT 23, pp. 47–58, 2019.
https://doi.org/10.1007/978-3-319-98557-2_5

Recently, observation with a low energy scanning electron microscope (LE-SEM) which is more accurate than C-DIC and PL has been put to practical use [2]. As a result, crystal information of just below the outermost surface can also be obtained. However, since image processing techniques targeting SEM images of SiC substrates have not existed so far, it has not been possible to efficiently and automatically extract defects from enormous amounts of data.

In this paper, we propose a method for detecting defects on SiC substrate by SEM and classifying them using deep learning. Furthermore, we develop a system based on the method and evaluate the performance by applying to actual images taken by SEM of SiC substrate.

2   Related Works

In this section, we describe the related studies on detecting defects on images.

Since one shooting range (we call it a tile in the rest of the paper) by SEM is extremely small as compared with an entire SiC substrate, it must be shot in multiple times. In addition, due to mechanical factors, overlapping or misalignment may occur between adjacent tiles. Therefore, it is necessary to accurately combine the tiles. It is called a stitching correction. There are some studies on stitching correction (ex. [3]). The reference [3] extracts feature points from the targeted image by using the amount of characteristic called SIFT and matches the feature points using RANSAC which is a type of robust estimation. The SIFT is effective for the panorama image of the landscape targeted by [3]; however, it is not appropriate to deal with the noise of the SEM image, because the SIFT outputs a lot of false detection of feature points.

There are some studies on classification of defects. The references [4] automatically generate defect classification rules from decision trees based on some predefined feature values for defects on semiconductor wafers taken by an optical microscope. When photographed by SEM, there are many overlapping of defects and difference of scale, which are the properties not seen in defects observed with the optical microscope. Therefore, it is not easy to apply the previous methods using some predefined feature values.

Deep learning is an effective approach to classification of defects. A convolution neural network (CNN) [5] is one of deep learning methods. Many models of CNN have been actively researched [6–9].

VGG [6] was developed by Visual Geometry Group of Oxford University. VGG has 16 layers and 19 layers models, which are called VGG-16 and VGG-19. It indicated excellent performance in Large Scale Visual Recognition Competition(ILSVRC) [10] held in 2014.

GoogLeNet [7] is also a model showing outstanding performance in ILSVRC2014. It is a model of CNN developed by Google, and it has 22 layers. Even after that, development has been continued under the name of Inception. Inception V3 [8] has succeeded in further improving performance.

ResNet [9] is a model called a residual network, developed by Microsoft. This aims to deal with problems called gradient loss caused by only deepening the layer. Since the gradient loss is caused by the fact that the difference between the input and the output becomes extremely small as the learning reaches to the deeper layer, ResNet directly uses the difference between the input and the output in layers.

There is the reference [11] as an example using deep learning in the medical field. This uses a technique called transfer learning. By using transfer learning, it is possible to construct a model even with only a small amount of data by using parameters of a model obtained from a large data set.

3 Method for Detection of Defects on SiC Substrate by SEM and Classification Using Deep Learning

3.1 System Architecture

In this section, we describe a method for detection of defects on SiC substrate by SEM and classification using deep learning.

First, the tile combining method is described in Sect. 3.2. Gap and overlap are detected by using linear defects as a clue. Since a tile includes much noise and difference of contrast, they are removed first and linear defects are extracted. When there are multiple linear defects and they are contained in multiple tiles, the overlap among the tiles is estimated by the distance between any pair of the linear defects, and the coordinates of the tiles are determined. Thus, all tiles are combined and an image of the entire SiC substrate is generated.

Next, a method of detecting and classifying defects is described in Sect. 3.3. The patterns of defects are difficult to geometrically characterize. Therefore, we aim to design a method of automatically extracting defects and positions from the entire SiC substrate image by using the deep learning.

We show the concept of the proposed system in Fig. 1.

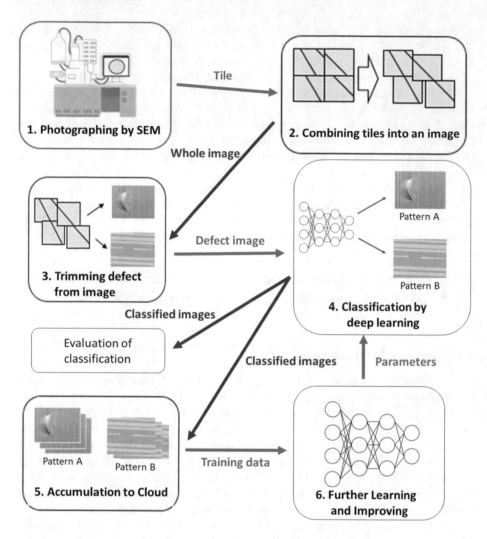

Fig. 1. Concept of the proposed system

3.2   Tile Combining Method

First, since a SEM image contains white noise, noise is smoothed using a Gaussian filter. Further, adaptive binarization is performed while reducing influence of contrast. Thus, linear defects are emphasized. Then, linear defects are detected by using Hough transformation.

Next, we describe an algorithm correcting horizontal gap of two vertically adjacent tiles. We define the gap between two tiles by the number of pixels. For the vertically adjacent two tiles, let T_1 be upper boundary of the lower tile and T_2 be the lower boundary of the upper tile. When the gap is s between T_1 and T_2, the number of pairs such that the position of a linear defect in T_1 is the same position of a linear defect in T_2 is counted, and the gap s such that the number of pairs is the maximum is found by increasing s from 0 to the size of the tile one by one. The idea is shown in Fig. 2.

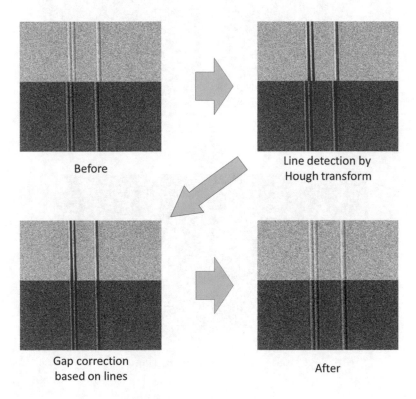

Fig. 2. Horizontal gap correction

Next, we describe an algorithm estimating the width of the overlapped part. For the horizontally adjacent two tiles, let T_l be the left boundary of the right tile and T_r be the right boundary of the left tile. Let $maxlap$ be the maximum overlap width between T_l and T_r, and the tile is cut by s' from the right side of the left tile and the left side of the right tile, respectively. Then, the cut-out parts are compared and the number of pairs such that the position of a linear defect in T_l is the same position of a linear defect in T_r is counted, and the gap s' such that the number of pairs is the maximum is found by increasing s' from 0 to the size of the tile one by one.

Thus, the relative positions of the tiles are determined. We show an example in Fig. 3.

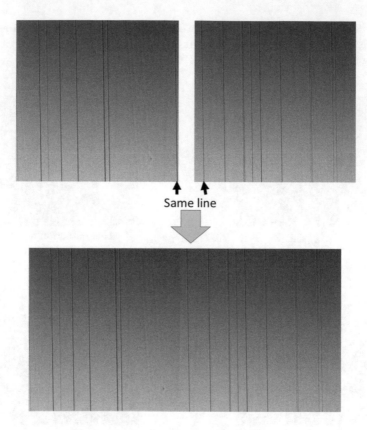

Fig. 3. Stitching correction

3.3 Method of Detecting and Classifying Defects

We describe a method of detecting defects on SiC substrate by SEM and classifying them using deep learning. The method automatically classifies defects included in a rectangular area cut out by an operator. We use a convolutional neural network (CNN) which is known to be able to classify defects highly accurately. We use Inception v3 [8] and ResNet [9] as the models of CNN.

Since there is not much training data for learning, we attempt an approach of transfer learning. In transfer learning, a model is learned in advance for a data set different from the data set of the target, and the model using a part of the learned model learns again for a data set of the target. We use only fully connected layers of Inception v3 and ResNet for transfer learning.

4 Performance Evaluation

In this section, we evaluate the system developed based on the proposed method in Sect. 3 by applying to actual images taken by SEM of SiC substrate.

First, we compare the image manually corrected by an operator and the image automatically corrected by the system.

The image of 4 × 4 tiles manually corrected by an operator is shown in Fig. 4.

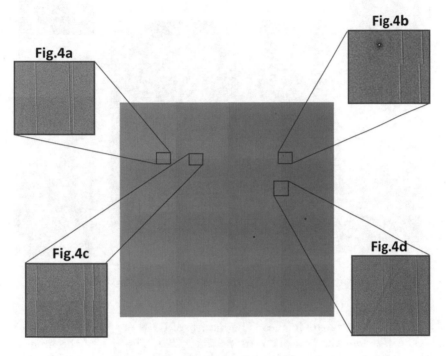

Fig. 4. Manually correction by operator

Next, the image automatically corrected by the proposed system is shown in Fig. 5.

Comparing the Fig. 4 with the Fig. 5, the contradictions of the positions of linearly defects are corrected. As shown in Fig. 4a in Fig. 4, there are some parts that are accurately combined even with manual correction, but there are many contradictions as shown in Fig. 4c. On the other hand, the contradiction can not be seen in Fig. 5.

However, since the proposed system corrects based on the linear defects, it is not possible to correct based on other defects. The overlap in Fig. 4b and d in Fig. 4 can be manually corrected; however, Fig. 5b and d in Fig. 5 are not corrected by the system. This is because non-linear defects are found by the system.

Next, we evaluate the performance of detection and classification of defects in the proposed system.

We used 736 images in total 7 defect classes are divided into 608 images for training and 128 images for test (validation). Furthermore, the number of these images are increased by rotating, vertically and horizontally inverting, slightly

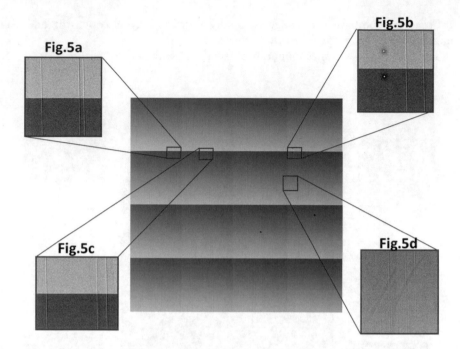

Fig. 5. Automatic correction by proposed system

enlarging or reducing the images. The models of CNN are learned by images in ImageNet [12] and the transfer learning is applied. The size of the input layer is 224 × 224, and the output layer uses the dense layer provided by Keras [13]. The dense layer is the usual fully connected layer. As the activation function of this part, softmax function is applied. The optimizer and the loss function in the back propagation method are the RMS prop and the cross entropy error method, respectively, which are often used in the transfer learning. For the mini batch learning, the number of batches is 32, the number of the samples in training data is 704, and the number of the samples of test data is 96. The number of epochs is 50.

We show an example of a typical SEM image in each defect class in Table 1.

First, we show the results of defect classification by the Inception v3. The training time is about 4.5 h by using Intel Xeon E5-2630V3 (CPU), Geforce GTX-1080 (GPU), and 32 Gbyte (RAM). The learning process for each epoch is shown in Figs. 6 and 7.

In Figs. 6 and 7, the horizontal axis represents the number of epochs and the vertical axis represents loss and accuracy. The accuracy converges at around 60%, which is not high.

Next, we show the results of ResNet50. The training time is about 9 hours by using Intel Xeon E5-2630V3 (CPU), Geforce GTX-1080 (GPU), and 32 Gbyte (RAM). The learning process for each epoch is shown in Figs. 8 and 9.

Table 1. Example in each defect class

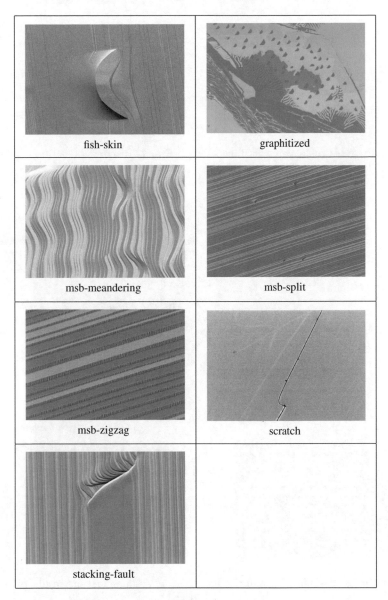

The accuracy is about 70%, which is better than Inception v3. Therefore, in the rest of the paper, we use the ResNet model.

We were able to correctly classify the fish-skin class, the graphitized class, scratch class, msb-meandering with high accuracy. We show an example of the graphitized class in Figs. 10 and 11.

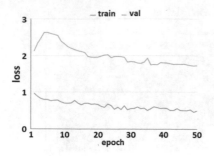

Fig. 6. Progress of learning in Inception v3 (loss)

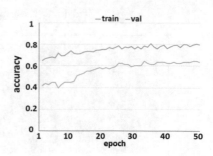

Fig. 7. Progress of learning by Inception v3 (accuracy)

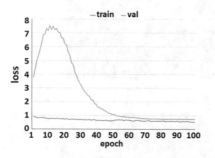

Fig. 8. Progress of learning by ResNet50 (loss)

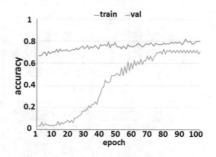

Fig. 9. Progress of learning by ResNet50 (accuracy)

Fig. 10. Image sample of graphitized class

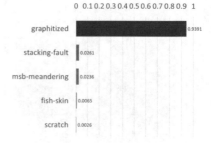

Fig. 11. Classification result of graphitized class

An example of msb-zigzag image is shown in Fig. 12 and in Fig. 13; An example of stacking-fault image is shown in Fig. 14 and in Fig. 15.

Although Fig. 14 is the stacking-fault image, the scratch has the highest score in Fig. 15, which is incorrect as a classification. This is because these defects are similar, and the classification is difficult even by experts. Similarly, although

Fig. 12. Image sample of msb-zigzag class

Fig. 13. Classification result of msb-zigzag class

Fig. 14. Image sample of stacking-fault class

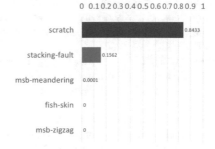

Fig. 15. Classification result of stacking-fault class

Fig. 12 is the msb-zigzag image, the msb-split has the highest score in Fig. 13, which is incorrect as a classification. However, it was found by some experts that the features of msb-split also appeared in this image, and it was found that clear classification is difficult.

5   Conclusions

In this paper, we proposed the method for detecting defects on SiC substrate by SEM and classifying them using deep learning. Moreover, we developed the system based on the method. This system consists of the tile combining function and the defect detection and classification function. The tile combining function has the purpose of accurately generating an entire image combined from multiple tiles photographed by SEM. In the defect detection and classification function, an operator cuts out a rectangle including a defect image from the entire image and a defect included in the rectangle are automatically detected and classified by using convolution neural network, a kind of deep learning.

As a result of the performance evaluation, the system can combine multiple tiles without inconsistency of linear defects and can detect and classify the seven defects classes, which achieved classification accuracy equivalent to experts.

References

1. Pushpakaran, B.N., Subburaj, A.S., Bayne, S.B., Mookken, J.: Impact of silicon carbide semiconductor technology in Photovoltaic Energy System. Renew. Sustain. Energy Rev. **55**, 971–989 (2016)
2. Ashida, K., Kajino, T., Kutsuma, Y., Ohtani, N., Kaneko, T.: Crystallographic orientation dependence of SEM contrast revealed by SiC polytypes. J. Vac. Sci. Technol. B **33**(4), 04E104 (2015)
3. Brown, M., Lowe, D.G.: Automatic panoramic image stitching using invariant features. Int. J. Comput. Vis. **74**, 59–73 (2007)
4. Shibuya, H., Okabe, T., Nakagawa, Y.: Automatic generation technique of defect classification rules using decision tree. J. Inst. Image Electr. Eng. Jpn. **36**, 731–737 (2007)
5. LeCun, Y., Bottou, L., Bengio, Y., Haffner, P.: Gradient-based learning applied to document recognition. Proc. IEEE **86**(11), 2278–2324 (1998)
6. Simonyan, K., Zisserman, A.: Very deep convolutional networks for large-scale image recognition. In: Advances in Neural Information Processing Systems (NIPS) (2012)
7. Szegedy, C., Liu, W., Jia, Y., Sermanet, P., Reed, S., Anguelov, D., Erhan, D., Vanhoucke, V., Rabinovich, A.: Going deeper with convolutions. In: Proceedings of the IEEE Conference on Computer Vision and Pattern Recognition (CVPR), pp. 1–9 (2015)
8. Szegedy, C., Vanhoucke, V., Ioffe, S., Shlens, J., Wojna, Z.: Rethinking the inception architecture for computer vision. In: Proceedings of the IEEE Conference on Computer Vision and Pattern Recognition (CVPR), pp. 2818–2826 (2016)
9. He, K., Zhang, X., Ren, S., Sun, J.: Deep residual learning for image recognition. In: Proceedings of the IEEE Conference on Computer Vision and Pattern Recognition (CVPR), pp. 770–778 (2016)
10. http://www.image-net.org/challenges/LSVRC
11. Shin, H.-C., Roth, H.R., Gao, M., Le, L., Ziyue, X., Nogues, I., Yao, J., Mollura, D., Summers, R.M.: Deep convolutional neural networks for computer-aided detection: CNN architectures, dataset characteristics and transfer learning. IEEE Trans. Med. Imaging **35**, 1285–1298 (2016)
12. Dong, W., Socher, R., Li, L.-J., Li, K., Fei-Fei, L.: ImageNet: a large-scale hierarchical image database. In: Proceedings of the IEEE Conference on Computer Vision and Pattern Recognition (CVPR) (2009)
13. Chollet, F.: Keras (2015). https://github.com/fchollet/keras

Semantic Analysis of Social Data Streams

Flora Amato[1]([⊠]), Giovanni Cozzolino[1], Francesco Moscato[2], and Fatos Xhafa[3]

[1] Department of Electrical Engineering and Information Technology,
University of Naples "Federico II", Naples, Italy
{flora.amato,giovanni.cozzolino}@unina.it
[2] Department of Scienze Politiche,
University of Campania "Luigi Vanvitelli", Caserta, Italy
francesco.moscato@unicampania.it
[3] Department of Computer Science,
Universitat Politècnica de Catalunya, Barcelona, Spain
fatos@cs.upc.edu

Abstract. Social Networks Analysis has become a common trend among scholars and researchers worldwide. A great number of companies, institutions and organisations are interested in social networks data mining. Information published on many social networks, like Facebook, Twitter or Instagram constitute an important asset in many application fields, overall sentiment analysis, but also economics analysis, politics analysis and so on. Social networks analysis comprehends many disciplines and involves the application of different methodologies and techniques to define the criteria for generating the analytics, according to the purpose of the study. In this work, we focused on the semantic analysis of the content of textual information obtained from social media, aiming at extracting hot topics from social networks. We considered, as case study, reviews from the Yelp social network. The same methodology can be also applied for social and political opinion mining campaigns.

1 Introduction

Nowadays, social network data are a valuable source for many application fields. Overall, users trends and preferences are used in "social media marketing" field to organize advertising campaigns: knowing the tastes and habits of users, companies can reach a bigger number of customers and can implement a novel concept of advertising, which is quicker, immediate and in topic with hot "hashtags".

Important information that analysts are interested in, mainly includes how people feel about an event occurred in their lives or near their geographical area, city, neighbourhood, etc. Previous works [17] assert that the use of emoticons or hashtags can improve the identification of sentiments that people feel, since they are the easiest way to share human feelings on the net. So, sentiment analysis has became an hot topic research trend also for ICT researchers because, mining information through forums, blogs, social media, etc. helps to understand how people react to a real life event (i.e. a new law, terrorist attack, and so on).

© Springer Nature Switzerland AG 2019
F. Xhafa et al. (Eds.): INCoS 2018, LNDECT 23, pp. 59–70, 2019.
https://doi.org/10.1007/978-3-319-98557-2_6

Therefore, in social networking literature there are many tools and projects designed to perform different kind of analysis on social data [2].

In this paper we focus our work on the *semantic* analysis of textual information obtained from social media streams. Our aim is to extract hot topics from social network streams. We consider, as case study and to exemplify the approach, reviews obtained from the social network Yelp [36].

1.1   Text Mining Procedures

Effective and efficient access to domain relevant information requires the ability to automatically process and organize the information especially if it is contained in Big Data repositories [3,4]. The most used approaches in Big Data processing are based on graph algorithms and parallel and distributed architectures. Some Big Data infrastructures use Apache Hadoop, Spark, Flink and other platforms [27] for data-intensive distributed applications [14]. They are based on functional languages programming models such as the popular MapReduce programming model and a HDFS (Hadoop Distributed File System). In MapReduce, a master node splits input datasets into independent subsets that are mapped in parallel to worker nodes. This step of mapping is then followed by reduction tasks, where outputs from map steps are used to compute the result of the original job. There are many open source tools for Big Graph mining, such as Pegasus, a big graph mining system built on top of MapReduce. Likewise, there are approaches to deal with very large graphs such as Pregel [33].

Using Big Graph mining allows to find patterns of interest, events, anomalies, etc. in massive real-world graphs. Apache promotes many projects focused on Big Data, scalable machine learning and data mining features. These projects aim to fast manage large-scale data at very low cost through a collection of hardware, software (mainly based on Hadoop, Spark and Flink), and design patterns. Examples of such software include BIDMat –an interactive matrix library that integrates CPU and GPU acceleration–, MLlib for machine learning and others.

Regarding text analysis, *morpho-semantic* approaches have been already proposed for many languages and applications are known for the medical domain. Relevant works include Pratt [26] on the identification and on the transformation of terminal morphemes in the English medical dictionary; Wolff [37] on the classification of the medical lexicon based on formative elements of Latin and Greek origin; Pacak et al. [26] on the diseases words ending in -*itis*; Norton and Pacaknorton1983morphosemantic on the surgical operation words ending in -*ectomy* or -*stomy*; Dujols *et al.* on the suffix -*osis*.

The automatic population of thesauri problem was faced by many researchers, such as Lovis *et al.* [18], that derives the meaning of the words from the morphemes that compose them; in Hahn et al. [13] authors segment the sub-words in order to recognise and extract medical documents; Grabar and Zweigenbaum [12] use machine learning methods on SNOMED thesaurus to analyse morphological data. Several other works present the definition and the implementation of a comprehensive architecture for information structuring, and for ensure the semantic interoperability of domain entities [15,20,25] by mapping the content of different corpora on a set of shared concepts [22,38].

On the other hand, considerable research is devoted to decision support systems, which in the literature are usually divided in two categories: knowledge-based and non knowledge-based [5,24] (for the former the reader can find details in [21]). AAPHelp, created in 1972, was an early attempt to implement automated reasoning under uncertainty. Other systems comprise Asbru, EON and PRODIGY [35]; PROforma, SAGE [34]; and the Clinical Reminder System [16]. The later is based on the [30] evidence-based medicine and provides evidence-based clinical guidelines. A more detailed and systematic overview on many other CDSS can be found in [11].

Recently, many studies focused on medical information extraction from structured or unstructured texts. Fette [10] proposes a clinical information system, that, through Information Extraction techniques, retrieve medical unstructured information, inserted by physicians, transforming them into a structured format stored in a data warehouse. Rink [29] proposes a method for the automatic extraction of information from electronic medical reports. Medical concepts, and their relationships, are extracted with supervised machine learning algorithms. The author uses several language resources (such as WordNet, Wikipedia and the General Inquirer lexicon) and processing resources (i.e. semantic role labeller, POS tagger, phrase chunk parser) for feature extraction. In [9] the authors introduce an automated system to extract medications and related information from discharge summaries through an integrated that combine some NLP tools. Finally, several modelling approach have been proposed [1,8,23] to define proper data model.

In this work, we focus on the semantic analysis of the content of textual information obtained from social media, aiming at extracting hot topics from social networks. The proposed methodology can be also applied for social and political opinion mining campaigns. We considered, as case study, reviews from the Yelp social network.

The rest of the paper is organised as follows. In Sect. 2, we overview the main approaches used in social network analysis including Text Analysis Tools Sect. 2.1 and methodology Sect. 2.2. Based on the identified methodologies, we designed and experiment, which is described in Sect. 3. Computational results are summarised in Sect. 4. We end the paper with some conclusions and remarks in Sect. 5.

2 A Social Network Analysis Methodology

Nowadays the most common approach to the social networks analysis make use of the graph theory to model the network. This theory provides a general model for studying networks of any kind. Applied to social networks, the single user or groups of users are represented as a vertex and their relationships are represented as edges. Such kind of modelling allows to represent the network data in matrix form that can be directly analysed, without the need of drawing the graph, which helps the management of large social network data sets. The edges can have a direction used to track which vertex influences the other, and a weight, that is a numerical value used to represent the strength of the relation.

2.1 Text Analysis Tools

2.1.1 TaLTac

TaLTaC (*Trattamento automatico Lessicale e Testuale per l'analisi del Contenuto di un Corpus* - Lexical and Textual automatic processing for analysing the Content of a Corpus) [31] is a software able to perform many different operations on documents and data written in natural language. Such operations include Text Analysis, Text Mining and Corpus Analysis. When using this environment, a first task to do is creating a work session, which is the file containing the information, next we build the Corpus, which is the main object to be analysed with various methods and operations. The Corpus is then divided in two parts called fragment and section. The first is identified by four asterisks (****) a name and some variables, if needed, identified themselves with an asterisk, a name and a value. The latter is defined with four plus signs (++++) and a name. Every Corpus can contain a number of fragments and sections. One of the main features of TaLTaC is the extraction of relevant information from the Corpus (Text Mining). TaLTaC Text Mining uses endogenous and exogenous resources. The endogenous resources are composed by a number of fragments, in which we can divide a document, and by categorical variables, which can be associated to the text in order to identify fractions of the corpus logically related. Through them TaLTaC is able to perform a Specificity Analysis. The exogenous resources are lists which contain the frequency of a term or lexical unit in a corpus, that allows to identify peculiar language of the text.

2.1.2 General Architecture for Text Engineering (GATE)

GATE (General Architecture for Text Engineering) [28] is a open source software, specialised for text analysis and information extraction. The GATE software is a family made up of an IDE (GATE Developer), a web app (GATE Teamware) and a framework (GATE Embedded). In addition, GATE comes with an information extraction system, called ANNIE (A Nearly New Information Extraction System), which provides many processing resources specialised for information extraction tasks. Some of such resources are the English Tokeniser (splits the text into annotations of type Token) and the POS Tagger (assigns to every token an annotation that describes its characteristic, i.e. NNP for Proper Noun in singular form).

2.2 The Methodology

In order to achieve a fine grain analysis of our data sources we considered different types of operations.

2.2.1   Text Pre-treatment

This is the first process to be applied to the text in order to obtain a clear analysis in the successive processes. It is made up of the following two phases.

Normalization: This phase normalizes the writings of names, acronyms and other entities. This result is achieved by executing the following tasks:

- Changing apostrophes into stresses (for the words that is needed), in order to determinate the right word;
- Labelling words/sequence of words in order to annotate their right meaning and avoid mistakes with others expressions (i.e. a name can be mistaken for a noun, Rose or rose);
- Changing capital letters depending on the eventual labels associated with the word. They are changed into lower-case for the words that are not labelled or depending on the label means: if it is a name (of person, of a city or a general proper name) then it will keep the capital letter, in other cases, it will not.

Correcting Spelling Errors: This phase consists of comparing a misspelled word with the system dictionary in order to correct and analyse it in the right way.

2.2.2   Lexical Analysis

Lexical analysis processes the segments of the Corpus, defined as sequences of graphic forms separated by a strong divider (i.e. punctuation such as ".",";"). Once segments are obtained, it can be estimated various analysis parameter such as the IS index to measure the level of absorption of the segment regarding the single elements that compose it. Other important operations are the Tagging, which links to every word a description of the grammatical or semantic characteristics [7], and the Lexation, which identifies the sequence of words defined during the pretreatment as a unique entity. Last we define the Corpus" key words by comparing the repetition rates, assuming that the ones that have a noticeable standard deviation (considering only the integers) are more meaningful.

2.2.3   Textual Analysis

The first step of Textual Analysis is the study of the Concordances that examine the context of every chosen word or segment. Through the calculation of TF-IDF rate we can sort the research's results according to the frequency and distribution of the search keyword in the documents provided (see Eq. (1)).

$$\text{TF-IDF} = tf \cdot \log \frac{N}{n} \tag{1}$$

where tf is the number of occurrences of an element, and the remaining part is the logarithm of the ratio between the number of documents building the Corpus (N) and the number of documents which present that element (n). Another import part of Textual Analysis is the co-occurrences identification, where with co-occurrences we identify those couples of near elements that repeat in the text. Thsis identification is useful to define the primary concepts contained in the Corpus.

3   Experimental Design

Our experimentation aims to process and analyse on-line social networks data sets in order to derive useful information. The data set analysed comes from the social network Yelp, which publishes crowd-sourced reviews about local businesses [36]. Reviews are made up of tuples structured as follows: anonymised user name; anonymised reviewed place; date of the review and review.

3.1   TaLTac Analysis

Prior starting the analysis we pass through the text pre-treatment, and after parsing of the Corpus we normalise it and compute the sub-occurrences (see Fig. 1 for a snapshot of graphical user interface).

The next phase is the Textual Analysis, in which we start with the identification of the segments, these are saved in two files: the former is "List of Segments (with index IS)" which contains the segments with their relative number of occurrences, number of elements forming the segment and the IS index; the latter, named "List of Significant Segments" containing significant segments.

Fig. 1. Text normalization window

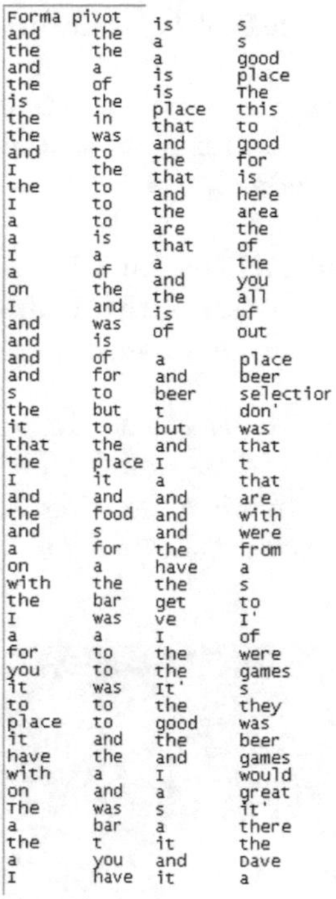

Forma	pivot	is	s
and	the	is	s
the	the	a	s
and	a	a	good
the	of	is	place
is	the	is	The
the	in	place	this
the	was	that	to
and	to	and	good
I	the	the	for
the	to	that	is
I	to	and	here
a	to	the	area
a	is	are	the
I	a	that	of
a	of	a	the
on	the	and	you
I	and	the	all
and	was	is	of
and	is	of	out
and	of	a	place
and	for	and	beer
s	to	beer	selectior
the	but	t	don'
it	to	but	was
that	the	and	that
the	place	I	t
I	it	a	that
and	and	and	are
the	food	and	with
and	s	and	were
a	for	the	from
on	a	have	a
with	the	the	s
the	bar	get	to
I	was	ve	I'
a	a	I	of
for	to	the	were
you	to	the	games
it	was	It'	s
to	to	the	they
place	to	good	was
it	and	the	beer
have	the	and	games
with	a	I	would
on	and	a	great
The	was	s	it'
a	bar	a	there
the	t	it	the
a	you	and	Dave
I	have	it	a

Fig. 2. Co-occurences window

Then we analyse the specificity of our Corpus together with the computation of the TF-IDF index. The last operation is the Textual Analysis with the concordances and co-occurrences computation (see Fig. 2).

The results of these operations show that the system can retrieve meaningful information, such as the TF-IDF index that shows how much a word is important in the document, or the co-occurrences, which show the main concepts of the text due to the couple of words that recur all the time. Unfortunately TaLTac cannot perform all the analysis on our data set due to the fact that it is in English and so our work is not totally complete nevertheless our results are very significant.

3.2   GATE Analysis

Our first operation with this software is the initialization of ANNIE with default configuration. Once all the processing resources are loaded we run ANNIE and start in a sequence:

- **Document Reset PR:** resets the document to its original state, by removing all the annotation sets and their contents;
- **ANNIE English Tokeniser:** splits the text into very simple tokens such as numbers, punctuation and words of different types;

- **ANNIE Gazetteer:** identifies entity names in the document text based on predefined lists;
- **ANNIE Sentence Splitter:** The sentence splitter is a cascade of finite-state transducers which segments the text into sentences;
- **ANNIE POS Tagger:** produces a part-of-speech tag as an annotation on each word or symbol;
- **ANNIE NE Transducer:** ANNIE named entity grammar;
- **ANNIE OrthoMatcher:** adds identity relations between named entities labelled by the semantic tagger, in order to perform co-reference.

Due to these operations we can overcome TaLTac limits and run a grammatical tag on our data set. It should be noted that our original data have been pre-processed to analyse only the review section due to the fact that in GATE the other information are not relevant for our purpose (Fig. 3).

Fig. 3. Annie pipeline

4 Computational Results

From our initial data set, made up of more than 5000 tuples, we extracted data samples of 50 tuples from which we obtained computational results, summarised next.

4.1   TaLTac Results

TaLTac functions enabled us to observe that our peculiar lexicon, shown in Fig. 4, is composed by words with highest occurrences and TF-IDF like: "games", "beer", "good", "great", "food", "place" so it is reasonable to assume that the reviews analysed reviews about a place to eat food, drink beer or play some games and that the main audience thinks that it is a good place. Other meaningful data are the co-occurrences, which are reported in Fig. 2, these elements strengthen the evidence about the place reviewed with expressions such as "the bar", "a good", "a place", "beer selection", "the games". Last, we took a meaningful word, "food", and studied its concordances through all the Corpus (see Fig. 5), in order to understand the context of the lemma.

Forma grafica	Occorrenze totali	Lunghezza	TFIDF	Forma grafica	Occorrenze totali	Lunghezza	TFIDF
was	83|03		3,37217	they	38|04		1,82534
s	62|01		2,72497	here	21|04		1,82342
for	71|03		2,61815	area	15|04		1,79898
games	34|05		2,46866	bar	24|03		1,79825
is	97|02		2,36757	get	20|03		1,78191
beer	24|04		2,19451	great	20|05		1,77350
are	34|03		2,10028	as	17|02		1,77066
to	132|02		2,07886	with	37|04		1,76399
t	37|01		2,06400	had	27|03		1,75748
good	37|04		2,06279	it	57|02		1,75409
selection	15|09		2,06025	place	44|05		1,75382
you	46|03		2,04468	out	27|03		1,72490
were	32|04		1,98688	I	131|01		1,71641
there	29|05		1,97350	fun	13|03		1,70398
that	57|04		1,95741	4.0	15|03		1,65914
	25|01		1,93592	It	18|02		1,65544
Great	7|05		1,92191	The	52|03		1,65391
in	56|02		1,91502	from	24|04		1,64308
on	47|02		1,90879	wings	9|05		1,64125
food	39|04		1,87488	be	18|02		1,62180
I	26|01		1,86714	&.	20|01		1,59426
of	91|02		1,64201	have	46|04		1,59094
1qCuOcks5HRv67OHovA	26|22		1,58947	but	48|03		1,57259
all	18|03		1,58031	can	12|03		1,55776
and	194|03		1,57490	5.0	10|03		1,55417

Fig. 4. Peculiar lexicon by occurences

ID Fr...	Intorno sinistro	Forma grafica	Intorno destro
fragm...	, and authentic . If you' re looking for good Irish	food	and a cold pint , you can' t go wrong at the Pour House
fragm...	worth seeking out . They have some of the best Irish	food	I' ve had in Pittsburgh- the colcannon is awesome and
fragm...	out of this world . If you' re not looking for Irish	food	, then try the grilled cheese- and make sure you ask
fragm...	there on a Saturday night with a mind to try the Irish	food	. Apparently , we were out of luck . I' ve always thought
fragm...	secrets of restaurant success is to actually stock	food	for people to eat . He told us before we ordered that
fragm...	were out ". At that point , realizing that the only	food	to be had in the place was what was crusted on the
fragm...	heoc96QX·TbecWVw333qhQ	2011-08-20 Best Irish	food	in the Burgh . Great bar food too . The service is
fragm...	2011-08-20 Best Irish food in the Burgh . Great bar	food	too . The service is maybe a bit surly and it' s not
fragm...	Excellent wings and sandwiches , generally good	food	otherwise , and fair prices- nice casual place . The
fragm...	1qCuOcks5HRv67OHovAVpg tPUGLIDZLF7HrOC46NqT...	food	here can actually be a little hit and Miss , but I
fragm...	wonderful character and ambiance of this place , but the	food	was average at best . We were there on a Pens play-off
fragm...	2 people working all the tables in both rooms . Our	food	came to us cold and unimpressive at that . Our orders
fragm...	were pretty good , also . If you are looking for good	food	, Homestead better choices at Blue Dust or Tin Front
fragm...	beer selection , but it won' t be for a while . The	food	has always been average at best , and the pizza sub
fragm...	op2Gve4sAMQ4qEzq2Tad0g	2013-09-15 I' m reading o...	food	is really hit or Miss . I' ve only gone to Duke' s
fragm...	Miss . I' ve only gone to Duke' s one time , but the	food	was good . It wasn' t too busy so our service was very
fragm...	very attentive and it didn' t take long to get our	food	at all . The place is separated between a bar area

Fig. 5. Concordances of the word "food"

4.2   GATE Results

By using the Tokeniser we can distinguish spaces from words, and due to the Gazetteer and POS Tagger every word in our data set has a description. At the end of our analysis with this software we can assert the grammatical features of our data and we can give even more meaning to the analysis described in Subsect. 4.1.

Steps in GATE can be pipelined. In addition, each work-flow can be replied on different documents. Hence, high scalability for large datasets is achieved by simply spawning different step managers on different documents, or even on parts of them. This is a behaviour similar to map-reduce frameworks, and we will implement map-reduce porting of our framework as well as through message-passing platforms such as Flink, leading to performance and scalability improvements.

5   Conclusions

Social networks data stream analysis has become a important research and development trend among scholars, researchers, software designer and data analysts from academia, business and industry. In literature, there are various tools and projects focusing on different kinds of social network data stream analysis. In our work, we focused on the analysis of the content of the text obtained from social media streams. Through our study it was possible to extract hot topics from different information sources, such as reviews. This study can constitute a starting point for further analysis, also on different domains such as marketing (through an analysis of the feedback of customers) or cyber-security [6,19,32] (detection of text containing malicious messages), crawling information from social networks data streams.

Acknowledgment. This research was funded by the European Commission through the project "colMOOC: Integrating Conversational Agents and Learning Analytics in MOOCs" (588438-EPP-1-2017-1-EL-EPPKA2-KA).

References

1. Amato, F., Moscato, F.: Pattern-based orchestration and automatic verification of composite cloud services. Comput. Electr. Eng. **56**, 842–853 (2016)
2. Amato, F., Moscato, F.: Exploiting cloud and workflow patterns for the analysis of composite cloud services. Future Gener. Comput. Syst. **67**, 255–265 (2017)
3. Balzano, W., Murano, A., Stranieri, S.: Logic-based clustering approach for management and improvement of vanets. J. High Speed Netw. **23**(3), 225–236 (2017)
4. Balzano, W., Murano, A., Vitale, F.: SNOT-WiFi: sensor network-optimized training for wireless fingerprinting. J. High Speed Netw. **24**(1), 79–87 (2018)
5. Coiera, E.: Guide to Health Informatics. CRC Press, London (2015)
6. Coppolino, L., D'Antonio, S., Mazzeo, G., Romano, L.: Cloud security: emerging threats and current solutions. Comput. Electr. Eng. **59**, 126–140 (2017)
7. D'Acierno, A., Moscato, V., Persia, F., Picariello, A., Penta, A.: iWIN: a summarizer system based on a semantic analysis of web documents. In: Proceedings of IEEE 6th International Conference on Semantic Computing, ICSC 2012, pp. 162–169 (2012)
8. Di Lorenzo, G., Mazzocca, N., Moscato, F., Vittorini, V.: Towards semantics driven generation of executable web services compositions. J. Softw. **2**(5), 1–15 (2007)

9. Doan, S., Bastarache, L., Klimkowski, S., Denny, J.C., Xu, H.: Integrating existing natural language processing tools for medication extraction from discharge summaries. J. Am. Med. Inform. Assoc. **17**(5), 528–531 (2010)
10. Fette, G., Ertl, M., Wörner, A., Kluegl, P., Störk, S., Puppe, F.: Information extraction from unstructured electronic health records and integration into a data warehouse. In: GI-Jahrestagung, pp. 1237–1251 (2012)
11. Garg, A.X., Adhikari, N.K.J., McDonald, H., Rosas-Arellano, M.P., Devereaux, P.J., Beyene, J., Sam, J., Haynes, R.B.: Effects of computerized clinical decision support systems on practitioner performance and patient outcomes: a systematic review. JAMA **293**(10), 1223–1238 (2005)
12. Grabar, N., Zweigenbaum P.: Automatic acquisition of domain-specific morphological resources from thesauri. In: Proceedings of RIAO, pp. 765–784. Citeseer (2000)
13. Hahn, U., Honeck, M., Piotrowski, M., Schulz, S.: Subword segmentation–leveling out morphological variations for medical document retrieval. In: Proceedings of the AMIA Symposium, p. 229. American Medical Informatics Association (2001)
14. Javanmardi, S., Shojafar, M., Shariatmadari, S., Ahrabi, S.S.: FR trust: a fuzzy reputation-based model for trust management in semantic P2P grids. Int. J. Grid Util. Comput. **6**(1), 57–66 (2015)
15. Jin, H., Sun, A., Zheng, R., He, R., Zhang, Q.: Ontology-based semantic integration scheme for medical image grid. Int. J. Grid Util. Comput. **1**(2), 86–97 (2009)
16. Kang, U., Chau, D.H., Faloutsos, C.: PEGASUS: mining billion-scale graphs in the cloud. In: 2012 IEEE International Conference on Acoustics, Speech and Signal Processing (ICASSP), pp. 5341–5344. IEEE (2012)
17. Kouloumpis, E., Wilson, T., Moore, J.D.: Twitter sentiment analysis: the good the bad and the OMG! In: ICWSM, vol. 11, no. 538–541, p. 164 (2011)
18. Lovis, C., Baud, R., Rassinoux, A.-M., Michel, P.-A., Scherrer, J.-R.: Medical dictionaries for patient encoding systems: a methodology. Artif. Intell. Medicine **14**(1), 201–214 (1998)
19. Mazzeo, G., Coppolino, L., D'Antonio, S., Mazzariello, C., Romano, L.: SIL2 assessment of an active/standby cots-based safety-related system. Reliabil. Eng. Syst. Saf. **176**, 125–134 (2018)
20. Mikkilineni, R., Morana, G., Zito, D., Keshan, S.: Cognitive application area networks. Int. J. Grid Util. Comput. **8**(2), 74–81 (2017)
21. Miller, R.A.: Medical diagnostic decision support systems past, present, and future. J. Am. Med. Inform. Assoc. **1**(1), 8–27 (1994)
22. Moore, P., Xhafa, F., Barolli, L.: Semantic valence modeling: emotion recognition and affective states in context-aware systems. In: Proceedings of 2014 IEEE 28th International Conference on Advanced Information Networking and Applications Workshops, IEEE WAINA 2014, pp. 536–541 (2014)
23. Moscato, F.: Exploiting model profiles in requirements verification of cloud systems. Int. J. High Perform. Comput. Netw. **8**(3), 259–274 (2015)
24. Musen, M.A., Middleton, B., Greenes, R.A.: Clinical decision-support systems. In: Biomedical Informatics, pp. 643–674. Springer (2014)
25. Pandey, M., Pathak, V.K., Chaudhary, B.D.: A framework for interest-based community evolution and sharing of latent knowledge. Int. J. Grid Util. Comput. **3**(2–3), 200–213 (2012). Cited by 6
26. Pratt, A.W., Pacak, M.: Identification and transformation of terminal morphemes in medical english. Methods Inf. Med. **8**(2), 84–90 (1969)
27. The Apache Hadoop project. Apache hadoop

28. The GATE project team. Gate
29. Rink, B., Harabagiu, S., Roberts, K.: Automatic extraction of relations between medical concepts in clinical texts. J. Am. Med. Inform. Assoc. **18**(5), 594–600 (2011)
30. Sackett, D.L., Rosenberg, W.M.C., Gray, J.A.M., Haynes, R.B., Richardson, W.S.: Evidence based medicine: what it is and what it isn't (1996)
31. Bolasco, A.M.S., Baiocchi, F.: TalTac
32. Staffa, M., Sgaglione, L., Mazzeo, G., Coppolino, L., D'Antonio, S., Romano, L., Gelenbe, E., Stan, O., Carpov, S., Grivas, E., Campegiani, P., Castaldo, L., Votis, K., Koutkias, V., Komnios, I.: An openNCP-based solution for secure ehealth data exchange. J. Netw. Comput. Appl. **116**, 65–85 (2018)
33. Steinbauer, M., Anderst-Kotsis, G.: DynamoGraph: extending the pregel paradigm for large-scale temporal graph processing. Int. J. Grid Util. Comput. **7**(2), 141–151 (2016)
34. Tu, S.W., Campbell, J.R., Glasgow, J., Nyman, M.A., McClure, R., McClay, J., Parker, C., Hrabak, K.M., Berg, D., Weida, T.: The sage guideline model: achievements and overview. J. Am. Med. Inform. Assoc. **14**(5), 589–598 (2007)
35. Veloso, M., Carbonell, J., Perez, A., Borrajo, D., Fink, E., Blythe, J.: Integrating planning and learning: the prodigy architecture. J. Exp. Theor. Artif. Intell. **7**(1), 81–120 (1995)
36. The Free Encyclopedia Wikipedia. Yelp
37. Wolff, S.: The use of morphosemantic regularities in the medical vocabulary for automatic lexical coding. Methods Inf. Med. **23**(4), 195–203 (1984)
38. Xhafa, F., Barolli, L.: Semantics, intelligent processing and services for big data. Future Gener. Comput. Syst. **37**, 201–202 (2014)

Exploring User Feedback Data via a Hybrid Fuzzy Clustering Model Combining Variations of FCM and Density-Based Clustering

Erind Bedalli[1,2(✉)], Enea Mançellari[2], and Esteriana Haskasa[3]

[1] University of Elbasan, Elbasan, Albania
erind.bedalli@uniel.edu.al
[2] Epoka University, Tirana, Albania
{ebedalli,emancellari}@epoka.edu.al
[3] UBS Investment Bank, Krakow, Poland
esteriana.haskasa@gmail.com

Abstract. In today's dynamic environments, user feedback data are a valuable asset providing orientations about the achieved quality and possible improvements of various products or services. In this paper we will present a hybrid fuzzy clustering model combining variants of fuzzy c-means clustering and density based clustering for exploring well-structured user feedback data. Despite of the multitude of successful applications where these algorithms are applied separately, they also suffer drawbacks of various kinds. So, the FCM algorithm faces difficulties in detecting clusters of non-spherical shapes or densities and moreover it is sensitive to noise and outliers. On the other hand density-based clustering is not easily adaptable to generate fuzzy partitions. Our hybrid clustering model intertwines density-based clustering and variations of FCM intending to exploit the advantages of these two types of clustering approaches and diminishing their drawbacks. Finally we have assessed and compared our model in a real-world case study.

1 Introduction

Basically, clustering is portrayed as the process of grouping the elements of a dataset into clusters (groups) such the elements within the same cluster are more similar to each other and less similar to the elements of the other clusters. This is an unsupervised procedure which is carried without any prior information about the structure of the dataset; it merely relies on the similarity/dissimilarity among the elements of the dataset. These attributes make clustering a very useful exploratory and data summarization tool with a wide variety of applications including, but not being limited to cognitive sciences, economy, medicine, image processing, web mining and social networks [1]. There are several approaches towards the clustering problem, like hard and fuzzy clustering, hierarchical and partition clustering, overlapping and non-overlapping clustering etc. One of the foremost categorizations of clustering algorithms is in crisp (rough, hard) clustering and fuzzy (soft) clustering [2, 3]. In the crisp clustering the instances are distributed in clusters where each instance belongs (having a full membership) to exclusively one of the clusters, while in the fuzzy clustering the

F. Xhafa et al. (Eds.): INCoS 2018, LNDECT 23, pp. 71–81, 2019.
https://doi.org/10.1007/978-3-319-98557-2_7

instances have partial memberships (a value between 0 and 1) into several clusters. The fuzzy approach to the clustering problem not only offers more flexibility, but it is also more realistic [4, 5]. Fuzzy c-means (FCM) is one of the most eminent representatives in the family of fuzzy clustering algorithms. It is devised as a generalization of the classical k-means clustering algorithm allowing partial memberships of the data elements in the clusters. Many other fuzzy clustering algorithms are developed as variations of the FCM algorithm, like the Gustafson-Kessel algorithm, the Gath-Geva algorithm, kernel-based fuzzy clustering algorithm etc. [6, 7]. On the other hand, in density-based clustering, clusters are conceived as regions of higher density and they are separated by regions of lowers density. The entities in these separating regions are either considered as border points (if they are in the proximity of the clusters) or they may be considered as noise, thus being excluded from the clusters. DBSCAN algorithm is one most frequently used density-based clustering algorithms. It reveals groupings of higher densities as clusters (classes/categories) while considering the data in low density regions as noise or outliers [8]. Compared to the FCM (and its variations) the DBSCAN algorithm has the advantage of being able to automatically detect the number of clusters, so it is not required to be specified as one of the parameters of the algorithm.

In our work we are going to combine density-based clustering with FCM and Gustafson-Kessel algorithms in order to set up a refined and more robust fuzzy clustering model, which will operate in two stages. In the first stage the DBSCAN algorithm will operate on the user feedback data generating a temporary set of clusters and noise points. In the later stage the FCM or Gustafson-Kessel operates using as initial prototypes the centers of the clusters generated in the initial stage. This model will offer information at different level of granularity which can be controlled by the parameters being tuned for the density based clustering. Thus different point of views can be provided for the same dataset of user feedback data, which offers a great flexibility in the utilization of these data. Although our model is versatile and can adapted to obtain fuzzy clusters on various data, in our work we are primarily focused on exploring well-structured user feedback data. These data represent a valuable asset providing orientations about the achieved quality and possible improvements of various products or services, thus supporting decision making processes.

2 The Fuzzy C-Means and Gustafson-Kessel Algorithms

The fuzzy c-means algorithm aims at partitioning a dataset into soft clusters, such that an instance does not have to exclusively belong to any of the clusters, but it may have partial memberships in several clusters simultaneously. The results generated by this algorithm are not only more flexible, but very frequently they are also more meaningful compared to the results of hard clustering algorithms. The algorithm operates in an unsupervised way; it doesn't have any prior information on the structure of the dataset, but it needs several parameters to be specified beforehand, like the number of clusters, the fuzzy exponent and the distance metrics. On the other hand, the initial values of the cluster centers may be either selected as predefined parameters or may be picked randomly [9]. The proper choice of these parameters remains a challenge, as they affect significantly the quality of the generated results.

Basically, the FCM algorithm is an iterative procedure which minimizes an objective function which is formally defined as:

$$J = \sum_{i=1}^{n} \sum_{j=1}^{c} \mu_{ij}^{\varphi} d^2 (x_i, c_j)$$

Here n is the cardinality of in the dataset, c is the number of the clusters (a pre-defined parameter), c_j is the center of the j-th cluster, x_i is the i-th element of the dataset, μ_{ij} is the membership value of the x_i instance in the c_j cluster, $d(x_i, c_j)$ is the distance between x_i and c_j regarding to the pre-defined distance metric and φ is the fuzzy exponent (a real value greater than 1, which is another predefined parameter). There are several possible choices for the distance metrics like the Euclidean distance, the Manhattan distance, the Minkowski distance, the maximum distance, the Pearson correlation etc. [10, 11]. The algorithm may be summarized by the given pseudo-code:

1. Initialize the centers of the clusters (according to the given parameters or randomly).
2. Initialize the U_0 partition matrix (assigning 0 to all its entries)
3. Set *nr* = 0 (iteration counter)
4. Evaluate the distance of each instance from each center of the actual clusters (employing the distance metric specified as parameter).
5. Update the partition matrix $U_{nr+1} = [u_{ij}]$, evaluating its entries according to

$$u_{ij} = d(x_i, c_j)^{-\frac{2}{\varphi-1}} \left(\sum_{k=1}^{c} d(x_i, c_j) \right)^{-1}$$

6. Evaluate the new cluster centers according to $c_i = \dfrac{\sum_{j=1}^{n} u_{ij}^{\varphi} X_j}{\sum_{j=1}^{n} u_{ij}^{\varphi}}$

7. $nr = nr + 1$ (the iteration counter is incremented)
8. If $\|U_{nr} - U_{nr-1}\| > \varepsilon$ continue to step 4.
9. END.

Although the FCM algorithm is generally efficient and has a decent accuracy when the dataset is characterized by hyper-spherical shapes of approximately equal sizes, its accuracy is significantly affected when the datasets contain clusters of various shapes, sizes or densities. Gustafson-Kessel is another important fuzzy clustering algorithm, which is delivered as an extension of the fuzzy c-means algorithm with a significant change in the distance metric which is applied. Instead of the classical distance metrics, in this case an adaptive distance is applied to enable the detection of clusters of various shapes (not only hyper-spherical ones) within the same dataset [7]. The objective function and the parameters of the Gustafon-Kessel algorithm remain the same as in the conventional fuzzy c-means algorithm, so the vital change is in the way the distance is evaluated [9]:

$$d^2 (x_j, c_i) = (x_j - c_i)^T V_i (x_j - c_i)$$

Here V_i is a matrix evaluated using a scaled inverse fuzzy covariance matrix for the respective cluster. The algorithm is briefly described by the given pseudo-code [10]:

1. Initialize the centers of the clusters (according to the given parameters or randomly).
2. Initialize the U_0 partition matrix (assigning 0 to all its entries)
3. Set $nr = 0$ (iteration counter)
4. Evaluate the covariance matrices: $F_i = \dfrac{\sum_{i=1}^{k} \sum_{j=1}^{n} \mu_{ij}^{\varphi} (x_i - c_j)(x_i - c_j)^{\mathrm{T}}}{\sum_{i=1}^{k} \sum_{j=1}^{n} \mu_{ij}^{\varphi}}$
5. Evaluate the distance of each instance from each center of the actual clusters:
   $d^2(x_j, c_i) = (x_j - c_i)^{\mathrm{T}} V_i (x_j - c_i)$, where $V_i = (\rho_i |F_i|)^{1/p} F_i^{-1}$
6. Update the partition matrix $U_{nr+1} = [u_{ij}]$, evaluating its entries according to
   $$u_{ij} = d(x_i, c_j)^{-\frac{2}{\varphi-1}} \left(\sum_{k=1}^{c} d(x_i, c_j) \right)^{-1}$$
7. Evaluate the new cluster centers according to: $c_i = \dfrac{\sum_{j=1}^{n} \mu_{ij}^{\varphi} X_j}{\sum_{j=1}^{n} \mu_{ij}^{\varphi}}$
8. $nr = nr + 1$ (increment the iteration counter)
9. If $\|U_{nr} - U_{nr-1}\| > \varepsilon$ continue to step 4.
10. END.

3 Density-Based Clustering and the DBSCAN Algorithm

Density-based clustering relies on the general idea that clusters consist of concentrated groups of data points (instances), separated by areas which are sparse or empty, i.e. they have a low or no concentration of data points. In this context a cluster is conceived as a maximal contiguous set with density exceeding a specified threshold. On the other hand some data points may be positioned in the separating areas not belonging to any cluster, thus these points remain uncategorized and they are considered as noise or outliers. Several density-based clustering are devised among which DBSCAN, LDBSCAN, HDBSCAN and OPTICS are some eminent representatives [12].

In our work we are making extensive use of the DBSCAN (Density-Based Spatial Clustering of Applications with Noise) algorithm. The central idea of this algorithm is the detection of the clusters as contiguous dense regions based on the neighborhood analysis of each data point. More specifically, the algorithm is driven by two key parameters which are *Eps* (ε) and *MinPts*, denoting respectively the neighborhood radius and the minimal number of points that a region of radius ε must contain to be considered as dense. These two parameters are used to partition the dataset into three types of points: core points, border points and noise. The $\varepsilon - neighborhood$ of a point x_i is formally defined as $N_\varepsilon(x_i) = \{x \in X | d(x_i, x) \le \varepsilon\}$. Then a point x_i is a core point if it satisfies the condition $card(N_\varepsilon(x_i)) \ge MinPts$. On the other hand a point x_i will be a border point if $card(N_\varepsilon(x_i)) < MinPts$ but there exists at least one core point y such that $d(x_i, y) \le \varepsilon$. Finally a point is considered as noise if $card(N_\varepsilon(x_i)) < MinPts$ and there are no core points y such that $d(x_i, y) \le \varepsilon$ [13].

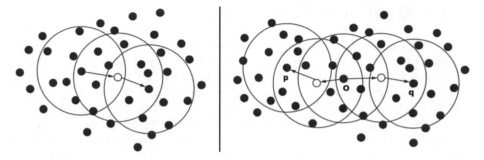

Fig. 1. Density edges and density-connected data points

A density edge is defined as an edge connecting two core points and furthermore a pair of two core points p and q is defined as density-connected, if there exists a path consisting of density edges connecting them as shown in the Fig. 1.

The concept of density-connected edges is the crux of the DBSCAN algorithm, based on which the clusters are delivered. Initially starting by picking an unassigned core point, the cluster is gradually expanded to potentially any direction clamping all the unassigned core points which are density-connected to the initially picked core point. This procedure is repeated with the core points which are unassigned yet, thus constructing the other core-clusters one by one. After all the core-clusters are constructed, the border points are also distributed into these clusters, thus completing the final clustering. The fundamental steps of the DBSCAN algorithm are summarized by the following pseudo-code [13]:

1. Categorize the instances of the dataset as core, border and noise points based of the neighborhood analysis of each data point.
2. The noise points are discarded (so they will not affect the clustering process anymore).
3. For each core point x which is not already assigned to one of the clusters:
   3.1 Create a new cluster consisting of the point x and all the other core points which are density-connected to x.
4. Assign each border point to the cluster containing the closest core point.

There are several significant advantages of the DBSCAN algorithm like its ability to detect clusters of various shapes and sizes, its ability to determine automatically the natural number of clusters and its resistance towards the presence of noise and outliers in the dataset. On the other hand it suffers several drawbacks like its sensitivity to parameters, difficulties in handling varying densities and high-dimensional data [14].

4 The Hybrid Fuzzy Clustering Model

In this work we combine density-based clustering with FCM or Gustafson-Kessel algorithms in order to set up refined and more robust fuzzy clustering models, which will operate in two stages. In the first stage density-based clustering will be applied on the dataset to generate the intermediate clusters. In the second stage we firstly discard the noise points from the dataset and then we apply the fuzzy clustering algorithm on the centers of clusters generated during the first stage by the density-based clustering algorithm. The key idea is to enhance the FCM and Gustafson-Kessel algorithms in several aspects. Firstly we can address two critical issues related to these algorithms: selecting the proper number of clusters and initialization of the cluster prototypes [15]. Moreover we avoid the noise and outliers in the dataset which may significantly affect the accuracy of the fuzzy clustering algorithm [16, 17]. Thus the model intends to exploit the advantages of the density-based and fuzzy clustering approaches and to diminish their drawbacks. On the other hand the main drawback of this model relies on the fact that it is more computationally expensive compared to the standard DBSCAN clustering, standard FCM and Gustafson-Kessel algorithm.

We have tested two versions of the model, the first version combining the DBSCAN and FCM algorithms, while the second version combines DBSCAN and Gustafson-Kessel algorithms. A complete view of the first model is provided by the following pseudo-code:

1. Categorize the instances of the dataset as core, border and noise points according to the predefined ε and *MinPts*.
2. The noise points are discarded from the dataset.
3. For each core point x which is not already assigned to one of the clusters:
   3.2 Create a new cluster consisting of the point x and all the other core points which are density-connected to x.
4. Assign each border point to the cluster containing the closest core point.
5. Calculate the centers of the generated clusters.
6. Initialize the centers of the new clusters using the centers of the previously generated clusters.
7. Initialize the U_0 partition matrix (assigning 0 to all its entries)
8. Set nr = 0 (iteration counter)
9. Evaluate the distance of each instance from each center of the actual clusters (employing the distance metric specified as parameter).
10. Update the partition matrix $U_{nr+1} = [u_{ij}]$, evaluating its entries according to
    $$u_{ij} = d(x_i, c_j)^{-\frac{2}{\varphi-1}} \left(\sum_{k=1}^{c} d(x_i, c_j) \right)^{-1}$$
11. Evaluate the new cluster centers according to $c_i = \dfrac{\sum_{j=1}^{n} u_{ij}^{\varphi} X_j}{\sum_{j=1}^{n} u_{ij}^{\varphi}}$
12. nr = nr +1 (the iteration counter is incremented)
13. If $\|U_{nr} - U_{nr-1}\| > \varepsilon$ continue to step 4.
14. END.

Moreover, to keep our model flexible we have prepared a module which designates the values of the parameters so that the total number of clusters is approximately equal to a given input. Our intention is to enable the generation of fuzzy clusters at different levels of granularity, which is a valuable tool providing important insights for the experts of the domain. A complete view of the second version of the model is provided by the following pseudo-code:

1. Categorize the instances of the dataset as core, border and noise points according to the predefined ε and *MinPts*.
2. The noise points are discarded from the dataset.
3. For each core point x which is not already assigned to one of the clusters:
   3.3   Create a new cluster consisting of the point x and all the other core points which are density-connected to x.
4. Assign each border point to the cluster containing the closest core point.
5. Calculate the centers of the generated clusters.
6. Initialize the centers of the new clusters using the centers of the previously generated clusters.
7. Initialize the U_0 partition matrix (assigning 0 to all its entries)
8. Set nr = 0 (iteration counter)
9. Evaluate the covariance matrices: $F_i = \dfrac{\sum_{i=1}^{k} \sum_{j=1}^{n} \mu_{ij}^{\varphi}\left(x_i - c_j\right)\left(x_i - c_j\right)^{T}}{\sum_{i=1}^{k} \sum_{j=1}^{n} \mu_{ij}^{\varphi}}$
10. Evaluate the distance of each instance from each center of the actual clusters: $d^2\left(x_j, c_i\right) = \left(x_j - c_i\right)^{T} V_i \left(x_j - c_i\right)$, where $V_i = \left(\rho_i |F_i|\right)^{1/p} F_i^{-1}$
11. Update the partition matrix $U_{nr+1} = [u_{ij}]$, evaluating its entries according to $u_{ij} = d\left(x_i, c_j\right)^{-\frac{2}{\varphi-1}} \left(\sum_{k=1}^{c} d\left(x_i, c_j\right)\right)^{-1}$
12. Evaluate the new cluster centers according to: $c_i = \dfrac{\sum_{j=1}^{n} \mu_{ij}^{\varphi} X_j}{\sum_{j=1}^{n} \mu_{ij}^{\varphi}}$
13. $nr = nr + 1$ (the iteration counter is incremented)
14. If $\|U_{nr} - U_{nr-1}\| > \varepsilon$ continue to step 4.
15. END.

5   Case Studies: Analysis and Interpretation of Results

In this section we are going to describe and make an interpretation of the results generated by the two versions of our model on a dataset of user feedback of a company. This dataset contains the user feedback data of 400 customers. These data are in tabular form containing the evaluation scores that the users have provided for various aspects of services offered by the company. We have employed both versions of the hybrid fuzzy clustering model on this dataset in order to get a complete fuzzy partitioning of the dataset. Moreover we have a prepared a module which designates the values of the parameters for the density-based clustering stage, in order to obtain results at different levels of granularity. Besides the overall summarizing table, the fuzzy memberships of each individual at different levels of granularity are also available. In the following series of tables we are summarizing the results of the general clustering by applying different versions of the clustering model and different levels of granularity on the dataset and also we are giving a partial view of the table containing the fuzzy memberships of the data points (individuals) into the generated clusters (Tables 1, 2, 3 and 4).

Table 1. The overall clustering results obtained using 1st version at 1st level of granularity

Clusters	Weighted score	Cluster size	Percentage
Cluster 1	89.6	114	28.5
Cluster 2	66.7	192	48
Cluster 3	38.2	60	15
Noise	58.4	34	8.5

Table 2. The overall clustering results obtained using 2nd version at 1st level of granularity.

Clusters	Weighted score	Cluster size	Percentage
Cluster 1	90.8	118	28.5
Cluster 2	68.5	186	46.5
Cluster 3	38.8	62	15.5
Noise	58.4	34	8.5

Table 3. The overall clustering results obtained using 1st version at 2nd level of granularity.

Clusters	Weighted score	Cluster size	Percentage
Cluster 1	92.4	71	17.25
Cluster 2	79.6	107	26.75
Cluster 3	62.4	94	23.5
Cluster 4	52.1	68	17
Cluster 5	35.6	41	10.25
Noise	58.4	19	4.75

Table 4. The overall clustering results obtained using 2nd version at 2nd level of granularity.

Clusters	Weighted score	Cluster size	Percentage
Cluster 1	93.1	68	17
Cluster 2	80.2	104	26
Cluster 3	61.0	96	24
Cluster 4	49.3	64	16
Cluster 5	36.7	49	12.25
Noise	58.4	19	4.75

As we notice through these tables, the results within the same level of granularity are generally close to each other. On the other hand by changing the level of granularity we notice significant changes as the data points are now re-distributed into the new fuzzy clusters. These tables provide useful results on the general user feedback and moreover we can manually label the generated clusters based on the domain expertise. On the other hand we can view the fuzzy memberships of various users into the generated fuzzy clusters. The following tables illustrate a partial view of these results (Tables 5 and 6).

Table 5. The membership values of a few users in the clusters obtained using 1st version at 1st level of granularity.

User	Membership in cluster 1	Membership in cluster 2	Membership in cluster 3
User 1	0.77	0.21	0.03
User 2	0.15	0.74	0.11
User 3	0.02	0.17	0.81
User 4	N/A	N/A	N/A
User 5	0.85	0.14	0.01

Table 6. The membership values of a few users in the clusters obtained using 2nd version at 1st level of granularity.

User	Membership in cluster 1	Membership in cluster 2	Membership in cluster 3
User 1	0.79	0.19	0.02
User 2	0.12	0.78	0.10
User 3	0.02	0.18	0.80
User 4	N/A	N/A	N/A
User 5	0.84	0.15	0.01

This type of tables provide useful insights if we are interested not only in aggregate results of clustering but also in categorizing particular users (data points) into the generated fuzzy clusters.

6 Conclusions and Future Work

In this study we have presented a hybrid fuzzy clustering algorithm which operates in two stages combining density-based clustering and fuzzy c-means or Gustafson-Kessel algorithms. This model intends to exploit the advantages of these two types of clustering approaches and to diminish their drawbacks. Although our model is versatile and can adapted to obtain fuzzy clusters on various data, in our work we are particularly focused on exploring well-structured user feedback data. This model was practically applied on a dataset of user feedback data of a certain company providing flexible results for different levels granularity. Besides the overall clustering, fuzzy membership values of each data point into the generated clusters are also available. These results assist the experts of the domain in gaining insights about the internal structures of the dataset. As a final remark we would point out that our model operates under the assumption that we are tolerant towards a higher computational complexity.

As a future work we may extend our project in a few different directions. Firstly we may offer more tuning options, for example to select among several distance metrics. Secondly we may involve local density-based clustering instead of just DBSCAN algorithm. Also we may employ cluster validation techniques for better orientation about the quality of the clusters.

References

1. De Oliveira, J.V., Pedrycz, W. (eds.): Advances in fuzzy clustering and its applications, pp. 32–49. Wiley (2007)
2. Nayak, J., Naik, B., Behera, H.S.: Fuzzy c-means (FCM) clustering algorithm: a decade review from 2000 to 2014. In: Computational Intelligence in Data Mining, vol. 2, pp. 133–149. Springer, New Delhi (2015)
3. Havens, T.C., Bezdek, J.C., Leckie, C., Hall, L.O., Palaniswami, M.: Fuzzy c-means algorithms for very large data. IEEE Trans. Fuzzy Syst. **20**(6), 1130–1146 (2012)
4. Ghosh, S., Dubey, S.K.: Comparative analysis of k-means and fuzzy c-means algorithms. Int. J. Adv. Comput. Sci. Appl. **4**(4) (2013)
5. Tsai, D.M., Lin, C.C.: Fuzzy c-means based clustering for linearly and nonlinearly separable data. Patt. Recogn. **44**(8), 1750–1760 (2011)
6. Bedalli, E., Ninka, I.: Exploring an educational system's data through fuzzy cluster analysis. In: 11th Annual International Conference on Information Technology and Computer Science (2014)
7. Yu, J., Yang, M.S.: Analysis of parameter selection for Gustafson-Kessel fuzzy clustering using Jacobian matrix. IEEE Trans. Fuzzy Syst. **23**(6), 2329–2342 (2015)
8. Ulutagay, G., Nasibov, E.: Fuzzy and crisp clustering methods based on the neighborhood concept: a comprehensive review. J. Intell. Fuzzy Syst. **23**(6), 271–281 (2012)
9. Graves, D., Pedrycz, W.: Kernel-based fuzzy clustering and fuzzy clustering: a comparative experimental study. Fuzzy Sets Syst. **161**(4), 522–543 (2010)
10. Bedalli, E., Mançellari, E., Asilkan, O.: A heterogeneous cluster ensemble model for improving the stability of fuzzy cluster analysis. Procedia Comput. Sci. **1**(102), 129–136 (2016)
11. Zhang, L., Lu, W., Liu, X., Pedrycz, W., Zhong, C.: Fuzzy c-means clustering of incomplete data based on probabilistic information granules of missing values. Knowl.-Based Syst. **1** (99), 51–70 (2016)
12. Liu, Q., Deng, M., Shi, Y., Wang, J.: A density-based spatial clustering algorithm considering both spatial proximity and attribute similarity. Comput. Geosci. **1**(46), 296–309 (2012)
13. Gan, J., Tao, Y.: DBSCAN revisited: mis-claim, un-fixability, and approximation. In: Proceedings of the 2015 ACM SIGMOD International Conference on Management of Data, 27 May 2015, pp. 519–530. ACM (2015)
14. Schubert, E., Sander, J., Ester, M., Kriegel, H.P., Xu, X.: DBSCAN revisited, revisited: why and how you should (Still) use DBSCAN. ACM Trans. Database Syst. (TODS) **42**(3), 19 (2017)

15. Tsiptsis, K.K., Chorianopoulos, A.: Data Mining Techniques in CRM: Inside Customer Segmentation. Wiley (2011)
16. Bedalli, E., Ninka, I.: Using homogeneous fuzzy cluster ensembles to address fuzzy c-means initialization drawbacks. Int. J. Sci. Eng. Res. 5(6), 465–469 (2014)
17. Silva Filho, T.M., Pimentel, B.A., Souza, R.M., Oliveira, A.L.: Hybrid methods for fuzzy clustering based on fuzzy c-means and improved particle swarm optimization. Expert Syst. Appl. 42(17–18), 6315–6328 (2015)

Expert Knowledge-Based Authentication Protocols for Cloud Computing Applications

Marek R. Ogiela[1(✉)] and Lidia Ogiela[2]

[1] AGH University of Science and Technology, 30 Mickiewicza Ave.,
30-059 Kraków, Poland
mogiela@agh.edu.pl
[2] Pedagogical University of Cracow, Institute of Computer Science,
Podchorążych 2 Street, 30-084 Kraków, Poland
lidia.ogiela@gmail.com

Abstract. In this paper will be presented new classes of authentication procedures, based on visual CAPTCHA solutions, which require special cognitive skills or expert-knowledge. Such authentication protocols can be used especially to differentiate access possibilities for particular group of persons, based of theirs expertise. For new authentication protocols some possible examples of applications will be presented with relation to Cloud computing and distributed authentication.

Keywords: Cryptographic protocols · User authentication
Cloud and fog computing · Cognitive CAPTCHA

1 Introduction

In modern authentication protocols an important role plays CAPTCHA codes. CAPTCHA has become a very popular challenge-response test, which are used on many websites, and protocols oriented for personal authentication. In simple CAPTCHA solutions the main purpose is to guarantee that the received response comes from a human, and not from computer systems. For such purposes there were proposed five different categories of CAPTCHAs i.e. text-based, audio-based, image-based, motion-based, and hybrid solutions combining previously mentioned types [1, 2].

What makes CAPTCHA safe is the randomness of distortion applied to the content (image, letter, sound etc.). This allow to change the shape or texture of visual CAPTCHA pattern, and make the recognition or understanding process more difficult.

The most common styles of distortions applies to the visual CAPTCHA are following [3]:

- Shadow, which for a bot could create a bit of a confusion, because may be interpreted as another letter or as a part of the original letter.
- Outline, which could mislead the pixel counting detection algorithms and cause a need to apply morphological segmentation operations.
- Grains, which results that histogram of this symbol can be different, what could potentially mislead the bot.

F. Xhafa et al. (Eds.): INCoS 2018, LNDECT 23, pp. 82–87, 2019.
https://doi.org/10.1007/978-3-319-98557-2_8

- Random outline degradation, what allow to create a distinct letter, which outline is severely influenced.
- Striped, where width, direction and brightness of placed stripes were randomly selected from a given range. This allow to arrange unique appearance to each character.
- Single and double outline, which allow to change the original shape of a letter.
- Rotation and tilting, which combine the rotation and shifting operations for the shape of symbols.

In modern security authentication protocols it is also possible to define procedures based on the perception abilities, or the specific knowledge possessing by particular person [4–6]. Such knowledge-based solutions can be realized in CAPTCHA creation. CAPTCHA codes can be oriented on specific group of participants, which possess specific expert knowledge, and can properly apply it for authentication processes. Personally oriented technologies can also play an important role in future IoT and Cloud Computing applications [7, 8].

The goal of this paper is to present selected ideas of application of cognitive CAPTCHA solutions for security purposes. Such solutions seem to be very promising for developing advanced cryptographic technologies, especially those oriented for personal authentication, which can evaluate specific human knowledge, expert information, or perception capabilities.

2   CAPTCHA for Knowledge-Based Authentication Protocols

Below will be described solutions based on cognitive CAPTCHA codes, which for authentication require expert or specific knowledge from a particular area or discipline.

Verification procedure using such knowledge-based CAPTCHAS can be done in distributed computer infrastructures with special accessing grants, and also in Cloud or Fog computer systems, in which participants (expert in the field) try to gain access to specific information or data [9].

In such knowledge-based verification procedures it is possible to gain access to data or computer systems, after proper selection by users CAPTCHA parts in proper manner or in determined order. Inappropriate selection during verification techniques can prevent to access the information or computer services. In this solution, it is necessary to correctly select the semantic combination of elements, which fulfil particular requirements or have a specific meaning. For example, we can select all or only a small number of visual parts presenting specific information (or its part). Figure 1 presents an example in which users are verified with different combinations of correct parts.

In such verification procedures, it is possible to specify more complex answers, which required expert knowledge or particular order in selection of correct parts.

Figure 2 presents another example with multi-level CAPTCHA answers. Depending on the whole image division on parts it is possible to find different possible answers. When we consider only biggest parts of this image it is possible to find only one answer (red rectangle). For more detailed division it is also possible to find other possible solutions based on yellow net or green parts. Different parts selection may be required for different participants, but it may be presented sequentially for only one user to deeply verify him on several stages [10, 11].

Fig. 1. Examples of visual CAPTCHA presenting different rare animals (A – kakapo, B – panda, C – platypus, D - echidna). Proper selection will depend on the question: where are animals (A, B, C, D), or where do you see egg-laying mammals (C, D), Where do you see critically endangered animals (A).

Considering expert knowledge-based CAPTCHA, it is also possible to create a verification procedure which required specific knowledge from particular area. It may be connected with history, medicine, engineering, art etc. In Fig. 3 is presented an example of such solution, based on medical visualization.

The meaning of such CAPTCHA can be different when using other visual content, but for presented example the proper answer for authenticating users can be connected with selection of images which show:

- brain structures (D)
- Chest structure (A, B)
- Neck (C)

It may also depend on proper answer for used modality:

- USG examination (C)
- RTG (A)
- MRI or tractography (B, D)

Fig. 2. Examples of visual CAPTCHA with multi-level answers for the question: on which part (s) do you see fish? Depending on the image division for particular parts it is possible to find only one answer (the red rectangle), four (yellow parts) or sixteen (green parts). The question may also be connected with the fishes having particular color, size or other features.

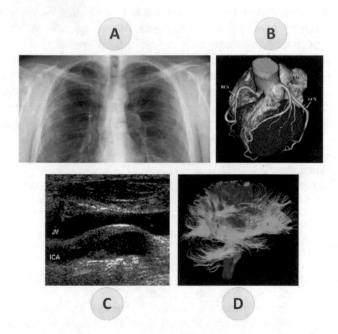

Fig. 3. Examples of cognitive CAPTCHA which required expert medical knowledge.

For verification by professionals we can also use selected medical structure with different lesions or disease stages. The proper answer in such cases can be find only by people having experiences or expert knowledge from particular area.

3   Cloud Applications

Cloud Computing offers unlimited storage and fast computing possibilities, which facilitate data acquisition and processing. It also provides various distributed services, which require user authentication. In such instances it is very important to use authentication procedures like CAPTCHA solutions, which may be applied especially in this areas or services which require proper authentication to prevent robot actions.

Also in Fog computing, we can perform information or data processing, and data analytics tasks. Both Cloud and Fog computing allow to move the data processing tasks from initial sensor layers, to these levels, which should be authorized, and available only for trusted users. For distributed access can be used multilevel or knowledge-based cognitive CAPTCHAs presented in this paper, with different possible answers generated for various accessing levels. Described CAPTCHA based on expert knowledge can be successfully used for such applications.

4   Conclusions

In this paper have been presented new approaches to creation of visual CAPTCHA codes, which security lays not only in difficulties in recognition of letters or image features, but especially on semantic meaning and expert knowledge.

It was described an idea of advanced knowledge-based CATCHAs which are based on using very specific or expert information for verification processes. Such solutions can be dedicated for groups of expert users with particular knowledge or mental analysis skills. If future research we'll try to extend presented solutions towards expert knowledge CAPTCHAs based on motion analysis and deep learning approaches [12].

Acknowledgments. This work has been supported by the AGH University of Science and Technology research Grant No 11.11.120.329.

This work has been supported by the National Science Centre, Poland, under project number DEC-2016/23/B/HS4/00616.

References

1. Alsuhibany, S.: Evaluating the usability of optimizing text-based CAPTCHA generation. Int. J. Adv. Comput. Sci. Appl. **7**(8), 164–169 (2016)
2. Bursztein, E., Bethard, S., Fabry, C., Mitchell, J., Jurafsky, D.: How good are humans at solving CAPTCHAs? a large scale evaluation. In: Proceedings of IEEE Symposium on Security and Privacy, pp. 399–413 (2010)
3. Krzyworzeka, N., Ogiela, L.: Visual CAPTCHA for data understanding and cognitive management. In: Barolli, L., Xhafa, F., Conesa, J. (eds.) Advances on Broad-Band Wireless Computing, Communication and Applications BWCCA 2017. Lecture Notes on Data Engineering and Communications Technologies, vol. 12, pp. 249–255. Springer (2018)

4. Ogiela, L.: Cognitive computational intelligence in medical pattern semantic understanding. In: Guo, M.Z., Zhao, L., Wang, L.P. (eds.) ICNC 2008, Fourth International Conference on Natural Computation Proceedings, Jian, Peoples Republic of China, vol. 6, 18–20 October 2008, pp. 245–247 (2008)
5. Ogiela, L.: Cognitive Information Systems in Management Sciences. Elsevier, Academic Press (2017)
6. Ogiela, L., Ogiela, M.R.: Data mining and semantic inference in cognitive systems. In: Xhafa, F., Barolli, L., Palmieri, F., et al. (eds.) 2014 International Conference on Intelligent Networking and Collaborative Systems (IEEE INCoS 2014) Salerno, Italy, 10–12 September 2014, pp. 257–261 (2014)
7. Ogiela, L., Ogiela, M.R.: Management information systems. LNEE, vol. 331, pp. 449–456 (2015)
8. Ogiela, L., Ogiela, M.R.: Insider threats and cryptographic techniques in secure information management. IEEE Syst. J. **11**, 405–414 (2017)
9. Ogiela, M.R., Ogiela, U.: Grammar encoding in DNA-like secret sharing infrastructure. LNCS, vol. 6059, pp. 175–182 (2010)
10. Ogiela, M.R., Ogiela, L.: On using cognitive models in cryptography. In: IEEE AINA 2016 - The IEEE 30th International Conference on Advanced Information Networking and Applications, Crans-Montana, Switzerland, 23–25 March 2016, pp. 1055–1058 (2016)
11. Ogiela, M.R., Ogiela, L.: Cognitive keys in personalized cryptography. In: IEEE AINA 2017 - The 31st IEEE International Conference on Advanced Information Networking and Applications, Taipei, Taiwan, 27–29 March 2017, pp. 1050–1054 (2017)
12. Osadchy, M., Hernandez-Castro, J., Gibson, S., Dunkelman, O., Perez-Cabo, D.: No bot expects the DeepCAPTCHA! Introducing immutable adversarial examples, with applications to CAPTCHA generation. IEEE Trans. Inf. Forensics Secur. **12**(11), 2640–2653 (2017)

Train Global, Test Local: Privacy-Preserving Learning of Cost-Effectiveness in Decentralized Systems

Jovan Nikolić[1], Marcel Schöengens[2], and Evangelos Pournaras[1(✉)]

[1] Professorship of Computational Social Science, ETH Zurich, Zurich, Switzerland
{jnikolic,epournaras}@ethz.ch
[2] Swiss National Supercomputing Centre, Zurich, Switzerland
schoengens@cscs.ch

Abstract. The mandate of citizens for more socially responsible information systems that respect privacy and autonomy calls for a computational and storage decentralization. Crowd-sourced sensor networks monitor energy consumption and traffic jams. Distributed ledgers systems provide unprecedented opportunities to perform secure peer-to-peer transactions using blockchain. However, decentralized systems often show performance bottlenecks that undermine their broader adoption: propagating information in a network is costly and time-consuming. Optimization of cost-effectiveness with supervised machine learning is challenging. Training usually requires privacy-sensitive local data, for instance, adjusting the communication rate based on citizens' mobility. This paper studies the following research question: How feasible is to train with privacy-preserving aggregate data and test on local data to improve cost-effectiveness of a decentralized system? Centralized machine learning optimization strategies are applied to DIAS, the *Dynamic Intelligent Aggregation Service* and they are compared to decentralized self-adaptive strategies that use local data instead. Experimental evaluation with a testing set of 2184 decentralized networks of 3000 nodes aggregating real-world Smart Grid data confirms the feasibility of a linear regression strategy to improve both estimation accuracy and communication cost, while the other optimization strategies show trade-offs.

1 Introduction

The optimization cost-effectiveness in decentralized networked systems is challenging, for instance, sensor networks making collective measurements for energy [3] or traffic monitoring [6]. Communication cost as well as convergence time required for data to propagate in a decentralized network are performance bottlenecks for their broader adoption. Instead, in centrally managed systems such as cloud computing infrastructures [22] in which performance data

© Springer Nature Switzerland AG 2019
F. Xhafa et al. (Eds.): INCoS 2018, LNDECT 23, pp. 88–102, 2019.
https://doi.org/10.1007/978-3-319-98557-2_9

from each system component can be locally available or remotely accessed, a broad spectrum of optimization and machine learning techniques can be used to tune performance [10]. Hibernation of idle system components to save energy and dynamic resource allocation based on varying computational demand are some examples. Nonetheless, the data based on which optimization of cost-effectiveness is performed can be personal and privacy-sensitive. Citizens may not be willing to share their data. As a result, the data-intensive operations of traditional optimization and machine learning techniques oppose the design of decentralized systems and in particular they are not easily applicable in a privacy-sensitive application context. The resolution of this discrepancy is the subject and focus of this paper.

This paper addresses the following research questions: (i) *How to optimize cost-effectiveness in decentralized systems using supervised machine learning that exclusively uses aggregate data for training and local data for testing?* (ii) *How to aggregate training data in a privacy-preserving way?* To address these questions, a centralized machine learning approach is introduced that relies on baseline and runtime performance data aggregated over the nodes in a privacy-preserving way using differential privacy [7] or homomorphic encryption [14]. The learning capacity of linear regression and neural network classifiers is studied given the information loss by the performed aggregation. This approach is compared to decentralized self-adaptive strategies that rely instead on local data. The optimization approaches are implemented in DIAS, the *Dynamic Intelligent Aggregation Service* [18] that computes aggregation functions in a fully decentralized fashion under continuously changing input data. The accuracy of the estimations and the communication cost are optimized by the strategies. Training data are generated from simulation test runs of 2184 decentralized networks with 3000 nodes aggregating real-word Smart Grid data. The linear regression classifier is the strategy that improves both accuracy and communication cost. The neural network classifier shows a trade-off of lower communication cost for a lower accuracy, while the self-adaptive strategies show the opposite trade-off: higher accuracy at a cost of higher communication load.

The contributions of this paper are summarized as follows: (i) A new privacy-preserving framework to apply a broad spectrum of existing machine learning techniques for the optimization of cost-effectiveness in decentralized systems. (ii) The qualitative and quantitative comparison of centralized machine learning optimization strategies using aggregate data with decentralized self-adaptive strategies using local data. (iii) The enhancement of DIAS with the machine learning and self-adaptive strategies that improve accuracy and communication cost.

This paper is organized as follows: Sect. 2 illustrates the challenge of cost-effectiveness optimization in decentralized systems. Section 3 introduces machine learning and self-adaptive optimization strategies. Section 4 shows their applicability on a decentralized sensing scenario. Section 5 outlines the experimental settings and the results. Finally, Sect. 6 concludes this paper and outlines future work.

2   Cost-Effectiveness Optimization in Decentralized Systems

A *decentralized system* is defined as a set of N (remote) autonomous nodes such as personal computers, smart phones, wearables or other Internet of Things devices that interact in a peer-to-peer fashion to achieve a collective goal. For instance, such devices can exchange numerical values, e.g. sensor data, to make collective measurements [18] known as in-network aggregation [5]: each device locally computes aggregation functions, for instance, the total load of the power grid [3] or the average vehicle traffic in a city [6]. Each node disseminates its values to other nodes whenever the values change or nodes join and leave the network. Without loss of generality and for a concrete illustration of the research challenge, *decentralized sensing* is the scenario studied in this paper.

The cost-effectiveness of decentralized systems is the focus of this paper. *Cost* is the amount of resources required to perform system operations. For instance, computational cost is the processing power consumed by nodes and communication cost is the number/size of exchanged messages between them. *Effectiveness* reflects on the quality of service and shows how well a decentralized system performs under a certain cost paid. For instance, the aggregation accuracy is indicator of effectiveness as it shows how 'close' the estimation of the aggregation functions is to the actual true values. Given that improvements of accuracy are a result of updates received by other nodes, i.e. input sensor data from joining nodes or updated data from connected ones, a higher communication cost can improve system effectiveness.

This paper addresses the following challenge: *self-management of trade-offs in the cost-effectiveness of decentralized systems*. A number of local parameters often regulate cost-effectiveness, for instance, the period of push-pull gossip requests [11] or Time-To-Live (TTL) in flooding [2]. In practice [20], the selection of these parameters is non-automated, system-wide (global) and made offline by system administrators/operators to control effectiveness given the available resources [21] in the deployed network infrastructure, i.e. network bandwidth or energy capacity.

Cloud computing virtualization [22] separates distributed processing and storage from centralized resource allocation. When universal access over distributed data is granted to a centralized authority, or when this authority collects these data, cost-effectiveness can be optimized online with existing (supervised) machine learning techniques [10]. However, this approach violates decentralization and raises privacy concerns over personal data that citizens may not be willing to share. This paper studies two types of privacy-preserving optimization strategies for the cost-effectiveness of decentralized systems: (i) *machine learning* and (ii) *self-adaptive* strategies. These two approaches are positioned in the design space of Table 1.

Unsupervised decentralized learning is highly complex and usually requires endogenous system redesign [1,15]. Decentralized systems impose partial data storage and exchange. Sharing regular updates of a full feature vector over the network is inefficient and often infeasible [13] considering biases by outliers

Table 1. The design space of cost-effectiveness optimization in decentralized systems: Centralized optimization with local data and decentralized optimization with aggregate data are excluded (x). The former is privacy-intrusive, while the latter requires complex and costly mechanisms integration. Instead, centralized optimization with aggregate data using machine learning and decentralized optimization with local data using self-adaptive strategies are studied (★).

Optimization	Local data	Aggregate data
Centralized	Privacy-intrusive (x)	*Machine learning* (★)
Decentralized	*Self-adaptive strategies* (★)	Complex & costly (x)

and initialization. Instead, the introduced supervised machine learning strategies (Fig. 1) perform training using aggregate data and testing with local data at each node. Two aggregate data types are required for training: (i) *baseline performance* and (ii) *cost-effectiveness* data. The former are used to compare system performance with an optimal performance. For instance, the actual aggregates of the sensor data can be used as a baseline against the aggregate estimations during system runtime. This comparison provides the effectiveness data. Both baseline and cost-effectiveness data are aggregated by the *optimizer* in a privacy-preserving way using differential privacy [7] or homomorphic encryption [14]. In differential privacy, nodes mask their data by adding a special noise, e.g. Laplace [7], that has the following property: when the masked data are summed up by the optimizer, the added noises cancel out and the aggregate data are revealed without revealing the individual data of the nodes. In homomorphic encryption this process is performed using cryptographic keys and is therefore more secure than differential privacy, though it usually requires key management by trusted third parties and more expensive computations [14].

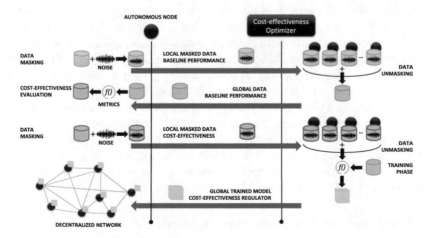

Fig. 1. The machine learning optimization approach with aggregate data.

The sequence of operations during the training phase is as follows: Nodes mask their local sensor data used as baseline performance and send them to the optimizer. The optimizer unmasks the data at an aggregate level and sends back the aggregate sensor data used as the global baseline performance. The nodes compare the estimates of the aggregates performed in a fully decentralized way (see Sect. 4) with the actual baseline aggregates sent by the optimizer using an error metric that measures the aggregation accuracy (effectiveness). These data together with the measured communication cost are masked before sent to the optimizer. The optimizer unmasks the cost-effectiveness data by aggregating them and feeds them in the learning model for training. The universally generated trained model is sent back to the nodes based on which they locally regulate the communication cost at low levels while maximizing effectiveness, i.e. accuracy.

On the other hand, nodes with self-adaptive strategies perform local adjustments over parameters that control cost-effectiveness using stimuli from other nodes they interact. No training is required. Communication cost is regulated by measuring the portion of the network from which updates are received or by monitoring the stability of the aggregates as an indicator of convergence, i.e. maximal effectiveness.

3   Machine Learning vs. Self-adaptive Optimization Strategies

Assume a system parameter that controls the resource utilization at each node. Resources have a cost paid to improve system effectiveness. For instance, a higher communication rate increases the speed of aggregation accuracy. Each node is given the autonomy to regulate such system parameters to *consume* or *save* resources. Resource utilization is controlled within the *flexibility range* $[B_L, B_U]$, where B_L is a lower and B_U an upper bound. Assume as well that the nodes of the decentralized system trigger and process events at discrete time steps referred to as *epochs*, e.g. sending, and processing received messages. Nodes choose at each epoch their resource utilization within the flexibility range $[B_L, B_U]$ as the means to control cost-effectiveness. Maintaining a maximum accuracy in decentralized sensing requires utilization of the maximum number B_U of exchanged messages per epoch. However, input data may cancel out each other in aggregation, e.g. $5 - 2 - 3 = 0$ and therefore the update of summation does not require these data records, whose exchange adds up communication and computational overhead. Saving resources from the nodes in which these records are originated is an optimization that can improve cost-effectiveness. The rest of this section illustrates the resource utilization of the machine learning and self-adaptive strategies in the flexibility range $[B_L, B_U]$.

3.1   Machine Learning Optimization Strategies

Each node is equipped with a classifier trained to distinguish between the two classes CONSUME and SAVE that indicate low and high cost-effectiveness respectively. During testing, CONSUME sets resource utilization to B_U, whereas SAVE to B_L. Training data are collected via (i) *simulations* or (ii) *pilot test runs*. In decentralized sensing, they concern the number of exchanged messages and the estimated aggregates among others. Formally, let $x_i \in \mathbb{R}^d$ be a d-dimensional feature vector with $i = 1, ..., n$, where n is the number of features. Labeling of the training data is performed by evaluating whether cost-effectiveness is above or below a fixed threshold, assuming it is a discrete variable. Formally, let $y_i \in \{0, 1\}$, $i = 1, ..., n$ be a label corresponding to the feature vector x_i, where 1 corresponds to CONSUME and 0 to SAVE. The data is then divided into training and validation set.

Two classifiers are studied: (i) *logistic regression* and (ii) *neural network*. The former is chosen for its non-linearity and learning efficiency since it directly learns posterior probabilities $P(y_i = 1|x_i)$. Given the optimization scope, ridge regularized logistic regression [12] is used that minimizes the following cost function:

$$\min_{\beta,C,b} \frac{1}{2}\beta^T\beta + C\sum_{i=1}^{n} \log(\exp(-y_i(x_i^T\beta + b)) + 1), \tag{1}$$

where β is a d-dimensional weight vector, b is the bias and C is the regularization strength parameter. Ridge regularization favors low values in β for irrelevant features and is chosen for its stability of the solutions and computational efficiency over L1 regularization, which favors a sparse weight vector β. Learning is performed via stochastic gradient descent. Let the learned weight vector β^* minimize Eq. 1 and $\tilde{x} \in \mathbb{R}^d$ be a new feature vector from the validation or test set. The probability of $\tilde{x}$ belonging to the class CONSUME for $\tilde{y} = 1$ is given as follows:

$$P(\tilde{y} = 1|\tilde{x}) = \frac{1}{1 + \exp(-(\tilde{x}^T\beta^* + b^*))}. \tag{2}$$

If $P(\tilde{y} = 1|\tilde{x}) < 0.5$, a node saves resources by consuming B_L. Otherwise, it expands the resource consumption to B_U. The logistic regression classifier is generalized to a *Multi-Layer Perceptron* (MLP) neural network, by adding h hidden layers, each with l_q, $q = 1, ...h$ hidden nodes. The activation function in the hidden layer is the ReLU function, chosen for its simplicity, ease of computation and reduced likelihood of vanishing gradients [9]. Learning is performed via backpropagation.

Both classifiers are validated on the validation set and then tested during the decentralized system runtime. The trained classifier is integrated into each node as a black box. It receives as input the locally estimated aggregates and provides as output the resource utilization that is the communication rate at each epoch.

3.2   Self-adaptive Strategies

The self-adaptive strategies are data-independent heuristics that predict cost-effectiveness by monitoring local system parameters, for instance, counting the number of nodes with which sensor data are exchanged. This measurement encodes both the (i) communication cost and (ii) aggregation accuracy. As the counter increases, a larger portion of the network receives the latest sensor data and estimations are more accurate. Therefore this counter can regulate resource allocation: a node joining the network or having new sensor data to share begins with a communication rate of B_U to minimize the time operating with low accuracy. As the number of data exchanges with different nodes increases, the communication rate decreases to save resources and eventually becomes B_L when data are disseminated to all nodes of the network.

Formally, let $r_j^{(t)} \in [0, 1]$ define the relative remaining communication cost that is required for a node j at epoch t to exchange the sensor data with all nodes in the network. Next assume a function $f(r_j^{(t)})$ that calculates the allocated resources such that $B_L \leq f(r_j^{(t)}) \leq B_U$. This paper studies the following functions:

Table 2. Functions used by the self-adaptive strategies.

Function name	Function	α
Linear	$B_L + \alpha \cdot r \cdot B_U$	$\frac{B_U - B_L}{B_U}$
Exponential	$B_L - 1 + e^{\alpha \cdot r \cdot B_U}$	$\frac{\ln(B_U - B_L + 1)}{B_U}$
Square root	$B_L + \sqrt{\alpha \cdot r \cdot B_U}$	$\frac{(B_U - B_L)^2}{B_U}$
Logistic regression	$B_U - \frac{B_U - B_L}{1 + 0.001 \cdot e^{14 \cdot r}}$	Parameters smoothly covering $[B_L, B_U]$

These self-adaptive strategies do not require data labeling and offline training. On the contrary, they assume apriori an exact relation between the relative remaining communication cost and the allocated resources.

4   Applicability on Decentralized Sensing

The optimization strategies are applied on DIAS,[1] the *Dynamic Intelligent Aggregation Service*. DIAS is a generic and highly dynamic service for fully decentralized and real-time computations of aggregation functions: Each DIAS node can share (data provider) and aggregate (data consumer) sensor data. DIAS can adapt and self-correct computations to improve accuracy under changing sensor data [18] or nodes joining and leaving the network [16,17]. A large family of aggregation functions, e.g. summation, average, maximum, minimum, count, standard deviation, etc., can be computed without any change in the core distributed algorithm.

DIAS nodes discover each other and disseminate their sensor data updates via the gossip-based peer sampling service [11]. Each node periodically updates its partial *view* that is a list of limited size c with randomly populated descriptors

---

[1] Available af http://dias-net.org (last access: May 2018).

of other nodes containing the IP address, the port number and other application-level information. The HEALING and SWAPPING gossip parameters determine the features of the discovered nodes: the latest nodes joined the network vs. random ones.

Real-time aggregation of highly violatile data is feasible in DIAS using the model of *possible states*. Let $P_j = (p_{j,u})_{u=1}^k$ be a sequence of k possible states generated by a node j using, for instance, historical sensor data [19]. For instance, smart meter power data can be abstracted by numerical representations corresponding to the low, medium and high consumption level. Moreover, let $s_j^t \in P_j$ be the local *selected state* of node j at epoch t that is the input in the aggregation functions of other remote nodes. Sensor data exchanges are performed within an *aggregation session* between an *aggregator* and a *disseminator*, locally or in two remote nodes. A node has a disseminator, an aggregator or both. The latter scenario is the most demanding one in terms of computational and communication resources. The disseminator initiates the aggregation session and sends its selected state s_j to the aggregator. The aggregator classifies it as a new input value, i.e. fist performed aggregation session, or as a replacing input value, i.e. an earlier aggregation session has been performed with another outdated selected state. The aggregator completes the aggregation session by sending an acknowledgment and the outcome back to the disseminator.

Aggregation accuracy improves by counting a new selected state or updating an outdated one. Any other scenario, i.e. duplicate selected states, is excluded using a distributed memory system based on probabilistic data structures, the *bloom filters* [4]. The memory system consists of two nested bloom filter layers as shown in Fig. 2: (i) the *interactions layer* verifies whether an aggregation session between an aggregator and a disseminator has been earlier performed while (ii) the *data layer* validates which latest selected state has been aggregated. Disseminators use the memory system to determine whether an aggregation session should be performed with an aggregator sampled from the gossiping service. The exact algorithms are out of the scope of this paper and can be found in earlier work [16–18].

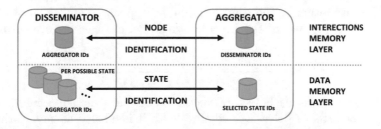

Fig. 2. The DIAS distributed memory systems with two nested layers of bloom filters: (i) *Interaction layer* - tracks the involved disseminators and aggregators of the aggregation sessions. (ii) *Data layer* - tracks the exchange of selected states.

Cost is measured by the total number of exchanged messages per epoch originated by the aggregation sessions. The effectiveness of DIAS at each epoch t is measured by the average estimation error $\tilde{S}_j^{(t)}$ of summation over N nodes, assuming each node has an aggregator:

$$\epsilon^{(t)} = \frac{1}{N} \sum_{j=1}^{N} \frac{|\tilde{S}_j^{(t)} - \sum_{v=1}^{N} s_v^{(t)}|}{\sum_{v=1}^{N} s_v^{(t)}}, \tag{3}$$

where $\tilde{S}_j^{(t)}$ is the estimated sum by each node j and $\sum_{v=1}^{N} s_v^{(t)}$ is the actual sum of all selected states. The aggregation function of summation is chosen as it is the most sensitive one to changes in cost-effectiveness, i.e. possible states. Recall from Fig. 1 that $\sum_{v=1}^{N} s_v^{(t)}$ is the baseline performance calculated via differential privacy schemes [7] to prevent leaking of the individual selected states to a third party.

The number of initiated aggregation sessions $b_j^{(t)}$ for each node j and epoch t, i.e. communication rate, is the system parameter (control variable) that regulates cost-effectiveness and the one controlled by the optimization strategies. Instead of fixing $b_j^{(t)}$ for all nodes, the strategies vary $b_j^{(t)}$ in the flexibility range $[B_L, B_U]$, where $B_L = 0$ determines no performed aggregation sessions (saving resources), whereas $B_U = c$ improves accuracy by maximizing[2] the number of aggregation sessions.

4.1 Optimization Stratagies in Decentralized Sensing

Machine Learning Strategies: Training is performed by varying in multiple test runs (i) the number of aggregation sessions and (ii) the possible states as well as the selected state of the nodes (different datasets). This process generates a 5-dimensional vector with the following metrics: (i) test run identifier, (ii) epoch number, (iii) number of aggregation sessions, (iv) average estimation of the sum over all nodes and (v) slope of this sum based on a time window of 5 epochs. The average estimation of the sum at epoch t is calculated as $\frac{1}{N} \sum_{j=1}^{N} \tilde{S}_j^{(t)}$, where $\tilde{S}_j^{(t)}$ is the estimation of the sum by node j at epoch t. The slope of the sum quantifies the dynamics of the selected states, i.e. changes performed during a test run:

$$\delta^{(t)} = \frac{Var[\frac{1}{N} \sum_{j=1}^{N} \tilde{S}_j^{(\tau)} | \tau = t, t-1, ..., t-4]}{Var[\tau | \tau = t, t-1, ..., t-4]}. \tag{4}$$

The optimizer labels each feature vector using the average estimation error (Eq. 3). If the error is below a fixed threshold, it is labeled as SAVE, otherwise, CONSUME. These labeled data are then used to train the classifier.

During the testing phase, each node j at epoch $t - 1$ uses the classifier to determine the number of aggregation sessions $b_j^{(t)}$ at epoch t. No interactions are required between nodes and optimizer. Autonomy and decentralization are

---

[2] $B_U = c$ assumes that gossiping updates the view at least once per epoch.

preserved. The test feature vector generated online during system operation contains the local sum estimation $\tilde{S}_j^{(t)}$ and the slope of this estimate instead of the average sum estimation and its slope over all nodes. The number of aggregation sessions are set to $b_j^{(t+1)} = B_U$ if the classifier determines CONSUME or $b_j^{(t+1)} = B_L$ if SAVE.

Self-adaptive Strategies: At each epoch $t - 1$, a disseminator j counts the aggregators $Q_j^{(t-1)}$ that have aggregated its current selected state $s_j^{(t-1)}$. If the selected state changes, the counter sets back to zero. The count aggregation function of DIAS provides an estimate of the total number of aggregators $\tilde{N}_j^{(t-1)}$ at each epoch $t - 1$ that is N if all nodes have an aggregator. The relative remaining aggregation sessions $r_j^{(t-1)}$ initiated by a disseminator j at epoch t are calculated as follows:

$$b_j^{(t)} = f(r_j^{(t-1)}) = 1 - \frac{Q_j^{(t-1)}}{\tilde{N}_j^{(t-1)}}, \tag{5}$$

where $r_j^{(t-1)} \in [0, 1]$ is fed in one of the functions of Table 2. The higher the rate of changes in the selected state, the higher the $r_j^{(t-1)}$.

5   Experimental Evaluation

DIAS and the peer sampling service on which DIAS relies on are implemented[3] in Java using the Protopeer distributed prototyping toolkit [8]. Training data are collected by deploying simulation test runs in the CSCS supercomputing infrastructure,[4] while DIAS is deployed in the Euler HPC cluster infrastructure[5] of ETH Zurich for execution of the testing phase and evaluation of the cost-effectiveness.

Real-world data from the *Electricity Customer Behavior Trial* project[6] are used for aggregation by DIAS. The data were collected in 2009 and 2010 (364 days) and concern the electricity consumption of 3000 residential consumers. They contain 30-min records of the power consumption. The possible states of each disseminator are the cluster centroids computed by k-means using Weka,[7] with $k = 5$. The dataset is randomly divided into two sets: 80% for training and 20% for testing.

The following parameters are used for the simulation test runs: 3000 nodes, 800 epochs, view size $c = 50$, a HEALING of 1 and a SWAPPING of 24. Each

---

[3] https://github.com/epournaras/DIAS   and   https://github.com/epournaras/PeerSamplingService (last access: May 2018).

[4] https://www.cscs.ch (last access: May 2018).

[5] https://scicomp.ethz.ch/wiki/Euler (last access: May 2018).

[6] http://www.ucd.ie/issda/data/commissionforenergyregulationcer/ (last accessed: May 2018).

[7] https://www.cs.waikato.ac.nz/ml/weka/ (last access: May 2018).

experiment runs with a fixed maximum number of aggregation sessions among the following values in different experiments: $\{5, 15, 20, 30, 40, 50\}$. Therefore, a total number of 6 sessions $* 364$ days $= 2184$ experiments are performed for training.

Each feature vector is labeled with 1 (CONSUME) if $\epsilon^{(t)} \geq 0.1$ and with 0 (SAVE) otherwise. The labeled feature vectors are fed into the two classifiers from *scikit-learn*.[8] The logistic regression uses the 'sag' solver and regularization strength of $C = 0.5$. The multilayer perceptron has $h = 2$ hidden layers with $l_1 = 6$ and $l_2 = 3$ hidden nodes. Learning rate is constant and equal to 0.001. Training is performed without regularization using the 'lbfgs' solver from the quasi-Newton family methods. Both classifiers are validated[9] on the training set. The trained classifiers are then integrated into each DIAS node by importing the weight vectors and biases.

During testing phase for Day 199, the flexibility range of $[B_L = 10, B_U = 40]$ is used by each node j to determine the number of aggregation sessions $b_j^{(t)}$ for both machine learning and self-adaptive strategies. The range $[B_L = 20, B_U = 40]$ is also evaluated for the machine learning strategies. Figure 3a illustrates how $b_j^{(t)}$ is determined by the self-adaptive strategies and also shows the actual summation estimated by each node during the testing phase.

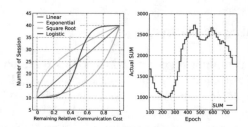

Fig. 3. Testing phase: Calculation of the aggregation sessions by the self-adaptive strategies and the actual summation estimated by each node.

The strategies are evaluated by comparing their cost-effectiveness against a fixed number of aggregation sessions for all nodes. This number is calculated by the average number of aggregation sessions observed in the compared optimization strategy. The former case is indexed with O and the latter with F. Comparison measurements are made in terms of *performance improvement* defined at each epoch t as follows:

$$g(v)^{(t)} = v_F^{(t)} - v_O^{(t)}, \tag{6}$$

---

[8] http://scikit-learn.org/stable/ (last access: May 2018).

[9] Linear regression has an average *precision, recall, f1-score* of 0.8 and 0.96 for neural network. 273662 occurrences appear for SAVE and 123557 for CONSUME in linear regression. The respective occurences are 274108 and 123111 for neural network. Validation metrics documentation: http://scikit-learn.org/stable/modules/generated/sklearn.metrics.precision_recall_fscore_support.html (last access: May 2018).

where v is the communication cost m or the estimation error ϵ. Similarly, the *mean relative performance improvement* over the DIAS runtime is introduced as follows:

$$\bar{g}(v) = \frac{1}{T_{max} - T_{min} + 1} \sum_{t=T_{min}}^{T_{max}} \frac{g(v)^{(t)}}{v_{\mathsf{F}}^{(t)}} \cdot 100\%, \tag{7}$$

where v is the communication cost m or the estimation error ϵ and $T_{max} = 800$, $T_{min} = 101$, excluding the first 100 epochs used for system bootstrapping.

5.1 Experimental Results

Figure 4 illustrates the mean relative improvement of the optimization strategies. The following observations can be made: The logistic regression with the narrower flexibility range of $[B_L = 20, B_U = 40]$ is the only strategy that achieves to decrease both the estimation error by 7.15% and the communication cost by 4.96%. The other strategies show two opposing trade-offs. The neural networks increase the estimation error by 12.35% on average but decrease communication cost by 7.37% respectively. In contrast, the self-adaptive mechanisms decrease the estimation error by 8.18% on average but increase communication cost by 11.25% respectively.

Figure 5 illustrates how performance improvement varies over runtime. The following observations can be made for the estimation error in Fig. 5a and b: The machine learning strategies and especially the logistic regression maximize the decrease of the estimation error during epochs 131 to 281 during which a significant change in summation is observed (Fig. 3). The self-adaptive strategies manage to maintain the improvement during the next epochs in contrast to the machine learning strategies that fall into a performance deterioration during epochs 281 to 800.

Figure 5c and d show the improvement of the optimization strategies in communication cost along with Fig. 6 that shows the number of aggregation sessions for each node over runtime. The machine learning strategies and especially the logistic regression consume a high number of messages during the first epochs in which large changes in the summation are observed. In this way they manage to decrease the estimation error during this period. During epochs 327 to

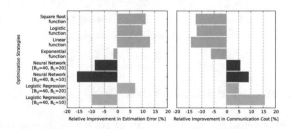

Fig. 4. Mean relative performance improvement of the optimization strategies.

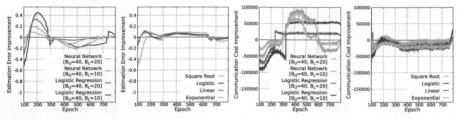

(a) Estimation error, machine learning strategies

(b) Estimation error, self-adaptive strategies

(c) Communication cost, machine learning strategies

(d) Estimation error, self-adaptive strategies

Fig. 5. Performance improvement of the optimization strategies over system runtime.

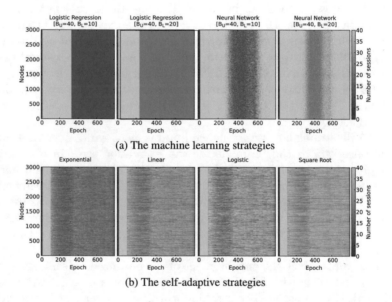

(a) The machine learning strategies

(b) The self-adaptive strategies

Fig. 6. The selections of the aggregations sessions by the optimization strategies.

800, all machine learning strategies reduce communication cost and as a consequence they do not capture the increase of summation. This justifies the negative improvement in the estimation error. The self-adaptive strategies decrease the estimation error by sacrificing communication cost, with the exponential function providing the best trade-off.

6 Conclusion and Future Work

This paper concludes that the optimization of cost-effectiveness in decentralized systems is feasible using machine learning without violating privacy. This is made possible by introducing a novel training scheme that relies entirely on aggregate data collected in a privacy-preserving way, while testing is locally performed in each node of the network using local data. A linear regression classifier manages to improve both accuracy and communication cost in an application scenario of decentralized sensing using real-world Smart Grid data, while the neural network classifier shows a trade-off: lower accuracy but higher savings in communication load. The opposite trade-off is observed in the self-adaptive strategies.

Future work includes the linking of the validation performance with the testing performance of the decentralized network as well as performance comparisons with the other two design options of Table 1. The feasibility of deep learning and other machine learning techniques empowered by cooperative data forwarding strategies is subject of future work as well.

Acknowledgements. This work is supported by the European Community's H2020 Program under the scheme 'ICT-10-2015 RIA', grant agreement #688364 'ASSET: Instant Gratification for Collective Awareness & Sustainable Consumerism' (http:// www.asset-consumerism.eu).

References

1. Azimi, R., Sajedi, H.: A decentralized gossip based approach for data clustering in peer-to-peer networks. J. Parallel Distrib. Comput. **119**, 64–80 (2018)
2. Barjini, H., Othman, M., Ibrahim, H., Udzir, N.I.: Shortcoming, problems and analytical comparison for flooding-based search techniques in unstructured P2P networks. Peer-to-Peer Netw. Appl. **5**(1), 1–13 (2012)
3. Bhattarai, B., de Cerio Mendaza, I.D., Myers, K.S., Bak-Jensen, B., Paudyal, S.: Optimum aggregation and control of spatially distributed flexible resources in smart grid. IEEE Trans. Smart Grid (2017). https://ieeexplore.ieee.org/document/7886280/
4. Broder, A., Mitzenmacher, M.: Network applications of bloom filters: a survey. Int. Math. **1**(4), 485–509 (2004)
5. Dagar, M., Mahajan, S.: Data aggregation in wireless sensor network: a survey. Int. J. Inf. Comput. Technol. **3**(3), 167–174 (2013)
6. Dietzel, S., Petit, J., Kargl, F., Scheuermann, B.: In-network aggregation for vehicular ad hoc networks. IEEE Commun. Surv. Tutorials **16**(4), 1909–1932 (2014)
7. Dwork, C., Roth, A., et al.: The algorithmic foundations of differential privacy. Found. Trends® Theor. Comput. Sci. **9**(3–4), 211–407 (2014)
8. Galuba, W., Aberer, K., Despotovic, Z., Kellerer, W.: ProtoPeer: a P2P toolkit bridging the gap between simulation and live deployement. In: Proceedings of the 2nd International Conference on Simulation Tools and Techniques, p. 60. ICST (Institute for Computer Sciences, Social-Informatics and Telecommunications Engineering) (2009)
9. Glorot, X., Bordes, A., Bengio, Y.: Deep sparse rectifier neural networks. In: Proceedings of the Fourteenth International Conference on Artificial Intelligence and Statistics, pp. 315–323 (2011)
10. Hameed, A., Khoshkbarforoushha, A., Ranjan, R., Jayaraman, P.P., Kolodziej, J., Balaji, P., Zeadally, S., Malluhi, Q.M., Tziritas, N., Vishnu, A.: A survey and taxonomy on energy efficient resource allocation techniques for cloud computing systems. Computing **98**(7), 751–774 (2016)

11. Jelasity, M., Voulgaris, S., Guerraoui, R., Kermarrec, A.M., Van Steen, M.: Gossip-based peer sampling. ACM Trans. Comput. Syst. (TOCS) **25**(3), 8 (2007)
12. Lee, A., Silvapulle, M.: Ridge estimation in logistic regression. Commun. Stat.-Simul. Comput. **17**(4), 1231–1257 (1988)
13. Luo, L., Liu, M., Nelson, J., Ceze, L., Phanishayee, A., Krishnamurthy, A.: Motivating in-network aggregation for distributed deep neural network training. In: Workshop on Approximate Computing Across the Stack (2017)
14. Parmar, P.V., Padhar, S.B., Patel, S.N., Bhatt, N.I., Jhaveri, R.H.: Survey of various homomorphic encryption algorithms and schemes. Int. J. Comput. Appl. **91**(8), 26–32 (2014)
15. Pilgerstorfer, P., Pournaras, E.: Self-adaptive learning in decentralized combinatorial optimization: a design paradigm for sharing economies. In: Proceedings of the 12th International Symposium on Software Engineering for Adaptive and Self-Managing Systems, pp. 54–64. IEEE Press (2017)
16. Pournaras, E., Nikolić, J.: On-demand self-adaptive data analytics in large-scale decentralized networks. In: 2017 IEEE 16th International Symposium on Network Computing and Applications (NCA), pp. 1–10. IEEE (2017)
17. Pournaras, E., Nikolic, J.: Self-corrective dynamic networks via decentralized reverse computations. In: Proceedings of the 14th International Conference on Autonomic Computing (ICAC 2017) (2017)
18. Pournaras, E., Nikolic, J., Omerzel, A., Helbing, D.: Engineering democratization in internet of things data analytics. In: Proceedings of the 31st IEEE International Conference on Advanced Information Networking and Applications-AINA-2017. IEEE (2017)
19. Pournaras, E., Vasirani, M., Kooij, R.E., Aberer, K.: Measuring and controlling unfairness in decentralized planning of energy demand. In: 2014 IEEE International Energy Conference (ENERGYCON), pp. 1255–1262. IEEE (2014)
20. Ramassamy, C., Fouchal, H.: A decision-support tool for wireless sensor networks. In: 2014 IEEE International Conference on Communications (ICC), pp. 7–11. IEEE (2014)
21. Wu, J., Jia, Q.S., Johansson, K.H., Shi, L.: Event-based sensor data scheduling: trade-off between communication rate and estimation quality. IEEE Trans. Autom. Control **58**(4), 1041–1046 (2013)
22. Xiao, Z., Song, W., Chen, Q.: Dynamic resource allocation using virtual machines for cloud computing environment. IEEE Trans. Parallel Distrib. Syst. **24**(6), 1107–1117 (2013)

Performance Evaluation of WMNs for Normal and Uniform Distribution of Mesh Clients Using WMN-PSOSA Simulation System

Shinji Sakamoto[1(✉)], Leonard Barolli[2], and Shusuke Okamoto[1]

[1] Department of Computer and Information Science, Seikei University,
3-3-1 Kichijoji-Kitamachi, Musashino-shi, Tokyo 180-8633, Japan
shinji.sakamoto@ieee.org, okam@st.seikei.ac.jp
[2] Department of Information and Communication Engineering,
Fukuoka Institute of Technology,
3-30-1 Wajiro-Higashi, Higashi-Ku, Fukuoka 811-0295, Japan
barolli@fit.ac.jp

Abstract. Wireless Mesh Networks (WMNs) have many advantages such as low cost and increased high-speed wireless Internet connectivity, therefore WMNs are becoming an important networking infrastructure. In our previous work, we implemented a Particle Swarm Optimization (PSO) based simulation system for node placement in WMNs, called WMN-PSO. Also, we implemented a simulation system based on Simulated Annealing (SA) for solving node placement problem in WMNs, called WMN-SA. In this paper, we implement a hybrid simulation system based on PSO and SA, called WMN-PSOSA. We evaluate the performance of WMN-PSOSA by conducting computer simulations considering Normal and Uniform distributions of mesh clients. Simulation results show that WMN-PSOSA performs better for Normal distribution compared with the case of Uniform distribution.

1 Introduction

The wireless networks and devises are becoming increasingly popular and they provide users access to information and communication anytime and anywhere [3,8,10–12,15,21,27,28,30,34]. Wireless Mesh Networks (WMNs) are gaining a lot of attention because of their low cost nature that makes them attractive for providing wireless Internet connectivity. A WMN is dynamically self-organized and self-configured, with the nodes in the network automatically establishing and maintaining mesh connectivity among them-selves (creating, in effect, an ad hoc network). This feature brings many advantages to WMNs such as low up-front cost, easy network maintenance, robustness and reliable service coverage [1]. Moreover, such infrastructure can be used to deploy community networks, metropolitan area networks, municipal and corporative networks,

© Springer Nature Switzerland AG 2019
F. Xhafa et al. (Eds.): INCoS 2018, LNDECT 23, pp. 103–115, 2019.
https://doi.org/10.1007/978-3-319-98557-2_10

and to support applications for urban areas, medical, transport and surveillance systems.

Mesh node placement in WMN can be seen as a family of problems, which are shown (through graph theoretic approaches or placement problems, e.g. [6,16]) to be computationally hard to solve for most of the formulations [38]. In fact, the node placement problem considered here is even more challenging due to two additional characteristics:

(a) locations of mesh router nodes are not pre-determined, in other wards, any available position in the considered area can be used for deploying the mesh routers.
(b) routers are assumed to have their own radio coverage area.

Here, we consider the version of the mesh router nodes placement problem in which we are given a grid area where to deploy a number of mesh router nodes and a number of mesh client nodes of fixed positions (of an arbitrary distribution) in the grid area. The objective is to find a location assignment for the mesh routers to the cells of the grid area that maximizes the network connectivity and client coverage. Node placement problems are known to be computationally hard to solve [13,14,39]. In some previous works, intelligent algorithms have been recently investigated [4,7,17,19,22–24,32,33].

In our previous work, we implemented a Particle Swarm Optimization (PSO) based simulation system, called WMN-PSO [25]. Also, we implemented a simulation system based on Simulated Annealing (SA) for solving node placement problem in WMNs, called WMN-SA [20,21].

In this paper, we implement a hybrid simulation system based on PSO and SA. We call this system WMN-PSOSA. We evaluate the performance of hybrid WMN-PSOSA system considering Normal and Uniform distribution of mesh clients.

The rest of the paper is organized as follows. The mesh router nodes placement problem is defined in Sect. 2. We present our designed and implemented hybrid simulation system in Sect. 3. The simulation results are given in Sect. 4. Finally, we give conclusions and future work in Sect. 5.

2   Node Placement Problem in WMNs

For this problem, we have a grid area arranged in cells we want to find where to distribute a number of mesh router nodes and a number of mesh client nodes of fixed positions (of an arbitrary distribution) in the considered area. The objective is to find a location assignment for the mesh routers to the area that maximizes the network connectivity and client coverage. Network connectivity is measured by Size of Giant Component (SGC) of the resulting WMN graph, while the user coverage is simply the number of mesh client nodes that fall within the radio coverage of at least one mesh router node and is measured by Number of Covered Mesh Clients (NCMC).

An instance of the problem consists as follows.

- N mesh router nodes, each having its own radio coverage, defining thus a vector of routers.
- An area $W \times H$ where to distribute N mesh routers. Positions of mesh routers are not pre-determined and are to be computed.
- M client mesh nodes located in arbitrary points of the considered area, defining a matrix of clients.

It should be noted that network connectivity and user coverage are among most important metrics in WMNs and directly affect the network performance.

In this work, we have considered a bi-objective optimization in which we first maximize the network connectivity of the WMN (through the maximization of the SGC) and then, the maximization of the NCMC.

In fact, we can formalize an instance of the problem by constructing an adjacency matrix of the WMN graph, whose nodes are router nodes and client nodes and whose edges are links between nodes in the mesh network. Each mesh node in the graph is a triple $v = <x, y, r>$ representing the 2D location point and r is the radius of the transmission range. There is an arc between two nodes u and v, if v is within the transmission circular area of u.

3   Proposed and Implemented Simulation System

3.1   PSO Algorithm

In PSO a number of simple entities (the particles) are placed in the search space of some problem or function and each evaluates the objective function at its current location. The objective function is often minimized and the exploration of the search space is not through evolution [18]. However, following a widespread practice of borrowing from the evolutionary computation field, in this work, we consider the bi-objective function and fitness function interchangeably. Each particle then determines its movement through the search space by combining some aspect of the history of its own current and best (best-fitness) locations with those of one or more members of the swarm, with some random perturbations. The next iteration takes place after all particles have been moved. Eventually the swarm as a whole, like a flock of birds collectively foraging for food, is likely to move close to an optimum of the fitness function.

Each individual in the particle swarm is composed of three $\mathcal{D}$-dimensional vectors, where $\mathcal{D}$ is the dimensionality of the search space. These are the current position $\vec{x}_i$, the previous best position $\vec{p}_i$ and the velocity $\vec{v}_i$.

The particle swarm is more than just a collection of particles. A particle by itself has almost no power to solve any problem; progress occurs only when the particles interact. Problem solving is a population-wide phenomenon, emerging from the individual behaviors of the particles through their interactions. In any case, populations are organized according to some sort of communication structure or topology, often thought of as a social network. The topology typically consists of bidirectional edges connecting pairs of particles, so that if j is in i's neighborhood, i is also in j's. Each particle communicates with some other particles and is affected by the best point found by any member of its topological neighborhood. This is just the vector $\vec{p}_i$ for that best neighbor, which we will denote with $\vec{p}_g$. The potential kinds of population "social networks" are hugely varied, but in practice certain types have been used more frequently.

In the PSO process, the velocity of each particle is iteratively adjusted so that the particle stochastically oscillates around $\vec{p}_i$ and $\vec{p}_g$ locations.

3.2   Simulated Annealing

3.2.1   Description of Simulated Annealing

SA algorithm [9] is a generalization of the metropolis heuristic. Indeed, SA consists of a sequence of executions of metropolis with a progressive decrement of the temperature starting from a rather high temperature, where almost any move is accepted, to a low temperature, where the search resembles Hill Climbing. In fact, it can be seen as a hill-climber with an internal mechanism to escape local optima. In SA, the solution s' is accepted as the new current solution if $\delta \leq 0$ holds, where $\delta = f(s') - f(s)$. To allow escaping from a local optimum, the movements that increase the energy function are accepted with a decreasing probability $\exp(-\delta/T)$ if $\delta > 0$, where T is a parameter called the "temperature". The decreasing values of T are controlled by a *cooling schedule*, which specifies the temperature values at each stage of the algorithm, what represents an important decision for its application (a typical option is to use a proportional method, like $T_k = \alpha \cdot T_{k-1}$). SA usually gives better results in practice, but uses to be very slow. The most striking difficulty in applying SA is to choose and tune its parameters such as initial and final temperature, decrements of the temperature (cooling schedule), equilibrium and detection.

In our system, cooling schedule (α) will be calculated as:

$$\alpha = \left(\frac{SA\ Ending\ temperature}{SA\ Starting\ temperature} \right)^{1.0/Total\ iterations}.$$

3.2.2   Acceptability Criteria

The acceptability criteria for newly generated solution is based on the definition of a threshold value (accepting threshold) as follows. We consider a succession t_k such that $t_k > t_{k+1}$, $t_k > 0$ and t_k tends to 0 as k tends to infinity. Then, for any two solutions s_i and s_j, if $fitness(s_j) - fitness(s_i) < t_k$, then accept solution s_j.

For the SA, t_k values are taken as accepting threshold but the criterion for acceptance is probabilistic:

- If $fitness(s_j) - fitness(s_i) \leq 0$ then s_j is accepted.
- If $fitness(s_j) - fitness(s_i) > 0$ then s_j is accepted with probability $\exp[(fitness(s_j) - fitness(s_i))/t_k]$ (at iteration k the algorithm generates a random number $R \in (0, 1)$ and s_j is accepted if $R < \exp[(fitness(s_j) - fitness(s_i))/t_k]$).

In this case, each neighbour of a solution has a positive probability of replacing the current solution. The t_k values are chosen in way that solutions with large increase in the cost of the solutions are less likely to be accepted (but there is still a positive probability of accepting them).

3.3 WMN-PSOSA Hybrid Simulation System

3.3.1 WMN-PSOSA System Description

Here, we present the initialization, particle-pattern, fitness function.

Initialization
Our proposed system starts by generating an initial solution randomly, by *ad hoc* methods [40]. We decide the velocity of particles by a random process considering the area size. For instance, when the area size is $W \times H$, the velocity is decided randomly from $-\sqrt{W^2 + H^2}$ to $\sqrt{W^2 + H^2}$. Our system can generate many client distributions. In this paper, we consider Normal and Uniform distribution of mesh clients.

Particle-pattern
A particle is a mesh router. A fitness value of a particle-pattern is computed by combination of mesh routers and mesh clients positions. In other words, each particle-pattern is a solution as shown is Fig. 1. Therefore, the number of particle-patterns is a number of solutions.

Fitness function
One of most important thing in PSO algorithm is to decide the determination of an appropriate objective function and its encoding. In our case, each particle-pattern has an own fitness value and compares other particle-pattern's fitness value in order to share information of global solution. The fitness function follows a hierarchical approach in which the main objective is to maximize the SGC in WMN. Thus, the fitness function of this scenario is defined as

$$\text{Fitness} = 0.7 \times \text{SGC}(\boldsymbol{x}_{ij}, \boldsymbol{y}_{ij}) + 0.3 \times \text{NCMC}(\boldsymbol{x}_{ij}, \boldsymbol{y}_{ij}).$$

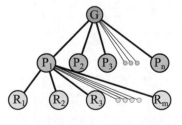

G: Global Solution
P: Particle-pattern
R: Mesh Router
n: Number of Particle-patterns
m: Number of Mesh Routers

Fig. 1. Relationship among global solution, particle-patterns and mesh routers.

3.3.2 WMN-PSOSA Web GUI Tool and Pseudo Code

The Web application follows a standard Client-Server architecture and is implemented using LAMP (Linux + Apache + MySQL + PHP) technology (see Fig. 2). Remote users (clients) submit their requests by completing first the parameter setting. The parameter values to be provided by the user are classified into three groups, as follows.

- Parameters related to the problem instance: These include parameter values that determine a problem instance to be solved and consist of number of router nodes, number of mesh client nodes, client mesh distribution, radio coverage interval and size of the deployment area.
- Parameters of the resolution method: Each method has its own parameters.

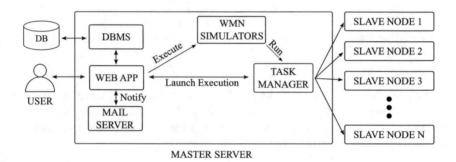

Fig. 2. System structure for web interface.

Simulator parameters, Particle Swarm Optimization and Simulated Annealing		
Distribution	Uniform ▼	
Number of clients	48 (integer)(min:48 max:128)	
Number of routers	16 (integer) (min:16 max:48)	
Area size (WxH)	32 (positive real number)	32 (positive real number)
Radius (Min & Max)	2 (positive real number)	2 (positive real number)
Independent runs	1 (integer) (min:1 max:100)	
Replacement method	Constriction Method ▼	
Starting SA Temperature value	10 (positive real number)	
Ending SA Temperature value	0.1 (positive real number)	
Number of Particle-patterns	10 (integer) (min:1 max:64)	
Max iterations	800 (integer) (min:1 max:6400)	
Iteration per Phase	4 (integer) (min:1 max:Max iterations)	
Send by mail	☐	

Run

Fig. 3. WMN-PSOSA Web GUI Tool.

- Execution parameters: These parameters are used for stopping condition of the resolution methods and include number of iterations and number of independent runs. The former is provided as a total number of iterations and depending on the method is also divided per phase (e.g., number of iterations in a exploration). The later is used to run the same configuration for the same problem instance and parameter configuration a certain number of times.

We show the WMN-PSOSA Web GUI tool in Fig. 3. The pseudo code of our implemented system is shown in Algorithm 1.

Algorithm 1. Pseudo code of PSOSA.

/* Generate the initial solutions and parameters */
Computation maxtime:= T_{max}, $t := 0$;
Number of particle-patterns:= m, $2 \leq m \in \mathbf{N}^1$;
Starting SA temperature:= $Temp$;
Decreasing speed of SA temperature:= T_d;
Particle-patterns initial solution:= $\mathbf{P}_i^0$;
Global initial solution:= $\mathbf{G}^0$;
Particle-patterns initial position:= $\mathbf{x}_{ij}^0$;
Particles initial velocity:= $\mathbf{v}_{ij}^0$;
PSO parameter:= ω, $0 < \omega \in \mathbf{R}^1$;
PSO parameter:= C_1, $0 < C_1 \in \mathbf{R}^1$;
PSO parameter:= C_2, $0 < C_2 \in \mathbf{R}^1$;
/* Start PSO-SA */
Evaluate($\mathbf{G}^0, \mathbf{P}^0$);
while $t < T_{max}$ **do**
   /* Update velocities and positions */
   $\mathbf{v}_{ij}^{t+1} = \omega \cdot \mathbf{v}_{ij}^t$
       $+C_1 \cdot \text{rand}() \cdot (best(P_{ij}^t) - x_{ij}^t)$
       $+C_2 \cdot \text{rand}() \cdot (best(G^t) - x_{ij}^t)$;
   $\mathbf{x}_{ij}^{t+1} = \mathbf{x}_{ij}^t + \mathbf{v}_{ij}^{t+1}$;
   /* if fitness value is increased, a new solution will be accepted. */
   if Evaluate($\mathbf{G}^{(t+1)}, \mathbf{P}^{(t+1)}$) ¿= Evaluate($\mathbf{G}^{(t)}, \mathbf{P}^{(t)}$) **then**
      Update_Solutions($\mathbf{G}^t, \mathbf{P}^t$);
      Evaluate($\mathbf{G}^{(t+1)}, \mathbf{P}^{(t+1)}$);
   else
      /* a new solution will be accepted, if condition is true. */
      if Random() ¿ $e^{\left(\frac{Evaluate(G^{(t+1)}, P^{(t+1)}) - Evaluate(G^{(t)}, P^{(t)})}{Temp} \right)}$ **then**
         /* "Reupdate_Solutions" makes particle back to previous position */
         Reupdate_Solutions($\mathbf{G}^{t+1}, \mathbf{P}^{t+1}$);
      end if
   end if
   $Temp = Temp \times t_d$;
   $t = t + 1$;
end while
Update_Solutions($\mathbf{G}^t, \mathbf{P}^t$);
return Best found pattern of particles as solution;

3.3.3   WMN Mesh Routers Replacement Methods

A mesh router has x, y positions and velocity. Mesh routers are moved based on velocities. There are many moving methods in PSO field, such as:

Constriction Method (CM)

   CM is a method which PSO parameters are set to a week stable region ($\omega = 0.729$, $C_1 = C2 = 1.4955$) based on analysis of PSO by M. Clerc et al. [2,5,36].

Random Inertia Weight Method (RIWM)

   In RIWM, the ω parameter is changing ramdomly from 0.5 to 1.0. The C_1 and C_2 are kept 2.0. The ω can be estimated by the week stable region. The average of ω is 0.75 [29,36].

Linearly Decreasing Inertia Weight Method (LDIWM)

   In LDIWM, C_1 and C_2 are set to 2.0, constantly. On the other hand, the ω parameter is changed linearly from unstable region ($\omega = 0.9$) to stable region ($\omega = 0.4$) with increasing of iterations of computations [36,37].

Linearly Decreasing Vmax Method (LDVM)

   In LDVM, PSO parameters are set to unstable region ($\omega = 0.9$, $C_1 = C_2 = 2.0$). A value of V_{max} which is maximum velocity of particles is considered. With increasing of iteration of computations, the V_{max} is kept decreasing linearly [31,35].

Rational Decrement of Vmax Method (RDVM)

   In RDVM, PSO parameters are set to unstable region ($\omega = 0.9$, $C_1 = C_2 = 2.0$). The V_{max} is kept decreasing with the increasing of iterations as

$$V_{max}(x) = \sqrt{W^2 + H^2} \times \frac{T - x}{x}.$$

Where, W and H are the width and the height of the considered area, respectively. Also, T and x are the total number of iterations and a current number of iteration, respectively [26].

4   Simulation Results

In this section, we show simulation results using WMN-PSOSA system. In this work, we consider Normal and Uniform distributions of mesh clients. The number of mesh routers is considered 16 and the number of mesh clients 48. The total number of iterations is considered 6400 and the iterations per phase is considered 32. We consider the number of particle-patterns 9. We conducted simulations 100 times, in order to avoid the effect of randomness and create a general view of results. We show the parameter setting for WMN-PSOSA in Table 1.

   We show the simulation results from Figs. 4, 5 and 6. In Figs. 4 and 5, we show results for Normal and Uniform distribution of mesh clients. For Normal distribution, the convergence of SGC is faster than Uniform distribution. Also, WMN-PSOSA performs better for the NCMC when the client distribution is Normal compared with the case of Uniform distribution. We show the visualized

Table 1. Parameter settings.

Parameters	Values
Clients distribution	Normal, Uniform
Area size	32.0×32.0
Number of mesh routers	16
Number of mesh clients	48
Total iterations	6400
Iteration per phase	32
Number of particle-patterns	9
Radius of a mesh router	2.0
SA starting temperature value	10.0
SA ending temperature value	0.01
Temperature decreasing speed (α)	0.998921
Replacement method	LDIWM

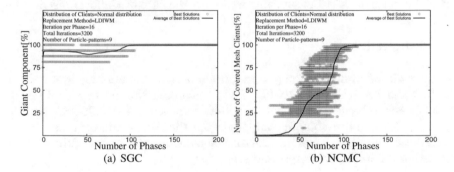

Fig. 4. Simulation results of WMN-PSOSA for Normal distribution.

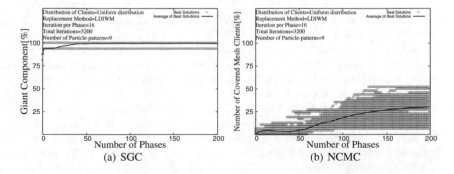

Fig. 5. Simulation results of WMN-PSOSA for Uniform distribution.

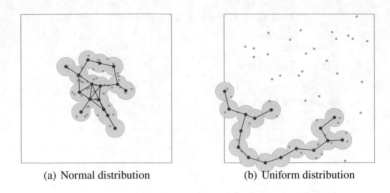

(a) Normal distribution   (b) Uniform distribution

Fig. 6. Visualized image of simulation results for different clients.

results for WMN-PSOSA in Fig. 6. As shown in Fig. 6(a), all nodes are covered for the Normal distribution. However, we see that many clients nodes are not covered for Uniform distribution (see Fig. 6(b)). Therefore, WMN-PSOSA performs better for Normal distribution compared with the case of Uniform distribution.

5   Conclusions

In this work, we evaluated the performance of a hybrid simulation system based on PSO and SA (called WMN-PSOSA) considering Normal and Uniform distribution of mesh clients. Simulation results show that WMN-PSOSA performs better for Normal distribution compared with the case of Uniform distribution.

In our future work, we would like to evaluate the performance of the proposed system for different parameters and scenarios.

References

1. Akyildiz, I.F., Wang, X., Wang, W.: Wireless mesh networks: a survey. Comput. Netw. **47**(4), 445–487 (2005)
2. Barolli, A., Sakamoto, S., Ozera, K., Ikeda, M., Barolli, L., Takizawa, M.: Performance evaluation of WMNs by WMN-PSOSA simulation system considering constriction and linearly decreasing Vmax Methods. In: International Conference on P2P, Parallel, Grid, Cloudand Internet Computing, pp. 111–121. Springer (2017)
3. Barolli, A., Sakamoto, S., Barolli, L., Takizawa, M.: Performance analysis of simulation system based on particle swarm optimization and distributed genetic algorithm for WMNs considering different distributions of mesh clients. In: International Conference on Innovative Mobile and Internet Services in Ubiquitous Computing, pp. 32–45. Springer (2018)
4. Barolli, A., Sakamoto, S., Ozera, K., Barolli, L., Kulla, E., Takizawa, M.: Design and implementation of a hybrid intelligent system based on particle swarm optimization and distributed genetic algorithm. In: International Conference on Emerging Internetworking, Data & Web Technologies, pp. 79–93. Springer (2018)

5. Clerc, M., Kennedy, J.: The particle swarm-explosion, stability, and convergence in a multidimensional complex space. IEEE Trans. Evol. Comput. **6**(1), 58–73 (2002)
6. Franklin, A.A., Murthy, C.S.R.: Node placement algorithm for deployment of two-tier wireless mesh networks. In: Proceedings of Global Telecommunications Conference, pp. 4823–4827 (2007)
7. Girgis, M.R., Mahmoud, T.M., Abdullatif, B.A., Rabie, A.M.: Solving the wireless mesh network design problem using genetic algorithm and simulated annealing optimization methods. Int. J. Comput. Appl. **96**(11), 1–10 (2014)
8. Goto, K., Sasaki, Y., Hara, T., Nishio, S.: Data gathering using mobile agents for reducing traffic in dense mobile wireless sensor networks. Mob. Inf. Syst. **9**(4), 295–314 (2013)
9. Hwang, C.R.: Simulated annealing: theory and applications. Acta Applicandae Mathematicae **12**(1), 108–111 (1988)
10. Inaba, T., Elmazi, D., Sakamoto, S., Oda, T., Ikeda, M., Barolli, L.: A secure-aware call admission control scheme for wireless cellular networks using fuzzy logic and its performance evaluation. J. Mob. Multimed. **11**(3&4), 213–222 (2015)
11. Inaba, T., Obukata, R., Sakamoto, S., Oda, T., Ikeda, M., Barolli, L.: Performance evaluation of a QoS-aware fuzzy-based CAC for LAN access. Int. J. Space Based Situat. Comput. **6**(4), 228–238 (2016)
12. Inaba, T., Sakamoto, S., Oda, T., Ikeda, M., Barolli, L.: A testbed for admission control in WLAN: a fuzzy approach and its performance evaluation. In: International Conference on Broadband and Wireless Computing, Communication and Applications, pp. 559–571. Springer (2016)
13. Lim, A., Rodrigues, B., Wang, F., Xu, Z.: k-center problems with minimum coverage. In: Computing and Combinatorics, pp. 349–359 (2004)
14. Maolin, T.: Gateways placement in backbone wireless mesh networks. Int. J. Commun. Netw. Syst. Sci. **2**(1), 44 (2009)
15. Matsuo, K., Sakamoto, S., Oda, T., Barolli, A., Ikeda, M., Barolli, L.: Performance analysis of WMNs by WMN-GA simulation system for two WMN architectures and different TCP congestion-avoidance algorithms and client distributions. Int. J. Commun. Netw. Distrib. Syst. **20**(3), 335–351 (2018)
16. Muthaiah, S.N., Rosenberg, C.P.: Single gateway placement in wireless mesh networks. In: Proceedings of 8th International IEEE Symposium on Computer Networks, pp. 4754–4759 (2008)
17. Naka, S., Genji, T., Yura, T., Fukuyama, Y.: A hybrid particle swarm optimization for distribution state estimation. IEEE Trans. Power Syst. **18**(1), 60–68 (2003)
18. Poli, R., Kennedy, J., Blackwell, T.: Particle swarm optimization. Swarm Intell. **1**(1), 33–57 (2007)
19. Sakamoto, S., Kulla, E., Oda, T., Ikeda, M., Barolli, L., Xhafa, F.: A comparison study of simulated annealing and genetic algorithm for node placement problem in wireless mesh networks. J. Mob. Multimed. **9**(1–2), 101–110 (2013)
20. Sakamoto, S., Kulla, E., Oda, T., Ikeda, M., Barolli, L., Xhafa, F.: A comparison study of hill climbing, simulated annealing and genetic algorithm for node placement problem in WMNs. J. High Speed Netw. **20**(1), 55–66 (2014)
21. Sakamoto, S., Kulla, E., Oda, T., Ikeda, M., Barolli, L., Xhafa, F.: A simulation system for WMN based on SA: performance evaluation for different instances and starting temperature values. Int. J. Space Based Situat. Comput. **4**(3–4), 209–216 (2014)
22. Sakamoto, S., Kulla, E., Oda, T., Ikeda, M., Barolli, L., Xhafa, F.: Performance evaluation considering iterations per phase and SA temperature in WMN-SA system. Mob. Inf. Syst. **10**(3), 321–330 (2014)

23. Sakamoto, S., Lala, A., Oda, T., Kolici, V., Barolli, L., Xhafa, F.: Application of WMN-SA simulation system for node placement in wireless mesh networks: a case study for a realistic scenario. Int. J. Mob. Comput. Multimed. Commun. (IJMCMC) **6**(2), 13–21 (2014)
24. Sakamoto, S., Oda, T., Ikeda, M., Barolli, L., Xhafa, F.: An integrated simulation system considering WMN-PSO simulation system and network simulator 3. In: International Conference on Broadband and Wireless Computing, Communication and Applications, pp. 187–198, Springer (2016)
25. Sakamoto, S., Oda, T., Ikeda, M., Barolli, L., Xhafa, F.: Implementation and evaluation of a simulation system based on particle swarm optimisation for node placement problem in wireless mesh networks. Int. J. Commun. Netw. Distrib. Syst. **17**(1), 1–13 (2016)
26. Sakamoto, S., Oda, T., Ikeda, M., Barolli, L., Xhafa, F.: Implementation of a new replacement method in WMN-PSO simulation system and its performance evaluation. In: The 30th IEEE International Conference on Advanced Information Networking and Applications (AINA-2016), pp. 206–211 (2016). https://doi.org/10.1109/AINA.2016.42
27. Sakamoto, S., Obukata, R., Oda, T., Barolli, L., Ikeda, M., Barolli, A.: Performance analysis of two wireless mesh network architectures by WMN-SA and WMN-TS simulation systems. J. High Speed Netw. **23**(4), 311–322 (2017)
28. Sakamoto, S., Ozera, K., Barolli, A., Ikeda, M., Barolli, L., Takizawa, M.: Implementation of an intelligent hybrid simulation systems for WMNs based on particle swarm optimization and simulated annealing: performance evaluation for different replacement methods. Soft Comput. 11 December 2017. https://doi.org/10.1007/s00500-017-2948-1
29. Sakamoto, S., Ozera, K., Barolli, A., Ikeda, M., Barolli, L., Takizawa, M.: Performance evaluation of WMNs by WMN-PSOSA simulation system considering random inertia weight method and linearly decreasing Vmax method. In: International Conference on Broadbandand Wireless Computing, Communication and Applications, pp. 114–124. Springer (2017)
30. Sakamoto, S., Ozera, K., Ikeda, M., Barolli, L.: Implementation of intelligent hybrid systems for node placement problem in WMNs considering particle swarm optimization, hill climbing and simulated annealing. Mob. Netw. Appl. **23**, 1–7 (2017)
31. Sakamoto, S., Ozera, K., Ikeda, M., Barolli, L.: Performance evaluation of WMNs by WMN-PSOSA simulation system considering constriction and linearly decreasing inertia weight methods. In: International Conference on Network-Based Information Systems, pp. 3–13. Springer (2017)
32. Sakamoto, S., Ozera, K., Oda, T., Ikeda, M., Barolli, L.: Performance evaluation of intelligent hybrid systems for node placement in wireless mesh networks: a comparison study of WMN-PSOHC and WMN-PSOSA. In: International Conference on Innovative Mobile and Internet Services in Ubiquitous Computing, pp. 16–26. Springer (2017)
33. Sakamoto, S., Ozera, K., Oda, T., Ikeda, M., Barolli, L.: Performance evaluation of WMN-PSOHC and WMN-PSO simulation systems for node placement in wireless mesh networks: a comparison study. In: International Conference on Emerging Internetworking, Data & Web Technologies, pp. 64–74. Springer (2017)
34. Sakamoto, S., Ozera, K., Barolli, A., Barolli, L., Kolici, V., Takizawa, M.: Performance evaluation of WMN-PSOSA considering four different replacement methods. In: International Conference on Emerging Internetworking, Data & Web Technologies, pp. 51–64. Springer (2018)

35. Schutte, J.F., Groenwold, A.A.: A study of global optimization using particle swarms. J. Glob. Optim. **31**(1), 93–108 (2005)
36. Shi, Y.: Particle swarm optimization. IEEE Connect. **2**(1), 8–13 (2004)
37. Shi, Y., Eberhart, R.C.: Parameter selection in particle swarm optimization. In: Evolutionary programming VII, pp. 591–600 (1998)
38. Vanhatupa, T., Hannikainen, M., Hamalainen, T.: Genetic algorithm to optimize node placement and configuration for WLAN planning. In: Proceedings of 4th IEEE International Symposium on Wireless Communication Systems, pp. 612–616 (2007)
39. Wang, J., Xie, B., Cai, K., Agrawal, D.P.: Efficient mesh router placement in wireless mesh networks. In: Proceedings of IEEE International Conference on Mobile Adhoc and Sensor Systems (MASS-2007), pp. 1–9 (2007)
40. Xhafa, F., Sanchez, C., Barolli, L.: Ad hoc and neighborhood search methods for placement of mesh routers in wireless mesh networks. In: Proceedings of 29th IEEE International Conference on Distributed Computing Systems Workshops (ICDCS-2009), pp. 400–405 (2009)

An Evaluation of Cooperative Communication in Cognitive Radio as Applied to Autonomous Vehicles

Jamal Raiyn[(✉)]

Computer Science Department, Al Qasemi Academic College,
Baka Al Gharbiah, Israel
raiyn@qsm.ac.il

Abstract. Autonomous vehicles are slowly being modified to be cognitive. A fully autonomous vehicle is a robotic vehicle that is designed to travel to destinations without human intervention. It uses vehicle-to-vehicle (V2V), vehicle-to-infrastructure (V2I), and vehicle-to-everything (V2X) communications in order to improve driving safety and traffic efficiency, and to provide information and entertainment to the driver. Its communication protocols are capable of driving without any human actions. V2X communications are designing and running in cellular networks. Some of the most commonly used methods for increasing the spectral efficiency of cellular systems involve resource allocation schemes. Various channel allocation schemes have been introduced to provide quality of service (QoS). To get high QoS, we have to examine the QoS requirements for vehicle cooperative communication systems. This paper discusses, analyzes, and evaluates the performance of a cooperative communication strategy for cellular systems, for applications in autonomous vehicle networks. The aim is to reduce delays in message exchange between vehicles that are caused by heavy traffic loads, a high blocking rate, and lengthy processing times. The major goal of our work is to design a robust message protocol for exchanging information among vehicles.

1 Introduction

In traditional traffic activity, eye contact, both between drivers and between drivers and other road users plays a critical role. Even pedestrians who want to cross the street without a crossing aid use eye contact to ensure that an approaching driver sees them. When a driver from a side street wants to merge onto a main street with heavier traffic, the merging driver also uses eye contact to adjust speed accordingly. Eye contact is a two-way form of communication. If one person looks away, it indicates the message that he or she has "not seen" the other person and does not intend to accept the message. It can be assumed that by the year 2030, 40% of vehicles will be equipped with V2V or V2X technology due to the major advantages these technologies offer, including reducing traffic jams and enhancing safety [14]. Autonomous vehicles will use V2X cooperation communication technology to reduce traffic congestion on urban roads [1, 2, 11, 15]. This can involve various communication standards, for example a cellular or WiFi network. It is essential in both cases that the latency and availability of the data

© Springer Nature Switzerland AG 2019
F. Xhafa et al. (Eds.): INCoS 2018, LNDECT 23, pp. 116–123, 2019.
https://doi.org/10.1007/978-3-319-98557-2_11

transfer be up to the necessary QoS requirements. One of the greatest challenges to supporting communication in cellular systems with applications in autonomous vehicle networks is to fulfill V2X demands within the constraints of limited radio bandwidth [4], and to utilize the limited spectrum available to meet the increasing demand for autonomous vehicle networks. Some of the most commonly used methods for increasing spectral efficiency are resource allocation schemes. Various channel allocation schemes have been introduced to provide QoS [3] and efficient channel utilization in cellular networks. There are many parameters for measuring the QoS of a network. These include throughput, latency, service availability, and so forth. The most important QoS parameters are blocking probability and radio coverage. This paper describes approaches to reducing message exchange latency in supporting communication applications in cellular systems, fulfilling vehicle–to–vehicle (V2V) demand under the constraints of limited radio bandwidth and utilizing limited spectrum availability to meet the increasing demand for autonomous vehicles. The channel allocation schemes that have been proposed were developed and designed to deal with a problem specific to one generation of mobile communications. Our research showed that cognitive radio provides the best results compared to fixed channel allocation (FCA), dynamic channel allocation (DCA) and distributed dynamic channel allocation (DDCA) [6]. The cognitive radio scheme achieves a high degree of channel usage and a low rate of call blocking in cellular systems [8], and it offers several features, such as, autonomy and negotiation. In addition, the use of cognitive radio affords greater autonomy to the base stations, which enables an increase in flexibility in dealing with new situations in the traffic load. Negotiation is used to avoid conflicts in resource allocation. In this paper, a cognitive radio scheme is used to deal with delays, and repeated unnecessary handoffs in the same cell [13]. Delayed response messages in cooperative communication cause traffic congestion and accidents. Cooperative communication is applied here in an autonomous vehicle network to improve its efficiency. Autonomous vehicles use satellite communication to determine vehicle position and use cellular systems to establish communication with other vehicles. The main challenge in cooperative communication is delays, which can be caused by the reception of several simultaneous request messages, failed communications, and cyber attacks [10]. To achieve efficient performance in cooperative communication, a new approach based on search-update-allocation has been proposed [5]. The search-update-allocation (SUA) scheme enables the system to collect information, and monitor neighboring vehicles. The allocation of the vehicle to an available lane is based on QoS requirements related to shortest delay. The rest of this paper is organized as follows: Sect. 2 introduces a cooperative communication strategy that can be applied to autonomous vehicles; Sects. 3 and 4 discusses the results of the implemented cooperation communication and conclude the paper.

2 Intelligent Cooperative Communications in AV Networks

The maturing of wireless communication technologies at affordable costs and with the right levels of security will enable the wide deployment of so-called connected vehicles. All of the ITSs based on cooperation and communication between vehicles or between vehicles and infrastructures are called cooperative systems [7, 9, 11].

It is usual to separate the various communication technologies supporting cooperative systems into two groups:

- vehicle to vehicle (V2V) communication systems, which enable safer transport, for example safe distance keeping, collision avoidance, and early warnings of unsafe conditions.
- vehicle to infrastructure (V2I) communication systems (and vice versa), which enable better use of existing infrastructure and provide valuable and consolidated information to intelligent vehicles for example improved information regarding travel times, ongoing roadwork, and weather and traffic conditions, and up-to-date information about parking availability or other means of transport.

Altogether, information, communication and positioning technologies will play key role in future transport systems and services. For instance, floating car data are mainly composed of position and/or speed information, and cooperative awareness messages (CAM) exchanged by ITS stations in V2V or V2V communications all contain the reference position of the station emitting the message. The success of positioning will depend on its capacity for better performance (including improvement and control), its tight integration with other ITS technologies on smart multi-service and multi-standards platforms, and the affordability of relevant services. Furthermore, to improve drivers' safety, V2V systems share information relating to traffic information and accident warning with nearby vehicles or road infrastructures. However, the incorporation of these new systems into vehicles increases the level of security risks. For example, some of the latest models can be hacked within 360 s; the actuators of modern vehicles can be remotely controlled; terrorists can potentially hack into V2V and V2I systems to cause chaotic traffic accidents (e.g., by hacking into an autonomous intersection system); and privacy information can be stolen from any driver [8]. AVs use vehicle–to-vehicle communication for cooperative merging on the highway, lane changes and intersection crossings as illustrates Fig. 1. An autonomous vehicle that would like to communicate with other vehicles must be in the same cluster. In a clustering scheme, we need to identify a single vehicle that will allocate a channel for communication. That is, when another autonomous vehicle joins a cluster area, the group must select one vehicle as the leader to manage the group and control the transfer of data among multiple vehicles. The success of this scheme depends on cooperative

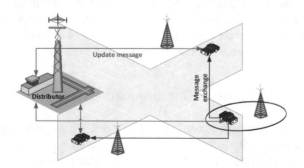

Fig. 1. Message exchange

communication between the leader and other vehicles in the cluster, and this cooperation should take place within the coverage area. In other words, the vehicles should maintain a minimum distance and should remain within the transmission range of the cluster. The area for the vehicles is calculated based on the formula given below in Eq. 1. It depends on transmission range, the message latency of autonomous vehicle, and the autonomous vehicles, and their speed.

2.1 Search-Update Strategy

In the previous sections, the analysis showed that the large numbers of message exchanges in traffic networks cause delays for autonomous vehicles. To reduce these delays, we propose a new intelligent cooperative communication based on a search-update strategy (SUS), which is described in detail in [5]. The SUS approach combines three modern techniques for vehicle communication in urban road traffic management, which are described in the three section that follow.

i. Searching

In the search phase, the vehicles use two kinds of messages to communicate with neighboring vehicles within a minimal distance. REQ (Req, node_name, ts, location), and RESP (Use, Free). Each autonomous vehicle can calculate the distance between itself and other vehicles based on the values of the x-axis and the y- axis obtained from their GPS. The proposed system is based on Eqs. 1 and 2, which calculate the distance and angle between two vehicles.

$$\text{Distance} = \sqrt{[(x_1 - x_2)^2 + (y_1 - y_2)^2]} \tag{1}$$

$$\text{angle} = \arctan(x_2 - x_1)/(y_2 - y_1) \tag{2}$$

where, (x_1, y_1) is the position of the first vehicle and (x_2, y_2) is the position of the second vehicle.

ii. Updating

Each section of the urban road network updates its status according to the capacity of the sections. Each vehicle in the system uses the following local variables. Use_i (k, time), whereby vehicle i reports a heavy traffic load in section k at time t. $Free_i$ (k, time), whereby, vehicle i reports a light traffic load in section k at time t. A node i retains information about the status of the neighboring nodes.

$$Available(i, t) = \begin{cases} 1 & \text{if } EMA(k, t) > Speed_{threshold} \\ 0 & \text{otherwise} \end{cases}$$

$Available(i, t)$ is an array that keeps track of the available sections from the starting node (start_node i) at time t until the end point (end_point). A node i keeps track of vehicles using the neighboring sections. HS_i is an array maintained by node i that denotes the number of sections under a heavy traffic load.

iii. Allocation

Vehicle i communicates with neighboring vehicles, N_i. This communication between vehicles is based on a message exchange protocol. Based on the information received from neighboring vehicles, the protocol decides which lane to allocate. Hence a vehicle may be source vehicle at time $t = 0$ to generate messages. The same vehicle can function as destination to receive packets sent at time $t = n$. These packets may be generated from other source vehicles and intermediate vehicles between the source and the destination, such as relay vehicles. The vehicles' iteration process ends when the specified time t is greater than t_{max}. The waiting time for a response message should be shorter than the threshold ε The algorithm minimizes delays in communication between vehicles by reducing unnecessary handoff in the same base stations.

$$t = 0$$
$$For \quad i = 1 \text{ to } k \qquad DO \qquad r_i(t) = choose \ (V)$$
$$REPEAT$$
$$For \ i = 1 \quad to \ k \quad DO \quad (\ t_{Ci} = 0 \ )$$
$$FOREACH \quad v \in V \quad DO$$
$$x = \arg\min_{i:i \in \{1,...,k\}} d(r_i(t), v)$$
$$ENDDO$$
$$For \ i = 1 \text{ to } k \quad DO$$
$$r_i(t) \quad = minimize \ (delay(t_{Ci}))$$
$$UNTIL \quad (\forall r_i : d(r_i(t), r_i(t\text{-}1)) \quad < \quad \varepsilon, \quad t > t_{max}$$
$$RETURN \ (\{r_1(t), \ ... \ , r_k(t)\})$$

3   Evaluation of the Cognitive Radio Approach

Several channel allocation schemes have been introduced in cellular systems, to meet new mobile service requirements. In general, a deficiency of current channel allocation strategies is that they were developed to deal a specific problem in cellular systems, and most of the solutions are the outcomes of an entirely reactive approach: a response to a series of events follows an algorithm that is designed to react to a specific situation. This limits their efficiency. Fixed channel allocation schemes are simple, however, they do not adapt to changing traffic condition and user distribution. The channels in centralized dynamic channel allocation schemes are assigned by a central controller, whereas in distributed dynamic channel allocation schemes a channel is selected according to update or search algorithms. In other words, distributed channel allocation is based on local assignments without involving a central controller. Dynamic channel allocation schemes provide flexibility and traffic adaptability; however, their main deficiency is that are less efficient than fixed channel allocation schemes under high load conditions. Due to high message complexity and the high channel acquisition time in dynamic channel allocation schemes, the call blocking rate is increased under a heavy traffic load. Real-time communication demands low delays (equal to zero). Some global optimization tools in cellular networks are of use; however, global optimization is resource intensive and costly, and several minutes are required to optimize a medium-sized cellular network communication. In other words, it is difficult for current cellular networks to react in real time to users' demands and difficult to optimize them. To fulfill the user's preferences and to improve the performance of the network, we have introduced an approach based on cognitive radio, which can handle more than one area that affected channel allocation in cellular systems. A social agent manages the resources in the cell; it is informed about resource states (free/ busy channels) by the distributor before it starts to communicate with its neighbor cells. The agent approach in the cognitive radio approach is deadlock free. A cell in a cognitive radio cannot wait indefinitely, because the negotiation between two cells is controlled and managed by the cluster distributor. When a cell needs to acquire a new channel, the distributor informs it about the state of the channels in its neighbor cells. Then the cell contacts the neighbor cell that has the greatest number of available channels. Every cell sends a response message in answer to the request message. Hence, in finite time (defined time) a cell receives all the response messages and it is either successful in acquiring a channel or not. Hence, a deadlock can occur when all of the cells are involved in cyclic waiting. The system cannot resolve the deadlock, because the cells by the rules of resource acquisition have no prior claim to the channels in the neighbor cells. Figure 2 describes the end-to-end delay that is caused due to repeated unnecessary handoffs in the same cell. The handoff process in the conventional scheme (DDCA) needs more time to finish. In the conventional approach the mobile station has to move the message to next neighbor cell any times, because the mobile station follows the strong SNR. The social agent in the cognitive radio scheme prevents repeated unnecessary handoffs, so that there is less delay. Figure 3 shows that, in cognitive radio, the number of requested source that are granted is higher than in the distributed dynamic channel allocation scheme, because the repeated unnecessary handoffs in the same cell in the distributed dynamic channel allocation scheme causes longer delays (see Fig. 3). These longer delays in the cellular system cause call blocking or call dropping. In other words, the

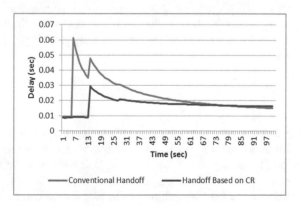

Fig. 2. End-to-end delay

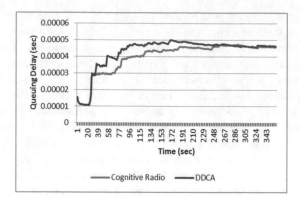

Fig. 3. Queuing delay

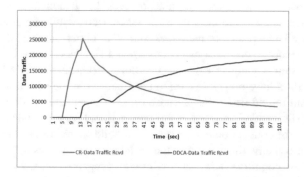

Fig. 4. Data traffic received

proposed cognitive radio scheme supports high quality compared to distributed dynamic channel allocation. Figure 4 shows that the received data traffic in cognitive radio is greater than that in the DDCA scheme. The cognitive radio approach avoids the repeated unnecessary handoffs that increase delays. Coverage planning depends on the received signal strength, that is, the covered area is almost only limited by the minimum signal strength at the cell range, while later capacity planning depends mainly on frequency allocation.

4 Conclusion

This paper has presented an overview of a cellular system in autonomous vehicle communication. Autonomous vehicles use V2 V, and V2X communications to update local information and to manage the traffic in multi-lane highways; however, delays in communication can cause traffic congestion and accidents. The search-update-allocation scheme discussed here is designed to reduce delays caused by an unnecessary handoff process in cellular systems.

Acknowledgement. The author wishes to thank the Center for Innovation in Transportation at the University of Tel Aviv for funding this research.

References

1. Jawhar, I., Mohamed, N., Usmani, H.: An overview of inter-vehicular communication systems, protocols and middleware. J. Netw. **8**(12), 2749–2761 (2013)
2. Alheeti, K.M.A., Al-ani, M.S., McDonald-Maier, K.: A hierarchical detection method in external communication for self-driving vehicles based on TDMA. PLoS ONE **13**(1), e0188760 (2018)
3. Seth, H.M., Momaya, K.: Quality of service parameters in cellular mobile communication. Int. J. Mob. Commun. **5**(1), 65–93 (2007)
4. Raiyn, J.: Handoff self-management based on SNR in mobile communication networks. Int. J. Wirel. Mob. Comput. **6**(1), 39–48 (2013)
5. Raiyn, J.: Road traffic congestion management based on search allocation approach. Transp. Telecommun. **18**(1), 25–33 (2017)
6. Groves, P.D., Matching, S.: A new GNSS positioning technique for urban canyons. J. Navig. **64**, 417–430 (2011)
7. Akyildiz, I.F., Won-Yeol, L., Vuran, M.C., Mohanty, S.: A survey on spectrum management in cognitive radio networks. IEEE Commun. Mag. **46**(4), 40–48 (2008)
8. Broggi, A., et al.: extensive tests of autonomous driving technologies. IEEE Trans. Intell. Transp. Syst. **14**(3), 1403–1415 (2013)
9. Kim, S.-W., et al.: Cooperative perception for autonomous vehicle control on the road: motivation and experimental results. In: 2013 IEEE/RSJ International Conference on Intelligent Robots and Systems (IROS), 3–7 November 2013, Tokyo, Japan (2013)
10. Petit, J., Shladover, S.E.: Potential cyberattacks on automated vehicles. IEEE Trans. Intell. Transp. Syst. **16**(2), 546–556 (2015)
11. Gora, P., Rüb, I.: Traffic models for self-driving connected cars. Transp. Res. Procedia **14**, 2207–2216 (2016)
12. Raiyn, J.: Real-time road traffic anomaly detection. J. Transp. Technol. **4**, 256–266 (2014)
13. Han, W., Li, J., Tian, Z., Zhang, Y.: Dynamic sensing strategies for efficient spectrum utilization in cognitive radio networks. IEEE Trans. Wirel. Commun. **10**(11), 3644–3655 (2001)
14. Jiang, L., Xi, W., Bai-Gen, C.: An improved GNSS receiver autonomous integrity monitoring method using vehicle-to-vehicle communication, pp. 249–260. Springer International Publishing Switzerland (2015)
15. Raiyn, J.: Real-time short-term forecasting based on information management. J. Transp. Technol. **4**, 11–21 (2014)

Method of Generating Computer Graphics Animation Synchronizing Motion and Sound of Multiple Musical Instruments

Amuro Takano, Jun-ya Hirata, and Hiroyoshi Miwa[✉]

Graduate School of Science and Technology, Kwansei Gakuin University,
2-1 Gakuen, Sanda-shi, Hyogo, Japan
{amr,miwa}@kwansei.ac.jp

Abstract. In recent years, some computer graphics (CG) animations in which musical instrument performance itself is important in the story, have attracted attention, and it is necessary to actually draw performance scenes by musical instruments. If we can automatically and semi-automatically generate CG animations including the performance of musical instruments, we can drastically reduce time and cost of production. For this reason, our research group has been advancing research and development of a system for generating CG animations of performance by musical instruments. In particular, the method of generating CG animations of performance by a plurality of musical instruments, such as a band performance and an orchestra performance, is necessary. For that purpose, we need to develop a method to detect onset times of each instrument from WAV format data separately recorded for each musical instrument and to synchronize the sound and the motion of all musical instruments. In this paper, we propose a method of generating a CG animation synchronizing motion and sound of multiple musical instruments.

1 Introduction

Many movies and animations including musical instrument's performance scenes have been produced. However, it had been practically impossible to create a computer graphics (CG) animation of a pianist's playing piano, because it is very difficult to draw accurate piano fingering. Therefore, there was no choice but to take solutions such as hiding hands and fingertips.

In recent years, some computer-generated animated movies in which musical instrument performance itself is important in the story, have attracted attention, and it is necessary to actually draw performance scene by musical instrument.

There are two approaches to make a CG animation of performance scene by musical instrument. One is to record the three-dimensional movement of a performer by using a motion capture system and generate a CG animation based on the movement data. The other is to directly generate the movement of fingers,

© Springer Nature Switzerland AG 2019
F. Xhafa et al. (Eds.): INCoS 2018, LNDECT 23, pp. 124–133, 2019.
https://doi.org/10.1007/978-3-319-98557-2_12

hand, and arm as a CG animation from musical score data. These approaches can enable to generate automatically and semi-automatically a CG animation including performance scenes by musical instrument; therefore, we can drastically reduce time and cost of production. For this reason, our research group has been advancing research and development of a system for generating CG animations of performance by musical instruments. We had developed the method of generating a CG animation of performance by a musical instrument and applied to the CG animation of the performance scenes of a pianist in the blockbuster Japanese TV animation "Nodame Cantabile". However, the method of generating a CG animation of performance by a plurality of musical instruments, such as a band performance and an orchestra performance, is also necessary.

When recording musical instrument performance data by a motion capture system, lack of data due to overlap of fingers and blind spots of cameras occur frequently; therefore, if the number of simultaneously recorded musical instruments increases, it becomes more difficult to correct the data. Consequently, it is not practical to simultaneously record performance of all instruments by a motion capture system. Moreover, when recording musical instrument's performance data by a motion capture system, since many markers are attached to the fingers, the performer hinders the high-quality performance. Therefore, it is difficult to simultaneously record motion data by a motion capture system and high quality sound source data, and it is necessary to separately record sound source data and motion data. For this reason, the gap between onset times in both data is inevitable, and it is necessary to correct the gap and to synchronize the motion and the sound. At the time of recording the sound source used here, an electronic musical instrument capable of recording the MIDI format including onset times cannot be used, because the high-quality sound with no compression must be recorded. To record high-quality sound, it must be recorded in WAV format not including the information of onset times by using an acoustic musical instrument. Therefore, to develop a CG animation of the performances by a plurality of musical instruments, it is necessary to detect onset times of each instrument from the WAV format data separately recorded for each musical instrument, and to synchronize the sound and the motion of all the musical instruments.

In this paper, we propose a method of generating a CG animation synchronizing motion and sound of multiple instruments. Furthermore, we evaluate the performance using actual performance data and demonstrate the effectiveness of the proposed method.

2   Related Works

We describe the related works of this research in this section.

As for the studies on the record of the performance motion, there is a study on the record of three-dimensional hand movement by a motion capture system [1]. It is difficult to capture the motion accurately, because the markers attached to the fingers cause misrecognition due to the short distance to other markers.

The reference [1] proposes the method to reduce the markers by attaching the markers only to the joints of the fingers. However, it is not useful for capturing musical instrument's performance, because the fingers moves very rapidly.

As for the studies on the detection of the onset times from WAV format data, there are some studies (ex. [2–4]). Since the WAV format data stores values obtained by synthesizing all frequencies for each time, the spectrogram is calculated by using Fourier transform and the peak of the intensity for each frequency is determined as the onset time of the frequency. However, since harmonic, reverb, and noise are included in a real sound data, it is not easy to detect the onset times by these methods for actual sound data.

To synchronize sound source and motion, it is necessary to change the speed of the sound and the motion so that the onset times of the multiple instruments are synchronized. As for the study on the variable speed of sound, there is the method of time stretch processing [5–7]. The sound source data is decompressed or compressed using a stretching technique that changes only the playing time without changing the pitch of the sound.

As for the study on the synchronization of the sound and the motion, the reference [8] proposes a method by using feature points of music and movement. However, the feature points are not always the onset times. Therefore, it is difficult to reduce the gap between the onset times for multiple musical instruments.

3   Method for Generation of CG Animation Synchronizing Motion and Sound of Multiple Instruments

In this section, we propose a method of generating a CG animation synchronizing motion and sound of multiple instruments.

We assume that sound source data and motion data are separately recorded for each musical instrument. The motion data consists of the collection of the time series of the position coordinates of the markers for each musical instrument. The data is captured by a motion capture system. The sound data is recorded in WAV format for each musical instrument. We assume that we have a musical score for each musical instrument and that the musical performance is coincident with the score.

We describe the basic idea of the proposed method as follows. First, the onset times of all the music notes are determined based on the musical score data for each musical instrument. We call the onset times the expected onset times. Next, the onset times are estimated in the sound data of the WAV format. We call the onset times the estimated onset times. Then, the estimated onset time is matched to the expected onset time for each music note by changing the speed of the music performance. Thus, the sound of the music is synchronized in total.

A method to estimate the onset times in the motion data depends on the musical instrument. For example, the onset times of the piano can be estimated by capturing the motion of the keys. Generally, the onset times can be estimated by using the electrical musical instrument recording MIDI data.

We show the concept of the proposed method in Fig. 1.

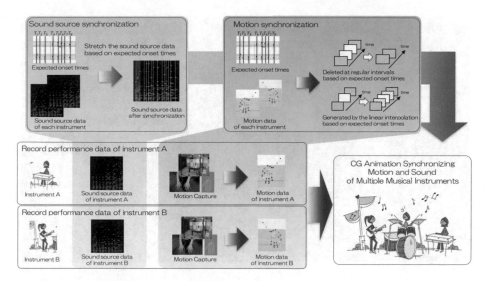

Fig. 1. Method of generating CG animation synchronizing motions and sounds of multiple instruments

3.1 Method for Synchronization of Sound of Multiple Instruments

In this section, we describe a method of detecting the onset times of a music instrument from the sound data recorded in WAV format.

First, a spectrogram is calculated by using the short-time Fourier transform. The spectrogram includes the pair of frequency and its signal intensity for all frequencies at each time. Figure 2 is an example of a spectrogram, where the horizontal axis indicates time, the vertical axis indicates frequency, and the contrasting density indicates the signal intensity of the frequency and the time.

Let $A(t)$ be the set of frequencies whose signal intensity is positive at time t. The note number i is assigned in the order of the notes in the musical score. Let H_i be the number of notes in note number i. Let $F_i = \{f_1, f_2, \ldots, f_{H_i}\}$ be the set of frequencies included in the note number i in $A(t)$. Let $S_i'(t)$ be the time series of the sum of the signal intensities of the frequencies in the range including the frequency corresponding each note for all the notes included in F_i. If the frequencies in the range including the frequency of k harmonics $(k = 2, 3, \ldots)$ of each note is included in $A(t)$, the value obtained by dividing the signal intensity by k to $S_i'(t)$ is added, and let the resulting value be $S_i(t)$.

We define the weight and the noise for each F_i. Let $w_i(t)$ be $dS_i(t)/dt$. $w_i(t)$ means the value indicating the possibility of the onset time of F_i at time t, because, when the signal intensity rapidly increase, the time corresponds to the onset time of the note. The noise $N_i(t)$ is defined as the time series of the sum of the signal intensities of the frequencies except the frequencies in $S_i(t)$.

Fig. 2. Example of spectrogram

We define $W_i(t) = w_i(t) - N_i(t)$. We call $W_i(t)$ weight. Thus, the weight set $\{W_i(t)|i = 1, 2, \ldots, n\}$ indicating the possibility of the onset time and noise set $\{N_i(t)|i = 1, 2, \ldots, n\}$ at each time are obtained for all the note numbers i included in the musical score data.

We discretize the time for each small constant time δ, and let the j-th time slice be time j. When n is the total number of the musical notes and E is the performance end time, we make $2n \times E$ partial lattice graph as follows. The vertices at row p and column q have the directed edges to the vertex at row p and column $q + 1$ and to the vertex at row $p + 1$ and column $q + 1$ (Fig. 3). Let $W_i(j)$ be the value of the vertex at the row $2i - 1$ and column j of the partial lattice graph and let $-N_i(j)$ be the value of the vertex at row $2i$ and column j.

By using the algorithm based on the dynamic programming, we find a path such that the sum of the weights of the directed edges in the path from the vertex at row 1 and column 1 to the vertex at row $2n$ and column E is the maximum (Fig. 4).

Let the vertices on row $2i - 1$ in the path be the vertex set including the onset time of the note i. Since the path includes vertices with large weight, the weights of the vertices in the vertex set is large. Since the weight of a vertex means the possibility of the onset time, a vertex in the vertex set means the time which is the onset time of the note i. Let the time at which the weight is the largest in the vertex set be the onset time of i. We show an example detecting the onset times by applying the method (Fig. 5). In the spectrogram, the detected onset time is the time corresponding to the white straight line. It can be seen that the straight line intersects the part where the signal intensity is strong.

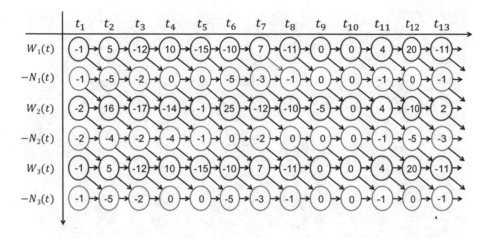

Fig. 3. Example of partial lattice graph

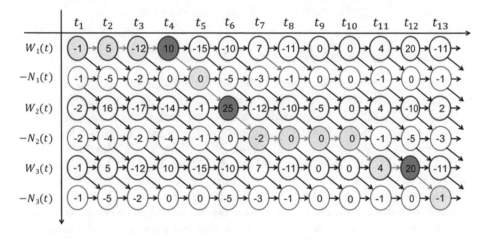

Fig. 4. Path with maximum weight

Furthermore, we improve the method so that, even if there is difference between the score and the sound data by mistakes in playing, the difference is detected. The above method can detect the onset times of the notes which are not included in the score. On the other hand, when the notes which are included in the score are not played, the method cannot the onset times correctly. Therefore, when $W_i(t)$ is less than a threshold, we determine that the note i is not played. Thus, we can expect that, even if the play is different from the score, the onset times can be detected.

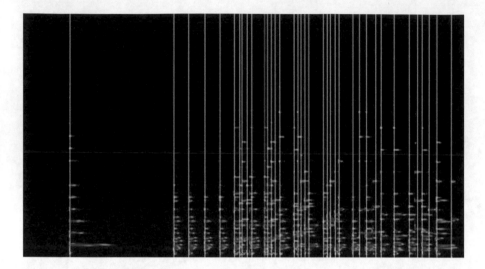

Fig. 5. Example of onset time detection

3.2   Method for Synchronization of Motion

We describe a method to estimate the onset times in the motion data. For example, the onset times of the piano can be estimated by capturing the motion of the keys. We detect the time that the marker attached on the key reaches the lowest point, and we determine the time as the onset time of the key. As for the guitar, the onset times of the motion can be detected by the motion of the markers of the thumb and forefinger of the right hand. In general, the onset times can be estimated by using the electrical musical instrument recording MIDI data.

Next, we describe a method to synchronize the motion of the musical instruments. The method matches the estimated onset time to the expected onset time by changing the speed of the motion. When the speed is reduced, the frames are deleted at a regular interval; when the speed is increased, the new frames are generated by the linear interpolation and inserted.

4   Performance Evaluation

In this section, we evaluate the performance of the proposed method.

We used the piano and the guitar for the performance evaluation in this paper.

First, the sound source data was recorded in WAV format. The sampling frequency is 44.1 kHz, stereo, 16 bit linear PCM.

The motion data was captured by the optical motion capture system with 8 cameras. One marker is attached to each keyboard on the piano, and one marker is attached to the end of each string and each fret of the guitar, respectively.

29 markers are attached to the fingers of one hand of the pianist. 24 markers are attached to the right hand except for the index finger and 23 markers are attached to the left hand except for the thumb of the guitarist. The onset times of the motion are detected by the motion of the marker on the key for the piano and the motion of the markers of the thumb and forefinger of the right hand for the guitar.

We show the results of the sychronization by applying the proposed method in Sect. 3 to the beginning part of "twinkle star" (Fig. 6). We used the same musical score for both piano and guitar.

Fig. 6. Musical score of Twinkle Stars

We show the onset times in Table 1.

Table 1. Onset times before synchronization

Note number	The onset times before synchronization		Gap between two instruments
	Piano (ms)	Guitar (ms)	(*ms*)
60	0.00000	0.00000	0.00000
60	603.71882	464.39910	139.31972
67	1160.99773	1021.67801	139.31972
67	1741.49660	1486.07710	255.41950
69	2298.77551	1996.91610	301.85941
69	2925.71428	2438.09524	487.61904
67	3575.87301	2948.93424	626.93877
65	4783.31065	3924.17234	859.13831
65	5387.02948	4435.01134	952.01814
64	5944.30839	4969.07030	975.23809
64	6524.80725	5433.46939	1091.33786
62	7174.96598	5944.30839	1230.65759
62	7778.68480	6501.58731	1277.09749
60	8428.84353	6965.98640	1462.85713

There is the gap of about 1.5 s between the last notes of the piano and the guitar within the musical performance of about 9.6 s. We show the results of the synchronization of the onset times of the piano and the guitar in Table 2.

Table 2. Onset times after synchronization

Expected onset times (*ms*)	The onset times after synchronization		Gap between two instruments (*ms*)
	Piano (ms)	Guitar (ms)	
0.0	0.00000	0.00000	0.00000
500.0	487.61905	487.61904	0.00001
1000.0	975.23810	998.45805	-23.21995
1500.0	1486.07710	1486.07709	0.00001
2000.0	1973.69615	1996.91610	-23.21995
2500.0	2484.53515	2507.75510	-23.21995
3000.0	2995.37415	2995.37415	0.00000
4000.0	3970.61225	3970.61224	0.00001
4500.0	4481.45125	4481.45124	0.00001
5000.0	4969.07030	4992.29025	-23.21995
5500.0	5479.90930	5503.12925	-23.21995
6000.0	5967.52834	5990.74830	-23.21996
6500.0	6478.36735	6501.58730	-23.21995
7000.0	6989.20635	6989.20635	0.00000

The gap is reduced to 23 ms at the maximum (Table 2). The method can synchronize the multiple instruments.

Since the estimated onset time is matched to the expected onset time for each music note, the onset time after synchronization must be the exact same as the estimated onset time. However, there is the small gap between the onset time after synchronization and the expected onset time in Table 2. This is because, since the onset times after synchronization are estimated in WAV format, the degree of the accuracy of the estimation causes the gap. In addition, when the speed of the music performance is changed, the small gap cannot be avoided at the joint. To improve the accuracy of the synchronization, it is necessary to improve the algorithm of detecting the onset times and the algorithm of changing the speed by considering the joints.

5   Conclusions

In this paper, we proposed the method for generation of CG animation synchronizing motion and sound of multiple instruments. The basic idea of the proposed method is described as follows. First, the expected onset times of all the music notes are determined based on the musical score data for each musical instrument. Next, the estimated onset times are determined in the sound data recorded in WAV format by using the algorithm based on the dynamic programming. The estimated onset times of the motion can be detected by a motion capture system or by a electrical musical instrument recording MIDI format. Then, the estimated onset time is matched to the expected onset time for each music note by changing the speed of the music and motion performances. Thus, the sound and the motion are synchronized in total.

Furthermore, we evaluated the performance using the data of the actual performance of the piano and the guitar. The gap is reduced to 23 ms at the maximum, even if the initial gap of 1.5 s. This indicates that the proposed method can synchronize the multiple instruments.

References

1. Miyata, N., Kouchi, M., Kurihara, T., Mochimaru, M.: Modeling of human hand link structure from optical motion capture data. In: 2004 IEEE/RSJ International Conference on Intelligent Robots and Systems (IROS), 28 September–2 October 2004
2. Patel, J.K., Gopi, E.S.: Musical notes identification using digital signal processing. Procedia Comput. Sci. **57**, 876–884 (2015)
3. Bello, J.P., Daudet, L., Abdallah, S., Duxbury, C., Davies, M., Sandler, M.B.: A tutorial on onset detection in music signals. IEEE Trans. Speech Audio Process. **13**(5), 1035–1047 (2005)
4. Marolt, M., Kavcic, A., Privosnik, M., Divjak, S.: On detecting note onsets in piano music. In: Electrotechnical Conference, MELECON 2002. 11th Mediterranean, 7–9 May 2002
5. Arfib, D., Verfaille, V.: Driving pitch-shifting and time-scaling algorithms with adaptive and gestural techniques. In: Conference on Digital Audio Effects (DAFx 2003), 8–11 September 2003
6. Kawai, T., Kitaoka, N., Takeda, K.: Acoustic model training using feature vectors generated by manipulating speech parameters of real speakers. In: Signal & Information Processing Association Annual Summit and Conference (APSIPA ASC), 3–6 December 2012
7. Mousa, A.: Voice conversion using pitch shifting algorithm by time stretching with PSOLA and resampling. J. Electr. Eng. **61**(1), 57–61 (2010)
8. Lee, H.C., Lee, I.K.: Automatic synchronization of background music and motion in computer animation. Comput. Graph. Forum **24**(3), 353–361 (2005)

Spectrum Trading in Wireless Communication for Tertiary Market

Anil Bikash Chowdhury[1], Fatos Xhafa[2], Rupkabon Rongpipi[3],
Sajal Mukhopadhyay[3(✉)], and Vikash Kumar Singh[3]

[1] Department of Master of Computer Applications, Techno India University,
Kolkata, India
abchaudhuri007@gmail.com

[2] Department of Computer Science, Universitat Politècnica de Catalunya,
Barcelona, Spain
fatos@cs.upc.edu

[3] Department of Computer Science and Engineering, NIT Durgapur, Durgapur, India
roopbon@gmail.com, sajmure@gmail.com, vikas.1688@gmail.com

Abstract. Number of devices needing wireless communication is more than ever. The main ingredient of wireless communication-the spectrum-is limited and due to the unprecedented growth of smart devices the demand for spectrum is high. To serve all the devices needing wireless communication, intelligent spectrum sensing and its trading has been the hot topic of research in this decade. In spectrum trading, so far in the literature, two layers are considered in terms of primary and secondary users. However, it may be the case, that the secondary users (may be NGOs) may redistribute their spectrum to some third party (downtrodden people of the rural areas) freely. To the best of our knowledge, this environment is not addressed in the literature so far. In this paper this three layer potentially demanding architecture is studied and algorithms are proposed based on the theory of mechanism design without money. Our algorithm is also simulated with a specially designed benchmark algorithm.

1 Introduction

The redistribution of the unutilized or underutilized spectrum to the secondary users held by the primary users has been an active area of research from decades in the field of wireless communications. Till date, the literature have addressed the issues revolving around the Government who is the owner, the primary users (say the wireless providers) and the secondary users (which may be any organization who are in need of some small range of frequencies), but not beyond that. At the outset we can say that the secondary users may be any entity like an organization, single atomic user, a Non-governmental organizations (NGOs) etc. We will refer them as agents wherever applicable. There may be two cases in which the secondary users may use the spectrum: (i) using for his purpose only or (ii) he may redistribute again. Moreover, the redistribution makes sense in several realistic situations. Say, for example, the secondary

© Springer Nature Switzerland AG 2019
F. Xhafa et al. (Eds.): INCoS 2018, LNDECT 23, pp. 134–145, 2019.
https://doi.org/10.1007/978-3-319-98557-2_13

users are the NGOs and their objective is to provide the Internet facility to the downtrodden people (located in different localities) free of cost and thereby creating another market for spectrum allocation, which we call in this paper the *tertiary market.*

In this paper, algorithms are proposed to redistribute the spectrum to the tertiary market from the market consisting of secondary users by the concept of mechanism design without money [1,7,9,11,12].

The main contributions of our paper are two fold: the tertiary market is addressed in spectrum trading in static environment and under strategic settings the redistribution algorithms are proposed and evaluated.

The rest of the paper is organized as follows: Sect. 2 discusses some literatures in this direction. Our proposed system model is presented in Sect. 3. In Sect. 4 the required economic properties are discussed. The proposed mechanism is illustrated in Sect. 5. In Sect. 6, the enhanced allocation system is presented. Experimental setup and results are discussed in Sect. 7. Section 8 concludes the work presented in this paper.

2   Literature Review

In literature, various related studies are done on the double auction spectrum trading like (*a*) [20] concentrating on spectrum reuse, (*b*) [2] that allows users to give their preferences for heterogeneous spectrum and in the same time satisfying all the economic properties. In [18] the spectrum heterogeneity in space, frequency and bandwidth was considered. There were other works regarding the sharing of spectrum among the various secondary users like [4,6,15–17,19]. In [6,13,14] instead of considering money, preferences of the interested participants are considered here. In this paper, we focus on the fact that if NGOs provide free of cost services (say providing Internet service for educational activities) to the community, then how the NGOs can be matched up with the localities who may be interested to have the service. This we call the tertiary market which is not addressed in the literature. We have investigated the problem in *static* environment.[1]

3   System Model

In our model, we have m number of **secondary users (NGOs)** given as $\eta = \{\eta_1, \eta_2, \ldots, \eta_m\}$ who are willing to provide access to the spectrum freely. There are n numbers of **tertiary users (say localities)** given by the set $\mathscr{L} = \{\mathscr{L}_1, \mathscr{L}_2, \ldots, \mathscr{L}_n\}$. It might be either case that $n = m$ or $m < n$ or in some cases it might be $n > m$. If $m < n$, one NGO can serve more than one localities. In our case, we are assuming that $n = m$, $m < n$ case is reserved for our future

---

[1] The *dynamic* environment version of the problem is reserved for our future work. By *dynamic* environment we mean that the agents may arrive and depart from the system on regular basis.

work. The strict preference is given by the set $\mathscr{P}_i^l = \{ \eta_1 \succ \eta_2 \succ, \ldots, \succ \eta_m \}$. The directed graph formed during the allocation process is $\mathscr{G}(\mathscr{V}, \mathscr{E})$, where $\mathscr{V} = \{\nu_1, \nu_2, \ldots \nu_{m+n}\}$ representing the NGO's $\{\eta_1, \eta_2, \ldots, \eta_m\}$ and the localities $\{\mathscr{L}_1, \mathscr{L}_2, \ldots, \mathscr{L}_n\}$, $\mathscr{E}$ is a set of directed edges which will represent that there may be an edge consisting of $(\eta_i, \mathscr{L}_i)$ for the random assignment of NGOs to the localities. Similarly, depending on the available preferences there may be an edge $(\mathscr{L}_i, \eta_i)$. Our idea is that we find the cycle and then we will allocate the NGO to the locality keeping in mind the preferences of the locality.

4   Required Definitions

Definition 1 (Strategyproof). A matching is Strategyproof if any participant, who is having a strict preference, do not lie about his preference. For, if he lies about his preference list, he will not be better off than the allocation done by the system (mechanism or algorithm) [10].

Definition 2 (Core Allocation). Core allocation means the allocation which is done in such a way that it gives the best available choice to the agents. The allocation must be such that even though if some agents tries to form a coalition (a subgroup form among themselves) would not make them better off [10]. This concept is utilized to measure the fairness of our proposed algorithms. For further fairness criteria such as [3,5] may be added in future work.

Definition 3 (Pareto Optimal). Pareto optimal is a property of an outcome which says that we can't make anyone better off without making someone else worse off [8].

5   Proposed Mechanism: STOM-OSM

We present a proposed mechanism namely *spectrum trading without money in one sided market* (STOM-OSM) motivated by [7,10]. It is to be noted that, in literature similar to the TTCA mechanism another greedy algorithm exists known as "The Draw" [8] It follows the serial dictatorship approach. The idea of STOM-OSM is a directed graph is formed by utilizing the preference list of the localities and initial random allocation of NGOs to the localities. Once the directed graph is formed, the goal is to determine the directed cycle in the directed graph. The allocation of NGOs to the localities is done by following the directed cycle in the graph. The detailed algorithms is depicted in Algorithms 1 and 2.

Algorithm 1. STOM-OSM $(\eta_m, \mathscr{L}_n, \mathscr{P}_j^l)$

---

1 **for** $i = 1$ *to* m **do**
2   **for** $j = 1$ *to* n **do**
3        $\eta_i \longrightarrow \mathscr{L}_j$ `// `η_i` points to any `$\mathscr{L}_j$` initially.`
4        $\mathscr{P}_j^l$ is the strict preference of the localities.
5        $\mathscr{L}_n : \mathscr{P}_n$ `// `$\mathscr{P}_j^l$`=`$\{\eta_1 \succ, \eta_2 \succ, \ldots, \succ \eta_n\}$
6        $\eta_j \longleftarrow \mathscr{L}_i$ `// `η` is preferred by `$\mathscr{L}$
7   **end**
8 **end**
9 Cycle $(\mathscr{C}) = 0$
10 Check if there exist a cycle.
11 $\mathscr{C} \leftarrow$ Cycle Detection Algorithm $(\eta_m, \mathscr{L}_n, \mathscr{P}_j^l)$
12 **if** $\mathscr{C} = 1$ **then**
13   For agents in $\mathscr{C}$ allot $\mathscr{L}_n$ the most preferred. Remove allocated agents.
14 **end**
15 **for** $i = 1$ *to* n **do**
16   If any unmatched NGO's/Localities, repeat step 1 to 15.
17 **end**
18 **return** $\mathscr{A}\mathscr{L}_n$

Algorithm 2. Cycle Detection $(\mathscr{G}, i)$

---

1 **Cycle** $(\mathscr{G}, u)$ `/* `$\mathscr{C}$` is the set representing the cycle found in the`
    `graph `$\mathscr{G}(\mathscr{V}, \mathscr{E})$` */`
2 **for** *all* $u \in \mathscr{G}.\mathscr{V}$ **do**
3   u.visit = FALSE
4 **end**
5 **for** *all* $u \in \mathscr{G}.\mathscr{V}$ **do**
6   **if** $u.visit = FALSE$ **then**
7   Cyclevisit $(\mathscr{G}, u)$
8   **end**
9 **end**
10 **Cyclevisit** $(\mathscr{G}, v)$ u.visit = TRUE **for** $v \in \mathscr{G}.Adj(u)$ **do**
11   **if** $v.visit = FALSE$ **then**
12        $\pi = u$ `// `π` is the parent/root of u`
13   **for** *all* i *in* π **do**
14   **if** i *repeat in* π **then**
15              $\mathscr{C}_n = i_n$ $\triangleright$ i_n = the vertex in the repeated range, starting and
                ending with the same vertex u
16   **end**
17   **end**
18   **end**
19 **end**
20 Cyclevisit $(\mathscr{G}, v)$
21 **return** $\mathscr{C}_n$

5.1   Analysis of STOM-OSM

The time complexity of STOM-OSM is $O(n^2) + O(1) + O(n) = O(n^2)$.

5.2   Illustrative Example

Let us take an example to understand our algorithm better. We have 4 NGO's $(\eta_4 = \eta_1, \eta_2, \eta_3, \eta_4)$ and 4 localities $(\mathscr{L}_4 = \{\mathscr{L}_1, \mathscr{L}_2, \mathscr{L}_3, \mathscr{L}_4\})$ with the strict preferences given as $\mathscr{L}_1$: $\eta_1 \succ \eta_3 \succ \eta_4 \succ \eta_2$, $\mathscr{L}_2$: $\eta_1 \succ \eta_3 \succ \eta_2 \succ \eta_4$, $\mathscr{L}_3$: $\eta_4 \succ \eta_2 \succ \eta_1 \succ \eta_3$, and $\mathscr{L}_4$: $\eta_1 \succ \eta_4 \succ \eta_3 \succ \eta_2$. In Fig. 1a all the NGO's pointing to some localities initially. Next, in Fig. 1b the localities also points to their most preferred NGO's. After that, we can see that a directed graph $\mathscr{G}(\mathscr{V}, \mathscr{E})$ is formed the 8 vertices. In Fig. 1b, we have found our first cycle that is formed between $\mathscr{L}_1$ and η_1 i.e; $\{\mathscr{L}_1 \rightarrow \eta_1 \rightarrow \mathscr{L}_1\}$. As $\mathscr{L}_1$ had got his best preferred NGO and η_1 is also allocated to one locality i.e; $\mathscr{L}_1$, they are removed from the system. Next, there will be 6 agents left in the system which are $\{\mathscr{L}_2, \mathscr{L}_3, \mathscr{L}_4\}$ and $\{\eta_2, \eta_3, \eta_4\}$. In Fig. 1c we have found our 2nd cycle that was formed between $\mathscr{L}_4$ and η_4 which is $\{\mathscr{L}_4 \rightarrow \eta_4 \rightarrow \mathscr{L}_4\}$. Now, $\mathscr{L}_4$ and η_4 are removed from the system. In Fig. 1d, we got our last cycle formed between $\mathscr{L}_2, \eta_2, \mathscr{L}_3$ and η_3 form as $\{\mathscr{L}_2 \rightarrow \eta_3 \rightarrow \mathscr{L}_3 \rightarrow \eta_2 \rightarrow \mathscr{L}_2\}$. So, our algorithm is complete. We can see that the final matching is done given as: $\mathscr{L}_1 : \eta_1, \mathscr{L}_2 : \eta_3, \mathscr{L}_3 : \eta_2, \mathscr{L}_4 : \eta_4$

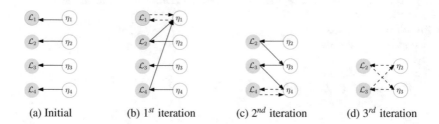

(a) Initial   (b) 1st iteration   (c) 2nd iteration   (d) 3rd iteration

Fig. 1. Illustrative example of STOM-OSM

5.3   Properties

Our proposed mechanism satisfies the following properties such as *Strategyproof*, *Unique Core allocation*, and *Pareto optimality* similar to TTCA.

Proposition 1. *The TTCA induces a DSIC mechanism [7, 10].*

Proposition 2. *For every house allocation problem, the allocation computed by the TTCA is the unique core allocation [7, 10].*

Proposition 3. *TTCA is Pareto Optimal [7, 10].*

Lemma 1. *STOM-OSM is Strategyproof.*

Proof. To prove any algorithm to be Strategyproof, we need to show that no one gets the incentive to lie. Here, the ranking of the preferences given by the localities is independent of the other locality ranked list. So, it will solely depend on his preference as how he gives his ranking. STOM-OSM gives the best available choice to the localities so, whenever a locality lie about his preference list he won't be better off. Here, $\mathscr{L}n_i$ denotes localities allocated in the $\mathscr{C}^{th}$ cycle or iteration of STOM-OSM when all localities are reporting truthfully (where, $\mathscr{L}n_1$ is the set of localities allocated in 1st iteration with $C = 1$, $\mathscr{L}n_2$ is the set of localities allocated in 2nd iteration with $C = 2$, and so on). Each locality of $\mathscr{L}n_1$ gets its best preferred choice and hence they have no incentive to lie. An NGO of $\mathscr{L}n_2$ is not pointed to by any localities belonging to $\mathscr{L}n_1$ else it would belong to $\mathscr{L}n_1$ instead of $\mathscr{L}n_2$. Thus, even if $\mathscr{L}_i$ will misreport, he would not get a house originally owned by a locality in $\mathscr{L}n_1$. Since, $\mathscr{L}_i$ gets its best choice other than the houses owned by $\mathscr{L}n_1$, it has no incentive to lie. We can say that, a locality $\mathscr{L}_i$ of $\mathscr{L}n_i$ is never pointed to in the first $i - 1$ iterations of the STOM-OSM by any locality in $\mathscr{L}_i \cup \ldots \cup \mathscr{L}_i - 1$. Thus, whatever it reports, $\mathscr{L}_i$ will not receive a house owned by an agent in $\mathscr{L}_i \cup \ldots \cup \mathscr{L}_i - 1$. Since the STOM-OSM gives $\mathscr{L}_i$ its favourite house outside this set, it has no incentive to misreport [8].

Lemma 2. *STOM-OSM gives unique core allocation.*

Proof. Due to limitation of space the detailed proof is not provided here.

Lemma 3. *STOM-OSM is Pareto Optimal.*

Proof. To prove that STOM-OSM gives a Pareto optimal outcome we will consider the same argument as given in [8]. Lets say, we term our present algorithm as $STOM - OSM_1$ and another similar algorithm $STOM - OSM_2$. We will assume that the algorithm $STOM - OSM_2$ doesn't make anyone worse off than in $STOM - OSM_1$ (else, the other algorithm would not be a threat to $STOM - OSM_1$). Our objective is to prove that the two algorithms are similar. Hence, any allocation different from $STOM - OSM_1$ allocation will make someone worse off and resulting in Pareto optimality. We will proof the above given statement by induction on k (first, we assume no one is worse off in $STOM - OSM_2$), the first k localities are allocated similarly in both the algorithms.

- **Base case** ($k = 0$): It is trivial, as both the empty allocation coincide.
- **Inductive step** ($k \geq 1$): By Induction hypothesis, the first $k-1$ are similarly in both the algorithms.

Therefore, the remaining NGO option in $STOM - OSM_2$ will be those options present in $STOM - OSM_1$ when k is considered. As proved in Propositions 1 and 2, $STOM - OSM_1$ allocate k his favourite NGO from the available NGO so, if $STOM - OSM_2$ allocate any different from $STOM - OSM_1$ it will result in making some k worse off. This completes the inductive step and the proof [8].

6   Enhanced Allocation System

In this section, we have considered the case where both the localities and the NGOs will be providing the strict preference ordering over the members of the NGOs and the localities respectively. For this set-up, we have proposed a mechanism namely; *spectrum trading with out money in two sided market* (STOM-TSM) mechanism motivated by [10,12].

6.1   Proposed Mechanism: STOM-TSM

In our algorithm STOM-TSM, we will assume that the localities will be the one proposing to the NGO's. So, it will be a locality-optimal algorithm. The general idea of the STOM-TSM is that, each member of the requesting party (say localities) requests to the highest ranked NGO on its list. Now, the NGOs accepts at most one request and rejects others. Each rejected locality removes the rejecting NGO from its list. If there were no rejections, then mechanism stops, otherwise the rejected members request to their second most preferred NGO. The algorithm iterates until no rejection takes place.

Algorithm 3. STOM-TSM $(\mathscr{P}_n^l, \mathscr{P}_m^\eta, \mathscr{L}_n, \eta_m, i)$

---

1 initialize $\mathscr{P}r_i^\eta \leftarrow 0$ /* $\mathscr{P}r_i^\eta$ is the number of proposal η is getting. */
2 **for** $i = 1$ *to* n **do**
3 | **if** η_i *is unmatched* **then**
4 | | $\eta_i \longleftarrow \mathscr{L}_i$ ▷ $\mathscr{L}_i$ point to his best preference.
5 | | $\mathscr{P}r_i^\eta = \mathscr{P}r_i^\eta + 1$ ▷ increment number of proposal for η.
6 | **end**
7 **end**
8 **for** $j = 1$ *to* m **do**
9 | **if** $\mathscr{P}r_i^\eta > 1$ **then**
10 | | $\eta_j \longrightarrow \mathscr{L}_j^\eta$ ▷ accept $\mathscr{L}_j$ which is the highest ranked holder, reject others.
11 | **end**
12 | **else**
13 | | $\eta \leftrightarrow \mathscr{L}$ ▷ accept temporarily until he gets a better choice.
14 | **end**
15 **end**
16 Remove matching pairs
17 $\mathscr{S}tm = \eta \leftrightarrow \mathscr{L}$
18 **for** $i = 1$ *to* n **do**
19 | If any unmatched $\mathscr{L}$ repeat 1 - 13.
20 **end**
21 **return** $\mathscr{S}tm_n$

6.2   Analysis of STOM-TSM

The running time is $O(n^2)$. As in our case $n = m$.

6.3   Illustrative Example

In Fig. 2a it can be seen that each of the localities $\mathscr{L}_1$, $\mathscr{L}_2$, $\mathscr{L}_3$, and $\mathscr{L}_4$ are requesting to the most preferred NGO from their respective preference list *i.e.* η_1, η_1, η_4, and η_1 respectively. In the next step, in Fig. 2b we will check any requested NGOs has got the multiple requests. The competitive environment between $\mathscr{L}_1$, $\mathscr{L}_2$, and $\mathscr{L}_4$ can be resolved by considering the strict preference ordering of η_1. From the strict preference ordering of η_1 it is clear that $\mathscr{L}_2$ is most preferred one. In Fig. 2c, $\mathscr{L}_1$ and $\mathscr{L}_4$ request to their second most prefered NGOs. In the similar fashion the final allocation is determined. The final allocation is given in Fig. 2f.

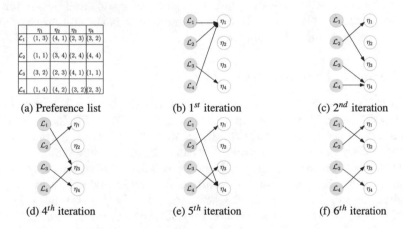

Fig. 2. Illustrative example of STOM-TSM

6.4   Properties

Our proposed mechanism satisfies the following properties such as *Strategyproof*, and *Perfect matching* similar to Stable matching algorithm.

Proposition 4. *Stable matching algorithm gives a Perfect Matching [7, 10].*

Proposition 5. *Stable Matching algorithm is Strategyproof [7, 10].*

Lemma 4. *STOM-TSM results in a Perfect Matching.*

Proof. Due to limitation of space the detailed proof is not provided here.

Lemma 5. *STOM-TSM is Strategyproof for the requesting party.*

Proof. Due to limitation of space the detailed proof is not provided here.

7 Experiments and Evaluation

We compared our proposed mechanisms against the benchmark mechanism (random mechanism). In this, each time a random locality is selected and randomly an NGO is selected from the locality preference list and allocated. Both the locality and the NGO are removed from the system. The process is repeated until the NGO list or the Locality list gets exhausted.

7.1 Performance Metrics

The performance of our proposed mechanisms $i.e.$ STOM-OSM and STOM-TSM is measured using two important parameters: (1) **Number of first preference (NFP):** It is the sum of the number of locality getting their first preference. Mathematically, the NFP is defined as: $NFP = \sum_{\mathscr{L}_i \in \mathscr{L}} f(\mathscr{L}_i)$; where, $\mathscr{L} = \{\mathscr{L}_1, \mathscr{L}_2, \ldots, \mathscr{L}_n\}$ is the set of total number of locality present in the system. Any locality $\mathscr{L}_i \in \mathscr{L}$ is allocated its first preference is captured by the function $f : \mathscr{L} \rightarrow \{0, 1\}$. (2) **Number of kth preference:** It is the sum of the number of locality getting their k^{th} preference.

7.2 Analysis of the Results

Our simulation result are analyzed by considering the two mechanisms namely; STOM-OSM and STOM-TSM separately. In STOM-OSM, Fig. 3, we have plotted the allocation graph taking 5, 10, 15 and 20 agents at a time for both STOM-OSM and random allocation. We can clearly see a huge difference in the number of localities getting their first preferred NGO. The algorithm is repeated 100 times for each number of the localities and after taking the average it is plotted in the graph. In Fig. 3b, 6 agents are taken both sides, i.e; 6 localities and 6 NGO's. we have compared not only for the 1st preference but for $n/2$ and n preference also. we have taken the average over 100 times. Localities getting the worst preference is also very high in random allocation whereas it is comparatively very low in STOM-OSM.

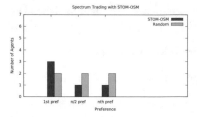

(a) Comparison of NFP   (b) Comparison of number of kth preference

Fig. 3. Evaluation of STOM-OSM

In Fig. 4a, we have plotted the allocation graph taking 4, 8, 10 and 12 agents at a time for both STOM-TSM and random allocation. We have considered the preference of the NGO as well. The NGO accepts only if they want or else they can reject the locality. We can clearly see a huge difference in the number of localities getting their first preferred NGO. The algorithm is repeated 100 times for each number of the localities and after taking the average it is plotted in the graph. In Fig. 4b, 12 agents are taken both sides, i.e; 12 localities and 12 NGO's. we have compared the preferences for 1st, $n/2$ and n preference. We have taken the average for over 100 times. Localities getting the worst preference is very high in random allocation whereas it is comparatively very low in STOM-TSM. In random mechanism the agents can gain by manipulation.

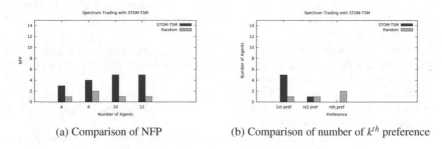

(a) Comparison of NFP   (b) Comparison of number of kth preference

Fig. 4. Evaluation of STOM-TSM

Overall, it is seen that our allocation algorithm gives us a better results than the *random* algorithm.

8 Conclusion and Future Works

In this paper, we have studied the problem of allocating the Internet services by some secondary users (say NGO's) to some downtrodden community of the society (say tertiary users) under *zero* budget environment. Designing an algorithm for the more general setup consisting of n NGO's and m tertiary users, where $m \neq n$ is our immediate future work. The work can further be extended to dynamic environment, where the NGO's and the tertiary users are arriving and departing from the system on a regular basis in a defined time (say *a day*).

References

1. Bandyopadhyay, A., Mukhopadhyay, S., Ganguly, U.: Allocating resources in cloud computing when users have strict preferences. In: 2016 International Conference on Advances in Computing, Communications and Informatics (ICACCI), pp. 2324–2328, September 2016
2. Feng, X., Chen, Y., Zhang, J., Zhang, Q., Li, B.: TAHES: a truthful double auction mechanism for heterogeneous spectrum. IEEE Trans. Wirel. Commun. **11**(11), 4038–4047 (2012)
3. Habes, M.R., Belleili-Souici, H.: Towards a fairer negotiation for dynamic resource allocation in cloud by relying on trustworthiness. Int. J. Grid Util. Comput. **8**(3), 185–200 (2017)
4. Hyder, C.S., Jeitschko, T.D., Xiao, L.: Bid and time strategyproof online spectrum auctions with dynamic user arrival and dynamic spectrum supply. In: 2016 25th International Conference on Computer Communication and Networks (ICCCN), pp. 1–9, August 2016
5. Köppen, M., Yoshida, K., Tsuru, M.: Relational approaches to resource-aware multi-maxmin fairness in multi-valued resource sharing tasks. Int. J. Space Based Situat. Comput. **3**(2), 91–101 (2013)
6. Mochaourab, R., Holfeld, B., Wirth, T.: Distributed channel assignment in cognitive radio networks:stable matching and walrasian equilibrium. IEEE Trans. Wirel. Commun. **14**, 3924–3936 (2015)
7. Roughgarden, T.: Lectures Notes on Algorithmic Game Theory. Cambridge university press, June 2014
8. Roughgarden, T.: Lecture #1: the draw and college admissions, 26 September 2016
9. Schummer, J., Vohra, R.K.: Mechanism design without money. In: Algorithmic Game Theory, pp. 243–267. Cambridge University Press, New York (2007)
10. Shapley, L.S., Scarf, H.: On cores and indivisibility. J. Math. Econ. **1**, 23–37 (1974)
11. Singh, V.K., Mukhopadhyay, S., Das, R.: Hiring doctors in e-healthcare with zero budget, pp. 379–390. Springer International Publishing, Cham (2018)
12. Singh, V.K., Mukhopadhyay, S., Sharma, A., Roy, A.: Hiring expert consultants in e-healthcare: a two sided matching approach. CoRR, abs/1703.08698 (2017)
13. Wang, X., Ji, Y., Zhou, H., Li, J.: A non monetary qos-aware auction framework toward secure communications for cognitive radio networks. IEEE Trans. Veh. Technol. **65**(7), 5611–5623 (2016)
14. Wang, X., Ji, Y., Zhou, H., Li, J.: Auction-based frameworks for secure communications in static and dynamic cognitive radio networks. IEEE Trans. Veh. Technol. **66**(3), 2658–2673 (2017)
15. Wang, Y., Yu, J., Lin, X., Zhang, Q.: A uniform framework for network selection in cognitive radio networks, pp. 3708–3713 (2015)
16. Yang, D., Xue, G., Zhang, X.: Group buying spectrum auctions in cognitive radio networks. IEEE Trans. Veh. Technol. **66**, 810–817 (2016)
17. Yi, C., Cai, J., Zhang, G.: Online spectrum auction in cognitive radio networks with uncertain activities of primary users. In: 2015 IEEE International Conference on Communications (ICC), London, pp. 7576–7581 (2015). https://doi.org/10.1109/ICC.2015.7249538

18. Zhan, S.-C., Chang, S.-C.: Double auction design for short-interval and heterogeneous spectrum sharing. IEEE Trans. Cogn. Commun. Netw. **2**(1), 83–94 (2016)
19. Zhang, Y., Song, L., Pan, M., Dawy, Z., Han, Z.: Non-cash auction for spectrum trading in cognitive radio networks: contract theoretical model with joint adverse selection and moral hazard. IEEE J. Sel. Areas Commun. **35**(3), 643–653 (2017)
20. Zhou, X., Zheng, H.: TRUST: a general framework for truthful double spectrum auctions. In: IEEE INFOCOM 2009, pp. 999–1007 (2009)

Collaborative All-Optical Alignment System for Free Space Optics Communication

Takeshi Tsujimura[1(✉)], Kiyotaka Izumi[1], and Koichi Yoshida[2]

[1] Saga University, Saga, Japan
tujimura@cc.saga-u.ac.jp
[2] Fukuoka Institute of Technology, Fukuoka, Japan

Abstract. This paper proposes collaborative laser beam alignment system of bilateral free space optics (FSO). Active free space optics apparatus is designed and prototyped which aligns transmission laser beam precisely. Adjustment statics is also investigated involving two techniques. A collaborative adjustment method makes it possible to connect the remote FSO apparatus using nothing but the transmission laser beam. Analytic estimation method of laser spot distribution is formulated based on the Gaussian beam optics. It helps the laser beam to reach the target receiver without feedback control.

1 Introduction

Free space optics (FSO) is an alternative telecommunication technology to the fiber optics or wireless radio wave. It emits collimated laser beam in the air, and exchanges optical signals between a pair of apparatus. FSO system is considered to save time and cost, especially in installation, compared with optical fiber system [1–3]. It also provides us a securer communication than the wireless radio transmission in terms of phone tapping. As conventional FSO has been developed as a fixed point-to-point telecommunication system, it seems inconsistent with ubiquitous communication [4–9].

The authors have newly proposed an active free space optics (a-FSO) technology [10–14] to apply to ubiquitous broadband user network. Figure 1 portrays sceneries of the active free space optics applications. A-FSO is capable of serving a rural area network or ad-hoc network in case of disaster as well as urban communication network. FSO is available to construct broadband network between buildings or hilltops more costly than optical fiber. Ad-hoc network is easily established thanks to its autonomous alignment function even in case of disaster.

Because it is important to retain line-of-sight for free space optics communication, conventional FSO apparatus is mounted on a stiff frame to prevent vibration from disturbing laser beam alignment. Nevertheless, the relative location between two confronting apparatus may shift by inches result from wind tremor, traffic quake, frame warp due to solar heat, and so on.

One of features of our active FSO system is the function of tracking a roaming FSO receiver. We have studied the feedback control of transmission laser beam to keep stable communication between restless transmission apparatus [12, 13]. The optical transmission line is established by emitting a thin laser beam to transit broadband

© Springer Nature Switzerland AG 2019
F. Xhafa et al. (Eds.): INCoS 2018, LNDECT 23, pp. 146–157, 2019.
https://doi.org/10.1007/978-3-319-98557-2_14

signals in the air, and by aligning bilateral transmission lines along a unique path by motorized mirrors.

This paper newly designs and prototypes an active FSO system which steers bidirectional transmission laser beams independently using motor-driven lens. It also proposes a collaborative tactic for laser beam adjustment using the transmission laser for both measuring adjustment error and exchanging control information. Estimation method is realized to determine the optical axis of laser beam based on the laser intensity only by distributed photodiodes.

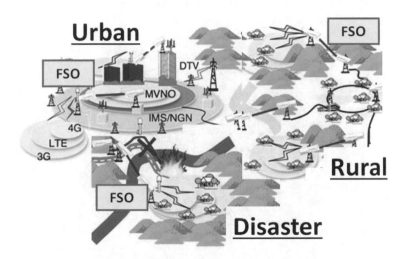

Fig. 1. Free space optics systems applied to various areas.

2 Active Free Space Optics System

2.1 Design of Active Free Space Optics System

The authors have designed an apparatus of active free space optics system as shown in Fig. 2. Our proposed active FSO system transmits a thin laser beam of 10 mm in diameter. Its optical system is designed to transmit 1 Gbit/s signals on 1550 nm wavelength laser beam in the air. It is equipped with two lens-barrels which contain collimator lenses for transmitter and receiver. Transmission laser pulses are introduced into the apparatus by an optical fiber and discharged from one of the collimator lenses to the opposite FSO apparatus. They are steered by a motor-driven lens installed on the path of the laser. The lens is suspended by a voice coil motor (VCM), and is precisely positioned in two-dimension by applying control voltage. Another collimator lens catches an arrived laser beam from the opposite apparatus. The laser beam travels through the lens-barrel and reaches the outlet to a single mode optical fiber (SMF).

Other lenses suspended in VCMs adjust position and orientation of the incident laser beam to effectively couple the aerial laser with the SMF. A beam splitter, installed in the midst of the optical path, leads a portion of the incident laser to a quadrant photodiode (QPD). Quadrant photodiode has four discrete components that are optically activated, and indicates intensities of laser beam incident on the corresponding components. If the branch laser beam divided by the beam splitter hits at the center of the QPD, the output voltages of four components are the same. If different, the offset of the laser beam axis can be estimated based on the proportion of output intensities. The VCMs are controlled so that the branch reaches the center of the QPD. Meanwhile, the main laser beam is precisely directed onto the outlet to SMF.

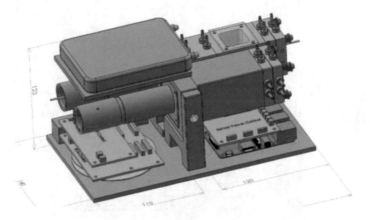

Fig. 2. Active free space optics apparatus.

The receiver is also equipped with positioning photodiodes to measure local intensities of diffused laser beam. They are distributedly arranged near the receiver lens to detect the off-target laser beam. The authors have designed a pentagon positioning PD set as shown in Fig. 3. It is made of five photodiodes, which detect intensity of 1550 nm wavelength, surrounding the receiver lens. Measured data is delivered to the control PC of the opposite FSO apparatus to create the proper control command for the VCM. Then, the controller adjusts the lens position to guide the laser beam within the receiver inlet window.

We have actually manufactured a prototype of active free space optics system as shown in the photograph of Fig. 4. It is equipped with two lenses. The right is a transmitter, which emits 1550 nm laser beam of 10 mm in diameter and transmits 1 G bit/s broadband data. Optical signals are supplied through a single mode fiber. The laser beam is steered by VCMs with a positioning accuracy of 7.54×10^{-6} m/V which is equivalent to a resolution of 1.15×10^{-9} m/bit.

The left is a receiver, which captures an arrived laser beam. The receiver lens is ringed by five InGaAs PIN photodiodes, whose wavelength range is from 1000 to 1600 nm, to find luminescence intensity of the stray beam. The PDs provide the

opposite transmitter five values of voltages representing optical intensity of the corresponding position of photodiodes. If a laser beam is successful in reaching the
receiver lens, it is guided by other VCMs to the outlet SMF. The VCMs adjust the
position and orientation of the laser beam to precisely couple with the SMF whose
mode field diameter (MFD) is around 10×10^{-6} m. Specification of the prototype
active FSO system is listed in Table 1.

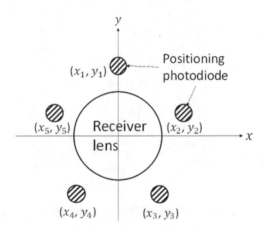

Fig. 3. Positioning photodiodes.

Fig. 4. Photograph of active free space optics apparatus.

Table 1. Specification of active FSO system.

Item	Parameter		Specification
Optical system	Wavelength		1550 nm
	Communication distance		10 m–100 m
	Communication speed		1 G/10 Gbps
	Transmission	Output power	10 dBm : 10 mW
		Laser class	10 mW : Class 1
		Beam angle	Maximum 1.2°
		Lens diameter	25 mm Φ
		Collimator light	10 mm Φ
		Beam divergence	0.02°
	Reception	Optical input power level	0 dBm–−30 dBm
		Telescopic QPD	Size : 3 mm Φ
		Field-of-view of data detector	1.2°
	External dimensions		140 mm × 140 mm × 250 mm

2.2   Fundamental Features of Active Free Space Optics System

The authors have conducted some experiments to confirm the fundamental features of the prototyped active FSO apparatus.

Optical directivity of the receiver system is investigated first. Coupling loss is measured by changing incident angle of laser beam. Experimental results are indicated in Fig. 5, where the horizontal and vertical axes represent the incident laser beam angle and the received power, respectively. Persistent and assiduous regulation of the optical system realizes wide directivity. The coupling loss is less than 5 dB within 0.4°.

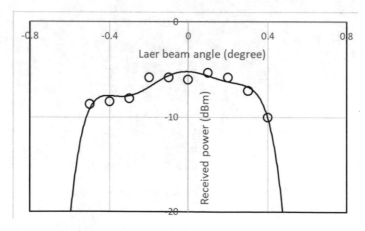

Fig. 5. Directivity of receiver optical system.

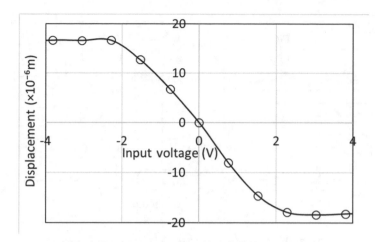

Fig. 6. Performance of voice coil motor motion.

Performance of the voice coil motor is evaluated next. It is required to accurately steer the lens to adjust optical path of the transmission laser beam. Figure 6 expresses the motion of VCM, where the horizontal and vertical axes represent the input voltage and displacement of VCM, respectively. Results show good linearity between 16×10^{-6} to -16×10^{-6} m while charging -2 to 2 V.

3   Alignment Strategy of Active Free Space Optics System

Bilateral optical communication is established between FSO apparatuses. Figure 7 represents a symmetrical block diagram of active free space optics communication by mutually discharging laser beam. Each apparatus contains a transmitter and a receiver as well as a controller PC. Broadband communication is performed by transferring the thin laser beam from the transmitter to the opposite receiver. It is necessary to keep the laser spot within the receiver lens even if it drifts along. Optical path of the transmission laser beam is adjusted by VCMs installed in the transmitter lens-barrel. The beam angle can be steered up to $1.2°$. The objective angle of laser beam is determined by the output data of the opposite positioning photodiodes.

The receiver has two types of laser beam alignment systems. Coarse adjustment is applied to perform between confronting transmitter and receiver in terms of its regional optical intensity. The laser luminescent distribution can be presumed based only on the dispersed regional intensity of the off-target laser beam. Measured data of the positioning PDs is transferred to the VCM in the opposite transmitter via reverse optical line, and is converted to feedback control commands. The VCM is controlled based on the positioning PD data to lead the laser beam within the receiver lens. Fine adjustment is conducted within the receiver. Around 10% of the incident laser beam is divided by the beam splitter and monitored by the QPD to evaluate positional shift of laser beam alignment. The feedback control algorithm maintains the monitor laser beam onto the

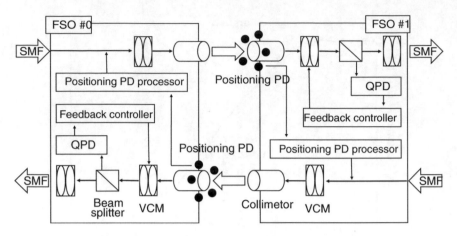

Fig. 7. Block diagram of bilateral FSO control system.

center of the QPD in real time, and the mainstream laser beam is consequently coupled to the outlet SMF at all time.

Coarse feedback control is executed between the transmitter and the receiver on both sides. In order to control the upstream laser beam which travels from FSO #0 to FSO #1, the arrived laser beam is detected by the positioning photodiodes attached around the receiver lens of FSO #1 at first. Their output data is forwarded to the positioning PD processor and is superposed on the transmission signals through the downstream transmission line from FSO #1 to FOS #0. The transferred information is evaluated by the positioning PD processor in FSO #0, and is converted to control commands to the VCM of the transmitter in FSO #0. Thus, the upstream line is adjusted after all. The downstream laser beam is also aligned in the same way.

Such a bilateral transmission involves a complementary and dependent feedback system. Coarse feedback control of either line can be accomplished as far as the opposite stream is connected. In case any line is interrupted, the other line simultaneously breaks.

When discussing tactics of laser beam alignment, we assume the situation of FSO communication into two phase in terms of transmission connection; the transient and steady phases. The transmission line is not connected in the transient phase, as the laser beam is unsuccessful in coupling to the receiver. Adjustment of the laser beam is required to transmit signals from the transmitter to the opposite receiver. In the steady phase, both receivers keep catching laser beams from corresponding transmitters. A tracking control is functioning in guiding the laser beam to the remote receiver lens. Once the FSO apparatus detects the laser beam, broadband communication is sustained all the time in principle.

The algorithm of laser beam alignment is shown in Fig. 8. Two tracking techniques of laser beam are applied with regard to the connection phases: coarse adjustment, and fine adjustment. Each of two tracking techniques is applied according to the condition of transmission line. Initial situation in installing FSO apparatus is the transient phase in general, because the laser beam is deviated from the receiver and the transmission line is disconnected at the beginning. The coarse adjustment is operated to find out the corresponding receivers. Off-line alignment technique looks for the line-of-sight from the transmitter to the receiver by emitting the laser beam over the region where the opposite FSO apparatus possibly exists. The positioning PDs detect luminescence of laser spot at dispersed points. Intensity distribution of the arrived laser luminescence suggests the positioning errors of the laser beam. The laser beam direction is corrected so that the optical axis of laser beam lies on the center of the receiver lens by evaluating the measured data.

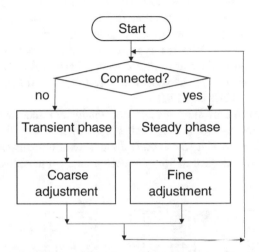

Fig. 8. Laser beam alignment procedure.

If the laser beam is introduced into the receiver, the fine adjustment algorithm drives it to track the restless receiver automatically in the steady phase. Relative motion between the transmitter and the receiver is estimated by the QPD output. PID controller adjusts the laser beam to hit the center of the QPD. The authors have already reported the tracking control of laser beam. As far as the system maintains such status with respect to both the upstream and downstream lines, the steady state continues.

If the laser beam chances to lose the receiver, the communication line is interrupted and the laser beam misses tracking the target receiver owing to incompletion of the feedback loop. Then the status changes to the transient phase, and the system begins seeking the receiver again.

3.1   Collaborative Alignment System

In case of such reconnection or at the outset of connection, some ingenious procedure is necessary to discover the opponent FSO apparatus and to exchange alignment information depending only on the transmission laser beam.

Figure 9 illustrates collaborative procedure of the initial alignment where the confronting apparatuses, FSO #0 and #1 are simplified with the transmitter, receiver, and controller. It represents situation of laser beam and flow of control information. A pair of FSO apparatus are remotely located facing each other.

At the onset of connection, laser beams are emitted toward the opposite FSO apparatus but they miss the receivers as shown in Fig. 9 (1). Neither transmission line is established yet. Positioning PDs in FSO #1 detect the intensity of off-target laser beam traveling from FSO #0.

Next, the laser beam haphazardly scans the area where the target receiver is expected to be as shown in Fig. 9 (2). Meanwhile, the laser beam carries both the transmitter orientation control data and the positioning PD data.

The moment the laser beam accidentally hits at the receiver by a fluke, it transfers those data to FSO #0 in a flash as shown in Fig. 9 (3). The controller in FSO #0 is informed the transmission laser direction of FSO #1 at the occasion of the correspondence as well as output values of the positioning PDs.

In the next instance, the laser beam passes by and goes out of the receiver. Then, FSO #0 controller adjusts the transmitter direction based on FSO #1 positioning PD data to reach the target receiver as shown in Fig. 9 (4). The downstream transmission line is connected from FSO #0 to #1.

Finally, the downstream line transfers positioning PD data of FSO #0 to FSO #1 controller, as it keeps connection ever after as shown in Fig. 9 (5). That makes it possible for FSO #1 controller to establish the upstream transmission from FSO #1 to #0. Bidirectional lines are successfully connected after all as shown in Fig. 9 (6).

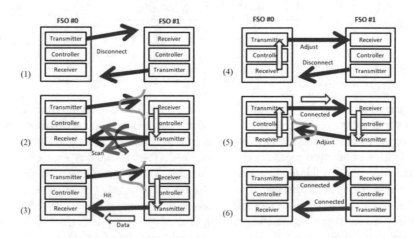

Fig. 9. Collaborative adjustment procedure of laser alignment.

3.2   Coarse Laser Beam Localization

Feedback system is unavailable during the transient phase, because the transmission lines are disconnected. That is why this section proposes an analytical method for coarse laser beam adjustment depending on the positioning photodiode information, namely, off-line laser beam localization. The method can estimate the distribution of laser intensity with some positioning photodiodes, and guide the laser beam to the target where the receiver is expected to place.

Coarse adjustment technique requires the line-of-sight of the laser beam in the transient state. On condition that the distribution of the laser beam intensity is hypothesized in advance, we can estimate the correct position of the laser beam based on local intensity data detected by several interspersed photodiodes. When the laser beam is subject to the Gaussian beam optics, analysis provides us with the optical axis of laser beam. Thus, we can calculate the control commands to adjust the laser beam onto the receiver accurately.

Note that our preceding experiments have revealed that the laser beam generated by the transmitter looks asymmetric distribution, though Gaussian beam optics claims that laser beam intensity forms a point symmetry. That is why a modified equation is used in this paper to analyze laser beam optics instead of the normal distribution.

Let us consider the formulation of the laser beam in the x-y-z coordinate system, assuming that the optical axis is parallel to the z-axis and a laser beam reached at (a, b) on the x-y plane. Optical intensity, E (x, y; E0, n, m, a, b) of a modified Gaussian beam at (x, y) on the x-y plane is expressed in principle as

$$E(x, y; E_0, n, m, a, b) = E_0 \exp\left(-\frac{(x-a)^2}{n^2} - \frac{(y-b)^2}{m^2}\right) \tag{1}$$

where E_0 denotes laser intensity on the optical axis of the laser beam. Parameters, a, and b represent x- and y-coordinate of the optical axis, and n, m are the standard deviations of the x- and y-direction.

Intensity, E (x, y; E_0, n, m, a, b) is actually output voltage measured by the positioning photodiodes. Because five positioning photodiodes are prepared enclosing the receiver in this paper, we obtain local intensity of diffused laser beam at five measuring point. Then five simultaneous equations are established.

Solution of simultaneous equations with respect to five local measuring points analytically brings out the coordinate values, (a, b) of the laser beam optical axis according to the Gaussian beam optics. Because Eq. (1) contains five unknown parameters, E_0, n, m, a, b, five independent conditions are necessary to solve the simultaneous equation in general.

Those photodiodes provide us five data sets. Let the data sets with regard to five photodiodes be (x_1, y_1, E_1), $(x_2\ y_2, E_2)$, (x_3, y_3, E_3), (x_4, y_4, E_4), (x_5, y_5, E_5), respectively, where the first two elements express x- and y-coordinates of each photodiode, and the last element stands for photodiode voltage of laser intensity.

With respect to the data sets, we obtain the following five simultaneous equations as

$$\begin{cases} E(x_1, y_1; E_0, n, m, a, b) = E_1 \\ E(x_2, y_2; E_0, n, m, a, b) = E_2 \\ E(x_3, y_3; E_0, n, m, a, b) = E_3 \\ E(x_4, y_4; E_0, n, m, a, b) = E_4 \\ E(x_5, y_5; E_0, n, m, a, b) = E_5 \end{cases}$$

The optical axis (a, b) of the laser spot is calculated by solving such simultaneous equations in terms of five variables as

$$a = \frac{a_1^u \log(E1) + a_2^u \log(E2) + a_3^u \log(E3) + a_4^u \log(E4) + a_5^u \log(E5)}{a_1^d \log(E1) + a_2^d \log(E2) + a_3^d \log(E3) + a_4^d \log(E4) + a_5^d \log(E5)} \qquad (2)$$

$$b = \frac{b_1^u \log(E1) + b_2^u \log(E2) + b_3^u \log(E3) + b_4^u \log(E4) + b_5^u \log(E5)}{b_1^d \log(E1) + b_2^d \log(E2) + b_3^d \log(E3) + b_4^d \log(E4) + b_5^d \log(E5)} \qquad (3)$$

In case of the prototyped PD set, we obtain the parameters of optical axis (a, b) as

$$a_1^u = -13.9, a_2^u = 13.9, a_3^u = -8.60, a_4^u = 0, a_5^u = 8.60$$
$$a_1^d = -0.74, a_2^d = 0.74, a_3^d = -0.46, a_4^d = 0, a_5^d = 0.46$$
$$b_1^u = -39.2, b_2^u = 39.2, b_3^u = -24.4, b_4^u = 0.1, b_5^u = 24.4$$
$$b_1^d = -2.46, b_2^d = 2.46, b_3^d = -1.51, b_4^d = 0.1, b_5^d = 1.51$$

4   Conclusion

This paper describes the active free space optics system and its alignment method.

The authors design a bilateral FSO apparatus using regulation lens mounted on a voice coil motors as well as dedicated positioning photodiode sets. A prototype of the active FSO system is manufactured and its fundamental characteristics are experimentally evaluated to confirm that the apparatus satisfies required features of containing sensors and actuators.

Alignment strategy of active free space transmission is investigated next. Distributed control system is established for bilateral remote laser transmission. Alignment procedure is designed with regard to line connection status; transient phase and steady phase. The procedure consists of both off-line laser beam distribution presumption for coarse adjustment and feedback tracking control of laser beam for fine adjustment.

Collaborative negotiation process is established for exploratory alignment in the transient phase. Off-line alignment method is also realized for coarse adjustment based on modified Gaussian beam optics. Simultaneous equations with respect to positioning photodiodes analytically determine the optical axis position of laser beam.

After all, the proposed system contributes to connect remote free space transmission line only by transmission laser beam, and to maintain high-quality broadband communication.

Acknowledgments. Research partially supported by Strategic Information and Communications R&D Promotion Program (SCOPE) of Ministry of Internal Affairs and Communications, Japan.

References

1. Pratt, W.K.: Laser Communication Systems, p. 196. Wiley, New York (1969)
2. Ueno, Y., Nagata, R.: An optical communication system using envelope modulation. IEEE Trans. Commun. **20**(4), 813 (1972)
3. Willebrand, H., Ghuman, B.S.: Free-Space Optics: Enabling Optical Connectivity in Today's Networks. Sams Publishing, Indianapolis (1999)
4. Nykolak, G., et al.: Update on 4x2.5 Gb/s, 4.4 km free-space optical communications link: availability and scintillation performance. In: Optical Wireless Communications II, Proceedings of the SPIE, vol. 3850, pp. 11–19 (1999)
5. Dodley, J.P., et al.: Free space optical technology and distribution architecture for broadband metro and local services. In: Optical Wireless Communications III, Proceedings of the SPIE, vol. 4214, pp. 72–85 (2000)
6. Vitasek, J., et al.: Misalignment loss of free space optic link. In: 16th International Conference on Transparent Optical Networks, pp. 1–5 (2014)
7. Dubey, S., Kumar, S., Mishra, R.: Simulation and performance evaluation of free space optic transmission system. In: International Conference on Computing for Sustainable Global Development, pp. 850–855 (2014)
8. Wang, Q., Nguyen, T., Wang, A.X.: Channel capacity optimization for an integrated wi-fi and free-space optic communication system. In: 17th ACM International Conference on Modeling, Analysis and Simulation of Wireless and Mobile Systems, pp. 327–330 (2014)
9. Kaur, P., Jain, V.K., Kar, S.: Capacity of free space optical links with spatial diversity and aperture averaging. In: 27th Biennial Symposium on Communications, pp. 14–18 (2014)
10. Tsujimura, T., Yoshida, K.: Active free space optics systems for ubiquitous user networks. In: 2004 Conferernce on Optoelectronic and Microelectronic Materials and Devices (2004)
11. Tanaka, K., Tsujimura, T., Yoshida, K., Katayama, K., Azuma, Y.: Frame-loss-free optical line switching system for in-service optical network. J. Lightwave Technol. **28**, 539–546 (2009)
12. Yoshida, K., Tsujimura, T.: Seamless transmission between single-mode optical fibers using free space optics system. SICE J. Control Meas. Syst. Integr. **3**(2), 94–100 (2010)
13. Yoshida, K., Tanaka, K., Tsujimura, T., Azuma, Y.: Assisted focus adjustment for free space optics system coupling single-mode optical fibers. IEEE Trans. Ind. Electron. **60**, 5306–5314 (2013)
14. Shimada, Y., Tashiro, Y., Yoshida, K., Izumi, K., Tsujimura, T.: Initial alignment method for free space optics laser beam. J. Appl. Phys. **55**, 8S3–08RB08 (2016)

Three-Dimensional Motion Tracking System for Extracting Spatial Movement Pattern of Small Fishes

Akira Senda[1], Miki Takagi[1(✉)], Hiroyoshi Miwa[2], and Eiji Watanabe[3]

[1] School of Science and Technology, Kwansei Gakuin University,
2-1 Gakuen, Sanda-shi, Hyogo, Japan
aparou0423@gmail.com, fzr46416@kwansei.ac.jp
[2] Graduate School of Science and Technology,
Kwansei Gakuin University, 2-1 Gakuen, Sanda-shi, Hyogo, Japan
miwa@kwansei.ac.jp
[3] National Institute for Basic Biology,
38 Saigou-naka, Meidaiji-cho, Okazaki-shi, Aichi, Japan
eijwat@gmail.com

Abstract. The mathematical properties of the spatial movement patterns of animals, humans, and insects have gradually become clear in recent years. Motion tracking is essentially necessary for the study of the spatial movement patterns. GPS telemetry is often used for large mammals, birds, and humans. However, it is difficult to track the migration paths of insects and small fish by using GPS telemetry. When the region of an object's movement is restricted, we can record the movement of the object by the video instead of GPS telemetry, but we must determine the position coordinates of objects from a video for the study of the spatial movement patterns. If this motion tracking can be executed not by manually but automatically, we can obtain and analyze Big data on motion. In this paper, we develop a system for motion tracking of one or more small fishes in an aquarium. This system solves the difficulties such as the overlap of fishes, the ghost image of the reflection, and outputs the trajectory of the 3-dimensional coordinates. Furthermore, we apply this system to actual videos and show that the detection of active and inactive phases is possible and that the spatial movement pattern follows Levy walk.

1 Introduction

The mathematical properties of the spatial movement patterns of animals, humans, and insects have gradually become clear in recent years [1–7]. For example, it is known that the human spatial movement pattern follows Lévy Walk (LW). In particular, in the case of the spatial dimension $d = 2$, the probability of a linearly moving distance to the next destination follows the power law $P_{LW}(l) = \mathcal{O}(l^{-(1+\beta)})$, where β $(0 < \beta \leq 2)$ is a power exponent in the moving distance [8]. The value of exponent β is 0.59 [1] for banknote movement data;

© Springer Nature Switzerland AG 2019
F. Xhafa et al. (Eds.): INCoS 2018, LNDECT 23, pp. 158–169, 2019.
https://doi.org/10.1007/978-3-319-98557-2_15

for the human mobility data, β is a value from 0.75 to 1.20 based on GPS data [2], a value from 0.75 ± 0.15 based on GSM data [3].

It is also known that the distribution of frequency that two persons meet each other follows the power law [9]. As the mobility model which is consistent with this property, Homesick Lévy Walk (HLW) model that a person moves based on Lévy Walk and it returns to a specified point (base) with a certain probability (return rate) has been proposed [9]. Similarly, the probability distribution of the time interval (Inter-Contact Time (ICT)) meeting the same person follows the power law and there is a cut-off at which the exponential decay attenuates [10–13]. The probability distribution of information collection time, which is the important measure in the store-carry-forward routing method in the research area of the information network, also has a similar property [14,15].

Movement patterns relating not only to humans but also other organisms are being investigated. For example, in the reference [16], vertical motion pattern of tuna was analyzed; data loggers were attached to tuna and the moving depth was collected at one second interval, and it was shown that it attenuated exponentially and was close to the random walk. In the reference [17], it was shown that the change of the activity amount before and after medicine was administered to a mouse was examined. The activity amount was measured by the change of the distance of the movement.

Motion tracking is essentially necessary for the study of the spatial movement patterns and the technology for the motion tracking is extensively required (ex. [18,19]). GPS telemetry is often used for large mammals, birds, and humans. However, it is difficult to track the migration paths of insects and small fish. When the region of an object's movement is restricted, we can record the movement of the object by the video instead of GPS telemetry, but we must determine the position coordinates of objects from a video for the study of the spatial movement patterns. If this motion tracking can be executed not by manually but automatically, we can obtain and analyze Big data on motion.

In this paper, we develop a system for motion tracking of one or more small fishes (Oryzias latipes) in an aquarium. This system solves the difficulties such as the overlap of fishes, the ghost image of the reflection, and outputs the trajectory of the 3-dimensional coordinates. Furthermore, we apply this system to actual videos and show that the detection of active and inactive phases is possible and that the spatial movement pattern follows Lévy walk.

2   Method for Three-Dimensional Motion Tracking for Extracting Spatial Movement Pattern of Small Fishes

In this section, we describe a method for motion tracking of one or more small fishes in an aquarium. There are many systems to track motion of a moving object in a video if the shooting condition is simple and the object and the movement route of the object can be easily extract; however, in general, the real shooting condition is not simple at all. Indeed, there are many difficulties as follows: the reflection of image of fishes on water surface, bottom surface, wall surface;

the shadow of fish; feces; bait; the overlapping due to denseness of fishes; the large change in shape upon direction change; the diversity of moving style and posture. Therefore, we cannot detect fishes using only a previous method such as a background subtraction method (or an inter-frame difference method) detecting a moving object from the difference between the current frame and a reference frame (background image).

We show the example of the video images in Fig. 1. We use two videos observing two or more fishes moving in an aquarium from the two directions at the same time; video shot from above the aquarium (Fig. 1(a)) and video shot from the side of the aquarium (Fig. 1(b)).

(a) Video image from above aquarium   (b) Video image from side of aquarium

Fig. 1. Example of video image

We describe the basic idea of the proposed method. First, we extract the candidates of the region of each fish from the video shot from above the aquarium by using the method described below (Fig. 2(a) and (b)). Then, we calculate the center of the image of the fish for each frame in the video and record the trajectory of the xz-coordinates of the center of each fish from the entire video. Similarly, we extract the trajectory of the xy-coordinates of the center of each fish from the entire video shot from the side of the aquarium (Fig. 2(c) and (d)). Next, we combine these trajectories and calculate the three-dimensional trajectory of the xyz-coordinates of each fish (Fig. 2(e)).

We describe the method of extracting the candidates of the region of each fish from the video from above the aquarium. First, we convert the video image to a grayscale image. A grayscale image is an image in which the brightness is represented by white, black, and gray 254 gray scales. The regions of fishes are extracted by using a threshold value according to the brightness of the body of the fishes in the video in order to distinguish the image of the real fishes from

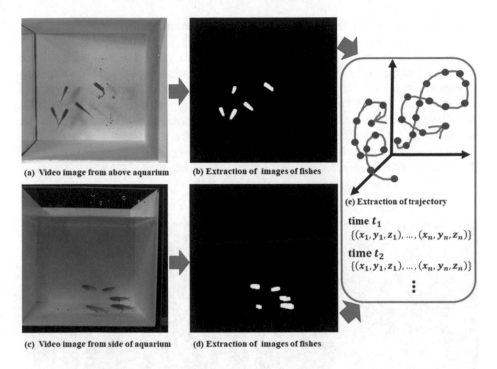

(a) Video image from above aquarium   (b) Extraction of images of fishes

(e) Extraction of trajectory

time t_1
$\{(x_1, y_1, z_1), ..., (x_n, y_n, z_n)\}$

time t_2
$\{(x_1, y_1, z_1), ..., (x_n, y_n, z_n)\}$

(c) Video image from side of aquarium   (d) Extraction of images of fishes

Fig. 2. Method for motion tracking of one or more small fishes in an aquarium

the reflection and the shadow, and then we convert the grayscale image to a binary image of white and black, where the regions of the fishes are white. Thus, we can extract the objects of the candidates of the fishes.

Next, we apply the contraction and expansion processing to the image. The contraction processing is a process of removing the pixels at the boundary between the object and the background by one pixel and is a technique used for removing noise such as small points; the expansion processing is a process of increasing the pixels of the boundary by one pixel and is a technique used for filling small holes and small irregularities. Even if the image of a fish is not connected, since the divided objects are not far apart, it can be repaired by the expansion processing. Since the feces and the bait of the fishes were often extracted at small points, it can be removed by the contraction processing. Thus, we repair the image and extract the objects, each of which is a fish.

Next, we calculate the center of the image of each object. We define the coordinates of the center as the position of the fish.

In each frame, we give each object the unique label. When the center of an object in a frame and the center of an object in the successive frame is the closest, we presume that these objects are the same. We give the same label to the same object in all frames.

Thus, we extract the trajectory of the xz-coordinates of the center of each fish from the entire video.

Similarly, we convert the video image from the side of the aquarium to a grayscale image; the regions of fishes are extracted; and then we convert the grayscale image to a binary image of white and black. Next, we apply the image the contraction/expansion processing. We repair the image and extract the objects, each of which is a fish. Then, we extract the trajectory of the xy-coordinates of the center of each fish from the entire video.

Since we have the trajectories of the xz-coordinates and the xy-coordinates of the center of each fish, we can make the xyz-coordinates of each fish. We show the example of the labeling to the fishes in Fig. 3.

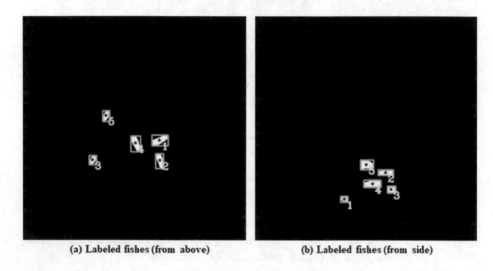

(a) Labeled fishes (from above)   (b) Labeled fishes (from side)

Fig. 3. Labeling of fishes

3   Experiments

We implemented the method for three-dimensional motion tracking for extracting spatial movement pattern of small fishes in Sect. 2, and we developed the motion tracking system for small fishes in an aquarium. First, we evaluate the performance of the method and the system in Sect. 3.1, and we analyze the spatial movement pattern of small fishes (Oryzias latipes) in an aquarium by using this system in Sect. 3.2

We use the videos shot from above and from the side of the aquarium; we show some video images in Fig. 4. The number of fishes is five. The length of the videos is about 51 s; the frame rate is 30 fps; the videos includes 1536 frames.

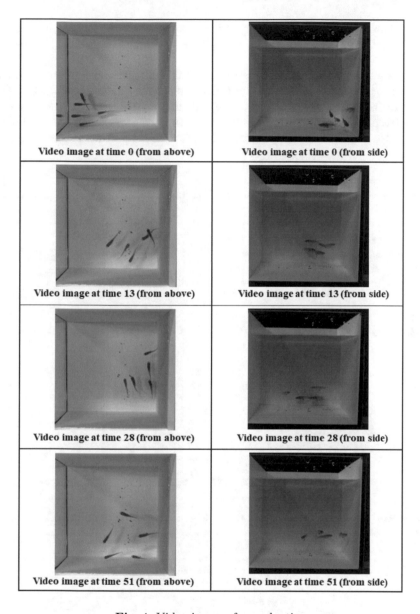

Video image at time 0 (from above)	Video image at time 0 (from side)
Video image at time 13 (from above)	Video image at time 13 (from side)
Video image at time 28 (from above)	Video image at time 28 (from side)
Video image at time 51 (from above)	Video image at time 51 (from side)

Fig. 4. Video images for evaluation

3.1   Performance Evaluation of Motion Tracking Method

The threshold of the brightness to extract the regions of fishes is determined in a heuristic manner so that the split of the image of a fish rarely occurs and that the reflection and the shadow are rarely detected.

We show the trajectories of the fishes extracted by the system in Figs. 5, 7, 9, 11 and 13 from the above the aquarium; Figs. 6, 8, 10, 12 and 14 from the side of the aquarium.

We evaluated the proposed method by the number of the errors of labeling to fishes. When the proposed method is applied to the video from the above the aquarium (resp. from the side of the aquarium), the labeling to the fishes in 4 frames (resp. 53) of 1536 frames failed. The number of the errors of the labeling in the video from the side of the aquarium is more than that in the video from

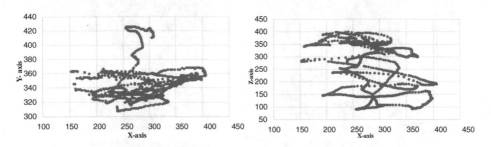

Fig. 5. Trajectory of fish #1 (from side)

Fig. 6. Trajectory of fish #1 (from above)

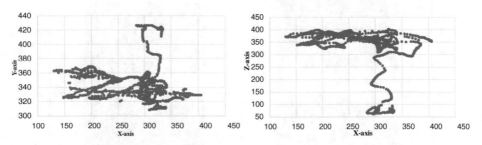

Fig. 7. Trajectory of fish #2 (from side)

Fig. 8. Trajectory of fish #2 (from above)

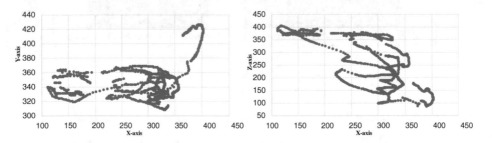

Fig. 9. Trajectory of fish #3 (from side)

Fig. 10. Trajectory of fish #3 (from above)

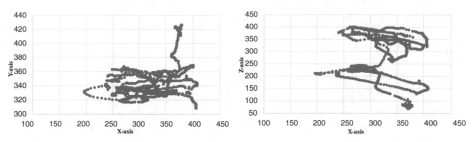

Fig. 11. Trajectory of fish #4 (from side)

Fig. 12. Trajectory of fish #4 (from above)

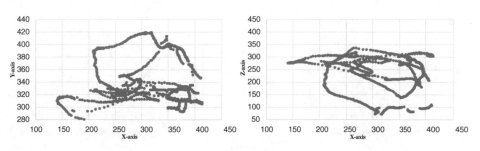

Fig. 13. Trajectory of fish #5 (from side)

Fig. 14. Trajectory of fish #5 (from above)

the above the aquarium, because the fishes often move at the same depth and the overlapping of the fishes often occurs.

The number of the errors by the proposed method is 4 frames of 1536 frames (0.3%) and 53 frames of 1536 frames (3.5%) and, since these numbers are not large, these errors can be corrected manually. Consequently, the motion tracking by the proposed method is useful from a practical viewpoint.

3.2   Analysis of Spatial Movement Pattern of Small Fish

In this section, we investigate whether the spatial movement pattern follows Lévy Walk. Furthermore, we investigate whether the active period and the inactive period can be detected.

We show the frequency distribution of moving distances per one second by double logarithmic chart in Figs. 15, 16, 17, 18 and 19.

Since the fishes move in the aquarium, the moving distance is restricted, and the video recording time is not long; however, these results indicate that the distribution follows the power law. Therefore, the spatial movement pattern follows Lévy Walk.

Next, we investigated whether the system can detect active and inactive phases.

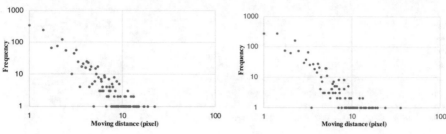

Fig. 15. Distribution of distance of fish #1

Fig. 16. Distribution of distance of fish #2

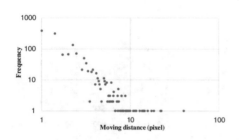

Fig. 17. Distribution of distance of fish #3

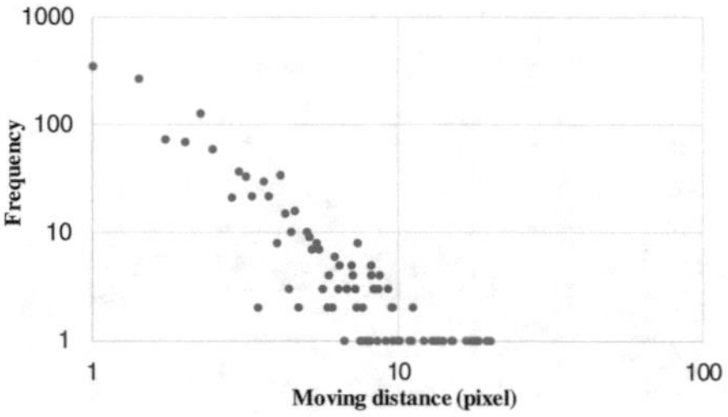

Fig. 18. Distribution of distance of fish #4

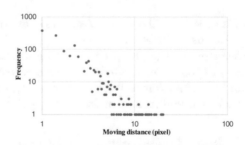

Fig. 19. Distribution of distance of fish #5

In the reference [17], the change of the movement amount of a mouse per unit time is observed to detect the active and inactive phases. Similarly, we observe the change of the movement amount of the fishes per one second by the proposed system (Figs. 20, 21, 22, 23 and 24).

In these figures, we can observe the active period and the inactive period. For example, in Fig. 20, the active period of fish #1 is about from 25 s to 29 s. If we determine the mathematical definition of the active period, we can detect the active period based on the definition. Thus, the proposed system can detect the active period and the inactive period.

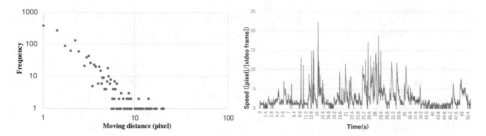

Fig. 20. Time-series change of speed of fish #1

Fig. 21. Time-series change of speed of fish #2

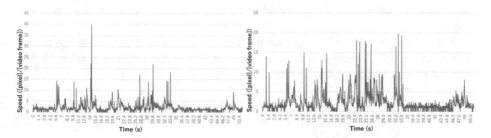

Fig. 22. Time-series change of speed of fish #3

Fig. 23. Time-series change of speed of fish #4

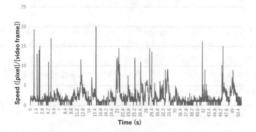

Fig. 24. Time-series change of speed of fish #5

4 Conclusions

Motion tracking is essentially necessary for the study of the spatial movement patterns, However, it is difficult to track the migration paths of small fishes, because GPS telemetry cannot be used. When the region of an object's movement is restricted, we can record the movement of the object by the video instead of GPS telemetry, but we must determine the position coordinates of objects from a video for the study of the spatial movement patterns. If this motion tracking can be executed not by manually but automatically, we can obtain and analyze Big data on motion.

In this paper, we proposed the method for motion tracking of one or more small fishes in an aquarium and developed the motion tracking system. This system solves the difficulties such as the overlap of fishes, the ghost image of the reflection, and outputs the trajectory of the 3-dimensional coordinates. We applied this system to the actual video and evaluated the performance of the system. As a result, the error rate of labeling to fishes is from 0.3% to 3.5%, which is small and the motion tracking by the proposed method is useful from a practical viewpoint. Furthermore, from the data of the trajectories of the fishes, we found that the spatial movement pattern follows Lévy Walk. In addition, we found that the detection of active and inactive phases is possible by the proposed system.

The developed system cannot detect the motion in real time. For the future work, it remains to design a method to realize a motion tracking in semi real time by accelerating the image processing.

Acknowledgement. This work was partially supported by the Japan Society for the Promotion of Science through Grants-in-Aid for Scientific Research (B) (17H01742) and JST CREST JPMJCR1402.

References

1. Brockmann, D., et al.: The scaling laws of human travel. Nature **439**, 462–465 (2006)
2. Rhee, I., et al.: On the levy-walk nature of human mobility: do humans walk like monkey? In: Proceedings of the 27th IEEE International Conference on Computer Communications, pp. 924–932 (2008)
3. González, M.C., et al.: Understanding individual human mobility patterns. Nature **453**, 779–782 (2008)
4. Viswanathan, G.M., et al.: Lévy flight search patterns of wandering albatrosses. Nature **381**, 413–415 (1996)
5. Viswanathan, G.M., et al.: Optimizing the success of random searches. Nature **401**, 911–914 (1999)
6. Edwards, A.M., et al.: Revisiting Lévy flight search patterns of wandering albatrosses, bumblebees and deer. Nature **449**, 1044–1048 (2007)
7. Sims, D.W., et al.: Scaling laws of marine predator search behaviour. Nature **451**, 1098–1102 (2008)
8. ben-Avraham, D., Havlin, S.: Diffusion and Reactions in Fractals and Disordered Systems. Cambridge University Press, Cambridge (2000)
9. Fujihra, A., Miwa, H.: Homesick levy walk: a mobility model having Ichi-go Ichi-e and scale-free properties of human encounters. In: Proceedings of 2014 IEEE 38th Annual Computer Software and Applications Conference (COMPSAC), Vasteras, Sweden, 21–25 September (2014)
10. Chaintreau, A., et al.: Impact of human mobility on opportunistic forwarding algorithms. IEEE Trans. Mobile Comput. **6**(6), 606–620 (2007)

11. La, C.-A., et al.: Characterizing user mobility in second life. In: Proceedings of the First Workshop on Online Social Networks (2008)
12. Karagiannis, T.: Power law and exponential decay of intercontact times between mobile devices. IEEE Trans. Mobile Comput. **9**(10), 1377–1390 (2010)
13. Galati, A., Greenhalgh, G.: Human mobility in shopping mall environments. In: Proceedings of the Second International Workshop on Mobile Opportunistic Networking (2010)
14. Fujihara, A., Miwa, H.: Efficiency analysis on an information sharing process with randomly moving mobile sensors. In: International Symposium on Applications and the Internet, pp. 241–244 (2008)
15. Fujihara, A., Miwa, H.: Scaling relations of data gathering times in an epidemically data sharing system with opportunistically communicating mobile sensors. In: Intelligent Networking, Collaborative Systems and Applications, pp. 193–206 (2010)
16. Kadota, M., Komeyama, K., Furukawa, S., Kawabe, R.: Analysis of the vertical movement of pacific bluefin tunas as non-levy random walk. Fisheries Eng. **50**(1), 7–17 (2013)
17. Shoji, H., Nakatomi, Y., Yokoyama, C., Hanai, K.: New index based on the physical separation of motion into three categories for characterizing the effect of cocaine in mice. J. Theor. Biol. **333**, 68–77 (2013). https://doi.org/10.1016/j.jtbi.2013.05.008
18. Ahmad, N., Imran, M., Khursheed, K., Lawal, N., O'Nils, M., Oelmann, B.: Model, placement optimisation and verification of a sky surveillance visual sensor network. Int. J. Space Based Situated Comput. **3**(3), 125–135 (2013)
19. Takano, K., Li, K.F.: A multimedia tennis instruction system: tracking and classifying swing motions. Int. J. Space Based Situated Comput. **3**(3), 155–168 (2013)

Proposing a System for Collaborative Traffic Information Gathering and Sharing Incentivized by Blockchain Technology

Akihiro Fujihara[✉]

Chiba Institute of Technology, 2-17-1 Tsudanuma, Narashino, Chiba 275-0016, Japan
akihiro.fujihara@p.chibakoudai.jp

Abstract. In the context of transportation technologies for smart city, recently, self-driving and connected cars have been studied with increased attention to realize more efficient and safer transportation. On the other hand, applications of blockchain and its Layer 2 technologies to IoT and cyber-physical systems have been considered for achieving sharing economy. The blockchain explores new incentive-based value exchange system using reliable data structure with resistance to information tampering despite the absence of reliable central control. In this paper, we propose a distributed system for collaborative traffic information gathering and sharing incentivized by blockchain technology. I consider that beacon devices are deployed along road segments by many and unspecified users. These devices collaboratively estimate traffic count and status, and also detect traffic jam and accidents to gather and share them coordinated by the incentive from its digital currency. After introducing an idea of how to realize this system, I demonstrate numerical simulations to evaluate some preliminary results for how the proposed system works and responds to real-time traffic conditions, such as traffic accident and jam in the road segment.

1 Introduction

After its potential and the speculative boom of Bitcoin [1] and other cryptocurrencies, the blockchain technology have been attracted with increasing attention [2–7]. Blockchain does not need to rely on any authority and reliable bystander, which enables to not only publicly issue currency, but also an existence proof of transactions or other digital information. The blockchain consists of block and chain: each block includes elements of transactions or information and the block is linked with the previous one like a chain to create one-dimensional ordered data structure. For instance, the blocks of Bitcoin includes transactions and newly issued coins. They are linked using the digest (hashed value) of the previous block. The blockchain usually maintains on a P2P network. The blocks also include a value called nonce. To add a new block into the blockchain to issue cryptocurrency, a miner must find a digest value of the block less than

© Springer Nature Switzerland AG 2019
F. Xhafa et al. (Eds.): INCoS 2018, LNDECT 23, pp. 170–182, 2019.
https://doi.org/10.1007/978-3-319-98557-2_16

a given threshold value called a target. The target is inversely proportional to difficulty. By changing the difficulty, the blockchain system controls a controlled computational effort to generate the new block once per about ten minutes. This mechanism is called Proof of Work (PoW). By PoW, it is difficult to not only add a block into the blockchain, but also to manipulate it. In order to manipulate the blockchain, they need to recalculate all the digests of blocks after the manupilated block less than the target. This recalculation *is* possible, but practically it is very hard because the new block is generated after competing with all the other computational power that all the peers have to find the correct nonce. The blockchain selects the longest chain generated by the most energy-consumed computation as the unique and correct one. This is the reason why the blockchain has a reliable data structure.

The blockchain technologies are generally categorized into two types: public and private. But, the private blockchain technologies assume an implicit leader peer as the reliable authority or bystander. Therefore, the public blockchain technologies are the only one that have a potential to create new applications that never exist before. Crowdsourcing is a promising application by blockchain. There exist some applications of crowdsourcing that has been considered. For example, FixMyStreet [8] is a representative application. The citizen users report broken roads nearby they live, therefore the government and community can gather and find unlawful or abnormal conditions around the road like a bump and unauthorized dumping collaboratively. But, the service have a weak incentive to report them. The report does not necessarily motivate the government and community to improve the situations because of cost issue for fixing the road conditions. Since the blockchain can issue cryptocurrency for the incentive, the combination of crowdsourcing and blockchain technologies has a potential solution to the problem.

On the other hand, in the recent years, automatic driving and connected car technologies have also been attracted attention to acheive smart transportation systems in smart city. One of the most important aims in smart transportation system is to reduce traffic congestion by controlling automatic driving and connected cars. To this end, it is important to consider how to measure traffic count and status on as many places as possible.

In Japan, ETC 2.0 [9] has started in 2014 to monitor city traffic conditions. Electronic toll collection (ETC) system originally started for tolling at the entrance and exit of express highway and tollway. But, ETC 2.0 also supports providing precise information about wide-area traffic congestion, suggesting dynamic route navigation for avoiding congestd routes, delivering useful information in service areas, and reporting disaster information at the time of disaster. It uses 5.8 GHz Dedicated Short Range Communication (DSRC) to provide cars traffic information using roadside equipments called ITS spots. Although ITS spots are more than 1,700 places and the number is gradually expanding throughout Japan, they are usually around urban expressways. Because the cost for deploying ITS spots is very high, the governmental agencies have the initiative to deploy them.

Recently, bus location services have also been in practical use to monitor where city buses are running using Wi-Fi [10]. Each bus has smartphone to emit Wi-Fi radiowave and roadside beacon modules detect Wi-Fi to monitor where buses are. This system is for location-based service, but it is similar to the ETC system. Car proving data using GPS sensor and cellular networks has also been in practical use within coveraged areas. Waze [11] is a successful crowdsourcing platform for general citizens to gather and share traffic information. Wazers try to report traffic congestion and road accidents by themselves while they are driving using smartphones. Because most information is reported manually and not so many people check the quality of the reports, it is difficult to remove misinformation and outdated information. Furthermore, if there is no wazer, no traffic jam and accident are reported. As far as I know, there is no idea for a crowdsourcing platform to reliably and automatically generate and share traffic information by general citizens to monitor traffic conditions in more road segments for traffic control. This is a big challenge to acheive the smart city.

In this paper, I propose a system for collaborative traffic information gathering and sharing incentivized by blockchain technology to generate reliable traffic information automatically. In the system, roadside beacons are deployed for each road segment and Wi-Fi signals emitted by cars passing in the road segment are scanned by the beacons. By doing this, the beacons can record car IDs to know which car was passed. These beacons also create blockchain which contains traffic information, such as car IDs and a status of the road segment in addition to issuing its cryptocurrency for supposed users like car and beacon owners. Because the blockchain is competitively and collaboratively generated by the interaction between cars and beacons, resulting traffic information is highly reliable if the number of cars and beacons are sufficiently large. Moreover, because the blockchain is open for public, anyone who own his or her car or beacon can join in the crowdsourcing platform to gather and share traffic information. I assume that the system provides its cryptocurrency to share a successful miner (a beacon owner) and car ID holders (car owners) whose car ID is recorded in the mined block as the incentive for generating traffic information. This incentive mechanism is expected to increase the number of users joining in the system and grows to build a traffic observation network by general citizens. I perform numerical simulations of car traffic and blockchain generation to evaluate the performance of the proposed system. I find that the blockchain successfully collect information of traffic volume. Moreover, the system can also detect abnormal road status such as traffic congestion and accident by observing whether the blockchain tends to be forked. Each road segment manages its unique blockchain, so the system can detect traffic volume and which road is congested or in accident automatically. Because forked blockchain is unstable to obtain the incentive, car owners avoid coming to congested roads, which results in the decrease of traffic volume. Therefore, the system uses these properties of blockchain technology fully for traffic control.

2   Related Works

Proximity-Based Service (PBS) is proposed as a concept of local production of information for local consumption [12]. For example, bluetooth beacons are used to wirelessly advertise information to nearby human-carried mobile devices, such as iBeacon. A system which consists of many bluetooth beacons deployed along corridors in a building is useful for not only location-based services without the help of GPS sensor, but also route navigation [13]. By saving the history of bluetooth scan providing information for location detection, an advertisement beacon located nearby the intersection provides the next direction to go if the system knows where a user want to go beforehand.

Smart Street Solution [14] is also proposed to provide useful information on their neighborhoood to cars and pedestrians using digital signage placed around the roadside. At the time of disaster, the signage shows disaster information and evacuation route to nearby refuges. However, in this solution, the system is managed by reliable authority or bystander to work. It may be difficult for general citizens to manage digital signage in public space.

Disaster evacuation guidance using opportunistic networking has been proposed as a disaster evacuation method for disaster victims evacuating to a nearby refuge collaboratively with the help of human-carried mobile devices like a smartphone and wearable device [15–19]. Each evacuee starts evacuation following a shortest-path-based evacuation route guided by his or her mobile device. If the evacuee encounters an impassable road by, for example, fire, building collapse, congestion, and other reasons, the evacuee record the road status manually or automatically using sensors to the device. Then, the device recalculates an alternative route to the refuge. When a couple of evacuees having the device happens to encounter, the device share the disaster information using opportunistic networking to reroute a path to the refuge. In this study, average and maximum evacuation times are evaluated by numerical simulations. The result shows that the guidance contributes to decrease both evacuation times well even if the traffic congestion by evacuees are taken into consideration. Furthermore, disaster information in affected areas are rapidly gathered to the refuges using the mobility of evacuees. Therefore, gathered information can be easily integrated by connecting refuges by disaster tolerant networking. This is a crowdsourcing of disaster information. However, it is known that at the time of disaster misinformation and fake news by unreliable sources can be spread easily. In the proposed system, however, no one can pledge that gathered disaster information is really true and when the information is outdated. This problem may be the cause of confusion for evacuees to evacuate properly.

3   System Proposal

I explain the system for collaborative traffic information gathering and sharing incentivized by blockchain technology to try to realize reliable traffic inforamtion generation only by general citizens without assuming any reliable authority or bystander.

3.1   Road Segment and Roadside Beacons

Firstly, I define a road segment as a continuous domain on the road network separated by the intersections as shown in Fig. 1. Cars basically go through the road segment from left to right or from right to left as shown in Fig. 2. Although they sometimes stop on a road segment, most of them go through, but no cars can vanish suddenly. On this road segment, a sufficiently large number of roadside beacons are deployed by anonimous citizens living around the segment, such as deploying Wi-Fi routers these days. In general, the beacons are owned by multiple citizen managers who do not necessarily have reliability to anyone. Since Wi-Fi is handy compared to DSRC, the beacons are assumed to scan devices emitting Wi-Fi radiowave. For example, the devices can be smartphones, Wi-Fi routers, and a car navigation system moving with car mobility. I assume that the beacons basically receive Wi-Fi radiowave without loss of information. If a beacon receives car IDs by Wi-Fi scan, it saves them as the traffic information in its memory pool.

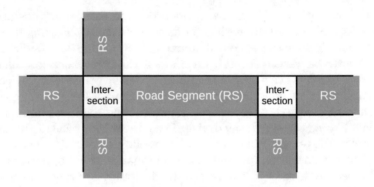

Fig. 1. Road segments on a road network

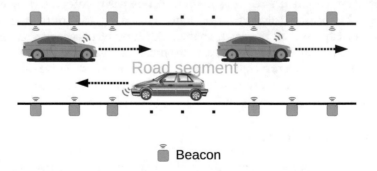

Fig. 2. Roadside beacons and cars on a road segment

3.2   How to Use Blockchain and Incentive Mechanism

The Bitcoin blockchain uses PoW, which is originally proposed in Hashcash by Back [20]. PoW is a mechanism to prove to a group of users that a certain amount of computational resources has been utilized for a period of time. For Bitcoin, a PoW function

$$F_d : (c, x) \rightarrow SHA256^2(c + x) < 2^{256}/d \in \{True, False\} \tag{1}$$

is used, where d, c, and x is called difficulty, challenge, and nonce, respectively. Double SHA256 is used for the hash function, and if the return hashed value is less than the target value $2^{256}/d$, the PoW function returns $True$. It is computationally difficult, but feasible to find x such that $F_d(c, x) = True$ for fixed parameters d and c, which is called PoW. For Bitcoin, full node users perform PoW to compete to find an appropriate nonce x and the winner who find the nonce has a right to make a block and also get a new coinbase. For the proposed system, PoW is supposed to be done by roadside beacons to create a block which contains car IDs and road status (for example, normal or congested) in the memory pools. The procedure of PoW in a beacon is listed in Algorithm 1.

A roadside beacon which find the correct nonce is allowed to create the next block containing car IDs gone through the road segment and the coinbase is shared with the winner of PoW and car ID holders for the incentive for creating the traffic information. The created block is automatically connected to the blockchain by every roadside beacon after checking PoW is correctly performed. The procedure of checking a candidate of the next block is shown in Algorithm 2. The proposed structure of block and block header is shown in Tables 1 and 2. In this proposal, the blockchain is managed for each road segment (RS). All the blockchains in the system is forked from the genesis block to create RS genesis blocks as shown in Fig. 3.

Algorithm 1. Procedure of PoW in a roadside beacon
***

1: Nonce $x = 0$, Contents to be contained in the next block c, and difficulty d
2: **while** $F_d(c, x) == False$ **do**
3:     x = x+1
4: **end while**
5: a candidate of the next block b and its nonce x is broadcast within a road segment.

Algorithm 2. Procedure of checking a candidate of the next block
***

1: A candidate block b is received
2: Check the longest length of the blockchain h_{max} and the terminal blocks b_{max}
3: Check b is correctly added to one of the blockchain terminals
4: Calculate the length of the connected block h_b
5: **if** $h_b > h_{max}$ **then**
6:     $h_{max} = h_b$
7:     $b_{max} = b$
8:     Check car IDs in the new block to remove them from the memory pool.
9: **end if**

Table 1. Block structure

Field	Description
Block size	Data size of the block
Block header	See Table 2
Traffic data	Traffic count and status of the road segment
Coinbase	Newly generated coin shared with the miner and detected car users
Car IDs	List of (hashed) car IDs

Table 2. Block header

Field	Description
Version	Data size of the block
Location data	Coordinates or road segment ID
Block digest	Digest of the previous block
Car IDs' digest	Digest (Merkle root) of hashed car IDs passed in the road segment
Timestamp	Unix time when the block is generated
Difficulty	Difficulty of PoW when the block is generated
Nonce	Number used at ONCE when PoW is solved

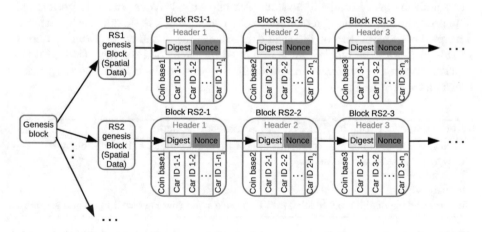

Fig. 3. Blockchain for traffic information system and its incentives for roadside beacons and cars.

Next, I explain how the road status is automatically recorded in the blockchain. As shown in Fig. 4, if a traffic accident occurs on the road, cars nearby the accident site stop to emerge traffic congestion. In this case, the blockchain makes the block contents eosiler to vary because detected car IDs tend to change for each roadside beacon. This is similar to eclipse attack to fragment the original Bitcoin network into two separate networks to fork the blockchain.

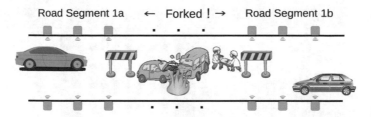

Fig. 4. When a traffic accident occurs, for example, the blockchain in the road segment is forked into two separate ones because the road becomes impassable.

Since beacons in the road segment share generated blocks with each other, they can detect whether the blockchain is forked or not. If car IDs in a block is diffirent from other blocks, the road status is estimated as congested automatically. This anormaly detection is possible to calcuate a similarity index, for example, Jaccard index of car IDs in two different blocks. After the traffic accident is recovered, cars go through the road segment smoothly. In this case, the blockchain tends not to be forked and stabilized. The roadside beacons automatically detect the road status is recovered to be normal automatically. Therefore, the proposed system actively utilize blockchain's fork for anomaly detection of road status, which is a characteristic point of the system. However, system's reliability decreases by the fork of blockchain, which is the weak point of the system. To avoid this, the condition that the number of roadside beacons are sufficiently large in the road segment is necessary to avoid attack to fork the blockchain by intention.

4 Performance Evaluation

I perform some numerical simulations of time evolution of traffic condition and the blockchain. A window of the simulator is shown in Fig. 5. In the simulations, I assume that the Wi-Fi radiowave emitted from cars can be detected by the roadside beacons without loss of car ID information. Cars arrive at the road segment that we focus in the simulation with a (Poisson) arrival rate α. As shown in Fig. 5, the blockchain is successfully generated with concensus of the new block between roadside beacons and record the traffic volume.

I also simulate the effect of traffic accident as shown in Fig. 6. In the simulation, the road becomes impassable and cars make line around the accident site. In this case, car IDs in the memory pools becomes different betweetn beacons, which results in the fork of the blockchain. In general, if the arrival rate of cars is stationary, the number of car IDs recorded in each block will also be stationary as shown in Fig. 7 on the left. However, after the traffic accident occurs, the number of car IDs becomes unstalbe and the recorded traffic count changes drastically as shown in Fig. 7 on the right. By using this instability to fork the blockchain, the system can detect the anomaly of road status.

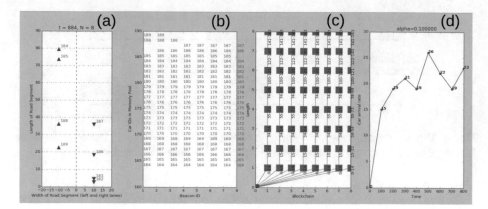

Fig. 5. Showing the traffic and blockchain simulator for evaluating the proposed system. (a) Cars are moving on two traffic lanes. The vertical dotted line indicates the separate line between two opposite lanes. The horizontal dotted lines show where roadside beacons are to scan car IDs using Wi-Fi. (b) Memory pools of roadside beacons. The numbers show the car IDs memorized in the memory pools. (c) Blockchains generated in each beacons. (d) Time evolution of the traffic volume (= the number of car IDs) contained in the mined block.

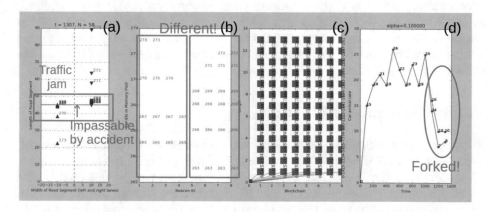

Fig. 6. Showing the simulator for evaluating the effect of impassable road. (a) The horizontail solid line shows the place where the road is impassalbe in the road segment. Cars stop in front of the line and make traffic congestion. (b) Car IDs in the memory pools are separated into two different types because of the impassable road avoids to generate the same block. (c), (d) Even though the blockchain is constantly generated, it is forked in general.

Next, I also perform numerical simulations where the traffic volume changes periodically. A result is shown in Fig. 8. The figure on the right shows the time evolution of the relation between the arrival rate of cars and the number of car IDs recorded in the blockchain. Because one of serious weak points of the

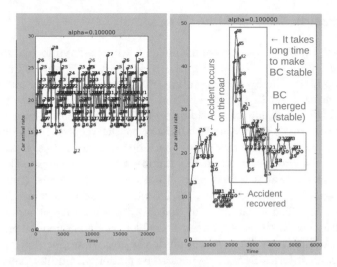

Fig. 7. Time evolution of the number of car IDs contained in the mined block when the car arrival rate is stationry (left) and that when the road becomes impassable for a while because of the traffic accident (right).

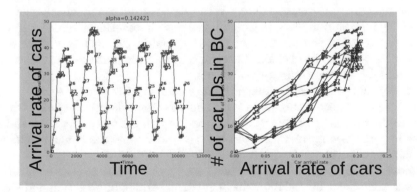

Fig. 8. Time evolution of the number of car IDs contained in the mined block when the car arrival rate is periodic (left). Time evolution of the relation between the arrival rate of cars and the number of car IDs in blockchain (right).

blockchain is delay to generate the block. For Bitcoin core, it takes ten minutes on average to generate the new block. Because of the effect of delay, simulation results shows that the peak of the number of car IDs delays one or two blocks compared to the peak of actual arrival rate of cars. Simulation data in the blockchain as shown in Fig. 8 on the right is used for the calculation. I also calculate the correlation coefficient between the arrival rate of cars and the number of car IDs. As a result, it reaches $R = 0.929223$. This result means that these values are highly correlated enough to estimate quasi-real-time traffic count using the number of car IDs recorded in the blockchain.

5   Discussion and Future Work

In this paper, I proposed a system for collaborative traffic information gathering and sharing incentivized by blockchain technology to generate reliable traffic information automatically. In the system, roadside beacons are deployed for each road segment and Wi-Fi signals emitted by cars passing in the road segment are scanned by the beacons. By doing this, the beacons can detect car IDs to record them in the blockchain. In addition to the record, the blockchain also issue its cryptocurrency for owners of roadside beacons and cars. This incentive mechanism is expected to increase the number of users joining in the system for crowdsourcing the traffic information. Because the blockchain is competitively and collaboratively generated by the interactions between cars and beacons, resulting traffic information is highly reliable if the number of cars and beacons are sufficiently large. Moreover, because the blockchain is open for public, anyone who own his or her car or beacon can join in the crowdsourcing platform to gather and share traffic information. I also performed numerical simulations of car traffic and blockchain generation to evaluate the performance of the proposed system. I found that the blockchain successfully collect information of traffic count and road status using the state of blockchain.

To realize the proposed system, there are many study to do as future works. I assume that roadside beacons detect Wi-Fi signals from the moving cars without loss of any information. But, in general, the signals sometimes loss, which results in the difference of the contents of memory pool in the roadside beacons, and the resulting fork of blockchain. It is important to simulate more realistic road conditions for the performance evaluation.

The system assumes a sufficiently large number of roadside beacons in each road segment to generate reliable traffic information. Maintaining roadside beacons needs some cost in general. It is necessary to estimate incentive and cost issues for supposed users of the system to consider some mechanism to increase the number of users.

What I finally want to consider with the proposed system is to reduce traffic congestion in a city by collaboratively gathering traffic information and sharing traffic status with cars. I did not consider this point well in this paper. An idea is that the beacons also run as API servers to provide traffic information they gathered. The information can be shared for free of charge or a user can buy the information using cryptocurrency that the sytem issues. It is important to perform agent-based simulations to show how owners of cars and beacons exchange the traffic information to reduce traffic congestion as a smart city solution. It may be interesting to control cars avoiding accident sites and congested areas by issuing cryptocurrency. To do this, showing road segments active for gathering traffic information on the map is necessary. Preparing the map using P2P technology might be a challenging task.

In this proposal, I assume the blockchain from Bitcoin core. On the other hand, similar but other type of technologies for enabling reliable information geneartion using Directed Acyclic Graphy (DAG), such as IOTA, Hashgraph, and BBc-1 have been proposed recently. By crossing the histories of multiple blockchains and other data structures, it may be possible to propose another reliable system without using PoW to collaboratively generate useful information for everyone.

The blockchain itself is developing now to solve many weak points. The delay of block generation is the fatal weak point to limit the applicability of blockchain when we consider applications. It is important to consider a method to minimize the delay to gather traffic information in real time. The most important thing to do is to implement the proposed system to do experiment in real situations.

Acknowledgements. This work was partially supported by the Japan Society for the Promotion of Science (JSPS) through KAKENHI (Grants-in-Aid for Scientific Research) Grant Numbers 17K00141 and 17H01742. The author thanks Dr. Shigeichiro Yamasaki for useful discussion and comments on the proposed system.

References

1. Nakamoto, S.: Bitcoin: a peer-to-peer electronic cash system (2008). https://satoshi.nakamotoinstitute.org/emails/cryptography/1/#selection-19.14-63.19, https://bitcoin.org/bitcoin.pdf
2. UK Government Chief Scientific Adviser, Government Office for Science, Distributed Ledger Technology: beyond block chain (2016). https://assets.publishing.service.gov.uk/government/uploads/system/uploads/attachment_data/file/492972/gs-16-1-distributed-ledger-technology.pdf
3. Narayanan, A., Bonneau, J., Felten, E.W., Miller, A., Goldfeder, S., Clark, J.: Bitcoin and Cryptocurrency Technologies: A Comprehensive Introduciton. Princeton University Press, Princeton (2016)
4. Tapscott, D., Tapscott, A.: Blockchain Revolution: How the Technology Behind Bitcoin is Changing Money, Business and the World. Portfolio Penguin, New York (2016)
5. Antonopoulos, A.M.: Mastering Bitcoin: Programming the Open Blockchain. O'Reilly & Associates Inc., USA (2017). https://github.com/bitcoinbook/bitcoinbook
6. Watternhofer, R.: The Science of the Blockchain. CreateSpace Independent Pub, USA (2016)
7. Watternhofer, R.: Distributed Ledger Technology. CreateSpace Independent Pub, USA (2017)
8. Fix My Street. https://www.fixmystreet.jp/
9. What is "ETC 2.0" ITS, Road Bureau - MLIT Ministry of Land, Infrastructure, Transport and Tourism. http://www.mlit.go.jp/road/road_e/p1_its.html#a2
10. Verification survey with the private sector for disseminating Japanese technologies for new location information system and traffic observation system for urban transport improvement in Vientiane city. http://www.openjicareport.jica.go.jp/pdf/1000028290.pdf
11. Waze, Free Community-based GPS, Maps & Traffic Navigation App. https://www.waze.com/
12. Fujihara, A.: Proximity-based service: an advanced way of extending human proximity awareness. In: Smart Sensors Networks, pp. 292–307 (2017). ScienceDirect

13. Fujihara, A., Yamagizawa, H.: Proposing an extended iBeacon system for indoor route guidance. In: 2015 International Conference on Intelligent Networking and Collaborative Systems (INCOS), pp. 31–37. IEEE (2015)

14. Panasonic's smart street solution (Japanese), Huawei Smart Street Solution for Dubai Silicon Oasis. https://www.youtube.com/watch?v=UqUGrd53U5g, https://www.youtube.com/watch?v=wvwFg36TylQ

15. Fujihara, A., Miwa, H.: Real-time disaster evacuation guidance using opportunistic communication. In: IEEE/IPSJ-SAINT 2012, pp. 326–331 (2012)

16. Fujihara, A., Miwa, H.: Effect of traffic volume in real-time disaster evacuation guidance using opportunistic communications. In: IEEE-INCoS 2012, pp. 457–462 (2012)

17. Fujihara, A., Miwa, H.: On the use of congestion information for rerouting in the disaster evacuation guidance using opportunistic communication. In: IEEE 37th Annual Computer Software and Applications Conference Workshop, ADMNET 2013, pp. 563–568 (2013)

18. Fujihara, A., Miwa, H.: Disaster evacuation guidance using opportunistic communication: the potential for opportunity-based service. In: Big Data and Internet of Things: A Roadmap for Smart Environments, Studies in Computational Intelligence, vol. 546, pp. 425–446 (2014)

19. Fujihara, A., Miwa, H.: Necessary condition for self-organized follow-me evacuation guidance using opportunistic networking. In: Intelligent Networking and Collaborative Systems (INCoS), pp. 213–220 (2014)

20. Back, A.: Hashcash - a denial of service counter-measure. http://hashcash.org/papers/hashcash.pdf, http://hashcash.org/papers/

Employment in Information and Communication Technologies in European Countries and Its Potential Determinants and Consequences

Peter Pisár[1]($\boxtimes$), Ján Huňady[1], and Peter Balco[2]

[1] Faculty of Economics, Matej Bel University in Banska Bystrica, Tajovskeho 10, 975 90 Banska Bystrica, Slovakia
{peter.pisar, jan.hunady}@umb.sk
[2] Faculty of Management UK, Odbojárov 10, P.O.BOX 95, Bratislava, Slovakia
peter.balco@fm.uniba.sk

Abstract. Information and communication technologies have important role in many different innovations and further positively affect the productivity and economic growth. Our aim was to examine examine the share of ICT sector in selected European countries and identify its potential determinants and consequences. We used mostly panel data for EU28 countries plus Norway and Switzerland and applied correlation and panel Granger causality tests. The employment in ICT was used as main examined variable. Our results suggest that there is a positive correlation between ICT skills as well as the share of people with ICT education and employment in ICT. The causal effect seems to be working in direction from education to employment. Moreover, variables capturing GDP per capita, trade openness, R&D expenditure, number of internet users, quality of regulation and political stability appear to be all positively correlated with employment in ICT and especially with employment in ICT services.

1 Introduction

The paper focuses its attention on Information and communication technologies (ICT) sector for serval reasons. ICTs in general have several positive effects on economy. They have been recognized as one of the key factors determining innovation, economic growth [1, 9, 12].

There are three main impacts of ICT on economy [5]:

- ICT industries contribute directly to productivity and growth through their own technological progress,
- ICT use improves the productivity of other factors of production,
- there are some spill-over effects' on the rest of the sectors in the economy as ICT.

Perhaps the most important effect of ICT is on productivity and economic growth. The most ICT-intensive industries appeared to experience significantly larger productivity gains than other industries [11].

© Springer Nature Switzerland AG 2019
F. Xhafa et al. (Eds.): INCoS 2018, LNDECT 23, pp. 183–194, 2019.
https://doi.org/10.1007/978-3-319-98557-2_17

Furthermore, ICT could affect economic growth both directly or indirectly. As mentioned direct effect of ITC sector on product is evident but there are several other effect. There are three channels through which ICT can affect economic growth [12]:

- fostering technology diffusion and innovation,
- enhancing the quality of decision-making by firms and households,
- increasing demand and reducing production costs, which together raises the output level.

Innovation drives economic competitiveness and sustained long-term economic growth. Hence, the positive effect of ICT [4]. As stated by Hall et al. [6] research and development (R&D) together with information and communication technology (ICT) investment have been both identified as main sources of relative innovation underperformance in Europe vis-à-vis the USA. In knowledge-based economy, ICT plays important role in many types of innovation which could increase productivity [2, 3]. According to Lall [7] ICT sector is the source of more than 50% of innovation worldwide and has an even greater role in developing countries. Thus, based on mentioned previous findings we can say that ICT sector is one of the key innovation leaders among all sectors of the economy. Hence, higher share of ICT sector on economy could mean higher innovation potential of the country.

In our paper we also examine factors potential determining the share of ICT sector. There are several factors, which could be seen as those having effect on ICT sector and its share. Ngwenyama and Morawczynski [8] argue that the deregulation is not enough for ICT sector expansion. They found that that existing there are several existing socio-economic determinants such as economic development, human capital, geography, and civil infrastructure. Especially these factors have to be taken into account when setting policy frameworks for ICT sector. Moreover, Rai and Kurnia [10] based on the example of Buthan economy found that especially foreign direct investment (FDI), policy, infrastructure and human resources affect the growth of ICT sector.

The paper aims to firstly examine the share of ICT sector in selected European countries and then identify its potential determinants and consequences.

2 Data and Methodology

As stated in the introduction, we want firstly want to examine the share of ICT sector in EU28 and two non-EU European countries. Next we identify its potential determinants and consequences. We used panel data with cross-sectional and period dimensions. The dataset consists of 28 EU member states plus Norway and Switzerland during the period 2005–2015. Together we get maximum of 330 observations. However, the missing observations pose a problem in the case of several variables. Some of the missing observations were replaced by the nearest available data from the same country. Consecutive missing observations were excluded from the sample during the analysis. In some graphs we used cross-sectional or period dimension separately. All variables used in the analysis are described in Table 1.

Table 2 shows basic descriptive statistics for variables used in the analysis.

Table 1. Short description of variables used in the analysis

Variable	Short description	Source
ICT employment	Total employment in ICT sector (% of active population)	Eurostat
ICT manufactur. employment	Employment in ICT manufacturing (% of active population)	Eurostat
ICT services employment	Employment in ICT services (% of active population)	Eurostat
ICT infrastructure	Proxy variable for ICT infrastructure - Secure Internet servers (per 1 million people)	World bank database
ICT education	Persons with ICT education on active population (%)	Eurostat
Internet users	Individuals using internet on population (%)	World bank
GDP per capita	GDP per capita in PPP	World bank
FDI inflows	Foreign direct investment, net inflow (% of GDP)	World bank
Openness	Import + Export (as % of GDP)	World bank
Political stability	Index of political stability – World bank Worldwide governance indicators	World bank
Regulations Quality	Index of regulatory quality – World bank Worldwide governance indicators	World bank
R&D expenditure	Intramural R&D expenditure (GERD) as % of GDP	Eurostat
Total tax rate	Total tax rate (% of company profit)	World bank

Source: Authors.

We firstly analysed the development of ICT employment in V4 countries during selected period and compared it to the EU28 average. Next we compared cross-sectional data for European countries. Based on the scatter plots and Pearson correlations coefficients we examine potential relationships between selected pairs of variables with the focus on the employment in ICT. Further we also used panel Granger causality test in order to identify the significance and direction of potential effect between selected variables. Despite our best effort our methodology do not allow us to make strong conclusions about the effect of each factor on ICT sector determinants due to potential problem of endogeneity. It is clear that in most cases there could be also some reveres causality arising from the share of ICT sector to selected variables. This reveres causality are to some extent tested using Granger causality test.

Table 2. Basic descriptive statistics for selected variables used in the analysis

Variable	Obs.	Mean	Std. Dev.	Min.	Max.
ICT employment	290	2,75	0,85	1,28	4,82
ICT manufactur. employment	275	0,49	0,40	0,03	1,89
ICT services employment	314	2,32	0,75	1,01	4,33
ICT infrastructure	330	637,15	673,5	5,44	3101
ICT education	321	1,03	0,45	0,21	2,65
Internet users	330	67,97	17,58	19,97	97,3
GDP per capita	330	102,41	44,39	35,00	270,0
FDI inflows	330	12,54	43,62	−58,32	451,7
Openness	330	118,1	64,66	45,61	410,2
Political stability	330	0,80	0,41	−0,47	1,59
R&D expenditure	330	1,55	0,89	0,37	3,75
Total tax rate	330	42,61	12,93	18,40	76,70

Source: Authors.

3   Results

In the first stage of our analysis we are focused on examining the variables capturing employment in ICT. Figure 1 shows the development of this employment in Visegrad 4 countries (Czech Republic, Hungary, Slovakia and Poland) during selected period from 2005 to 2015. As we can see the employment in this sector is ether stable during the time. However, there is a significant drop during the 2008 or 2009, which seems to be related to economic crisis during these years. This decrease was partly corrected during the next several years. The employment in ICT is the highest in Hungary followed by Czech Republic and Slovakia.

We further compared total employment in ICT and employment in ICT services in European countries included in the sample. As mentioned above total ICT employment can be divided into employees in ICT services and employees working in ICT manufacturing. We are focused on ICT services. However, the difference between total ICT employment and employment in ICT services represents the employment in ICT manufacturing. Data for whole ICT employment are not available in Cyprus, Ireland, Luxembourg, Netherlands and Portugal. However, as it can be seen in Fig. 2 Ireland and Luxemburg have both very high employments in ICT services. Together with Malta, Sweden and Denmark are these five countries leaders in this statistic. On the other hand ICT sector is developed significantly less in Greece, Romania, Poland and Portugal. Slovakia is in the eleventh place. So we can say that the share of ICT sector in Slovakia is rather high compared to other European countries.

In the next part of our analysis we examine potential factors determining the share of ICT sector in the country. Firstly we focused on IT skills of population in the country and its relation to Employment in ICT sector. We take to the account two different types of ICT skills. Coping or moving a file represents a rather basic IT skill. Due to the lack of available data we used only cross-sectional data in this case. On the other hand writing a code in programming language was used as a proxy for significantly more advanced IT

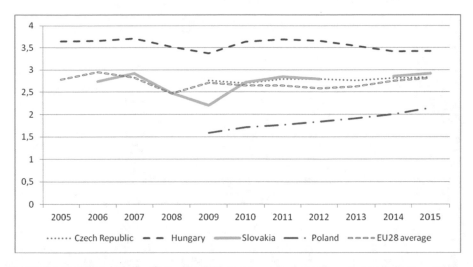

Fig. 1. Total employment in ICT sector (% of active population) in selected countries Source: Authors based on data from Eurostat database. Note: Missing lines represents missing values in the sample

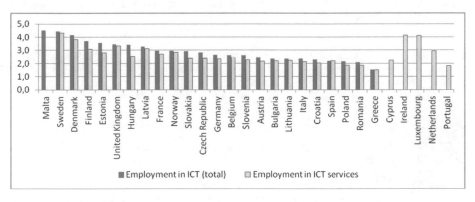

Fig. 2. Total employment in ICT sector and ICT services (% of active population) in European countries Source: Authors based on data from Eurostat database. Note: Missing columns represents missing values in the sample

skills. We used both of them in two separate Scatter plots on X axis together with the employment in ICT sector on Y axis (see Figs. 3 and 4).

As we can see basic IT skills are the most developed in population of Denmark followed by Latvia and Norway. There seems to be slightly positive correlation between basic IT skills and employment in ICT sector in the country. Similar relationship is also evident in the case of more advanced IT skills as it can be seen in Fig. 4. However, in this case the slope of the fitted line appears to be mostly affected by two top performing countries which are Finland and Sweden. Slovakia is approximately at the average of all countries in the case of basic IT skills, but performs bellow the average in advanced IT

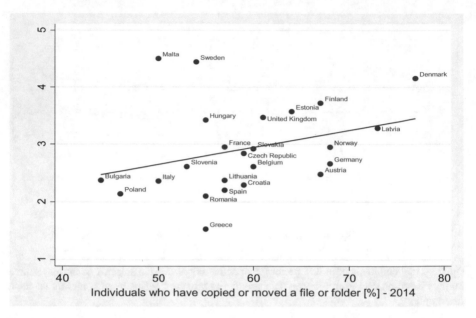

Fig. 3. Potential relationship between employment in ICT and basic IT skills Source: Authors based on data from Eurostat database

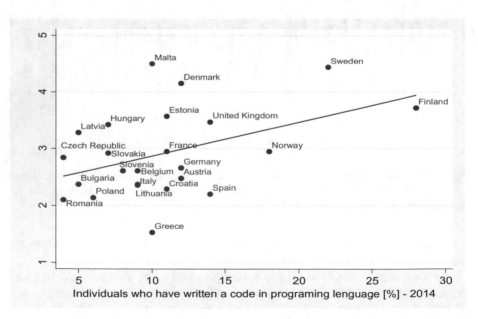

Fig. 4. Potential relationship between employment in ICT and advanced IT skills   Source: Authors based on data from Eurostat database

skills. Thus, it skills of population could be potentially important factor determining the share of ICT sector in the country. On the other hand, it is also possible that higher employment in ICT causes better IT skills in the population.

We also take to the account other potential determinants besides IT skills. One of them is ICT education, which is of course closely related to IT skills. We can assume that higher share of population with ICT education is related to higher share of ICT sector in the country. On one hand, more people with ICT education may lead to new domestic firms in this sector as well as attract more foreign ICT firms. On the other hand, higher demand for ICT employees on labour market may cause in the longer run the increase in ICT education within population. We also take to the account the level of R&D expenditure, which could be also either result of ICT industry or prerequisite for the development of ICT sector in the country. Both of these variables are shown in Fig. 5 in relation to employment in ICT sector. This time we are able to use panel data. This means that each dot on the graphs represents the observation for single European country and single year from 2005 to 2015. As it can be seen there seems to be a positive correlation in both cases. However, this correlation appears to be more intensive between intramural R&D expenditure and employment in ICT sector.

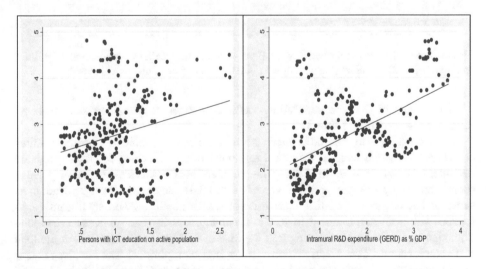

Fig. 5. Potential relationship between the share of population with ICT education (left), R&D expenditure (right) and employment in ICT   Source: Authors based on data from Eurostat database

Finally, we selected another two indicators related mostly to external economic environment, namely GDP per capita and regulatory quality in the country. We believe that both of these indicators could be positively related to the share of ICT sector. Economically more developed countries with better quality of institutions and regulations could represent appropriate external environment for the development of ICT sector. This assumption seems to be true according to our empirical data. Figure 6 shows motioned relationships based on panel data and we can see that there seem to be relatively strong positive correlation between both economic variables and employment in ICT sector. However in order to test the intensity and statistical significance of these correlations we further applied Pearson correlation coefficients.

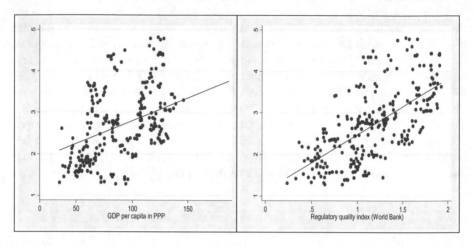

Fig. 6. Potential relationship between GDP per capita (left), quality of regulations (right) and employment in ICT sector Source: Authors based on data from Eurostat database

The results of Pearson correlation coefficients together with their statistical significance are shown in Table 3.

As we can see in this table there seem to be several relatively strong positive correlations between selected variables. As expected there is strong positive correlation between total employment in ICT and employment in ICT services. However, what is more important from our point of view is that GDP per capita, R&D expenditures, share of internet users and ICT infrastructure are all positively and significantly correlated with both employment in ICT and employment in ICT services. These relationships appear to be even stronger in the case of ICT service rather than in the total ICT employment. The positive correlations between political stability, regulatory quality and trade openness on one hand and employment in ICT on another hand are all slightly less intensive but still statistically significant at 5% level of significance. All mentioned variables have weaker positive correlations with employment in ICT manufacturing, rather than with employment in ICT services.

Thus, based on the results, we can say that countries with higher GDP per capita and higher R&D per capita have mostly higher share of ICT sector on total economy. The same is true for those having more internet users, better regulatory quality and political stability. However, we are still not able to exactly identify the cause and the consequence based on this methodology. In order to identify the direction of the causality we further applied panel Granger causality tests. The results can be seen in Table 4.

Table 3. Results of Pearson correlation coefficients for selected variables (panel data)

	Employment in sectors		
	ICT total	ICT manufacturing	ICT services
FDI inflows	0.26***	0.14**	0.19***
GDP per capita	0.44***	0.01	0.70***
ICT education	0.23***	−0.05	0.24***
ICT employment	1.00***	0.56***	0.87***
ICT infrastructure	0.51***	−0.12**	0.64***
ICT manufacturing employment	0.56***	1.00***	0.15***
ICT services employment	0.87***	0.15**	1.00***
Internet users	0.58***	0.03	0.74***
Openness	0.37***	0.28***	0.42***
Political stability	0.55***	0.42***	0.47***
Regulations quality	0.56**	0.25***	0.55***
R&D expenditure	0.67**	0.22***	0.72***
Total tax rate	0.04	0.23***	−0,22**

Source: Authors calculations
Note: */**/*** means significance at the 10%/5%/1% levels. %.

Our results suggest several statistical significant causalities in Granger sense. With respect to potential consequences of ICT sector we can say that employment in ICT services appears to have significant effect on GDP per capita and as far as we know from our previous analysis this effect spouse to be positive. However, in the case of total ICT employment we fail to prove statistically significant effect on GDP by using the Granger causality test. On the other hand, we found positive and significant effect of ICT employment on ICT infrastructure measured by number of servers per inhabitant. This is of course in line with our expectation, because higher share of ICT sector could be related to better ICT infrastructure. Moreover, there is positive and significant effect of both total ICT employment and employment in ICT manufacturing on trend openness of the economy. This effect is highly significant on both cases, but it is even more intensive for employment in ICT manufacturing. Thus we can say that countries higher share of ICT sector in the economy could positively affect the GDP per capita as well as increase the trade openness of the economy. First may be due to higher productivity and innovation in this sector and second could be the consequences of potentially higher import needs and more export orientation of firms in this sector. There is also some slightly weaker evidence for potential bi-directional effect of internet users and employment in ICT manufacturing.

With respect to potential determinants of ICT sector employment there is also some evidence that higher openness of the economy could have also some backward positive effect on the ICT employment in the country. We can also conclude, that share of people with ICT education may be the determining factor. Hence, education in ICT field of study seems to be important for increasing the employment in this sector, which could be due to either attracting foreign ICT companies or increasing the number of home-grown enterprises.

Table 4. Results of panel Granger causality tests

	F-Statistic (2 lags)	F-Statistic (3 lags)
ΔICT employment does not Granger Cause ΔGDP per capita	1.73	1.07
ΔGDP per capita does not Granger Cause ΔICT employment	0.99	1.95
ΔICT services employment does not Granger Cause ΔGDP per capita	2.77*	7.56***
ΔGDP per capita does not Granger Cause ΔICT services employment	0.72	0.64
ΔICT employment does not Granger Cause ΔICT education	0.58	0.41
ΔICT education does not Granger Cause ΔICT employment	2.48*	3.21**
ΔICT manufact. employment does not Granger Cause ΔICT education	0.01	0.03
ΔICT education does not Granger Cause ΔICT manufact. employment	1.07	0.72
ΔICT employment does not Granger Cause ΔICT infrastructure	6.22***	0.24
ΔICT infrastructure does not Granger Cause ΔICT employment	0.91	0.21
ΔICT employment does not Granger Cause ΔInternet users	1.03	1.11
ΔInternet users does not Granger Cause ΔICT employment	1.14	0.03
ΔICT manufact. employment does not Granger Cause ΔInternet users	0.76	4.93**
ΔInternet users does not Granger Cause ICT manufact. employment	8.95***	2.08
ΔICT employment does not Granger Cause ΔOpenness	19.40***	13.01***
ΔOpenness does not Granger Cause ΔICT employment	3.41**	0.89
ΔICT manufact. employment does not Granger cause ΔOpenness	31.43***	26.36***
ΔOpenness does not Granger cause ΔICT manufact. employment	0.55	0.11

Source: Authors calculations

Note: */**/*** means significance at the 10%/5%/1% levels. %. Numbers of lag was selected based on the results of Akaike and Schwarz criterion.

4   Conclusion

ICT sector is seen as one of the key sources of innovation in the economy. Hence, it can accelerate the productivity and economic growth by several ways. Higher share of ICT sector could therefore be beneficial for a country with respect to increasing its innovation potential and economic level. The relative size of ICT sector was in our case proxied by the employment in the sector.

We further examine the employment in ICT services and ICT manufacturing in European countries. We found that the highest share of ICT sector is in Ireland followed by Luxembourg, Malta, Sweden and Denmark. Furthermore, we are focus on examining the potential determinants and consequences of the relative size of ICT sector in the country. There is some positive correlation between ICT skills and employment in ICT sector in the country.

Based on the panel data analysis we further found that the employment in ICT sector is significantly positively correlated also with several other country characteristics such as its GDP per capita, trade openness, R&D expenditure, number of internet users, number of servers per inhabitant, quality of regulation and political stability. Policies dealing with support of ICT sector should take to the account these findings. However, based on correlation analysis we were of course not able to identify casual relationships and its direction. Hence, we also applied panel Granger causality test. The results suggest that there is some positive effect of higher employment in ICT services on GDP. However, we fail to prove the same for the share of total employment in whole ICT sector. Moreover, higher share of ICT sector seem to have certain positive effect on trade openness and the relative number of servers in the country. On the other hand, more people with ICT education seems to have positive impact on the relative size of ICT sector in economy.

Acknowledgments.   This work has been supported by the Scientific Grant Agency of Slovak Republic under project *VEGA No.* 1/1009/16 "Innovation potential of the regions of Slovakia, its measurement and innovation policy at the regional level".

References

1. Bauer, J.M., Shim, W.: Regulation and innovation in telecommunications. In: Quello Center Working Paper 01–2012, Presented at Scientific Seminar Entitled Communications & Media Markets: Emerging Trends & Policy Issues, pp. 1–26 (2012)
2. Brynjolfsson, E., Saunders, A.: Wired for Innovation. How Information Technology is Reshaping the Economy. MIT Press, Cambridge (2010)
3. Brynjolfsson, E.: Innovation and the E-economy. MIT, Sloan School of Management (2011, Unpublished paper)
4. Cardona, M., Kretschmer, T., Strobel, T.: ICT and productivity: conclusions from the empirical literature. Inf. Econ. Policy **25**(3), 109–125 (2013)
5. Doucek, P.: Human resources in ICT–ICT effects on GDP. In: Information Technology– Human Values, Innovation and Economy, IDIMT-2010, pp. 97–105 (2010)

6. Hall, B.H., Lotti, F., Mairesse, J.: Evidence on the impact of R&D and ICT investments on innovation and productivity in Italian firms. Econ. Innov. New Technol. **22**(3), 300–328 (2013)
7. Lall, S.: Learning from the Asian Tigers: Studies in Technology and Industrial Policy. Springer (1996)
8. Ngwenyama, O., Morawczynski, O.: Factors affecting ICT expansion in emerging economies: an analysis of ICT infrastructure expansion in five Latin American countries. Inf. Technol. Dev. **15**(4), 237–258 (2009)
9. Pohjola, M.: The adoption and diffusion of ICT across countries: patterns and determinants. In: The New Economy Handbook, pp. 77–100 (2003)
10. Rai, D., Kurnia, S.: Factors affecting the growth of the ICT industry: the case of Bhutan. In: International Conference on Social Implications of Computers in Developing Countries, pp. 728–739. Springer, Cham (2017)
11. Stiroh, K.J.: Information technology and the US productivity revival: what do the industry data say? Am. Econ. Rev. **92**(5), 1559–1576 (2002)
12. Vu, K.M.: ICT as a source of economic growth in the information age: empirical evidence from the 1996–2005 period. Telecommun. Policy **35**(4), 357–372 (2011)

Automatic Process of Continuous Integration of Web Application

Tomasz Krym[1], Aneta Poniszewska-Marańda[1($\boxtimes$)], Erich Markl[2],
and Remy Dupas[3]

[1] Institute of Information Technology, Lodz University of Technology, Łódź, Poland
aneta.poniszewska-maranda@p.lodz.pl
[2] University of Applied Sciences Technikum, Wien, Austria
erich.markl@technikum-wien.at
[3] Laboratoire d'Intgration du Matriau au Systme CNRS, Universite de Bordeaux,
Bordeaux, France
remy.dupas@gmail.com

Abstract. Along with the evolution of software development models, the role of tests in the whole process has been increasing, mainly in the direction of earlier detection of errors than at the time of final product delivery by the customer. The innovativeness and volatility of the Internet applications market leads to a situation where there is great pressure to quickly implement a good quality solution. It is crucial to ensure both high quality and effective monitoring, which will allow to decide on the implementation of the application at the right time. The paper concerns the problem of software testing and its automation, with particular emphasis on web applications created in Java.

1 Introduction

Software testing is considered an indispensable part of the manufacturing process right from the beginning of software engineering. Along with the evolution of software development models, the role of tests in the whole process has been increasing, mainly in the direction of earlier detection of errors than at the time of final product delivery by the customer. The main reason for this was the observation that the earlier the error was detected, the lower the costs of its removal. Currently, it is estimated that from 40% (for typical projects), up to 90% (for innovative projects), expenditure on software development is the cost of testing. It is so high, because tests are carried out repeatedly in the program's life cycle, from the specification to the maintenance of the system [1]. The automation of the testing process has a great potential to reduce these costs, which allows to more effectively and economically verify the quality of the produced software [17].

Currently, Internet applications are a very popular and relatively young category of programs. The innovativeness and volatility of the internet applications market leads to a situation where there is great pressure to quickly implement a good quality solution. The decisive factor in a failure of a project may be a month

F. Xhafa et al. (Eds.): INCoS 2018, LNDECT 23, pp. 195–206, 2019.
https://doi.org/10.1007/978-3-319-98557-2_18

of delay against competition or improperly functioning functionality. Therefore, it is crucial to ensure both high quality and effective monitoring, which will allow to decide on the implementation of the application at the right time. For this reason, it is required that testing is effective and efficient, that is, it is effective in finding errors and as fast and as cheap as possible. By automating the testing of web applications, we can significantly reduce costs and also perform a greater number of test cases at the same time. Execution of test variants can take several minutes, not hours, as in the case of manual tests [2,6].

The problem presented in the paper concerns the process of software testing and its automation, with particular emphasis on web applications created in Java. It implements and uses the methods of software development, such as continuous integration and test-driven production.

The presented paper is structured as follows: Sect. 2 presents the elements of construction and testing of web application. Section 3 describes the selected issues of automatic process of building Java EE application while Sect. 4 deals with automatic testing of Java EE web application.

2   Construction and Testing of Web Application

Nowadays more and more emphasis is put on the fastest possible embodiment of an idea for an Internet application. Therefore, incremental models of project management are chosen, where the first prototypes are made available to end users as beta versions, which implement only a part of the target functions. After launching the application on the target environment, the project team starts implementing further functionalities from the list of so-called "product backlogs". At the same time, feedback about the application from end users is collected, based on which the expectations for the product are developed, to which the initial requirements and specifications are adjusted. Subsequent iterations can last very shortly, which means that tests of the whole application must be carried out equally often. Therefore, in projects developed in an incremental model according to modern agile methodologies, very much emphasis is put on the creation of automated tests before or along the implementation of functionality. The source code is also subjected to continuous integration, which allows to obtain current information on the current quality and stability of the product, so we can evaluate the product's readiness for acceptance testing and implementation. The result of this is that we can provide new functionalities to users faster, more often and less risky [1,2,18].

The unquestionable usefulness of continuous integration, i.e. the process of automatic building and testing of source code, is the cause of its growing popularity, and its effective implementation should be considered as a factor building a competitive advantage over software producers who do not use it. According to Martin Fowler, the implementation of the continuous integration process should meet the following guidelines [1,6]:

- source code is stored in the central repository,
- process of building the application is automated,

- automated tests are part of the process,
- every programmer makes changes to the repository every day,
- every change in the repository starts the construction process,
- construction process is carried out in a short time,
- testing takes place in an environment as close as possible to the target one,
- the last stable version of the application is easily accessible,
- installation of the new version in the production environment is automatic.

The following section shows how the automatic process of continuous application integration should look like, having first of all in mind what features of the application can be tested during it so as to get the fullest possible information on the quality of the created software. Libraries and tools of the Open Source type have been proposed that can be used for this purpose. First of all, they are tools dedicated to software created on the Java EE (JEE, Java Enterprise Edition) platform with a network profile. In this work, Open Source software is used for applications and libraries made available under a license that meets the definition of Open Source Software according to OSI (Open Source Initiative) and is included in the Open Source software license list on the OSI website. The choice of the JEE platform for the presentation of the solution is motivated above all by its constant development and high availability of Open Source solutions available to it.

On the JEE platform, the application is run on an application server that provides services defined in the JEE6 application server specification, such as transaction management, Object-Relational Mapping (ORM) or dependency injection, and we can use these services using the application programming interface (API). Both the application server and the application itself must fulfill the set of specifications that make up the JEE standard.

During the continuous integration process, the basic element is the systematically repeated process of building and testing applications. The proposed steps and their sequence that make up this process are [2]:

1. initialization of the process after changes to the repository by the programmer or at a defined time,
2. compilation of program sources,
3. static tests,
4. unit tests,
5. creating a network application,
6. launching a web server with a downloaded network application in order to perform acceptance tests,
7. acceptance tests,
8. disabling the web server after acceptance tests,
9. publication of results and, in case of successful completion of the process, a stable version of the application has been built.

The continuous integration server is responsible for process initialization and reporting its result, while the remaining steps are the domain of the script automatically building the application, which may use other tools to perform individual operations. Such a division of roles is caused by the fact that most of the remaining actions are also performed through a script on the programmers' machines while they are working on the created software.

Continuous integration server (CI server) is an application that manages the process of continuous integration, dedicated to the process of building applications and publishing its results. Requirements for the application managing the continuous integration process are [2]:

- monitoring the source code repository,
- launching the program that builds the web application after introducing changes to the repository,
- publication of the results of the construction process on the website,
- providing the last stable version of the application.

There are many CI server solutions available that meet the above requirements and provide many other additional functionalities. Among the most popular Open Source products are Apache Continuum, Cruise Control, Luntbuild, Jenkins and Hudson [3,5,10]. All CI servers run under many operating systems and work with most of the code control systems (SCM, Source Code Management), and also integrate with tools for automatic Java project development, so for initialization the construction process can be used any of them, although the amount of functionality and the active project community speak for Jenkins [3].

3   Automatic Process of Building Java EE Application

The built-in network application created on the JEE platform is a ZIP-compressed archive with a .war extension, which contains compiled class code, libraries, and other resources, such as static web content, configuration files, and style sheets. The minimum structure of such an archive consists of the root directory and subdirectories META-INF and WEB-INF. The META-INF subdirectory stores information about the zip archive itself, while WEB-INF stores libraries, compiled classes and configuration files. Both of these subdirectories are not directly accessible from outside. Static HTML files, CSS style sheets, graphics or dynamic Java Server Pages (JSP) pages are located in the root directory or its subdirectories, excluding META-INF and WEB-INF [11]. The following steps are necessary to create such an archive:

- building application components,
    - establishing external libraries (dependencies) necessary for compilation,
    - compiling the component code,
    - creating libraries from compiled code and resources,
- establishing dependencies that are necessary for the compilation and functioning of components, excluding libraries provided by the application server,
- compiling a non-partitioned code for components,
- creating an archive with an application that has the correct structure, contains compiled code, components, libraries and other resources.

In order to perform the above steps, we can write a system shell script, but the JEE platform is so popular that there are ready-made tools to avoid this. One of them is the Apache Ant and Apache Maven projects developed by the Apache Software Foundation [4].

Apache Ant is a tool written in Java that builds the application according to the steps placed in the procedure defined in the XML file. The default file is build.xml, where we specify goals that contain a list of tasks to perform. Apache Ant has a large number of tasks built in, such as compilation, calling another purpose, packing the archive directory, and many more, and you can implement your own tasks in Java [3,11].

Apache Maven is another tool that is used to build applications, but it is implemented using a different concept than in Apache Ant, where a procedure for a specific project is created, what actions are to be performed. The Apache Maven project assumes the existence of certain operations (e.g. compilation) and their sequences that are common to many projects, and differ only in parameter values. Instead of creating a precise procedure, an object model project (POM, Project Object Model) is defined, which has an XML file format (XML file) and has information about the design attributes and what operations are to be performed during construction [4].

4   Automatic Testing of Java EE Web Application

TDD (TestDriven Development) and ATDD (Acceptance TestDriven Development) techniques recognize automatic quality verification as an integral part of the software development process. However, the source code of tests serves not only as a tool to control the correctness of application operation, but also as a documentation of the program's operation, because, under conditions of rapidly changing requirements, it is often the only complete form of product specification. An important aspect is also that in TDD and ATDD, tests are written by programmers who know the technical aspects of the application, and those responsible for quality assurance only supervise that all key areas have been properly checked. In subsequent iterations, previously created automatic tests are used to detect the resulting regression in functionality [7].

Having automated tests, they should be included in the process of continuous integration, where they will be performed on a dedicated machine during program construction. The environment on which the CI process is run is usually similar to the target production environment, and in practice is at least more similar than the environments on the programmers' machines. Therefore, the aggregated results of tests performed on such a machine are more reliable than the results obtained from each programmer. Even more important is the fact that the testing process is carried out on the current version of the application downloaded from the repository, which allows obtaining information on the current quality and stability of the product, so you can evaluate the product's readiness for acceptance testing and implementation.

4.1   Unit Tests

Unit tests as a basic element of software quality assurance and the foundation of TDD methodologies are widely used in JEE projects. As for most popular aspects of application development in Java, also for unit tests, there are tools that make it easier to write and run tests, and report their results. The most popular non-commercial platforms for unit tests dedicated to Java are the JUnit and TestNG libraries.

The *JUnit* library according to Koskela is actually a standard when it comes to unit testing in Java [7]. Creating unit tests using this library requires a minimum configuration in the unit test code. All we need to do is annotate (from version 4 JUnit and 5 Java) the test method so that we can start it using JUnit. We can also define methods that will be started before and after each test or test class, which makes it much easier to isolate the test, i.e. to prepare the data and the environment for the test case to be performed, and to recreate the initial conditions after it.

The biggest advantages of JUnit are simplicity and ease of use as well as popularity, and thus integration with the majority of Integrated Development Environments (IDE) and a large number of solutions that extend its functionality. The disadvantage of JUnit, resulting from the concept of complete isolation of each unit test, is the inability to define a different order of test methods and defining dependencies between them, which is a disadvantage mainly when using the library for automatic acceptance testing, where it can often be pointless to run a test to check the functionality, which is known from previous tests that it is not available [8,9].

A good example of such a situation may be a critical error in logging into the web application that blocks any verification of functionality available only after logging in. In this situation, JUnit will execute and mark all functional tests as running and failed, which will distort the image of the application's quality status, because probably a significant part of these tests could be successful, if not for a single error in the critical login function [9].

The *TestNG* library is modelled after JUnit, which is very similar in use. Also, most of the tools dedicated to JUnit extensions can be used with TestNG. The most important advantage of TestNG, compared to JUnit, is the elimination of the described defect JUnit.

In TestNG there is an extensive possibility of configuration in the XML file of test groups, order and dependencies between tests. Therefore, in the case of ATDD implementation, the selection of TestNG as a unit test platform can be considered beneficial, whereas in the case of the TDD itself, the TestNG and JUnit libraries are equivalent.

During unit testing, in addition to the results of the tests themselves, an important metric is covering the code coverage by unit tests. Several solutions are available for Java, such as Cobertura, CodeCover, EMMA and JaCoCo [13]. Cobertura, CodeCover and EMMA are tools that have been available for a long time and are based on Java code instrumentation, permanently changing it before

Contacts manager   Sessions

Contacts manager

Element	Missed Instructions	Cov.	Missed Branches	Cov.	Missed	Cxty	Missed	Lines	Missed	Methods	Missed	Classes
pl.lodz.p.ics.tomaszkrym.contact.form		36%		15%	31	39	46	61	10	16	1	2
pl.lodz.p.ics.tomaszkrym.contact.controller		0%		0%	7	7	16	16	5	5	1	1
pl.lodz.p.ics.tomaszkrym.contact.dao		0%		0%	5	5	7	7	4	4	1	1
pl.lodz.p.ics.tomaszkrym.contact.service		0%		n/a	4	4	6	6	4	4	1	1
pl.lodz.p.ics.tomaszkrym.contact.filter		72%		86%	3	11	7	15	2	4	1	2
Total	279 of 402	31%	47 of 66	29%	50	66	82	105	25	33	5	7

Created with JaCoCo 0.5.6.201201232323

Fig. 1. HTML report generated by *JaCoCo*

or after compilation. Therefore, they are not suitable for use during the continuous integration process, because the created stable version of the web application would contain unnecessary code, used only to calculate the code coverage with unit tests, which would lower the application's performance in the target environment. *JaCoCo* [13] is based on the instrumentation of "on-the-fly" code, while loading classes into JVM (Java Virtual Machine). Thanks to this, the built-in stable version of the application does not include the instrument code, so using JaCoCo is the only choice if the code coverage report is to be obtained during the continuous integration process. JaCoCo generates reports in the form of HTML pages (Fig. 1), which can be viewed in a web browser, as well as CSV files (Comma Separated Values) and XML files, from which data can be presented in other tools [13].

4.2 Static Tests

Static analysis of the source code allows detection of many potential errors that may not be found in unit tests. In addition, it allows to obtain a more transparent code with better quality and to avoid duplicate program parts which make it difficult to maintain the application.

Findbugs is a tool that inspects bytecode Java for logical errors. In addition to logic errors independent of the programming language, like infinite loops, it also searches for Java-specific code fragments that may run counter to the programmer's intention by overlooking or not knowing some of the methods. Findbugs' error detection rules are constantly updated, the project is run by the University of Maryland and the main donor is Google, which uses this tool to search for errors in its projects [14]. Findbugs is available as a library that generates reports in the form of XML files as well as a desktop application with a graphical user interface (Fig. 2).

Another tool is *PMD*, which aims to search for code fragments that violate the principles of good Java programming practices. These are primarily unused, duplicated or excessive complex parts of the program, and also warns against using the code marked as obsolete. PMD is a configurable tool that allows to both manage existing rules and add your own. Unlike Findbugs, PMD does not

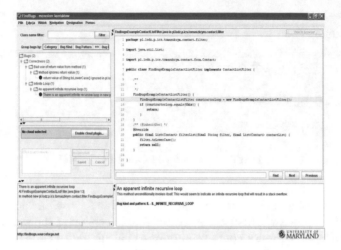

Fig. 2. View of the *Findbugs* graphical user interface

Fig. 3. View of HTML report generated by PMD

have a GUI, running the application is possible only from the command line. The results of the analysis are published as XML files and HTML pages that can be opened in a web browser (Fig. 3).

Checkstyle is a program that verifies the source code style against Java conventions, good practices and user rules. It allows to keep the code transparent and to enforce one standard in the formatting of the source code, for example the same size of indents and characters used to obtain them, or the length of the line. The triviality of these rules does not mean their unsuitability; for example, developers using differently configured formatting rules in IDE will automatically make a lot of changes to the files they edit, which later makes it difficult to track the changes of individual lines of code in the file history in the version control system. Just like PMD, Checkstyle has only a textual user interface, and the generated reports are published as XML files and HTML pages (Fig. 4).

Fig. 4. View of the HTML report generated by *Checkstyle*

4.3   Acceptance Tests

Automated acceptance testing of web applications is very desirable in connection
with reducing the workload of quality assurance, and adding them to the integration process adds to its value. When starting to automate acceptance tests
and attach them to the automatic construction process, we should consider some
basic problematic issues, independent of the business logic of the application:

- instrumentation of a web browser to simulate user behaviour,
- number of potential user environments in which the tests should be carried
  out,
- time needed to perform all tests on all user environments.

These problems are solved by *Selenium*, a tool for automatic testing of websites, which instruments the browser and provides the API for testing [15].
From version 2, the WebDriver project has been incorporated into the Selenium project, which provides its own API and instruments the browser directly,
without using JavaScript and Selenium RC. Both programming interfaces support the majority of popular web browsers, including Internet Explorer, Firefox
and Chrome [15]. In order to run tests on a remote machine, it is necessary to
install a remote Selenium server on it, to which test instructions are sent in sele-
nese or using the remote WebDriver API. Figure 5 shows how to perform tests
on a single remote machine.

The basic functionality of Selenium allows for the instrumentation of one
browser, while the Selenium Grid subproject has been created to run parallel
tests on many environments. It introduces a central Selenium server (Selenium
Hub) to which belongs remote Selenium servers that target specific browsers.
The test, instead of to a specific Selenium server, connects to the Selenium Hub
specifying the requirements for the user's environment, for example the name
and version of the browser. The central server selects one of the connected servers
from the pool of connected servers and provides test instructions to it, and also
balances the load balancing of connected servers. Figure 6 shows support for
many different user environments using Selenium Grid [16].

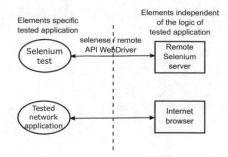

Fig. 5. Performing *Selenium* tests in a single remote web browser

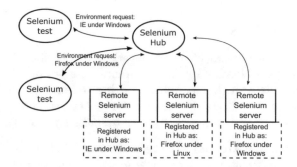

Fig. 6. Support for multiple user environments in *Selenium Grid*

From the point of view of the Selenium Grid client, i.e. a specific test, there is no difference whether it connects to a remote single Selenium server or to the Selenium Hub, the interface remains the same. This makes it much easier to develop the tests, because the test source code can be tested on a local machine treated as a remote one, and running it later with Selenium Grid does not require any changes. The great advantage of this solution is also the fact that the test infrastructure, i.e. the central server and the environment with remote servers, is completely independent of the logic of the tested application and can be used in many projects in a given organization. The use of Selenium Grid enables significant acceleration of test execution, as well as facilitates running tests in various user environments, which can be created even with the use of virtual machines enabled only for the time of testing. Figure 7 shows how to perform Selenium tests using Selenium Grid [16].

Selenium provides libraries for writing tests in many programming languages, including Java. Thanks to this, we can use the same test platforms that are used during unit testing, but we must configure the acceptance tests so that this happens when the archive with the web application is already created and "loaded" on the network server. Activities related to the web server, of course, can be solved with a dedicated shell script, but this is so typical problem that the *Cargo* library was created, which allows manipulation of many types of network

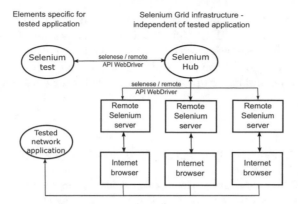

Fig. 7. Performing Selenium tests in many remote web browsers with the use of Selenium Grid

servers and applications using one API, which is integrated with both Apache Ant and Apache Maven [12].

The Open Source community undoubtedly recognizes the importance of the testing process and its automation, which is confirmed by the tools described in this section. In order to maximize their effective use, they should be integrated in a single process of continuous integration, which should result in synergy through the possibility of obtaining in one place up-to-date information on many aspects of the quality of software.

5   Conclusions

The presented elements of automatic process of continuous integration shows how to implement the methods and techniques of software development. The solution shows how to create a platform for automatic building and testing of applications. A particularly important element of it is the method of performing the acceptance tests, written using the Selenium library. The Selenium Grid application allows to verify the operation in user environments.

The automatic continuous integration process allows to regularly obtain a lot of useful information about the quality of the application being produced, making it easier to identify and repair the errors at the lowest possible cost, i.e. a short time after their creation. The solution also promotes the test-driven methods. Quick detection and error repair along with access to code quality reports should help the project management decide whether to release the application on demanding Internet market.

Thanks to the conventions promoted by the Apache Software Foundation, the implemented process can easily be adapted to any network application created for Java EE platform. The presented solution can be extended in the future with additional elements, such as automatic management of used tests of user environments and automatic performance tests that run daily as part of a continuous integration process that can detect the regression in the field of a performance. The possibility of easy adding the additional elements and standardizing the approach to building and testing the projects cause that the implementation of the solution in an organization should bring its long-term benefits by promoting good programming and design practices, and thus better quality of the software.

References

1. Meyers, G.J., Badgett, T., Sandler, C.: The Art of Software Testing, 3rd edn. Wiley, New York (2015)
2. Rossel, S.: Continuous Integration, Delivery, and Deployment: Reliable and Faster Software Releases with Automating Builds, Tests, and Deployment. Packt Publishing, Birmingham (2017)
3. Leszko, R.: Continuous Delivery with Docker and Jenkins: Delivering Software at Scale. Packt Publishing, Birmingham (2017)
4. Zyl, J.V.: Maven: The Definitive Guide. O'Reilly Media, Sebastopol (2008)
5. Laster, B.: Jenkins 2: Up and Running: Evolve Your Deployment Pipeline for Next Generation Automation. O'Reilly Media, Sebastopol (2018)
6. Fowler, M.: Continuous integration (2006). https://martinfowler.com/articles/continuousIntegration.html. Accessed May 2018
7. Koskela, L.: Test Driven: Practical TDD and Acceptance TDD for Java Developers. Manning Publications, Greenwich (2007)
8. Appel, F.: Testing with Junit.Packt Publishing (2015)
9. Gulati, S., Sharma, R.: Java Unit Testing with JUnit 5: Test Driven Development with JUnit 5. Apress, Berkeley (2017)
10. Oracle, Java JEE Enterprise Application Technologies, Oracle Technology Network (2018). http://www.oracle.com/technetwork/java/javaee/tech/entapps-138775.html
11. Apache Software Foundation, Apache Ant, Apache Ant (2018). http://ant.apache.org/. Accessed Apr 2018
12. Apache Software Foundation, Maven Documentation. Apache Maven Project (2018). http://maven.apache.org/guides/. Accessed Jan 2018
13. EclEmma, JaCoCo – Mission (2018). http://www.eclemma.org/jacoco/trunk/doc/mission.html. Accessed Jan 2018
14. University of Maryland, Findbugs (2018). http://findbugs.sourceforge.net/. Accessed Jan 2018
15. OpenQA, Selenium Documentation. SeleniumHQ (2018). http://seleniumhq.org/docs/01_introducing_selenium.html. Accessed Jan 2018
16. OpenQA, Selenium Grid (2018). https://www.seleniumhq.org/docs/07_selenium_grid.jsp. Accessed Jan 2018
17. Pawlak, M., Poniszewska-Marada, A.: Towards the testing management process in information technology projects. WULS Press, Information Systems in Management (2018)
18. Kryvinska, N.: Building consistent formal specification for the service enterprise agility foundation. the society of service science. J. Serv. Sci. Res. 4(2), 235–269 (2012)

Heuristic Min-conflicts Optimizing Technique for Load Balancing on Fog Computing

Muhammad Babar Kamal, Nadeem Javaid$^{(\boxtimes)}$, Syed Aon Ali Naqvi,
Hanan Butt, Talha Saif, and Muhammad Daud Kamal

COMSATS University, Islamabad 44000, Pakistan
nadeemjavaidqau@gmail.com
http://www.njavaid.com

Abstract. Cloud is an on-demand centralized global internet service provider to the end-user. Cloud computing, however, faces problems like high latency and low degree of security and privacy. For low latency, better control of the system and high-security fogs are integrated in the architecture. The cloud-fog based architecture provides security, control of data, quick response and processing time. Recently one of the emerging research areas of Smart Grid (SG) is the integration of Internet-of-Things (IoTs) with SG services to improve its capability. IoTs are interrelated digital machines, objects, and computing devices which have the ability to transfer information over the internet without human interaction with the system. SG is a modern energy management grid for smart use of resources and to optimize Peak Average Ratio (PAR) of energy consumption. Cloud-fog based architecture is integrated with SG for efficient utilization of resources and better management of the system. In cloud-fog computing, the load balancer allocates requests of end-user to Virtual Machines (VMs). In this paper a load balancing scheduling algorithm is presented namely; Min-conflicts scheduling algorithm. The algorithm takes a heuristic approach to solve a Constraint Satisfaction Problem (CSP).

Keywords: Fog computing · Cloud computing
Load balancing algorithm · Resource allocation · Virtual machines
Optimization algorithm

1 Introduction

A repaid growth of the Internet of Things (IoTs) is observed in recent years dominantly in appliances of daily use. [1] integration of IoTs in Smart Grid (SG), which helps to enhance the capability of SGs. This field has a large scope of expansion. Cloud computing is a model which provides facilities and services to the user. Cloud computing is rapidly progressing. However, it faces challenges

© Springer Nature Switzerland AG 2019
F. Xhafa et al. (Eds.): INCoS 2018, LNDECT 23, pp. 207–219, 2019.
https://doi.org/10.1007/978-3-319-98557-2_19

like high latency, lack of mobility, less reliable, security and awareness of location which needs to be resolved [2,3].

In [4] authors introduce architecture of decentralized cloud computing for power generating grid. In [5] the authors consider the behavior of load shifting using optimizing techniques for scheduling of charging and discharging of Electric Vehicles in fog computing.

In [6] the authors use fog computing which provides the network of cloud-fog based architecture for storage, to achieve low latency, control of data and administration of the system in one hop proximity to the end-user. [7] examine the fog computing on SG data management system while [8] uses fog computing to achieve parallel processing system, real-time information retrieval and secure access to data.

In this paper, a heuristic load balancing algorithm on fog computing is presented. The load balancer allocates Virtual Machines (VMs) to the user request for processing of data. The Min-conflicts algorithm is implemented to achieve quick Response Time (RT) and higher processing by resolving conflicts of Constraint Satisfaction Problem (CSP).

1.1 Motivation

In recent years, the exponential growth of the IoT is creating millions of new endpoints [9]. As discussed in [10] IoTs services include electric vehicles, wireless sensors and SG. Cloud extended fog computing is introduced for the handling of resources providing low latency, heterogeneity, mobility and location awareness. Fog computing provides a rich platform for portfolios, applications and new services working as edged-cloud in one-hop proximity.

In [11] the authors propose a systematic way of cloud computing services integrated into SG, three aspects are considered, i.e. information-management, security and energy management in SG architecture. In [12] uses heuristic algorithm, where electricity management controller is used in SG to minimize Peak-to-Average Ratio (PAR), maximize customer comfort and efficient integration of renewable energy-source to the system [13,14].

[15] proposes a novel bio-inspired technique which is used for load management of VM in the cloud computing to find an optimal solution for efficient utilization of resources. Inspired by [10–12,15] load balancing technique for fog resource utilization is introduced in this paper.

1.2 Contribution

Fog architecture provides many benefits to the end-user. Optimal resources are allocated to end-user through load balancing, following are the contributions of this paper:

- Achieving better load optimization in on-peak hours.
- Low latency due to fog less distance from end-user.
- Fog implementation reduces the processing time of the end-user request.

- The Min-conflict algorithm implementation provides better performance results as compared to Throttled and Round Robin (RR) algorithms.
- Simulation of the scenario on Min-conflicts scheduling algorithm achieves effective scheduling under PAR constraint, considering consumer preferences.

2   Related Literature Review

This section comprises of recent related literature. In [16], authors work is based on a bio-inspired technique for load balancing namely firefly algorithm. In [17], authors implements different load balancing techniques RR, Throttled, First Come First Serve (FCFS) and equally spread current execution. Authors compare these algorithms and concluded that RR shows better results in their proposed scenario.

Authors in [18], introduce a novel technique in cloud computing; namely Advance Max Sufferage (AMS) algorithm. Implementation of AMS achieve better results of load balancing. In [19], authors shows an analytical comparison of load balancing algorithms. Authors implemented central queuing, equally spread current execution, honeybee, Throttled and RR algorithms. The simulation results show that equally spread current execution load balancing algorithm achieves better client satisfaction and optimal resource utilization by reducing processing time.

In [20], authors consider a distributed auction model in which buyer's place bids to the seller, i.e., mobile devices for resource allocation and the sellers makes the auction decisions locally acting as an auctioneer. An auction system is designed to allocate resources to smart phones, and the price paid for each resource is determined. Efficient evaluation of payment is implemented to detect any untruthful and dishonest activity of sellers is detected. The proposed Auction system has many desirable properties, i.e., individual rationality, budget balancing, computation efficiency and truthfulness.

In paper [21], the authors proposed a novel multi-tenant computing framework for cloud computing. Businesses are considered in the paper which provide services of cloud computing such as resources of VMs to customers.

An Analytical model is proposed which simultaneously investigate and address RT to end-user which covers two factors of RT i.e., percentile of RT which gives the maximum time required for a sub-application to execute, mean of RT which gives mean RT.

The workload is allocated and optimization of service-performance is done in the proposed max-min cloud algorithm through load balancing of VMs according to their capability. Heterogenous VMs are utilized in virtual infrastructure where VMs have different numbers of virtual CPUs which are allocated to subtasks.

The proposed probabilistic framework which implements cloud computing services. The model based on framework determines important features and removes concern in cloud environment which include multi tenancy. The characteristics of multi-tenant model are heterogeneous VMs and Stochastic Response Time (SRT) which serves the request with a general probability distribution.

Two features of SRT are analytically characterized for requests in closed form, namely mean of RT and percentile of RT. These two metrics are mainly used for evaluating the efficiency of services in cloud infrastructure.

The allocation of workload, i.e., scheduling of tasks is critical in the heterogeneous computing environment of the cloud. Here, the workload is efficiently allocated which improve service performance. However, to optimize load allocation in the heterogeneous cloud computing environment is generally categorized as NP hard problem. The proposed framework for cloud services derives resource distribution algorithm, namely max min cloud for allocation of sub-requests to the optimal VMs in polynomial time.

The algorithm is implemented on the bases of load-balancing where execution time for sub-task is deterministic. The condition for getting less RT for the task such as execution times of the sub-tasks are close as much as possible. The proposed algorithm uses the greedy approach to customize m heterogeneous sub-tasks to fix set of VM n.

The authors in [22] introduces cloudlets in Mobile Cloud Computing (MCC). Smoother interaction of smart mobile with the cloud in one-hop proximity is achieved. Cloudlets provide high quality of service to customers. To motivate the cloudlet for sharing of Resources Incentive based system is introduced namely Auction model.

In this model buyer's bids, for resources and Sellers provide resource pool for particular ask, Auctioneers are the intermediate agents in the process where resources are auctioned. Authors proposed algorithm based on the Auction model called Incentive-Compatible-Auctions-Mechanism (ICAM) to serve nearby smart phones. Resources are efficiently utilized and latency is reduced and workload on the centralized cloud is balanced. ICAM provides these desirable qualities live truthfulness, budget balancing, efficient resource allocation, computational efficiency and payment and clearing price in polynomial time.

Table 1. Analysis of scheduling techniques.

Algorithm(s)	Objective(s)	Contribution(s)	Shortcoming(s)
Energy management system [23]	Peak to average ratio and load minimization	Better performances of Energy management system	Base function could be improved
Ant colony optimization [24]	Improve the simulation of fuzzy controller	Better performance achieved than original Ant colony optimization	More parameters of fuzzy system needs to be defined
Particle swarm optimization-pattern search [25]	Reliability, Power utilization and environmental friendly	Improvement of cost and better results for scenarios	Higher emission of Hydrogen fuel cells can be replaced

The paper utilizes an auspicious paradigm of cloudlets in MCC to provide resources to the nearby smartphones. Heterogeneous valuation for smartphones in cloudlet is utilized. The smartphone employ services of cloudlet. A double auction algorithm ICAM is introduced to achieve better latency and incentive capable system. The system calculate the price and ensure payment. Unlike other auction mechanisms, ICAM utilizes heterogenous VMs.

Analysis of different scheduling techniques is shown Table 1.

3   System Model

In this section a three-layered architecture of cloud-fog computing is presented in Fig. 1. The proposed model which is comprised of three-layers, i.e., end-user layer, core-cloud layer, and fog layer. Core-cloud is centralized internet network consisting of remote servers for processing, storage, and management of data. Fog computing is an extension of core-cloud at the edge of the network. Fog and end-user communicate in one-hop proximity to reduce latency. There are a limited number of VMs in a fog which provide services to end-users i.e. processing, storage and management of data.

There are two regions, i.e. Europe and Asia, in our proposed system model. Europe and Asia are selected as region 1 and region 2 respectively because a dense concentration of population lives in these two regions. Efficient energy management is required in this part of the world. For better energy management,

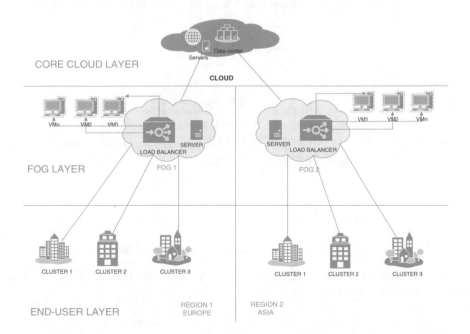

Fig. 1. System model

the load of peak hour is transferred to off-peak hours. Each region has one fog and three clusters of buildings. The Number of VMs vary on these fogs from 30–45. There are average 50 and 200 users from each cluster in on peak and off-peak hour respectively. Clusters are connected to fog and macro grid. The user requests for resources from fog in response fog allocates resources to the end-user.

The fog communicates with cloud if the fog does not have sufficient resources to fulfill the user request. The deployment of fog in layered architecture gives us less latency, high processing speed and less RT. VMs are installed on fogs for handling user request, storage, and memory requirement. VMs are allocated through load balancing, different optimizing algorithms are utilized for this purpose.

3.1   Problem Formulation

Fog resources consist of VMs. Requests are assigned to VMs for processing. Main performance parameters for fog computing includes RT, cost and processing time. Load scheduling techniques are used for optimizing of the workload of fog by assigning VMs accordingly.

Let C be the set of clusters $C = \{c_1, c_2, \ldots, c_c\}$ and have N numbers of users per cluster $N = \{n_1, n_2, \ldots, n_n\}$. There are r requests from a user per hour and $r \epsilon R = \{r_1, r_2, \ldots, r_m\}$. Set of VMs is represented as $VM = \{vm_1, vm_2, \ldots, vm_y\}$. R requests are sent to fog which assigns VMs to request through load balancing. The total number of requests from a cluster of N users are given as:

$$R_{cluster} = \sum_{i=1}^{m} (r_i) \tag{1}$$

And total request:

$$R_{total} = \sum_{cluster=1}^{c} (R_{cluster}) \tag{2}$$

By mapping R_{total} to y number of VMs performance parameters are affected [26] where heavy load of requests cause delay.

3.1.1   Cost
The cost can be calculated for VMs as

$$Cost_{VM} = Cost_{server} + Cost_{storage\ array} + Cost_{operating\ system} \tag{3}$$

The amount of data is considered in GigaBytes (GBs) where the cost of data transfer can be calculated as:

$$Cost_{DT} = Cost\ of\ Data_{(GB)} * Data_{(GB)} \tag{4}$$

3.1.2 Processing Time

Processing time T_p can be calculated as start time minus end time.

$$T_p = T_{start} - T_{end} \tag{5}$$

Time for allocating r request to vm_y is processing time of r request at vm_y i.e., P_{ry}:

$$P_{ry} = \frac{Length\ of\ r}{vm_y} \tag{6}$$

Where λ_{ry} defines the status of request r at vm_y as either assigned or not. The conflict for constraint implies non-allocation of VMs where the number of allocated VMs is known. Reduction of conflicts gives us better processing time. Objective of the algorithm is to minimize processing time:

$$Min\_T_p = \sum_{j=1}^{y} \sum_{i=1}^{r} (\lambda_{ji} * P_{ji}) \tag{7}$$

3.1.3 RT

RT is the sum of request processing time and delays time.

$$RT = T_{end} - T_{start} + T_d \tag{8}$$

Where T_d is delay time when the request r_m reaches to the fog.

3.2 Load Balancing Algorithm

VMs are installed on fog servers. Allocation of requests to VM is done through load balancing algorithms, namely Throttled algorithm and RR algorithm. The Throttled maintains the index of VMs and their status either busy or available.

Algorithm 1. Min-conflicts

1: Min-conflicts parameters initialization: CSP
2: Input: VM-present-state, initial assignment of random state to VM.
3: Input: VM-maximum-steps, the maximum number of steps permitted until scrambling.
4: Output: VM_solution_state or failure.
5: **for** n = 1:maximum-steps **do**
6:      **if** VM-present-state has no conflicts **then**
7:          **return** VM-present-state
8:      **end if**
9:      variable<− scrambled variable of conflicted VMs-CONFLICTED(CSP)
10:      val = The minimize 'v' in CONFLICTS(variable,v,VM-present-state,CSP).
11:      **return** Variable = val in VM-present-state.
12:      **return** Failure.
13: **end for**

The table of indexes is scanned from top first to the bottom for available VMs and allocate it to the respective task. RR uses time slicing or quantum where resources are allocated for a time interval to a particular task. A novel technique Min-conflicts for load balancing is implemented in this paper.

3.2.1   Min-conflicts Algorithm

Min-conflicts is optimizing algorithm to CSP [27]. The algorithm randomly assigns VMs after assignment conflicts are checked if there are no conflicts then the VMs are assigned to requests. Otherwise, it minimizes the conflicts by solving the nastiest conflicts. Min-conflicts is shown in Algorithm 1.

4   Simulation and Results

Simulation and their results are presented in this section. The simulations are based on performance parameters which are calculated through Eqs. (2), (3), (4), (5), (7) and (8). The simulations for these parameters are done on cloud-analyst which depicts the real-world environment, from which we evaluate numbers of experiments to get and optimize results according to a real-world scenario. Graph plots comparison of Throttled, RR, and Min-conflicts is presented in the following subsections. The plots show Cost, RT, and processing time for algorithms where Min-conflicts shows better results.

4.1   RT

In Figs. 2 and 3 RT is shown from the simulation. Figure 2 shows the overall RT of each algorithm total RT of Throttled in the scenario is 65.25 ms, RR RT is

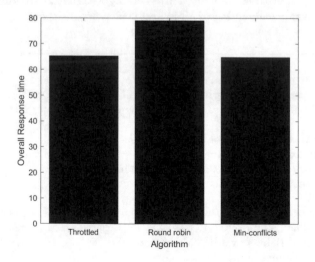

Fig. 2. Overall RT of fogs

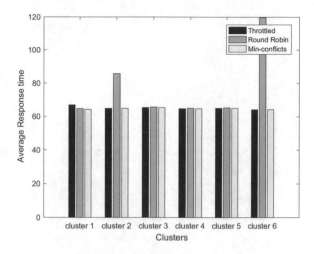

Fig. 3. RT of clusters

78.95 ms, whereas, the proposed algorithm Min-conflicts overall RT is 64.74 ms. Min-conflicts performs better than other two algorithms by giving less RT and high efficiency in fog computing.

In Fig. 3 RT to all six clusters of fog is given where Min-conflicts algorithm performs better.

4.2 Processing Time

Figures 4 and 5 display the results of simulations for processing time of fogs. Figure 4 shows the processing time 15.60 ms, 28.24 ms, and 15.14 ms for Throttled, RR, and Min-conflicts respectively.

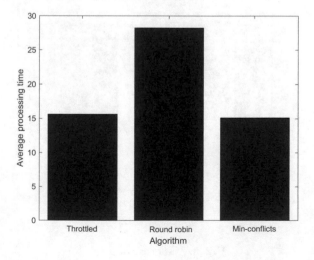

Fig. 4. Overall processing time

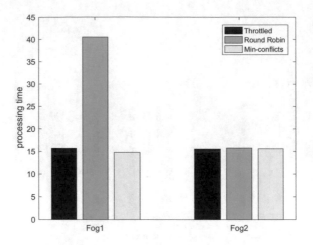

Fig. 5. Processing time of fogs

Min-conflicts perform better than other two algorithms where overall processing time required for processing is less. Figure 5 shows the processing time of fogs.

4.3   VM Cost

In Fig. 6 the cost of fogs is shown for Min-conflicts, RR and Throttled algorithms. The Min-conflicts shows same cost as RR and Throttled. The total data transfer cost in dollars is the shown in Fig. 7 where the Min-conflicts algorithm gives a slightly better cost for data transfer as compare to RR and Throttled.

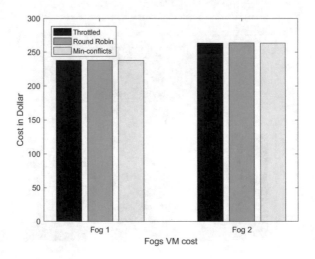

Fig. 6. VMs cost of fogs

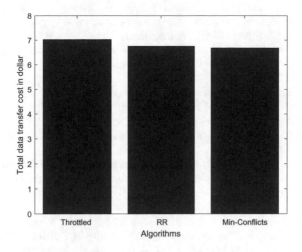

Fig. 7. Total data transfer cost

5   Conclusion

A cloud-fog based architecture is used in given scenario. The proposed architecture comprises of three layers i.e., cloud, fog and end-user layer. The cloud-analyst tool is used for the simulation where two regions Asia and Europe are considered. In this scenario, one fog is installed in one region and three clusters of buildings are connected to it. The fog communicates with cloud and end-user.

Min-conflict load balancing algorithm is implemented to allocate VMs to the requests. This algorithm provides better simulation results of RT and processing time then RR and Throttled.

The proposed heuristic Min-conflicts algorithm satisfy constrain problem by allocating the VMs to minimal conflicted request for processing. Optimal resources are allocated to requests where small task gets less resources and heavy task gets more computational resources. Three different performance matrices are derived in this paper, i.e., cost, RT and processing time.

References

1. Blanco-Novoa, O., Fernandez-Carames, T.M., Fraga-Lamas, P., Castedo, L.: An electricityprice-aware open-source smart socket for the internet of energy. Sensors **17**(3), 643 (2017)
2. Xia, Z., Wang, X., Zhang, L., Qin, Z., Sun, X., Ren, K.: A privacy-preserving and copy-deterrence content-based image retrieval scheme in cloud computing. IEEE Trans. Inf. Forensics Secur. **11**(11), 2594–2608 (2016)
3. Fu, Z., Ren, K., Shu, J., Sun, X., Huang, F.: Enabling personalized search over encrypted outsourced data with efficiency improvement. IEEE Trans. Parallel Distrib. Syst. **27**(9), 2546–2559 (2016)

4. Luo, F., Zhao, J., Dong, Z.Y., Chen, Y., Xu, Y., Zhang, X., Wong, K.P.: Cloud-based information infrastructure for next-generation power grid: conception, architecture, and applications. IEEE Trans. Smart Grid **7**(4), 1896–1912 (2016)
5. Xing, H., Fu, M., Lin, Z., Mou, Y.: Decentralized optimal scheduling for charging and discharging of plug-in electric vehicles in smart grids. IEEE Trans. Power Syst. **31**(5), 4118–4127 (2016)
6. Chiang, M., Zhang, T.: Fog and IoT: an overview of research opportunities. IEEE Internet Things J. **3**(6), 854–864 (2016)
7. Okay, F.Y., Ozdemir, S.: A fog computing based smart grid model. In: 2016 International Symposium on Networks, Computers and Communications (ISNCC), pp. 1–6. IEEE (2016)
8. Reka, S.S., Ramesh, V.: Demand side management scheme in smart grid with cloud computing approach using stochastic dynamic programming. Perspect. Sci. **8**, 169–171 (2016)
9. Javaid, N., Sher, A., Nasir, H., Guizani, N.: Intelligence in IoT based 5G networks: opportunities and challenges. In: IEEE Communications Magazine, June 2018
10. Bonomi, F., Milito, R.: Fog computing and its role in the Internet of Things. In: Proceedings of the MCC Workshop on Mobile Cloud Computing (2012). https://doi.org/10.1145/2342509.2342513
11. Bera, S., Misra, S., Rodrigues, J.: Cloud computing applications for smart grid: a survey. IEEE Trans. Parallel Distrib. Syst. (2014). https://doi.org/10.1109/TPDS.2014.2321378
12. Rahim, S., Javaid, N., Ahmad, A., Khan, S.A., Khan, Z.A., Alrajeh, N., Qasim, U.: Exploiting heuristic algorithms to efficiently utilize energy management controllers with renewable energy sources. Energy Build. **129**, 452–470 (2016). ISSN 0378-7788. https://doi.org/10.1016/j.enbuild.2016.08.008
13. Javaid, N., Ahmed, A., Iqbal, S., Ashraf, M.: Day ahead real time pricing and critical peak pricing based power scheduling for smart homes with different duty cycles. Energies **11**(6), 1464 (2018). ISSN 1996-1073. https://doi.org/10.3390/en11061464
14. Aslam, S., Javaid, N., Khan, F.A., Alamri, A., Almogren, A., Abdul, W.: Towards efficient energy management and power trading in a residential area via integrating grid-connected microgrid. Sustainability **10**(4), 1245 (2018). ISSN: 2071–1050. https://doi.org/10.3390/su10041245.
15. Dam, S., Mandal, G., Dasgupta, K., Dutta, P.: An ant-colony-based meta-heuristic approach for load balancing in cloud computing. In: Intelligence and Soft Computing in Engineering, pp. 204–232. IGI Global (2018)
16. Vivek, M.S., Manohar. P.: A load balancing model using bio inspired firefly algorithm in cloud computing. Int. J. Eng. Technol. **7.1.1**, 671–674 (2018)
17. Kaur, P.: A comparison of popular heuristics for load balancing in cloud computing (2018)
18. Chiang, M.L., Hsieh, H.C., Tsai, W.C., Ke, M.C.: An improved task scheduling and load balancing algorithm under the heterogeneous cloud computing network. In: 2017 IEEE 8th International Conference on Awareness Science and Technology (iCAST), Taichung, pp. 290–295 (2017)
19. Rohith, K.P.S.S., Anand, A.: Analytical Study of different Load balancing algorithms. Int. J. Adv. Studi. Comput. Sci. Eng. **7**(1), 21–26 (2018)
20. Wang, X., et al.: A distributed truthful auction mechanism for task allocation in mobile cloud computing. IEEE Trans. Serv. Comput. (2018)
21. Wang, Z.: Optimizing cloud-service performance: efficient resource provisioning via optimal workload allocation. IEEE Trans. Parallel Distrib. Syst. **28**(6), 1689–1702 (2017)

22. Jin, A.-L., Song, W., Zhuang, W.: Auction-based resource allocation for sharing cloudlets in mobile cloud computing. IEEE Trans. Emerg. Topics Comput. **6**(1), 45–57 (2018)
23. Varela Souto, A.: Optimization and energy management of a microgrid based on frequency communications (2016)
24. Olivas, F., Valdez, F., Castillo, O., Gonzalez, C.I., Martinez, G., Melin, P.: Ant colony optimization with dynamic parameter adaptation based on interval type-2 fuzzy logic systems. Appl. Soft Comput. **53**, 74–87 (2017)
25. Gabbar, H.A., Labbi, Y., Bower, L., Pandya, D.: Performance optimization of integrated gas and power within MG using hybrid PSO-PS algorithm. Int. J. Energy Res. **40**(7), 971–982 (2016)
26. Wickremasinghe, B., Buyya, R.: CloudAnalyst: a CloudSim-based tool for modelling and analysis of large scale cloud computing environments. MEDC Project Rep. **22**(6), 433–659 (2009)
27. Minton, S.: Minimizing conflicts: a heuristic repair method for constraint satisfaction and scheduling problems. Artif. Intell. **58**(1–3), 161–205 (1992)

State Based Load Balancing Algorithm for Smart Grid Energy Management in Fog Computing

Muhammad Junaid Ali, Nadeem Javaid$^{(\boxtimes)}$, Mubariz Rehman,
Muhammad Usman Sharif, Muhammad KaleemUllah Khan,
and Haris Ali Khan

COMSATS University, Islamabad 44000, Pakistan
nadeemjavaidqau@gmail.com
http://www.njavaid.com

Abstract. The use of the traditional grid with Information and Communication Technology (ICT) gave birth to Smart Grid (SG). New services and applications are built by utility companies to facilitate electricity consumers which generate a huge amount of data that is processed on the cloud. Fog computing is used on the edge of the cloud to reduce the load on cloud data centers. In this paper, a four-layered architecture is proposed to reduce electricity shortage between consumer and electricity providers. Clusters layer consist of clusters of buildings which are connected to Micro Grid (MG) layer. MG layer is further connected to fog and cloud layer. Three load balancing algorithms Round Robin (RR), Honey Bee Optimization (HBO) and State-Based Load Balancing (SBLB). Results demonstrate that SBLB outperforms RR and HBO in terms of Response Time (RT) and Processing Time (PT).

Keywords: Smart Grids · Round Robin · Micro Grids
Processing time · Response time · Cost · Cloud computing
Fog computing

1 Introduction

Smart Gird (SG) is the modernization of electric grid that includes variety of measures such as Renewable Energy Sources (RES), smart appliances and meters. Utility companies developed many applications to monitor electricity usage. SG uses two-way communication between utility and consumer to get necessary information to maintain electricity load. Many Demand Side Management (DSM) systems have been developed for handling electricity usage in SG.

Demand Response (DR) provides an opportunity for consumers to reduce electricity usage during peak usage periods [2]. Advancement in Internet of Things (IoT) technology enables to transfer consumer's electricity usage on cloud. Different approaches have been proposed for using cloud with SG [3]. Cloud computing is the collection of software and hardware resources that are available over the Internet. Cloud provides three types of services.

© Springer Nature Switzerland AG 2019
F. Xhafa et al. (Eds.): INCoS 2018, LNDECT 23, pp. 220–232, 2019.
https://doi.org/10.1007/978-3-319-98557-2_20

- Infrastructure as a Service (IaaS): IaaS refers to hardware services available online via remotely, that includes storage, VMs etc.
- Platform as a Service (PaaS): PaaS provides development environment to the end user. Typically web servers, operating systems and development environments.
- Software as a Service (SaaS): SaaS is also known as software distribution model. It is deployed on internet. Customers or users get access through this service.

Cloud computing stores data in centralized data center and requires huge computation and cost. However, fog computing provide services on the edge of network and depends on personal devices or local servers that are further connected to cloud [4]. Private data of consumers can be held in fog devices and only relevant information is sent to cloud, which increases security and reliability of SG.

Fog computing provides better solution for handling large amount of data. It provides low latency, as it is placed near network edge. Cloud and fog computing have been used with SG to reduce cost of computation tasks. Allocating task to desired VM is very important issue in cloud computing. In case of SG based cloud-fog architecture, allocating user request to desired VM in given time constraint and low latency is challenging issue. To overcome this, different load balancing algorithms have been proposed [12,13].

In this paper, a four-layered SG based architecture is proposed for better communication between consumer and electricity company. Our model cover a large residential area. Regions are selected on the basis of smart homes and electricity usage.

1.1   Motivation

The SG is combination of traditional grid that includes information and communication technology. DSM techniques are used in order to control electricity usage using different DR techniques, user involvement is very necessary in order to fulfill his comfort. The authors in [1] gave the idea of using cloud computing with SG, where as in [6] Yigit *et al.* discussed the integration of cloud computing in different SGs applications.

An effective approach for computer applications to handle DSM is proposed in [3], The Authors present a model to allocate cloud computing resources in an efficient and cost-oriented way. By Increasing the number of consumers it becomes difficult for cloud computing to handle user requests, to overcome this issue, fog computing is incorporated as a middle layer between consumer and cloud environment. Fog computing reduces the load on cloud and minimizes the latency [4]. Resource management is an issue in cloud computing. To solve this problem, different load balancing algorithms are developed in [6,7]. Motivated from this concept, we use different load balancing techniques for effective utilization of cloud-fog based architecture in SG network.

1.2 Our Contribution

In this paper, an efficient model for communication between cloud-fog based architecture and SG is proposed.

- Fog server provides low latency, as it is placed near to end user. It responds faster to the request of user as compared to cloud.
- Fog computing is used in distributed areas, as we have different areas and requests of buildings cannot be fulfilled by cloud, requests are first sent to fog which handle the data to cloud.
- In cloud-fog based architecture, to deal with load balancing, State Based Load Balancing (SBLB) algorithm is used and compared with Round Robin (RR) and Honey Bee Optimization (HBO).
- A four layered framework is proposed. End user layer consist of group of buildings called clusters. Each cluster is connected to fogs in fog layer, fogs are connected with Micro Grids (MG) in MG layer and centralized cloud. VMs are installed on each fog. The consumers request electricity from fog that contains electricity usage and location information of consumer. However, necessary information is send to the cloud by fog for further request. Additional electricity is provided to consumer by MGs.
- MGs are integrated in order to fulfill extra energy demand of customers for electricity request allocation a centralized structure is proposed.

The paper is composed as follows. In Sect. 2 the related work is described. Section 3 covers the complete description of proposed system model. In Sect. 4 load balancing algorithms are discussed. Section 5 covers the service broker policies. In Sect. 6 results and discussions are discussed and the final Sect. 7 contains the conclusion and the future work.

2 Related Work

In the last few years, various load balancing algorithms are proposed. Being SG as a complex system that requires huge storage and computation, for that purpose cloud computing is used for effective utilization of resources. Yiğit *et al.* discussed an architecture for effective communication between SG and cloud computing. Challenges that can be faced while incorporating cloud based environment with SG's and open research areas are presented [8]. Distributed generation of electricity also take attention in recent years.

To get environment friendly generation of electricity, MG's are small electricity distribution systems that are attached with utility. The key challenges in enhancing SG's with MG's are discussed in [9]. Author's discuss MG architecture and how it can be used with SG to outcome electricity shortage issues.

A Novel approach for Electric Vehicles (EV's) charging and discharging is presented in [10]. In order to schedule EV's charging and discharging, two priority attribution algorithms are used in cloud computing. The waiting time of EV is considered in these algorithms and scheduling is done on the basis of peak hours.

In [11], authors proposed energy efficient way for handling electricity shortage in smart homes using MG. They proposed an efficient model for handling users electricity requests. Kazmi *et al.* proposed a model for demand and supply of consumers [12]. They user meta-heuristic algorithms for handling load between them.

Honey Bee Load Balancing (HBB-LB) algorithm is proposed in [13], in which they used HBB-LB for balancing load among VM's in cloud computing. In [14], a game theory based task scheduling algorithm for cloud computing is proposed. In which they used game strategy to calculate the allocation rate of each task, based on their weights resources are allocated to the tasks. An Augmented Shuffled Frog Leaping Algorithm (ASFLA) is proposed in [15], which is inspired from frog behavior used for work-flow scheduling in cloud computing.

In [16], authors proposed a Hybrid algorithm of Gravitational Search Algorithm (GSA) and Heterogeneous Earliest Finish Time (HEFT) that is used for bi-objective work-flow scheduling in cloud computing. Two objectives: Budget Constraint and Schedule length are considered for task scheduling. Direct Adjacency Graph (DAG) is used for VM representation. However, many approaches have been used for handling DSM in SG and how to effectively balance electricity load [17,18]. Task allocation in dynamic cloud is a challenging topic as load on cloud fluctuates randomly. In [19], a new approach is used for task allocation in open and dynamic cloud environment. Two new phases are added on propagation based technique: pruning and decomposition. Undirected Graph Model (UGM) is used for VM representation.

From the literature review, it is analyzed that a load balancing technique is required for resource utilization between SG and cloud-fog based system. Multiple algorithms are proposed for load balancing in cloud computing. By using various algorithms in SG and cloud-fog system Response Time (RT) and Processing Time (PT) is enhanced.

3 System Model

In this section, the architecture of proposed four layered framework is discussed. Graphical representation of system model is shown in Fig. 1. The detail of each layer is discussed. The end user layer consists of clusters of buildings, each cluster have a centralized controller, that communicate with fog using controller attached to each cluster of building. Two-way communication can be done between cloud and controller. Controller sends the data to fog on hourly basis and fog processes the data, which include electricity usage information, location etc.

In case of electricity shortage, the fog communicates with nearby MG in MG Layer which provides electricity to desired building in a cluster. There is no Power flow between Fogs and MG's. MG's consist of RES that are environmental friendly and have low cost. Fog sends the relevant information to cloud servers in cloud layer which stores data permanently. Cloud layer is connected with utility and service provider.

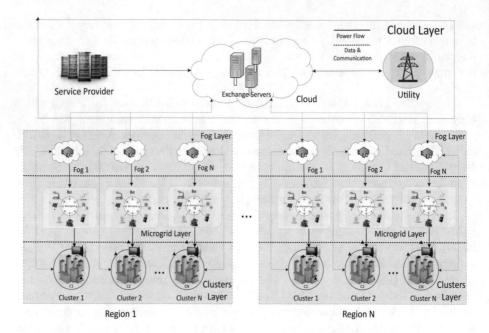

Fig. 1. Proposed system model.

The proposed scenario consists of five regions. In each region we have three clusters of buildings, each cluster contains 50 buildings. These clusters are connected to the fog layer that contains 3 fogs per region, that consist of 25 VM's. Three MG's are used in every region. A centralized cloud is connected with all regions that handle all the requests coming from fogs. Cloud is further attached with utility and cloud service provider. Cloud servers are used to handle all the communication between fogs and SG and provide communication between utility and consumers through Fogs, which is an intermediate layer between cloud and consumer. Different requests of buildings are allocated to different VM's in Fogs. For resource allocation load balancing algorithms are used. Load balancing algorithms used in above scenario are discussed below.

3.1 Problem Formulation

Following components are considered for allocation of resources in cloud and fog based architecture: set of regions, clusters and fogs. Let B be the set of buildings

$$B = \{b_1, b_2, ..., b_n\} \tag{1}$$

Where B is overall sum of buildings while $b_1, b_2, ..., b_m$ are number of buildings. Each cluster of building c_i represents the set of buildings which is represented as:

$$C = \{c_1, c_2, ..., c_n\} \tag{2}$$

Where c_1, c_2 represents the set of clusters. Requests from clusters C to fogs F is represented as

$$R_F^C = \{r_1, r_2, ..., r_n\} \tag{3}$$

where $r_1, r_2, ..., r_n$ are requests. Total requests is given by n clusters:

$$R = \sum_{j=1}^{p} r(j) \tag{4}$$

and each fog contains k number of VM's that handles requests coming from clusters to fogs. $F_1^N = \{vm_1, vm_2, ..., vm_k\}$ each request consists of tasks that can be assigned to specific VM by fog which we can say $T_R = \{T_1, T_2, ..., T_N\}$. For the performance evaluation of system some parameters are used for assessment of system. Following Parameters are used:

3.1.1   Processing Time
The Allocation of task i to VM j of processing time is represented as $P_{i,j}$ and status of each VM is VM_{State}. VM's cannot share resources with each other.

$$VM_{State} = \begin{cases} 1 \; if \; VM = Available \\ 0 \; if \quad VM = Busy \end{cases} \tag{5}$$

$$PT = \sum_{k=1}^{m} \sum_{j=1}^{n} (P_{i,j} * VM_{State}) \tag{6}$$

$$P_{i,j} = \frac{Length \, of \, task \, at \, ith \, place}{Capacity \, of \, VM \, at \, jth \, place} \tag{7}$$

3.1.2   Response Time
The time required to send a task from cluster of building to data center where $sum_{j=1}^{o} R(j)$ is the processing time of all VM's.

$$RT_k = \frac{\sum_{j=1}^{p} R(j)}{makespan * Number \, of \, VMs} \tag{8}$$

3.1.3   Cost
Cost of each VM is calculated by where $Cost_{DataTrans}$ is the cost of transfer of data from VM to Fog. $Cost_{VM}$ is the cost of all VM's during 24 h.

$$Cost_{Total} = Cost_{DataTrans} + Cost_{VM} \tag{9}$$

$$Cost_{VM} = \sum_{t=1}^{h} \sum_{l=1}^{p} (VM_i StartTime - VM_i EndTime) * VM_i HourlyPrice \tag{10}$$

Minimization function is defined as:

$$Minimize : F = \sum_{j=1}^{p} \sum_{i=1}^{n} \left\{ VM_{state} * Delay * Cost * P_{i,j} \right\} \tag{11}$$

where

$$P_{i,j} = \frac{\text{Length size of task in ith place of VM}}{\text{Space of VM at jth place}} \tag{12}$$

4   Load Balancing Algorithms

Load balancing is used to utilize resources in cloud and fog for better performance. VM's are allocated to desired resources on the basis of memory and storage. It is used to avoid overloading and to get better RT. In Cloud Analyst, load balancing algorithms used are as follow: RR, HBO and SBLB.

4.1   Round Robin

RR algorithm is used to allocate equal time slice to each host. RR develop a table of VM assignment of tasks. It allocates the job to desired VM one by one till it reached to the last VM in circular order without priority.

4.2   Honey Bee Optimization

Honey bees have three different behaviors known as forging, mating and breeding behavior. By inspiring from this behavior, HBO is proposed. HBO is one of the optimization algorithms for load balancing, which contains forger bees, scout bees and employee bees. Firstly, the forging process begins by searching the food sources by scout bees [7]. After getting the food source the bee's communication begins with other bees, known as waggle dance which provides information about food source quality and distance between the source and hives. The bees that take the food sources are known as employee bees. HBO tends to find optimal solution based on the fitness value of the agent.

4.3   State-Based Load Balancing Algorithm

SBLB algorithm dynamically allocates requests to the obtainable hosts. The host which are in queue are also assigned [5]. That avoids the host to be heavy loaded. SBLB algorithm carries two different tables lean on their VM states that's why the algorithm is named as SBLB. [6] VM have different threshold values based on their load. When the VM count reaches the threshold value, it change the state to busy state. However, if the threshold is not achieved then it is placed in available state table. RK Naha et al. in [6] propose SBLB algorithm and compare it with other algorithms. We use SBLB algorithm for load balancing in SG using cloud based infrastructure.

Algorithm 1. SBLB

```
1:  Input:List of VM requests, List of VM'S
2:  Output:Desired VM for task scheduling
3:    Initialize VmId,VmAvaliableStates and VmStatesList
4:  DC Controller receives request
5:  CheckVmAvalability()
6:  if (VmAvaliableStates is NULL) or (VmBusyStates is NULL) then
7:      for i=1 to VmStatesList.size() do
8:          if i is Busy then
9:              Add in VmBusyStates
10:         else if i is Avaliable then
11:             Add in VmAvaliable
12:         end if
13:     end for
14: else if VmAvaliable.size >0 then
15:     vmsize=VmStesList
16:     randVm=random(VmStesList)
17:     return VmId
18: else
19:     return VmId=-1
20: end if
```

5   Service Broker Policies

Service broker policy is at the helm of which fog should provide response of request coming from the user bases. The requests between clusters of buildings and fog are controlled by service broker policy. A service broker policy consists of several rules to decide which fog would be allocated to fulfill the request. Service broker policies used in this paper are discussed below.

5.1   Closest Data Center

This service broker policy maintains a table of all data centers in the region. [20] It maintains the index of all the regions and the fogs with minimum latency are calculated for the given region by the policy. It selects the fog located in the same or nearest region. If we have multiple regions in same area, then fog is selected randomly.

5.2   Dynamic Service Broker

This algorithm maintains the data centers list and another list of fogs with minimum RT. [20] This algorithm is an extension of closest data center. However, it also uses RT as a parameter for selection of fog. It also checks the list of RT of fogs and change the number of VM's count on the basis of RT.

6   Simulation Results and Discussions

In this article, different simulations are carried out for comparison between various load balancing algorithms. Service broker policies used in this paper are: closest data center and dynamic service broker. Results are compared between these two policies. Load balancing algorithms are: RR, SBLB and HBO. These algorithms are used for handling user requests and resource allocation. Performance parameters considered in this paper are: RT, cost (VM, MG, transfer and total cost) and PT.

Total five regions are considered based on number of Smart Buildings (SB) in each region and cloud facility is available. Each region represents the continent of the world. Every region is connected with three MG's that provide electricity in case of Consumer electricity demand (Table 1).

Table 1. Region distribution

Region	Region id
North America	0
South America	1
Asia	2
Europe	3
Australia	4

The average RT comparison of each fog is shown in Table 2 using three load balancing algorithm and service broker polices, where SBLB performs better as compared to other algorithms in closest Data center policy.

Table 2. RT of fogs using different algorithms are broker policies

Broker policy	Algorithm	Avg (ms)	Min (ms)	Max(ms)
Dynamic service	RR	52.39	37.86	140.81
	SBLB	51.15	37.36	132.07
	HBO	53.05	37.36	70.58
Closest datacenter	RR	51.21	37.32	67.62
	SBLB	50.67	37.38	67.36
	HBO	51.91	37.32	68.49

Figure 3 shows the PT comparison of RR, SBLB and HBO. Figure 2 shows dynamic service broker policy results and Fig. 3 shows closest data center policy results. Where in Fig. 2 we can see that RR with 4.3, 2.3 and 2.5 respectively, however with HBO and SBLB PT is 4.3, 2.7, 2.9 and 1.7, 1.2 and 1.3 respectively. We can see a slight difference among HBO, RR and SBLB, while SBLB outperforms these two techniques.

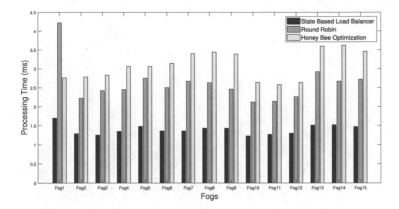

Fig. 2. PT of all fogs in dynamic service proximity policy

In Fig. 3 we can see that PT of RR with 0.7, 1.3 and 1.80 respectively, however with HBO and SBLB the PT is 0.6, 0.8, 0.9 and 1.2, 1.8 and 1.2. From the results we can conclude that SBLB works better as compared to HBO and RR.

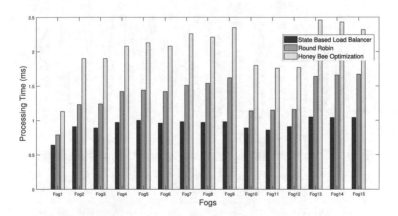

Fig. 3. PT of all fogs using closest data center

Figures 4 and 5 show cost wise comparison between HBO, RR and SBLB algorithms with respect to closest data center policy and dynamically reconfigure, which shows that cost is almost same. However, there is some changes of values in different fogs but overall the average cost remains same.

Similarly, in Figs. 6 and 7 RT of clusters of Buildings with respect to service broker policies and tested on different algorithms. In both scenario's it can be seen that RT of SBLB is better than RR.

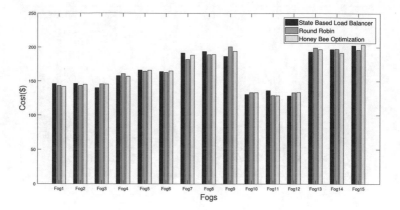

Fig. 4. Cost of all fogs in closest data center

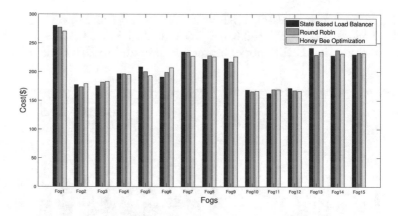

Fig. 5. Cost of all fogs in dynamically reconfigure data center

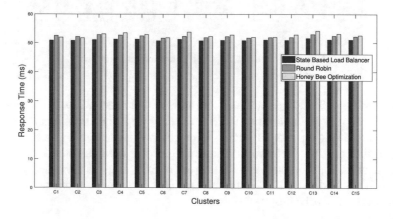

Fig. 6. RT of all clusters of buildings in reconfigure dynamically data center policy.

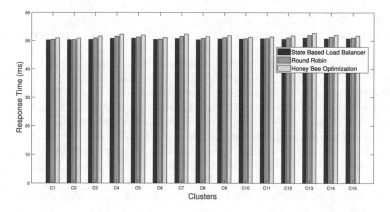

Fig. 7. RT of all clusters of buildings in closest data center policy.

7 Conclusion

In this article, a cloud-fog based SG architecture is proposed for effective task allocation. Three load balancing algorithms: RR, SBLB and HBO are used for VM allocation. Service broker policies used for simulations are: dynamically reconfigure and closest data center. Simulations are carried out on Cloud Analyst tool [20]. Results show that SBLB gives more performance by minimizing the RT and PT than other techniques.

References

1. Luo, F.: Cloud-based information infrastructure for next-generation power grid: conception, architecture, and applications. IEEE Trans. Smart Grid **7**(4), 1896–1912 (2016)
2. Gelazanskas, L., Gamage, K.A.A.: Demand side management in smart grid: a review and proposals for future direction. Sustain. Cities Soc. **11**, 22–30 (2014)
3. Cao, Z.: Optimal cloud computing resource allocation for demand side management in smart grid. IEEE Trans. Smart Grid **8**(4), 1943–1955 (2017)
4. Faruque, A., Abdullah, M., Vatanparvar, K.: Energy management-as-a-service over fog computing platform. IEEE Internet Things J. **3**(2), 161–169 (2016)
5. Bhardwaj, S., Jain, L., Jain, S.: Cloud computing: a study of infrastructure as a service (IAAS). Int. J. Eng. Inf. Technol. **2**(1), 60–63 (2010)
6. Naha, R.K., Othman, M.: Cost-aware service brokering and performance sentient load balancing algorithms in the cloud. J. Netw. Comput. Appl. **75**, 47–57 (2016)
7. Krishna, P.V.: Honey bee behavior inspired load balancing of tasks in cloud computing environments. Appl. Soft Comput. **13**(5), 2292–2303 (2013)
8. Yigit, M., Gungor, V.C., Baktir, S.: Cloud computing for smart grid applications. Comput. Netw. **70**, 312–329 (2014)
9. Yoldaş, Y.: Enhancing smart grid with microgrids: challenges and opportunities. Renew. Sustain. Energy Rev. **72**, 205–214 (2017)

10. Chekired, D.A., Khoukhi, L.: Smart grid solution for charging and discharging services based on cloud computing scheduling. IEEE Trans. Ind. Inform. **13**, 3312–3321 (2017)
11. Aslam, S.: Towards efficient energy management and power trading in a residential area via integrating a grid-connected microgrid. Sustainability **10**(4), 1245 (2018)
12. Kazmi, S., et al.: Towards optimization of metaheuristic algorithms for IoT enabled smart homes targeting balanced demand and supply of energy. IEEE Access (2017)
13. Chen, S.-L., Chen, Y.-Y., Kuo, S.-H.: CLB: a novel load balancing architecture and algorithm for cloud services. Comput. Electr. Eng. **58**, 154–160 (2017)
14. Yang, J., et al.: A task scheduling algorithm considering game theory designed for energy management in cloud computing. Future Gener. Comput. Syst. (2017)
15. Kaur, P., Mehta, S.: Resource provisioning and work flow scheduling in clouds using augmented Shuffled Frog Leaping Algorithm. J. Parallel Distrib. Comput. **101**, 41–50 (2017)
16. Choudhary, A.: A GSA based hybrid algorithm for bi-objective workflow scheduling in cloud computing. Future Gener. Comput. Syst. **83**, 14–26 (2018)
17. Khalid, A.: Towards dynamic coordination among home appliances using multi-objective energy optimization for demand side management in smart buildings. IEEE Access **6**, 19509–19529 (2018)
18. Hussain, H.M.: An efficient demand side management system with a new optimized home energy management controller in smart grid. Energies **11**(1), 190 (2018)
19. Kong, Y., Zhang, M., Ye, D.: A belief propagation-based method for task allocation in open and dynamic cloud environments. Knowl. Based Syst. **115**, 123–132 (2017)
20. Wickremasinghe, B., Buyya, R.: CloudAnalyst: a CloudSim-based tool for modelling and analysis of large scale cloud computing environments. MEDC Project Rep. **22**(6), 433–659 (2009)

Voting Process with Blockchain Technology: Auditable Blockchain Voting System

Michał Pawlak, Jakub Guziur, and Aneta Poniszewska-Marańda[✉]

Institute of Information Technology, Lodz University of Technology, Łódź, Poland
{michal.pawlak,jakub.guziur}@edu.p.lodz.pl,
aneta.poniszewska-maranda@p.lodz.pl

Abstract. There are various methods and approaches to electronic voting all around the world. Each is connected with different benefits and issues. One of the most important and prevalent problems is lack of auditing capabilities and system verification methods. Blockchain technology, which recently gained a lot of attention, can provide a solution to this issue. This paper presents Auditable Blockchain Voting System (ABVS), which describes e-voting processes and components of a supervised internet voting system that is audit and verification capable. ABVS achieves this through utilization of blockchain technology and voter-verified paper audit trail.

1 Introduction

In 2008, an individual or a group under the pseudonym Satoshi Nakamoto introduced a new and promising digital currency called Bitcoin [1]. It is based on blockchain technology, which since then became widely considered to have potential to revolutionize many fields not only in financial sector [2]. The idea of a blockchain is simple: it is a distributed system of ledgers stored in a chain of connected blocks and an algorithm that collectively negotiates and validates contents of the blocks in a peer-to-peer network [3,4]. The technology attracted a lot of attention and is now being thoroughly researched. Many potential and practical applications have already been identified by various researchers [2,5]. One of the very promising avenues of research is electronic voting (e-voting).

Voting is a basis of democracy and despite complex security measures, it is not free from frauds [6,7]. Modern voting systems are slow because in order to determine results ballots must be firstly collected from different locations and then counted by a single central institution. Moreover, the results are not verifiable. Voters have no way of assuring that their votes were included in the results or whether their ballots were tampered with [8]. This sparked interest in electronic voting, which resulted in the creation of many e-voting solutions [9,10]. E-voting solves many of the mentioned problems affecting voting processes (e.g. quick determination of results). However, the electronic voting systems used

© Springer Nature Switzerland AG 2019
F. Xhafa et al. (Eds.): INCoS 2018, LNDECT 23, pp. 233–244, 2019.
https://doi.org/10.1007/978-3-319-98557-2_21

today are not ideal and must face different issues like authentication, privacy or data integrity [6,11].

Blockchain technology can provide a solution to these problems. One of the properties of blockchain is an ability to create a platform for a public verification of data stored inside. This creates a possibility of creation of an e-voting system which can be audited by common voters instead of dedicated institutions and officials. Recognizing a great opportunity, some countries already started researching and implementing blockchain-based e-voting systems [11,12].

The existing blockchain-based voting systems provide its users with many advantages mostly connected with possibility of secure, anonymous and verifiable voting through an internet connection. These properties are obtained by utilization of various cryptographic techniques in addition to blockchain. However, the existing systems leave voter identification and authentication to election officials or depend only on cryptography. This can remove benefits of remote voting and create possibility of impersonating voters respectively.

In this paper, a full end-to-end blockchain-based e-voting system is presented. It is intended to be an enhancement and improvement of the existing voting process in Poland. The main goal of the system is to provide voters with the ability to follow and verify their votes. Furthermore, due to the distributed nature of blockchains the system provides an easy way of verifying voting results.

The paper is organised in the following way: Sect. 2 presents technical aspects of blockchain technology, Sect. 3 discusses a context of e-voting and its applications. Section 4 deals with an overview of works related to this field, while Sect. 5 describes the original e-voting system based on blockchain technology.

2   Blockchain Technology

Blockchain technology is a combination of two main elements [3]:

- blockchain data-structure,
- blockchain algorithm.

Data in blockchain is organized in units called blocks, which are connected to each other in a chronologically ordered chain-like structure. Each block is made of: (i) a block header, which stores block metadata (e.g. creation timestamp), a hash reference to the previous block header and a hash reference to transaction data; (ii) transaction data, for instance, transfer of money. Figure 1 presents an exemplary blockchain and how its elements are connected with each other. Ovals represent hash references in the block headers and arrows point to referenced elements.

It is important to note that the references are obtained through hashing of the data contained within each block with a cryptographic hash function. It is a one-way function that maps data of arbitrary size to a unique bit string of fixed size (hash value). One of the most important properties of this function is that any change in the input data results in a change in the hash value [13].

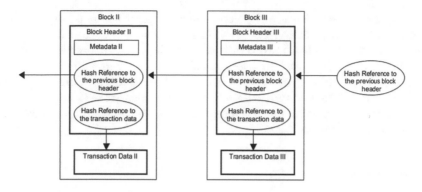

Fig. 1. Structure of a blockchain, adapted from [3]

The blockchain algorithm defines a sequence of instructions, which goal is to negotiate the information content of blockchains in a distributed peer-to-peer network of nodes. Each node maintains their own version of the whole blockchain and takes part in collective verification of each new block. In order for a new block to be appended to the chain, more than 50% of all nodes must give consent [3,4].

Blockchain technology extensively uses another cryptographic technology, namely *asymmetric cryptograph* [3,4]. It uses two keys: public and private. Both can be used to encrypt a message. However, a message encrypted with one key can be only decrypted with the other key. In other words, a message encrypted with a public key can be decrypted only with a corresponding private key and vice versa.

In general, there are two ways of using the public-private key pair [3]:

- *public-to-private*, where a message is encrypted with the publicly available public key and then decrypted with the private key, similarly to a mailbox where anybody who knows the address can send a message but only the addressee can open it
- *private-to-public*, where the author encrypts the message with private key and publishes it, so anyone with the corresponding public key can read the message, which is a way of proving ownership.

The blockchain uses asymmetric cryptography to identify accounts through public-to-private method. The accounts are differentiated by their addresses which are also cryptographic public keys. Initiators encrypt their transactions with the receiver's account number so only the transaction receiver can decrypt it. On the other hand, the private-to-public approach is used as a way of authorizing transactions. The process is simple: the message is encrypted and broadcasted to all nodes in the system, which check the authorisation by decrypting the message with the public key of the initiator.

Through a combination of all these technologies, blockchain technology provides many advantages over traditional storage systems. It is censorship-poof as there is no single central authority controlling the network. It is very secure and resilient on both transaction and system level. The blockchain delivers transparent data as transactions are conducted, verified, and instantly added to the blockchain, which is stored in every node of the system and can be viewed by anyone. An additional possible benefit is obtaining data in almost real time (depending on a given system). All these characteristics are desirable for e-voting systems, due to the fact that such systems need to be transparent, auditable and resistant to modification.

3   E-Voting

Electronic voting (e-voting) can be defined as employing electronic means or information technologies to conduct voting procedures [10, 11]. In general, every e-voting system consists of registration, authentication and authorization, vote casting, vote counting and vote verification [14]. E-voting is a broad term which covers a variety of different systems, solutions and implementations [11]. E-voting systems provide numerous benefits which include [15]:

- prevention of fraud, by reducing human involvement,
- acceleration of results processing,
- reduction of spoilt ballots by improved presentation and automatic validation of ballots,
- reduction of costs due to decreasing voting overhead,
- increase involvement in democratic process due to easer availability (remote voting),
- potential for more direct democracy.

Introducing an e-voting system is inherently connected with many challenges which are not only technical but also procedural and legislative. One the most substantial of weaknesses of e-voting systems are [15]: (i) lack of transparency and understanding of the system which leads to lack of trust in the solution and undermines its whole sense; (ii) lack of widely accepted standards for e-voting systems; (iii) risk of fraud and manipulation by privileged insiders or hackers; (iv) increased costs of voting infrastructure with regard to power supply, communication technology, etc.

What all e-voting systems have in common are basic requirements that must be met before a system can be considered valid. These requirements are the following [6, 16–18]:

- voter identification and authentication,
- voter privacy assurance,
- correctness,
- transparency and verifiability,
- ballot integrity,
- availability.

Voter identification and authentication requirement refers to ensuring that only people eligible will cast their votes. *Voter privacy assurance* describes a need for keeping voters anonymous and their choices secret and not connected with each other. *Correctness* is correct counting of all votes, disregarding invalid ones and enforcing each voter has a single vote. *Transparency and verifiability* are properties necessary to build trust which is essential for any electronic voting system. *Ballot integrity* refers to a requirement of protecting votes from being modified, removed or duplicated. Additionally, it also covers detection of attempts at such manipulation. *Availability* is making the system accessible and usable by any eligible citizen.

In general, e-voting systems can be differentiated by two key characteristics. The first is whether a system is remote (ballots are transmitted through a communication channel to some central location) or non-remote (ballots are stored locally on some a type of medium). The second distinction is whether a system is supervised (voting takes place in a supervised and controlled by election officials locations) or unsupervised (voting is not managed by election officials). These characteristics can be combined in different ways to create many distinct forms of e-voting [11, 19].

Additionally, there exists many different approaches and solutions for e-voting. In general, they can be divided into four groups [11, 19, 20]: electronic dedicated voting machines, optical scanning voting machines, electronic ballot printers, internet voting.

Electronic dedicated voting machines or direct-recording electronic (DRE) voting machines are devices that through use of keyboards, touch-screens, mouse devices and other interactive devices allow voters to directly record their votes. *Optical scanning voting machines* are machines that use an optical scanner to record and count votes from machine-readable paper ballots. *Electronic ballot printers* are machines that record the voters' choices and produce paper recipes or tokens which can be used in ballot boxes or as an input to some electronic counting machine.

As can be seen, electronic voting has numerous significant advantages which cannot be overlooked. However, before any e-voting system can be implemented and used, it must first overcome many challenges, some directly connected to e-voting concept itself and some connected to a specific implementation.

4   Related Works

Electronic voting is still a relatively unexplored field despite numerous publications [9]. In [12] authors describe research on possible applications of blockchain technology in e-governance and e-voting conducted in the Digital 5 (D5) countries. From the point of view of this work, cases of Estonia and South Korea are particularly interesting as those two countries actively research implementation of blockchain technology in electronic voting. In 2017 South Korea conducted a successful community vote using this technology.

One of the requirements of e-voting systems is transparency and verifiability. With that in mind, the authors of [21] analyse in detail what that requirement means for e-voting in general and examine methods of auditing and of such systems. Additionally, the publication provides a brief overview of some existing e-voting systems, i.a. Helios and WAVE. More practical considerations are presented in [22], where a working e-voting system using *Paillier homomorphic encryption algorithm* is described. The algorithm allows reading encrypted messages without decrypting them. This enables the system to verify voters by requiring their identification information without revealing their choices.

Another e-voting system called SAVE is described in [17]. The SAVE system is intended for medium to large scale voting, for example elections on universities. The work covers processes and components involved in SAVE system. The SAVE itself is a supervised voting with voting machines in a form of commonly available personal computers or smartphones. In addition, the system requires printers for generating VVPAT. Finally, the SAVE utilizes symmetric encryption for software signing, RSA asymmetric (2048 bit key length) encryption for data encryption and signing and HMAC-SHA256 for message authentication. Furthermore, blockchain-based e-voting solutions also exist. *Agora* [23] is a customizable multi-layer system which architecture enables high throughput and low overhead, which make Agora ideal for low bandwidth devices. The system allows conducting unsupervised and supervised internet voting with public *permissioned-and-permissionless blockchain hybrid*. It is important to note, that Agora leaves the identification and authentication to the voting administrators but offers a possibility of using digital signatures for this process.

The voting system presented in [24] is an e-voting system based on Bitcoin and is intended to be end-to-end verifiable. It utilizes *Anonymous Kerberos* protocol for authentication and transactions between a voter and a candidate for voting. The solution uses "tokens" to represent votes, which are smallest possible quantity of bitcoins that can be transferred plus transaction fee. In its basic form, the system fulfils most requirements of the electronic voting with excerption of voter's privacy, due to the possibility of linking voters with their transactions in blockchain. The presented in the paper system also requires the voters to register in person with election officials before they can participate in the voting process.

FollowMyVote [25] is an Ethereum-based voting platform for remote and unsupervised voting. It utilizes elliptic curve cryptography on the stored transactions for improving the security. The system uses webcams and official IDs for remote voter authentication and identification. Furthermore, the system allows users to watch the progress of elections in real time and switch their votes during the election.

5   Auditable Blockchain Voting System

Auditable Blockchain Voting System (ABVS) is intended to be a non-remote and supervised electronic voting system that utilizes internet connection to transmit votes and store them in a blockchain. The system is being in development stage and is intended to enhance and improve existing voting process in Poland. The following subsections present the ABVS process overview, components and implementation details.

5.1   Auditable Blockchain Voting System Process Overview

The voting process of the ABVS system is divided into three subsequent stages (Fig. 2):

1. election setup,
2. voting,
3. counting and verification.

Election setup is the first stage of the process. It starts with a selection of trusted public institutions which will act as nodes in the blockchain system of ABVS, by providing computing power and storage capabilities. The next step consists of a preparation of ABVS software and hardware. Election officials must sign and certify software and hardware for the ABVS procedure. In addition, an application for counting should be created and also certified. The next step is to generate unique Vote Identification Tokens (VITs) for voters, which are used for identification and verification of votes in later stages. The VITs are split and assigned to polling stations. Finally, the VITs and the ABVS equipment are distributed between polling stations.

The *Voting* stage begins right before the actual voting and ends with the closing of the polling stations. The election officials at each of the polling stations launch the ABVS hardware and software. The trusted nodes will only accept votes from the signed and certified equipment and locations. When the actual election begins, voters identify themselves with the election officials (in case of Poland, by presenting ID cards and signing in a registry). Next, each voter selects a random identification token containing a unique numeric code (VIT). Having completed these steps, voters proceed to voting stations which are equipped with printers and computers connected to the Internet. The voters cast their votes through application interface and confirm their choices using the codes form the VITs. The votes are sent through secured and encrypted communication channel to the trusted nodes where they are verified and processed by the blockchain algorithm. Approved blocks are added to a vote blockchain. At the same time, voters receive VVPATs and depose them into ballot boxes. Each VVPAT is mapped to a corresponding block in the blockchain via VIT present in both of them. The voters leave the polling stations with their VITs. Using the number that is provided on them, the voters can follow their vote and find it in the blockchain. It is important to note that at this stage it is impossible to check a "value" of the votes. Only existence of the votes can be verified. ABVS does not change the existing voting procedure in its general sense and due to its supervised and non-remote design, ABVS prevents possibility of voter impersonation.

Counting and verification stage starts after a defined time of the election elapses. The election officials at each polling station deactivate the system and open containers with unused VITs. A list of the remaining tokens is sent to the national electoral central authorities for vote verification. Moreover, the election staff at each node creates a backup of the blockchain stored at their nodes. From that moment, each of the voters can verify not only presence and correctness of their votes (assuming they still possess their VITs) but can also examine the whole chain. This, with the addition of VVPATs, allows vote audition and verification. Using the vote tokens it is possible to compare votes in the blockchain with their corresponding paper trail. At the same time, votes in the blockchain are assessed whether they contain a valid VIT (not on the list of unused) and whether they came from their assigned polling stations.

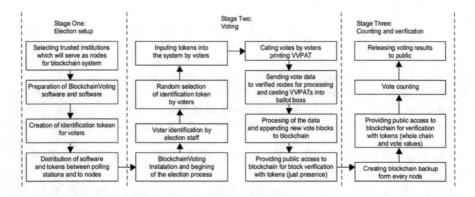

Fig. 2. Schema of Auditable Blockchain Voting System stages

5.2   Auditable Blockchain Voting System Components

ABVS is composed of five major components:

1. network of trusted nodes and polling stations,
2. Vote Identification Tokens,
3. ABVS blockchain technology,
4. counting application,
5. voter-verified paper audit trail,
6. vote error notification module.

Figure 3 presents a schema of relations between components of ABVS. Each component is represented by a rectangular box. The components are connected by certain relations shown by labelled arrows.

Network of trusted nodes and polling stations is made of two parts. The first one consists of a trusted super-node representing national electoral central authorities (National Electoral Commission for Poland) and pre-selected and verified public institutions (e.g. universities). All of the nodes will act as providers of computing power and storage space for the ABVS blockchain system. Furthermore, they will be responsible for verification of the transactions

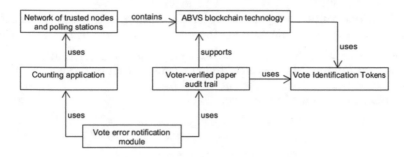

Fig. 3. Schema of relations between components of Auditable Blockchain Voting System

and of the whole blockchain, with the super-node taking precedence. The second part is composed of the polling stations understood as the ABVS software and hardware associated with the actual locations of the polling stations. The voters use them to cast their votes, which will be broadcasted to the nodes for verification and processing in accordance with the blockchain paradigm. Figure 4 illustrates the concept of this network.

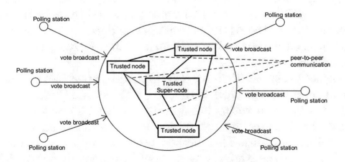

Fig. 4. Schema of the Auditable Blockchain Voting System network

Vote Identification Tokes (VITs) are alphanumerical codes which are used as means of authentication and authorization of the voters within the system. In addition, the VITs enable vote following and identification during and after the election. They can be stored on scratch cards, paper sheet in envelopes or any other mean that allows random selection of by the voters without revealing their contents in advance. The VITs are created in the first stage of the election and are assigned and distributed among the polling stations. A database pairing VITs with polling stations is maintained in the trusted nodes.

ABVS blockchain technology is the core of the whole system. It follows the blockchain paradigm with one exception. The nodes that verify and process new votes (transactions) are restricted to only verified and certified public institutions not just anyone willing to participate in the network. The ABVS blockchain data-structure is made of blocks that store the following transaction data (Fig. 5):

- Vote Identification Token, for a vote identification and verification,
- polling station identifier for identifying origin station of the votes,
- vote value, which contains the actual voters choice,
- vote timestamp, for additional security.

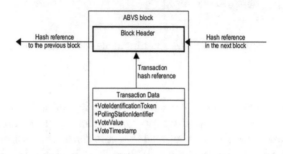

Fig. 5. Auditable Blockchain Voting System block model

Counting application is a signed and certified application which task is to iterate over the ABVS blockchain and count the votes. In order to provide fault tolerance, each node of the ABVS network is equipped with its own copy of the counting application for the software to create multiple reference points for the election result verification.

Voter-verified paper audit trail (VVPAT) is a paper ballot containing the same information as a single block in the ABVS blockchain (Fig. 5). The VVPATs are printed by printers in voting booths after the voters cast their e-votes. The paper ballots are disposed by the voters into traditional ballot boxes before leaving the polling stations. This method provides additional audit and verification capabilities to the whole system. ABVS assumes that the VVPATs take precedence over the blockchain contents in case of inconsistencies.

Vote error notification module is a dedicated application for error notifications. The voters who find inconsistencies with their votes can anonymously notify the election officials by providing their VITs and error explanation. Complaints with valid VITs are processed further, which results in comparison of the corresponding blocks from the ABVS blockchain with the analogous VVPATs.

The main benefit of ABVS is the possibility of its end-to-end verification thanks to the usage of blockchain technology. The voters are able follow and control their votes and due to the use of the Vote Identification Tokens, VVPATs and Vote error notification module can remain anonymous.

6   Conclusions and Future Works

There exist four main approaches to electronic voting: electronic dedicated voting machines, optical scanning voting machines, electronic ballot printers and internet voting. Each method has its own benefits and issues. Furthermore, due to a lack of standardisation and norms, each implemented e-voting system is different causing fragmentation of the e-voting field.

In this paper, Auditable Blockchain Voting System (ABVS) is described. ABVS is a non-remote internet voting system. It utilizes blockchain technology for storing cast votes in order to provide ballot integrity and vote correctness. Furthermore, voter-verified paper audit trails (VVPATs) are used to ensure verifiability of the results stored in the blockchain, thus further enforcing ballot integrity and correctness. The final component of ABVS is Vote Identification Token (VIT) which is a unique alphanumerical code randomly selected by the voters to identify their votes in the blockchain and in the audit trail while remaining anonymous.

AVBS fulfils all the requirements of the electronic voting systems and delivers an e-voting solution which provides voters with full audit and verification capabilities. In the near future, ABVS should be fully implemented and tested on all stages. Especially extensive software and system tests are necessary in order to adapt and adjust the system to available infrastructure and tasks at hand. Furthermore, the voter privacy assurance, which is an e-voting property not fully satisfied by ABVS, should be examined in order to find a possible solution.

References

1. Satoshi, N.: Bitcoin: a peer-to-peer electronic cash system (2008). https://bitcoin.org/bitcoin.pdf. Accessed 2018
2. Zhao, J.L., Fan, S., Yan, J.: Overview of business innovations and research opportunities in blockchain and introduction to the special issue. In: Financial Innovation, pp. 2–28. Springer, Heidelberg (2016)
3. Drescher, D.: Blockchain basics: a non-technical introduction in 25 steps, 1st edn. Apress, Frankfurt am Main (2017)
4. Karame, G., Audroulaki, E.: Bitcoin and Blockchain Security. Artech House Inc, Norwood (2016)
5. Risius, M., Spohrer, K.: A blockchain research framework - what we (don't) know, where we go from here, and how we will get there. Bus. Inf. Syst. Eng. **59**(6), 385–409 (2017)
6. De Faveri, C., Moreira, A., Arajo, J.: Towards security modeling of e-voting systems. In: Proceedings of IEEE 24th International Requirements Engineering Conference Workshops (REW), Beijing, China (2016)
7. Lehoucq, F.: Electoral fraud: causes, types, and consequences. Ann. Rev. Polit. Sci. **6**(1), 233–256 (2003)
8. Qi, R., Feng, C., Liu, Z., Mrad, N., Qi, R.: Blockchain-powered Internet of Things, E-Governance and E-Democracy. In: E-Democracy for Smart Cities, pp. 509–520. Springer, Singapore (2017)
9. Mpekoa, N., van Greunen, D.: E-voting experiences: a case of Namibia and Estonia. In: Proceedings of IST-Africa Week Conference (IST-Africa), Windhoek (2017)
10. Willemson, J.: Bits or paper: which should get to carry your vote? J. Inf. Secur. Appl. **38**, 124–131 (2018)

11. Braun Binder, N., Driza, A., Krimmer, R., Serdlt, U., Vinkel, P.: Focus on E-Voting, ACE Electoral Knowledge Network (2014). http://aceproject.org/ace-en/focus/e-voting/about. Accessed 22 Jan 2018
12. Ojo, A., Adebayo, S.: Blockchain as a next generation government information infrastructure: a review of initiatives in D5 countries. In: Government 3.0 – Next Generation Government Technology Infrastructure and Services, pp. 283–298. Springer, Cham (2017)
13. Stallings, W.: Chapter 11 cryptographic hash functions. In: Cryptography and Network Security: Principles and Practice, 6th edn, pp. 313–354. Pearson Education, Inc. (2013)
14. Naidu, P.S., Kharat, R., Tekade, R., Mendhe, P., Magade, V.: E-voting system using visual cryptography & secure multi-party computation. In: Proceedings of International Conference on Computing Communication Control and automation (ICCUBEA), Pune, India (2016)
15. Wolf, P., Nackerdien, R., Tuccinardi, D.: Introducing electronic voting: essential considerations. International Institute for Democracy and Electoral Assistance, 1 December 2011
16. Zhou, Y., Zhou, Y., Chen, S., Wu, S.S.: MVP: an efficient anonymous E-voting protocol. In: Proceedings of Global Communications Conference (GLOBECOM), Washington, DC, USA (2016)
17. Ochoa, X., Pelez, E.: Affordable and secure electronic voting for university elections: the SAVE case study. In: Proceedings of 4th International Conference on eDemocracy & eGovernment (ICEDEG), Quito, Ecuador (2017)
18. Caarls, S.: E-Voting Handbook: Key Steps in the Implementation of E-Enabled Elections. Council of Europe, Strasbourg (2010)
19. National Democratic Institute: Common Electronic Voting and Counting Technologies. https://www.ndi.org/e-voting-guide/common-electronic-voting-and-counting-technologies. Accessed 22 Jan 2018
20. United Nations Development Programm: Feasibility study on Internet Voting for the Central Electoral Commission of the Republic of Moldova: Report and Preliminary Roadmap (2016)
21. Mello-Stark, S., Lamagna, E.A.: The need for audit-capable e-voting systems. In: Proceedings of 31st International Conference on Advanced Information Networking and Applications Workshops (WAINA), Taipei, Taiwan (2017)
22. Anggriane, S.M., Nasution, S.M., Azmi, F.: Advanced e-voting system using Paillier homomorphic encryption algorithm. In: Proceedings of International Conference on Informatics and Computing (ICIC), Mataram, Indonesia (2016)
23. Agora Technologies: Agora_Whitepaper_v0.2.pd (2015). https://agora.vote/Agora_Whitepaper_v0.2.pdf. Accessed 20 Apr 2018
24. Bistarelli, S., Mantilacci, M., Santancini, P., Santini, F.: An end-to-end voting-system based on bitcoin. In: Proceedings of Symposium on Applied Computing, SAC 2017, New York, USA (2017)
25. Follow My Vote: The Online Voting Platform of The Future – Follow My Vote, Follow My Vote. https://followmyvote.com. Accessed 26 Jan 2018

Towards the Analysis of Self-rated Health Using Supervised Machine Learning and Business Intelligence

Laia Subirats[1,2(✉)], Estefania Piñeiro[2], Jordi Conesa[2], and Manuel Armayones[2]

[1] Eurecat, Centre Tecnològic de Catalunya, Unitat de eHealth,
C/ Bilbao 72, 08005 Barcelona, Spain
laia.subirats@eurecat.org

[2] eHealth Center, Open University of Catalonia, Rambla del Poblenou,
156, 08018 Barcelona, Spain
{epineiroc,jconesac,marmayones}@uoc.edu

Abstract. The perception that every person has about her health is very important for health, health evolution and therefore for enjoying a healthy life. So, it is important to know what the factors that most influence self-rated health are. If these factors are known, changes that have long term impact on people's health can be proposed, even for unhealthy people. The present paper performs a preliminary study in that direction, analyzing the causes of self-rated health from a couple of datasets: one private with 4848 people and one public with 802 people. These datasets have been analyzed to find out what socioeconomic, biological and environmental factors have more influence on health status and a dashboard has been created to allow analyzing the data interactively. Results show some factors that influence self-rated health such as chronic diseases, limitation in daily activities and depression were important.

Keywords: Quality of life · Supervised machine learning
Self-rated health · Business intelligence

1 Introduction

Self-rated health is an important measure as it has influence on mortality and provides very important information on health status and is an independent predictor of mortality in a large number of studies, [1]. There are several measures for self-rated health, perhaps the most popular one in Europe is EuroQoL [2], that is the one used in the health survey of Catalonia [3]. Doctors or practitioners can use this information at distance in order to follow the evolution of health both in a temporal and population manner becoming a useful tool to take public health decisions.

There are several longitudinal analysis of self-rated health in Spain. It was found that self-rated health is a predictive measure of mortality in both men and women, and it acts as a mediator between socio-economic, demographic

© Springer Nature Switzerland AG 2019
F. Xhafa et al. (Eds.): INCoS 2018, LNDECT 23, pp. 245–254, 2019.
https://doi.org/10.1007/978-3-319-98557-2_22

and health conditions [4]. Moreover, self-rated health is a predictor of hospital services use [5]. Another important issue to have in mind is that factors associated with self-rated health differ across age groups, particularly for lifestyle [6]. Among older people, inequalities in self-perceived health due to socioeconomic status and gender have persisted in time [7]. Therefore, it is important that governments develop health policies in this sector of the population. In addition, there are also self-reported health studies in Spain of disabilities of neurological [8,9] and cardiac origin [10]. In this case they use the World Health Organization (WHO) standard of the International Classification of Functioning, Disability and Health.

In Europe, several studies about self-rated health have also been done. They found that psychosocial conditions, material conditions and lifestyle factors are independently related with poor self-rated health [11]. They also found that there is also positive relationship between physical activity and self-rated health [12]. Finally, there are also studies that include countries outside Europe [13]. They found that there are differences in self-rated health among countries but could not be explained with differences in status, health condition or functional ability.

The goal of this research is to find out what variables influence most in the self-perception of health. To do so, two datasets resultant from the self-rated health surveys from different Spanish regions have been gathered and analyzed. Then, a dashboard has been created in order to allow users to analyze the obtained data. The paper presents the knowledge obtained during the analysis of the datasets, the differences found in the two regions and the dashboard created to analyze the data interactively.

The rest of the paper is organized as follows. Section 2 describes the materials and methods for their evaluation. Section 3 presents the results obtained and its usefulness. Finally, conclusions and future work are drawn in Sect. 4.

2   Materials and Methods

This paper presents the results of the study of two self-rated health datasets: one from Catalonia [3] and another one from Madrid [14].

The dataset from Catalonia corresponds to the results of the *Enquesta de Salut de Catalunya (ESCA)*, promoted by the Ministry of Health of the Catalan Government. It consists of thousands of personal interviews per year, done at the home of interviewed. These interviews have been conducted continuously since 2010 and collect information about the health of Catalonia inhabitants, without age restrictions, their behavior related with health and their use of healthcare services. Its goal is to provide relevant information for the definition and evaluation of healthcare policies. The dataset is divided into the following sections: socio-demographic data, health coverage, health state and quality of life, chronic diseases and accidents, consumptions of medicines, mental health, limitations, disabilities and personal autonomy, preventive practices, social support, psychological wellbeing, lifestyle and medical visits. The dataset was of 4818 people.

The dataset from Madrid corresponds to the result of the *Encuesta de Salud de la Ciudad de Madrid*. It consists of over 800 interviews performed to citizens from Madrid who are 16 years old or older. The interview assesses questions about health, health habits, healthy (or unhealthy) lifestyles and the main health problems of citizens. The dataset from Madrid is divided into the following sections: demographic data, home and housing, self-rated health and quality of life, mobility and activity limitation, functional dependency, mental health, health habits (such as physical state and leisure), preventive activities, work health and work situation. The dataset was of 802 people.

One of the goals of this work is to create a model that helps to estimate whether an observation corresponds to a good, bad or regular perceived health. This model, together with the dashboard, will provide information about certain patterns that may lead one person to a bad, regular or good perceived health.

Before applying classification algorithms to the data, some pre-processing has been done in order to improve the quality and suitability of data. In this process, some errors have been arranged, some variables have been joined, other irrelevant variables have been detected and deleted and the most suitable variables have been found. Some examples of actions conducted in this stage are removing variables that are too similar to the self-perceived health status (for example perceived self-status or quality of life in a short period of time) and converting categorical values to numbers (an attribute has been created for each category). The datasets analyzed suffer from the curse of dimensionality [15], since they have too many variables and not enough observations. For example, the Madrid dataset has 115 variables and just 802 observations. In order to deal with this problem we redefined the 5 value scale of the classifier variable (self-perceived health) from 5 (very bad, bad, regular, good, very good) to 3 (bad - includes very bad and bad-, regular, good -includes good and very good-).

Principal component analysis (PCA) was also used to reduce dimensionality (see Fig. 2). The first two components of the PCA are displayed using as colors the self-perceived health status. This representation helps to understand people's characteristics with reference to the self-rated health.

In both studies we will use random forest that it is a machine learning method often used to predict classes with a considerable number of attributes. "Random forest (rf) classifiers are a collection of decision trees [16], each of which is trained on a random subset of the training data and only allowed to use some random subset of the features. There is no coordination in the randomization (a particular data point or feature could randomly get plugged into all the trees, none of the trees, or anything in between). The final classification score for a point is the average of the scores from all the trees. The hope is that the different trees will pick up on different patterns and each one will only give confident guesses when its pattern is present. When it comes time to classify a point, several of the trees will classify it correctly and strongly while the other trees give answers that are on the fence, meaning that the overall classifier slouches toward the right answer. The individual trees in a random forest are subject to overfitting, but they tend to be randomly overfitted in different ways. These largely cancel each other out, yielding a robust classifier."

Random forest follow the following steps. Repeat: Take a bootstrap sample from the data, fit a classification regression tree. At each node: (1) Select m variables at random out of all M possible variables (independently at each node) (2) Find the best split on the selected m variables and (3) Grow the trees big. We can combine it by voting (classification) or averaging (regression). Random forest have several properties: same idea for regression and classification; handle categorical prediction naturally; quick to fit, even for large problems; no formal distributional assumptions; automatically fits highly non-linear interactions; automatic variable selection; handle missing values through proximities; not very easy to interpret if the tree is small; the terminal nodes does not suggest a natural clustering; the picture cannot give valuable insight into which variable are important and where.

Afterwards 80% of the data was used to create the training set and 20% of the data was used to create the test set. Then the grid search has been performed standardizing the data, performing 10-cross fold validation, and computing the best f1-weighted score trying the following parameters: classifier_n_estimators: [100, 500, 700], classifier_max_features: [auto, sqrt, log2], classifier_class_weight: [balanced, None]. It should be pointed out that the number of classifier_n_estimators in the Catalonia data was limited to 100 due to memory restrictions of the computer.

Finally, we used Tableau to create a dashboard to visualize the information [17]. The goal of the created dashboard has been:

1. to know the variables that affect self-perceived health. Variables are grouped in demographical characteristics, familiar structure, work situation, educational level, work uncertainty, living conditions, health state, preventive activities, lifestyles and the use of healthcare services.
2. to analyze the self-perceived health from different perspectives. Allowing people to analyze self-perceived health by combining different variables at different abstraction levels.

The dashboard provides color-coding code of traffic lights to visualize quickly potential problems.

The Madrid dataset is openly available so Madrid experiments are reproducible.

3   Results

The result section is divided in two subsections: one that analyzes data about self-rated health of the population in Catalonia and another that analyzes data of the population in Madrid.

3.1   Self-rated Health in Catalonia

The best parameters for maximizing f1_weighted score were classifier_class_weight: balanced, classifier_max_features: sqrt, classifier_n_estimators:

100. In Table 1 we can see the values of accuracy, precision, sensitivity or recall and specificity. The total accuracy is of 0.53. We can observe that bad, regular and excellent categories have a lower f1-score and are more difficult to predict. These categories have less instances in the dataset as can be seen in the column support.

Table 1. Performance of the prediction of self-rated health in Catalonia

Category	Precision	Sensitivity	Specificity	f1-score	Support
Bad	0.67	0.22	0.99	0.33	27
Regular	0.53	0.31	0.96	0.39	132
Good	0.52	0.62	0.67	0.56	348
Very good	0.55	0.66	0.68	0.60	355
Excellent	0.48	0.12	0.98	0.19	101
Total	0.53	0.52	0.75	0.51	963

The ranking of the 10th most important attributes for the random forest algorithm are (they are ordered from most important to less important): (1) sex, (2) month, (3) no chronic disease, (4) adulthood disorder, (5) chronic disease, (6) limitations, (7) medicaments adulthood, (8) year of the questionnaire, (9) limited daily activities and (10) depression.

In Fig. 1 we can see that self-perceived health can be filtered according to age and gender, social class and studies. In the upper left table in Fig. 1 it can be seen that men and women between 15 and 44 have usually health coverage. In the upper right bar figure we see that the different self-rated health levels are distributed in all social classes. The same happens with self-rated health and studies in the bottom-left part of the figure. The dashboard consists of a single dynamic web page that shows a series of variable menus depending on the active user. Its objective is to present on a screen as much information as possible to analyze the information of the population living in Catalonia on the state of health, behaviors related to health and the use of health services. Through the options available in these menus, a series of text boxes with the necessary input data in each case. And from these, a window with the result of the selected action is displayed.

3.2 Self-rated Health in Madrid

The best parameters for maximizing f1_weighted score were classifier_class_weight: balanced, classifier_max_features: sqrt, classifier_n_estimators: 100. In Table 2 we can see the values of accuracy, precision, sensitivity or recall and specificity. The total accuracy is of 0.57. We can observe that very bad and bad categories have a 0 f1-score and are more difficult to predict. These categories have less instances in the dataset as can be seen in the column support.

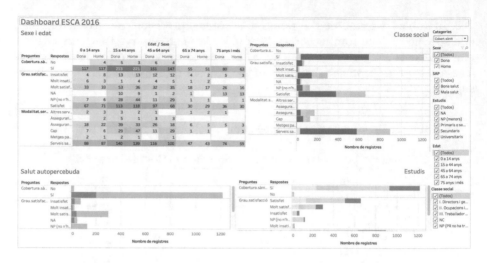

Fig. 1. Dashboard of the self-rated health datasets in Catalonia in 2016

Table 2. Performance of the prediction of self-rated health in Madrid

Category	Precision	Sensitivity	Specificity	f1-score	Support
Very bad	0	0	1	0	4
Bad	0	0	1	0	8
Regular	0.47	0.40	0.92	0.43	20
Good	0.58	0.94	0.30	0.72	68
Very Good	0.71	0.15	0.98	0.24	34
Total	0.54	0.57	0.63	0.49	134

The ranking of the 10th most important attributes for the random forest algorithm are (they are ordered from most important to less important): (1) age, (2) section, (3) during the last 12 months, I had to limit or reduce the performance of your usual activities because of this or these chronic problems I mentioned, (4) during the last two weeks, my physical health and emotional state did not limit my social activities with family, friends, neighbors or groups, (5) during the last two weeks I did not have difficulty when doing usual activities or tasks both inside and outside the home, because of my physical health or emotional problems, (6) I do not have any chronic disease, (7) during the last two weeks I had a lot of difficulty when doing usual activities or tasks both inside and outside the home, because of my physical health or emotional problems, (8) I have a chronic disease, (9) during the last two weeks I did very few physical activity and (10) district.

In Fig. 2 we describe in colors the self-rated health in the first and second component of the principal component analysis (PCA). The first two components

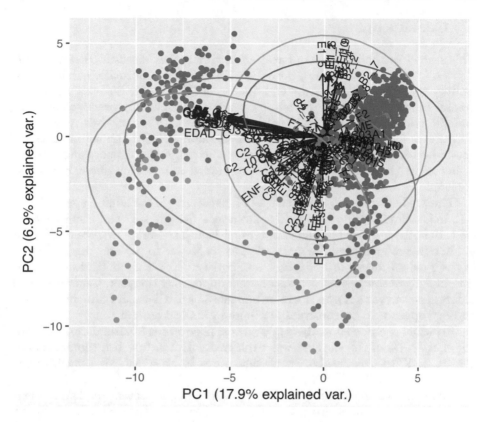

Fig. 2. Principal component analysis (PCA) representing with colors the self-rated health in Madrid

of the PCA explain 24.8% of the variance. In the figure we can see that self-rated health is described in blue, green and red colors:

- Blue: Regular self-rated health, these individuals are characterized by being the majority of the individuals in the third quadrant. These individuals are older, suffer from some type of disease or chronic disorder both self-perceived and recognized by the doctor, and poor mental health. They are characterized by having a poor physical condition and few hours of sleep.
- Green: Bad or very bad self-rated health, these individuals are very similar to those of the blue group, although the difference is that there are more individuals than belong to the fourth quadrant, characterized by having had some accident, or reduced sleep hours.
- Red: Good and very good self-perceived health, these individuals are mainly in quadrant 1, characterized by people living in the home, and children who live in the home, the time it takes to travel to work, the fact of working and the type of contract.

4 Conclusions

This study has applied random forest for the prediction of self-rate health status in both Catalonia and Madrid. Both datasets contained information about socio-demographic data, health state and quality of life, chronic diseases, accidents, mental health, limitations, disabilities, personal autonomy, preventive activities and lifestyle (such as physical health and leisure). However, there were some differences, for example the dataset from Catalonia also contained information about health coverage, consumptions of medicines, social support and medical visits; while the dataset from Madrid contained information about home and housing.

The prediction in a 5-level scale has a sensitivity of 0.52 and a specificity of 0.75 in Catalonia, while in Madrid it has a sensitivity of 0.57 and a specificity of 0.63. In both cases people who have a worst health self-rate are more difficult to predict. When analyzing the most relevant attributes to perform the prediction in the random forest, in both cases attributes such as chronic diseases, limitation in daily activities and depression were important. However, they have some differences, such as sex appears to be one relevant attribute in Catalonia dataset and age appears to be a relevant attribute in Madrid dataset.

Principal component analysis (PCA) is applied to the Madrid dataset obtaining that 24.8% of the variance was explained with the first two components of the PCA. With this new representation, three levels of self-rated health were displayed using colors.

Regarding the dashboard, Tableau appears to be a relevant method to visualize and extract value from the self-rated health status and other relevant attributes. The use of traffic lights colors help the user to detect problems and summarize the information. The visualization of the information in a dashboard helps in the challenge of filtering the information relevant and focus on the relevant interest and avoiding excess information. It helps to interpret visual information faster than written information and it facilitates the understanding. Creating attractive visualizations that are easy to digest enhance knowledge.

As future work, prediction performances could be improved applying com class imbalance methods such as cost-sensitive learning or Synthetic Minority Over-Sampling Technique (SMOTE). Those techniques could improve the performance of low-represented classes with health problems. In the future, a user study will also be performed in the dashboard to improve it and give value to potential users. Furthermore, other machine learning techniques could be applied and compared with random forest. Finally, some feature selection techniques could be applied such as filter, wrapper or embedded methods. From the data point of view, further work will imply the inclusion of datasets collected by other regions in the analysis. If Big Data is obtained, libraries such as MLlib could be used. The goal behind this would be to get a better ground of the variables that influence self-perceived health and to find out what of them are geographically independent.

Acknowledgments. We would like to thank to the *Departament de Salut de la Generalitat de Catalunya* for sharing the ESCA dataset. Without it, this research would have been impossible. This research have been partially funded by Recercaixa ("la Caixa" Foundation) and by the Catalonia Competitiveness Agency (ACC1Ó) and the eHealth Center from the UOC.

References

1. Idler, E.L., Benyami, Y.: Self-rated health and mortality: a review of twenty-seven community studies. J. Health. Soc. Behav. **38**, 21–37 (1997)
2. The Euroqol Group: EuroQol - a new facility for the measurement of healthy-related quality of life. Health Policy **16**(3), 199–208 (1990)
3. Resultats de l'enquesta de salut de Catalunya (ESCA) (2017). http://salutweb.gencat.cat/ca/el_departament/estadistiques_sanitaries/enquestes/esca/resultats_enquesta_salut_catalunya
4. Tamayo-Fonseca, N., Quesada, J.A., Nolasco, A., Melchor, I., Moncho, J., Pereyra-Zamora, P., López, R., Calabuig, J., Barber, X.: Self-rated health and mortality: a follow-up study of a Spanish population. Public Health **127**(12), 1097–1104 (2013)
5. Tamayo-Fonseca, N., Nolasco, A., Quesada, J.A., Pereyra-Zamora, P., Melchor, I., Moncho, J., Calabuig, J., Barona, C.: Self-rated health and hospital services use in the Spanish National Health System: a longitudinal study. BMC Health Serv. Res. **15**(1), 492 (2015)
6. Girón, P.: Determinants of self-rated health in Spain: differences by age groups for adults. Eur. J. Public Health **22**(1), 36–40 (2012)
7. Morcillo Cebolla, V., Lorenzo-Cáceres Ascanio, A., Domínguez Ruiz de León, P., Rodríguez Barrientos, R., Torijano Castillo, M.J.: Desigualdades en la salud autopercibida de la población española mayor de 65 años. Gac Sanit **28**(6), 511–521 (2014)
8. Subirats, L., Ceccaroni, L., Lopez-Blazquez, R., Miralles, F., García-Rudolph, A., Tormos, J.M.: Circles of Health: towards an advanced social network about disabilities of neurological origin. J. Biomed. Inform. **46**(6), 1006–1029 (2013)
9. Subirats, L., Lopez-Blazquez, R., Ceccaroni, L., Gifre, M., Miralles, F., García-Rudolph, A., Tormos, J.M.: Monitoring and prognosis system based on the ICF for people with traumatic brain injury. Int. J. Environ. Res. Public Health **12**(8), 9832–9847 (2015)
10. Calvo, M., Subirats, L., Ceccaroni, L., Maroto, J.M., de Pablo, C., Miralles, F.: Automatic assessment of socioeconomic impact in cardiac rehabilitation. Int. J. Environ. Res. Public Health **10**(11), 5266–5283 (2013)
11. Molarius, A., Berglund, K., Eriksson, C., Lambe, M., Nordström, E., Eriksson, H.G., Feldman, I.: Socioeconomic conditions, lifestyle factors, and self-rated health among men and women in Sweden. Eur. J. Public Health **17**(2), 125–133 (2007)
12. Abu-Omar, K., Rütten, A., Robine, J.M.: Self-rated health and physical activity in the European Union. Soc. Prev. Med. **49**(4), 235–242 (2004)
13. Bardage, C., Pluijm, S.M.F., Pedersen, N.L., Deeg, D.J.H., Jylhä, M., Noale, M., Blumstein, T., Otero, Á.: Self-rated health among older adults: a cross-national comparison. Eur. J. Ageing **2**(2), 149–158 (2005)

14. Encuesta de Salud de la Ciudad de Madrid (2013). http://datos.madrid.es/portal/
    site/egob/menuitem.c05c1f754a33a9fbe4b2e4b284f1a5a0/?vgnextoid=77e22cbf3ee
    07510VgnVCM1000001d4a900aRCRD&vgnextchannel=374512b9ace9f310VgnVC
    M100000171f5a0aRCRD&vgnextfmt=default
15. Köppen, M.: The curse of dimensionality. In: 5th Online World Conference on Soft
    Computing in Industrial Applications (WSC5), pp. 4–8, September 2000
16. Cady, F.: The Data Science Handbook. Wiley, Hoboken (2017)
17. Piñeiro, E.: Diseño de una herramienta business intelligence para el análisis de
    encuestas de salud. Master thesis of Universitat Oberta de Catalunya (2018)

Economic Interpretation of eHealth Implementation in Countrywide Measures

Peter Balco[1]($\boxtimes$), Helga Kajanová[2], and Peter Linhardt[2]

[1] Faculty of Management UK, Odbojárov 10, P.O. BOX 95, Bratislava, Slovakia
peter.balco@fm.uniba.sk
[2] FINESOFT, s.r.o., Popradská cesta 68, 040 11 Košice, Slovakia
helga.kajanova@gmail.com, peter.linhardt@outlook.com

Abstract. The eHealth represents a system, collecting data related to the evidence of the health care. Ensures the interoperability of patient's health history, medication and medical records, related to the care providers. Due precise evidence and data accessibility, this system definitely supports the fraud blocking, costs cutting and contributes to improvement of condition for better care provisioning by care providers, EU countries do not provide public data, targeted to clarify the results of eHealth in form of structured data set, representing the cost savings. This paper represents an alternative view to estimate the results of eHealth implementation into large measures health care system.

1 Introduction

Today, the need for having proper data for health care providers forces also health care authorities to define technology and legislation, followed by building electronic system targeted to store and process health related data of patients. Main issue is ensuring the data interoperability and regulated access, with a guarantee of the patient' right to maintain the access to her/his health related data. Patient is entitled to have full access (some diagnoses might be subject of discussion) to medical records, prescription and medication, suggestion and examination results. The physicians' access to those data provides the opportunity to evaluate the possible relation of current patient's medical problem to medical history, including the impact of medication. In most system, patients have possibility to introduce the personal record, related to observation, consumption of over the counter medication, health related activities like diet, exercises, remarks and understanding from alternative sources study, relevant to acute health status.

This evidence based approach is intensively extended with step-by-step introduction of AI modules, focused to provide physician decision support in diagnostics, healing proposals, medication. This process will have more weight in the future, is under a strong development, but it is inherently dependent on quality and complexity of all health related data of individual patient.

© Springer Nature Switzerland AG 2019
F. Xhafa et al. (Eds.): INCoS 2018, LNDECT 23, pp. 255–261, 2019.
https://doi.org/10.1007/978-3-319-98557-2_23

2 Direct and Indirect Evaluation of eHealth Implementation

European countries started the introduction of eHealth approaches 25–30 years ago, mostly in Nordic countries like Denmark, Sweden, the UK or Austria. Regardless long period of developing and building the digital and communication system, supporting the evidence and storage of information, related to the patients' health status, healing and/or prescription, countries and authorities do not strive to publish the full cost of infrastructure, database solution and application, as well as complex communication environment of eHealth systems. It is extremely difficult to find description of health segment reliable math model, which could provide evaluation of system changes, care costs, saving, caused by implementation and utilization of digital services, supporting the health care.

Research institute Empirica GmbH, Germany, in close cooperation with EU authorities has done plenty of research work, analyzes and have published hundreds of papers, dealing with comparison of status of eHealth solutions, their targets and result. Key researcher of Empirica is Dr. Stroetmann [1] led this work, which is further developed by Dr. Veli N. Stroetmann M.D. Ph.D., who obtained a doctorate in health informatics from the Bulgarian Academy of Sciences and the Academy of Medical Sciences. Booth researchers concentrate on the application of information and communication technologies (ICT) in clinical research, healthcare delivery, and public health. They analyze precisely eHealth policy and solution strategy development, implementation of applications in EU countries at national but also care providers and/or health system stakeholder levels. Empirica's researcher provide regular consultations to the European Commission in research policy, patient safety and on various topics including comparative effectiveness research, impact assessment, cross-border interoperability, uptake and sustainability of integrated care.

Results of the analyses is, that all countries has understood the urgent need of digital transformation of patient health records, EMR and EHR. Countries health representation are building and developing diverse eHealth systems, targeted to the main goal, which can be formulated as EHR interoperability and ensuring physicians' and patients' access to the proven, complex and well-arranged information, related to the patients' health history, status, medication, and laboratory results.

Most of the countries is not in favor of publication of structure, overview and evaluation of eHealth development and operation costs. The main reason is, that savings are evident in first stage on the health insurance accounts, which are ensuring financing of health care providers activities. Next reason is the absence of respected and widely accepted methodology of savings computation and evidence.

As a very rare analytic result, we can present illustration [2] of yearly costs and gains related to the implementation and operation of eHealth system in 10 selected health care organizations within the EU (Fig. 1), published by Prof. Stroetmann, Empirica GmbH, Germany:

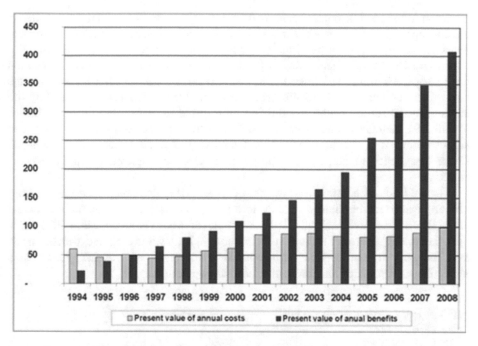

Fig. 1. The example explicitly show the positive economic impact of eHealth tools and solutions implementation in the process of care provisioning support in Mil €.

3 Indirect View on Economic Benefits of eHealth

While the digitalization and complex ITC support of healthcare services changes the landscape of health care in the national view, it influences and forces to change the legislation and process of care. It is essential to explore also unconventional evaluation of eHealth impact to the sector on larger scale.

When we use the crest view on country productive segments, we can see the health care as one of "industrial" segments, targeted to the regeneration, renewal and improvement of labor power of the population. The breakdown of this approach to the individual citizen is an indicator, the QALY [3], the WHO (The World health organization) is trying for dozens of years to introduce QALY as the national wide statistic indicators. QALY – Quality Adjusted Life Years is one of proper indicators to evaluate the cost effectivity of health care system. It represents number of years, the citizen live in good health conditions, without restrictions due illnesses, disability, invalidity, pain and discomfort as result of an illness or other health disorder. We can use QALY to express the time slot, the eHealth intervention extends the healthy life span of citizens.

The Disability-Adjusted Life-Year (DALY) [4] is a metric that combines the burden of mortality and morbidity (non-fatal health problems) into a single number. It is the primary metric used by the World Health Organization to assess the global burden of disease. It was developed in the 1990s as a way of comparing the overall health and life expectancy of different countries. The DALY is becoming increasingly common in the field of public health and health impact assessment (HIA). It "extends the concept of potential years of life lost due to premature death… to include equivalent years of 'healthy' life lost by virtue of being in states of poor health or disability."

In year 2000 World Health Report [5], the primary summary measure of population health used is Disability-Adjusted Life Expectancy, or DALE. DALE measures the equivalent number of years of life expected to be lived in full health, or healthy life expectancy. There are methods and data sources used to prepare the DALE estimates for the 191 member countries of WHO. In constructing the estimates, we sought to address some of the methodological challenges regarding comparability of the health status data collected and make a proposal of eHealth implementation in selected country.

On the basis of a simple survivorship curve, SMPH (summary measures of population health) can be divided broadly into two parts: health expectancies and health gaps. The bold curve in Fig. 2 is an example of a survivorship curve S(x) for a hypothetical population. The survivorship curve indicates, for each age x along the x-axis, the proportion of an initial birth cohort that will remain alive at that age. The area under the survivorship function is divided into two components:

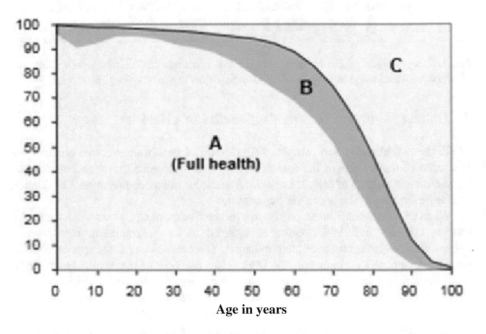

Fig. 2. Survivorship function for a population. Christopher J. L. Murray, David B. Evans, Organisation mondiale de la santé - 2003 - Health & Fitness

- Area A which is time lived in full health
- Area B which is time lived at each age in a health state less than full health.

The familiar measure of life expectancy at birth is simply equal to A + B (the total area under the survivorship curve. A health expectancy is generally of the form:

$$\text{Health expectancy} = A + f(B) \tag{1}$$

where f(.) is a function that weights time spent in B by the severity of the health states that B represents. In contrast to health expectancies, health gaps quantify the difference between the actual health of a population and some stated norm or goal for population health, for example after full implementation and utilization of eHealth. The health goal implied by Fig. 2 is for everyone in the entire population to live in ideal health until the age indicated by the vertical line enclosing area C at the right. In the specific example shown, the normative goal has been set as survival in full health until age 100. By selecting a normative goal for population health, the gap between this normative goal and current survival, area C, quantifies premature mortality.

We strongly believe that minimal set of desirable properties for summary measures that will be used to compare the health of populations – eHealth will cause age-specific mortality decreases in any age-group until everything else will remain being the same, then a summary measure should improve, resulting in health gap decreasing and a health expectancy increasing.

4 eHealth Effectivity

Effectivity of eHealth influence to the population health can be evaluated by saved years of life (DALY and/or QALY). The cost effectiveness can be later expressed, when we estimate the saved, or extended DALY years of citizens. In US for a long period the calculation were made by value of 50 000 USD/QALY.

During the last century and later around 2008 the researchers of Business Stanford University has stated the QALY value around 129 000 USD. British sources count with value 20–30 thousand of £, Swedish sources count with reference level od 60 000 €. In Slovakia we do not have available computation, but some Czech sources operates with QALY value of 1 000 000 Kcz. This corresponds to 40 000 €, what can be taken as median for Central European countries.

Let assume, that eHealth will improve the population health status by 1%. This can be visualized as gap B in Fig. 1. According the Info stat (Statistical office of SR) we can count with 2.54 mil. Working population in Slovakia. When we multiply the working population by QALY, it results approximatively 106 billion of EUR. When the assumption of 1% QALY improvements is acceptable, than the eHealth effectiveness is more than 1000 million EUR. Compared to the cost of eHealth development, implementation and integration and maintenance (estimated in Slovakia close to value of 100 mil of EUR), it makes all discussion of eHealth effectiveness non relevant, because ROI is deep under one year.

Some western resources implicates, the eHealth saves about 30% costs of medicaments prescription, about 10% of unnecessary laboratory examination. But this might be valid for the first, may be for the second year of eHealth utilization. But slight improvement of QALY results by 1% represents definitely much better savings, and we propose, that the eHealth effect should be presented to the population and professionals on the basis of QALY and/or DALE [6].

5   Summary

Until the authorities will accept some methods of eHealth implementation effectiveness, we have to point out the experience, that we see the main result of eHealth in changing the behavior and understanding of health role for the life among the patient, based on results of the information revolution.

The eHealth in form of good designed Patient health portal, electronic health book, ePrescription and eMedication details and summaries should lead the citizen to the better care of their personal health, to the prevention. The citizen will pay more attention to the healthy life style, to the reading and understanding the medical articles, professionals' suggestion and it should result in the citizens' effort, striving to remain better healthy instead of being treated and healed by the health providers.

Interoperability of health related information for the physicians would lead to the improvement of health care, will force physicians to cope better with patients. They will provide care based on deeper understanding of history and current problems and on cooperation of patients and professionals. We have to bear in our minds, that current phase of eHealth is basically the transformation process result only, transformation of media from paper to the digital media for evidence and reporting of care provisioning. The real expansion of eHealth will start with deeper integration and utilization of AI for evidence based medicine, for digital support of physicians decision by AI modules like Dr. Watson by IBM and other grandiose research projects.

Regardless the possibility to have quite applicable estimation and computation of eHealth development and operation costs, with fine-tuned procedures, we still call for development of a mathematical model of health segment, health care, where we can for the future provide much more accurate modeling, trend analyses, evaluation of influence of medicaments development, development of processes and other system activities.

6   Conclusion

Since the January 1st 2018, according to the Slovak eHealth law (the only pure eHealth legislation in Europe) the connectivity and integration of care providers information systems with the national eHealth solution represents a must for the care providers. The central EHR storage, applications are operated, and the system started to collect the real eHealth related data. E-Recepts, health records, EMR, created by physicians are collected in the central database, which will form key data set for monitoring and measuring the effectiveness of the whole eHealth building project. Health related data and databases would serve as a source of new added value creation for its users, the health professionals and management authorities. Currently, we are negotiating on agreement to have access to the anonymized data in the databases in order to participate in data analysis and to address new assignments that will move the research subject more closely to the main eHealth users, to the country population.

References

1. Stroetmann, K.A., Hammerschmidt, R., Stroetmann, V., Moldenears, I.: eHealth in action – Good practice in European countries, Good eHealth report January 2009 European Commission. Office for Official Publications of the European Communities, Luxembourg (2009)
2. Stroetmann, K.: Achieving the integrated and smart health and wellbeing paradigm. Int. J. Med. Inform. **82**(4), e29–e37 (2012)
3. Kříž, J.: Prevention and the economy. Hygiena **56**(3), 89–94 (2011)
4. https://en.wikipedia.org/wiki/Disability-adjusted_life_year
5. World Health Organization: World Health Report 2000. World Health Organization, Geneva (2000)
6. Inštitút informatiky a štatistiky výskumné demografické centrum, prognóza pracovnej sily v krajoch SR do roku 2025, Bratislava, June 2006

Development Methodology of a Higher Education Institutions Maturity Model

João Vidal Carvalho[1(✉)], Rui Humberto Pereira[1], and Álvaro Rocha[2]

[1] Polytechnic of Porto/CEOS.PP - R. Dr, Jaime Lopes de Amorim,
S. Mamede de Infesta, 4465-004 Porto, Portugal
{cajvidal, rhp}@iscap.ipp.pt
[2] Department of Informatics Engineering, University of Coimbra,
Pinhal de Marrocos, Coimbra, Portugal
amrocha@dei.uc.pt

Abstract. Maturity models have been introduced, over the last four decades, as guides and references for Information System (IS) management in organizations from different sectors of activity. In the educational field, maturity models have also been used to deal with the enormous complexity and demand of Educational Information Systems. This article presents a research project that aims to develop a new comprehensive maturity model for the Higher Education Institutions (HEI) area. The HEIMM (Higher Education Institutions Maturity Model) will be developed to help HEI to address the complexity of there IS, as a useful tool for the demanding role of the management of these systems, and institutions as well. The HEIMM will have the peculiarity of congregating a set of key maturity influence factors and respective characteristics, enabling not only the assessment of the global maturity of the HEI IS but also the individual maturity of its different dimensions. In this article, we present the second phase of our project by discussing the methodology for the development of maturity models that will be adopted for the design of the HEIMM and the underlying reasons for its choice.

Keywords: Stages of growth · Maturity models · Higher education institutions
Management

1 Introduction

In recent decades, we have seen many changes in the HEI sector. In the past, higher education was highly restricted in terms of the capacity of students. The expansion of this kind of education has put HEIs under enormous pressure for providing the required services and capacity. In addition, new teaching paradigms (Bologna Treaty), viewing the student as a "customer", opening up universities to enterprises, in terms of knowledge transference, and HEI rankings and competition have forced the HEI sector to reinvent higher education and adopt agile management methodologies, in order to adapt to constant changes in the market environment. In the same sense, the operational Information Technologies (IT) have grown in complexity to respond to the demands of this sector of activity. This increased complexity has led to the introduction of many

F. Xhafa et al. (Eds.): INCoS 2018, LNDECT 23, pp. 262–272, 2019.
https://doi.org/10.1007/978-3-319-98557-2_24

new systems, processes and approaches to academic integration, as well as the emergence of new platforms that provide services in this area. In this scenario, it is difficult to know if we are doing a good job managing these changes and monitoring progress on an ongoing basis. In addition, it is not easy to manage the interactions of systems and processes that are constantly evolving, as it is not easy to manage the impact of processes of low interoperability, security, reliability, efficiency and effectiveness. Thus, the benefits of modern technology in education, supported by better methods and better tools, cannot be obtained through undisciplined and chaotic processes. For this reason, we consider that the IS management in educational institutions should be carried out with the help that maturity models can provide. Such maturity model could help HEI in their organizational analytics processes, by enabling these institutions to obtain data regarding the maturity of their IS, in several dimensions, in terms what needs to be improved. The external organizations, such national evaluation agencies and certification agencies, also could benefit of this organizational analytics tool to determine the maturity of the implementation of the HEI's processes.

Several maturity models have been proposed over time, both for the evolution of people, for the general evolution of organizations and for the particular evolution of the IS function. These models differ mainly in the number of stages, evolution variables and focus areas [1, 2]. Each one of these models identifies certain characteristics that typify the target at different stages of growth or maturity and are applied in different organizations.

In the case of organizations in the area of education, several maturity models are also proposed. This finding stems from a systematic literature review that we conducted at a preliminary stage of our research. This literature review has been carried in order to identify any eventual gaps in existing educational maturity models. After following a systematic approach to our literature review methodology, we have considered a selection of thirty-five maturity models associated to educational ecosystems. Of these models, only five ([6–10]) are focused (or related) on the HEI IS management. After a detailed analysis of these models, it was verified that there is no comprehensive and detailed model that assesses the maturity of the IS of HEI, in its various dimensions. In fact, this analysis also revealed the inexistence of maturity models with dimensions or influence factors, with different weights considering its relative importance. It was verified that most of the maturity models are still in the early stages of development and in premature phases of affirmation and consolidation, on account of being proposed by their authors in the course of exploratory studies. Additionally, most of these models are not sufficiently explicit, considering the way they were developed and validated. Furthermore, because they are poorly detailed, they do not provide effective means to determine the maturity stage nor identify the characteristics to reach a stage of higher maturity. In the case of the adoption of a tool to assess a system's maturity, it was found that most of the models, besides of focusing on the assessment of the system's maturity, pay attention to an improvement path of such maturity. According to Bruin et al. [5] the reasons for these sometimes ambiguous results of maturity models stem from the insufficient focus on model validity, reliability and generalization testing, as well as poor documentation on how to develop and design such a model.

In view of these constraints, it was considered opportune to develop a research project that would contribute to an increase of knowledge about the maturity models applied to HEI IS, in order to spread an improvement in the practice of evaluating and promoting the maturity of their IS. Based on the description of the problem, the following research question was formulated:

▶ *Is there a model, which consists of several maturity-influencing factors and maturity stages, that can be applied to ISs in HEIs?*

In this paper, we report the progress of our work by discussing the chosen approach for the development of our maturity model in the next phase of our project.

In the next Section, we present a short state of the art of the methodologies for the development of Maturity Models. In Sects. 3 and 4, respectively, we discuss one of these methodologies and our arguments to apply it in the development of our model. Finally, we present our conclusions and the next steps of our project.

2   Methodologies for the Maturity Models' Development

Several studies have shown that over a hundred different maturity models have been proposed over time in the IS area [3, 5]. The recurrent publication of new maturity models, often for very similar applications, suggests a certain arbitrariness, as well as a lack of contextualization and clear definition of requirements that can distinguish them in a unique way [11]. In fact, authors rarely reveal their motivations and how they develop the model or method of procedure, as well as the results of their evaluation [3, 4]. Of course, in order to ensure that the development of a new model is recognized as solid and potentially relevant, both in academia and in society in general, there must be a concern to demonstrate that the model was developed with rigor and that it is subject to debate and verification.

So far, no standard approach describing how to rigorously construct new maturity models has been found [4]. In fact, there are few studies that refer to the process of designing maturity models. In the same direction, Poeppelbuss et al. [12] argue that the development of new models is often based on existing models rather than on development methodologies.

In the few methodological approaches applied in the development of maturity models that were found in the literature, the dominant approach is the *Design Science Research* (DSR) [13]. Indeed, some researchers argue that such models should be conceptualized and constructed as conventional IT artefacts [14]. In this context, it will be necessary to consider two iterative steps: (1) design - to describe the design of the maturity model in a transparent and traceable way; and (2) evaluation - to prove the utility and ability to solve the problem addressed [15]. It is in this sense that a robust and recognized research method, namely the DSR becomes essential for the successful development of this type of models.

Based on the literature review carried out by Elmaallam and Kriouile [13], a number of important methodologies have been identified [3, 5, 11, 16, 17]. In the same study carried out by these authors [13], it was also mentioned the guide for the development and implementation of "*maturity grids for assessing organizational capabilities*" developed by Maier *et al.* [18], as well as the maturity development approach "*focus area model*" [19].

3 Methodology Proposed by Mettler [17]

Mettler carried out a comparative study of the development of three maturity model methodologies [3, 5, 20]. As a result, a new approach emerged that introduced the so-called "decision parameter" elements (Fig. 1). This approach is based on an interactive design process that consists of five steps or design activities (white boxes). Several decisions must be made within each step (black lozenges), that is, at each stage of the model-building process the designer must choose a number of elements before proceeding with the process.

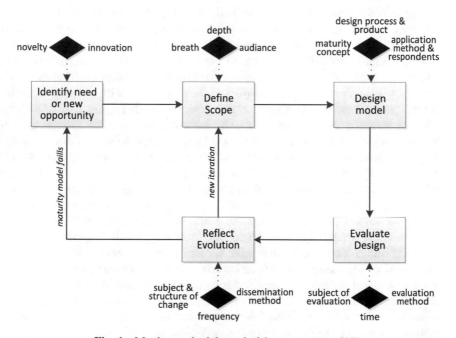

Fig. 1. Mettler methodology decision parameters [17]

To identify these decision parameters, Mettler relied on his own experience, as well as on the literature review he conducted in this area [17]. Next, we present the various decision parameters and the possible options that each one can offer for each one of the following five steps of the development process of a maturity model.

1. *Identify need or new opportunity* - During this first step, two parameters are considered. On the one hand, the novelty of the topic covered by the maturity model plays an important role, since it determines if there is a need to be filled by this model. The appearance of this new model should be duly substantiated by theoretical assumptions about maturity levels. On the other hand, innovation is another parameter of decision to consider, before beginning the development of a maturity model. For example, the designer has to decide whether to design a completely new model, a variant, or a version of an existing model.

2. *Define scope* - With regard to the definition of the scope, the author of the model is confronted with a very important decision-making. In the first decision, the breadth of the phenomenon to be studied must be defined, *i.e.*, it must be defined if the model addresses a generic topic (*e.g.* educational organizations) or a more specific (*e.g.* e-learning). In addition, the detail conditions of the maturity model should be considered. Finally, respecting one of the constituent guidelines of the DSR [21], some reflections on the potential "audience" of the maturity model should be made. There might be a need to adapt the model to the particular necessities of an audience, whether it has a technological orientation or a management orientation, or both.

3. *Design model* - Once the scope is defined, the design of the model itself begins. In this step, one of the major decisions to be made has to do with the definition of "Maturity" in the context of the model that is intended to be developed. In the most recent literature three different concepts of maturity appear [17], being more focused on the process, the object or the people. After clarifying the concept of maturity, the functional objective of the model (that is, how maturity will progress) is tacitly defined. For example, efficiency is almost always the underlying goal of process-oriented maturity, while increased satisfaction could be a goal of a people-centred maturity model. However, a maturity model can be targeted for multiple purposes. Therefore, another important decision to deliberate is related to the advancement of maturity if it is one-dimensional (that is, only focusing on a measure aimed at efficiency) or multidimensional (*i.e.* focusing on several, sometimes divergent, goals). Subsequently, the nature of the design process (theory-driven vs. practitioner-based or a combination of both) has to be determined, in order to determine the knowledge base to identify maturity levels, metrics and corresponding improvement recommendations. Another important decision parameter is the format of the model. This is particularly relevant in defining the development team (*e.g.* good writers are needed when the maturity model is disseminated in manual form, specialized software developers are needed when the maturity model is instantiated as software). This decision certainly has an effect on the choice of the method of application (*i.e.* whether data collection is based on a self-assessment or third-party evaluation such as outsourcing by certified professionals), as well as interviewees for the collection of data (*e.g.* managers, employees, business partners or a combination) [5].

4. *Evaluate design* - This step, is concerned with the verification and validation of the developed maturity model. According to Conwell *et al.* [22], verification is the process of determining that a maturity model *"represents the developer's conceptual description and specifications with sufficient accuracy"* and validation is the degree to which a maturity model is *"an accurate representation of the real world from the perspective of the intended uses of the model"*. Therefore, it is very important to define a strategy to define "what", "when" and "how", that is, to test the process of developing the model or the model itself, at what time (ex-ante or ex-post or both) and in what form, whether artificial (*e.g.*, laboratory experimentation, simulation) or natural (*e.g.*, case study, reflection groups).

5. *Reflect evolution* - Finally, in this last step, the designer has to decide on the mutability of the maturity model. Although often neglected, this circumstance is

particularly important. In fact, the maturity of the phenomenon under study, can be dynamic and therefore, the stages of the model and improvement activities should be rectified periodically (modify the requirements to reach a certain level of maturity, due to the appearance of new and better practices and technologies). On the other hand, changes in the form and function of the model may be required to ensure its standardization and global acceptance. Finally, in the possibility that the maturity model becomes increasingly mature, it must be decided whether the model will be available for free or if its dissemination and the dissemination of its results will be restricted to an exclusive group of persons and/or organizations.

4 Methodology for the Development of the Maturity Model HEIMM

Considering the methodological approaches adopted in the development of maturity models, and after analysis and thorough reflection, we selected the Mettler [17] methodology for the development of the HEIMM. This choice was based on a set of assumptions that included the following.

- This methodology is the result of a comparative study and systematization carried out by its author, work that is among some of the most referenced methodologies in this area.
- The methodology (or mental model) presented by this author is consistent with the DSR guidelines.
- This methodology respects the iterative nature of the development process of the maturity model.
- This methodology takes into account the need to combine theoretical and empirical research, as recommended by other maturity model researchers [5, 23, 24].
- This methodology is consistent with the type of maturity model intended for this project.

Table 1 presents the decisions (identified) made when developing the design for our Maturity Model for HEI IS Management.

The options indicated in Table 1, refer to the HEIMM development and will be justified in each one of the following steps of Mettler's development methodology.

Identify Need or New Opportunity
In the "identify need or new opportunity" step, two parameters are considered. First, the novelty of the topic covered by the maturity model plays a key role, since it determines if there is a need that this model is able to fulfil. In fact, the development of a new model (*emerging*) is justified by the fact that the existing maturity models in this field have weaknesses, both in their affirmation and their adoption by the HEI. Second, innovation is another parameter that needs to be considered before beginning the development of a maturity model. In the context of the HEIMM, we consider it to be completely new (*new*), because none of the models identified in the literature referred previously is fully focused in the IS of the HEI.

Table 1. Decisions made when developing the HEIMM (Adapted from Mettler [17])

Design activity	Decision parameter	Characteristic			
1. Identify need or new opportunity	Novelty	Emerging	Pacing	Disruptive	Mature
	Innovation	New	Variant	Version	
2. Define scope	Breadth	General issue		Specific issue	
	Depth	Individual/Group	Organization	Inter-organizational	Global/Society
	Audience	Management oriented	Technology oriented	Both	
3. Design model	Maturity concept	Process-focused	Object focused	People focused	Combination
	Goal function	One-dimensional		Multi-dimensional	
	Design process	Theory-driven	Practitioner based	Combination	
	Design product	Textual description of form	Textual Description of form and functioning	Instantiation (software)	Combination
	Application method	Self-assessment	Third-party assisted	Certified professionals	
	Respondents	Management	Staff	Business partners	Combination
4. Evaluate design	Subject of evaluation	Design process	Design product	Both	
	Point of time	Ex-ante	Ex-post	Both	
	Evaluation method	Naturalistic	Artificial	Combination	
5. Reflect evolution	Subject of change	None	Form	Functioning	Form and functioning
	Frequency	Non-recurring		Continuous	
	Structure of change	External/open		Internal/exclusive	
	Dissemination	Open		Exclusive	

Define Scope

In the second step, the first decision involves defining the magnitude of the phenomenon to be studied, *i.e.,* we must decide whether the model addresses a generic or a more specific topic. In the case of the HEIMM, the model is only applied to HEI and for this reason, the choice is "*specific issue*". The detailed conditions of the maturity model must then be considered. The HEIMM model focuses on *intra-organizational,* as well as *inter-organizational* aspects. In fact, this model incorporates aspects that are related to the internal processes of organizations and also aspects that represent the processes of cooperation with external organizations. This decision is based on the observation that many of the HEI's processes, such as research, relationships with enterprises and other HEI, involve external entities.

Finally, considering one of the DSR guidelines [21], the potential "*audience*" of the maturity model must be considered. In the case of the HEIMM, the target audience includes HEI managers who have decision-making authority. These can either be CEOs or department directors with responsibilities in the management field, such as CIOs or the directors of information systems and technologies (IST) area of HEI. In this case, the choice of the *"audience"* parameter falls under the option *"both"*.

Design Model

The construction of the model itself begins in the "design model" step. Here, one of the major decisions involves the definition of "maturity" in the context of the intended model. Mettler [17] resorts to the literature to justify the emergence of three different maturity concepts according to whether they focus more on the process, the object or the people. The HEIMM presents a multi-faceted approach (*combination*) to measure the maturity, in order to increase the overall efficiency of the HEI in terms of processes (*process-oriented*) and people (*people-oriented*), as well as the technologies that support them. In addition, the HEIMM assesses analytical capabilities, both organizational and technical.

A maturity model can have multiple goals, as the case with the HEIMM. Therefore, another important decision relates to the maturity level: *one-dimensional* (that is, focusing on one measure as an efficiency target) or *multi-dimensional* (*i.e.*, focusing on several, sometimes divergent, objectives). In fact, in the case of the HEIMM, the objective of the maturity measurement encompasses several influence factors (*multi-dimensional*), being able to measure the global maturity of the IS (calculated according to the different weights of each influence factor), but also the maturity of each sub-area of HEI represented by its influence factor. It should be noted that the identification of the main influence factors that will be part of HEIMM will be done in a first phase, based on a literature review and later, complemented by a questionnaire survey carried out to Portuguese IS specialists in the education area (mainly HEI managers and IT managers). Subsequently, the nature of the design process (e.g., *theory-driven* vs. *practitioner-based* or a combination of both) has to be determined, in order to identify the knowledge base with regard to the maturity levels, the metrics and the corresponding improvement recommendations. The HEIMM model uses a *combination* of both.

Another important decision parameter concerns the model format. Here too, there is a combination of two options, namely, the *"instantiation software"*, in respect of developing a tool to evaluate the HEI maturity, and the *"textual description of form"* as the HEIMM will be available in text format with a description of its applicability. This decision certainly affects the selection of the application method (*i.e.*, whether data collection is based on self-assessment or third-party assessment, such as outsourcing by certified professionals). It is our understanding that this model would essentially be implemented by managers of the health units whose maturity is to be assessed (*self-assessment*) since they are the ones who best know the reality of their organization.

Finally, in the data collection process for HEI maturity assessment, it is important to define the actors (*respondents*) in this collection. In the HEIMM model, data collection can be diversified and although it is essentially carried out by managers, other educational and IS professionals may be involved. Thus, the *"combination"* option is the most suitable for this final parameter of the *"design model"*

Evaluate Design

The "*evaluate design*" step concerns the verification and validation of the developed maturity model. With regard to the HEIMM, two parameters ("*design process*" and "*design product*") were initially considered in the context of the object to be evaluated (*subject of evaluation*). The HEIMM should be evaluated in terms of form and content and, therefore, the choice lies with "*design product*". Additionally, this new model should be evaluated before it was implemented, that is, in the "*point of time*" option, the choice falls to "*ex-ante*". Finally, the evaluation method must be "*naturalistic*" since the evaluation of the HEIMM should be based on the experience and reflection of real users. Semi-structured interviews were conducted with a diverse group of Portuguese HEI managers to validate the HEIMM. We ensure that they had considerable experience in HEI management. Additionally, different types of HEIs also should be represented, both private and public.

Reflect Evolution

In the final step, "*reflect evolution*", the designer has to decide on the mutability of the maturity model over time. In the context of this research project, this step will be performed after the HEIMM validation is completed. Later, we intend to implement the model in three HEIs (these should be representative of the private and public sectors), and the feedback obtained from its implementation will serve as input for the accomplishment of this final step.

5   Conclusion and Future Work

This paper presents a research project for the conception of a comprehensive maturity model for the assessment of the IS of HEI. The project started with a systematic literature review, in order to find the state of the art in the field of this kind of models. In this first phase of our exploratory work, we found five maturity models, but none of them fulfilled all the requirements. As far as we know, none of the models identified in the literature are sufficiently focused on the capability of the IS support complex, diversified, interoperable and dynamic organizational processes in the HEI sector. From this perspective, a new model to fill the gap should be designed. This new model should include the main influence factors with different weights depending on their relative importance, while its development should be informed by rigorous scientific methods of conceptualization and validation. This model should also identify key IS strategic areas for HEIs and apply international standards of management to the IS and HEI context. Such a maturity model will enable the evaluation of HEIs in terms of how their IS practices and strategies are supporting their institutional processes at all levels: that is, organizational strategies, management, operative management, teaching and research. Thus, it can empower the HEIs and their human capital (administrators, staff, teachers and students). Those previous works, however, are inspiring for us for the continuity of our research work.

We intend to develop our model, the HEIMM, following the principles enumerated above, applying rigorous scientific methods, in order to achieve a systematic methodology. Additionally, it will apply international standards for management of IS and HEI, as well. We are convinced that the success of such maturity model in terms of recognition, strength and relevance, in both the academic sector and in the society in general, requires formal methodology on all steps of its conception. Thus, in this paper we present the results of the second phase of our project, in which we discuss our options in terms of which methodology will be applied for the conception phase of the HEIMM.

To support our decision regarding the methodology to be applied, we conducted another literature review, in order to find the state of the art on methodologies to support the conception of maturity models. The approach proposed by Mettler was our option due to its characteristics, which we consider that are appropriated for our model.

As designers of our model, we discuss the decision parameters of the Mettler's methodology that must be taken before initiating this interactive methodology.

Our work is now at that third step of Mettler's methodology, the design stage of the maturity model. We expect to have a first version of the HEIMM in the next months, in order to be submitted to an evaluation as proposed by Mettler.

In the perspective of the acceptance and adoption of the model by HEI, we intend to develop an automatic tool for assessing HEI maturity. This tool should be built based on the principles established in HEIMM and should be made available on the Internet, enabling managers to perform HEI maturity assessments and simultaneously make comparisons with their competitors, as well as understand the evolution of their maturity over time.

References

1. Rocha, Á.: Evolution of information systems and technologies maturity in healthcare. Int. J. Healthc. Inf. Syst. Inf. **6**(2), 28–36 (2011)
2. Mettler, T., Rohner, P.: An analysis of the factors influencing networkability in the health-care sector. Health Serv. Manage. Res. **22**(4), 163–196 (2009)
3. Becker, J., Knackstedt, R., Pöppelbuß, J.: Developing maturity models for IT management – a procedure model and its application. Bus. Inf. Syst. Eng. **1**(3), 213–222 (2009)
4. Mettler, T., Blondiau, A.: HCMM – a maturity model for measuring and assessing the quality of cooperation between and within hospitals. In: 25th IEEE International Symposium on Computer-Based Medical Systems (CBMS) (2012)
5. de Bruin, T., et al.: Understanding the main phases of developing a maturity assessment model. In: 16th Australasian Conference on Information Systems (ACIS 2005) (2005)
6. Solar, M., Sabattin, J., Parada, V.: A maturity model for assessing the use of ICT in school education. J. Educ. Technol. Soc. **16**(1), 206–218 (2013)
7. Manjula, R., Vaideeswaran, J.: A new CMM-quality education (CMM-QE) framework using SEI-CMM approach and calibrating for its process quality and maturity using structural equation modeling-PLS approach. Int. J. Softw. Eng. Appl. **6**(4), 117–130 (2012)
8. Gu, D., Chen, J., Pu, W.: Online course quality maturity model based on evening university and correspondence education (OCQMM). In: 2011 IEEE 3rd International Conference on Communication Software and Networks (ICCSN), pp. 5–9. IEEE (2011)
9. Bass, J.M.: An early-stage ICT maturity model derived from Ethiopian education institutions. Int. J. Educ. Dev. Inf. Commun. Technol. **7**(1), 5–25 (2011)

10. Rossi, R., Mustaro, P.N.: eQETIC: a maturity model for online education. Interdisc. J. e-Skills Lifelong Learn. **11**, 11–24 (2015)
11. Mettler, T., Rohner, P.: Situational maturity models as instrumental artifacts for organizational design. DESRIST09, Malvern, PA, USA (2009)
12. Poeppelbuss, J., et al.: Maturity models in information systems research: literature search and analysis. Commun. Assoc. Inf. Syst. **29**(27), 505–532 (2011)
13. Elmaallam, M., Kriouile, A.: Toward a maturity model development process for information systems (MMDePSI). IJCSI Int. J. Comput. Sci. Issues **10**(3), 118–125 (2013)
14. Donellan, B., Helfert, M.: Applying design science to IT management: the IT-capability maturity framework. In: 16th Australasian Conference on Information Systems (AMCIS) - Lima, Peru (2010)
15. March, S.T., Smith, G.F.: Design and natural science research on information technology. Decis. Support Syst. **15**(4), 251–266 (1995)
16. Gottschalk, P., Solli-Sæther, H.: Towards a stage theory for industrial management research. Ind. Manage. Data Syst. **109**, 1264–1273 (2009)
17. Mettler, T.: Thinking in terms of design decisions when developing maturity models. Int. J. Strateg. Decis. Sci. (IJSDS) **1**(4), 76–87 (2010)
18. Maier, A., Moultrie, J., Clarkson, P.J.: Developing maturity grids for assessing organisational capabilities: practitioner guidance. In: 4th International Conference on Management Consulting Conference (2009)
19. Steenbergen, M.V., et al.: The design of focus area maturity models. In: Proceedings of the 5th International Conference on Design Science Research in Information Systems and Technology, pp. 317–332 (2010)
20. Mettler, T.: Supply Management im Krankenhaus: Konstruktion und Evaluation eines konfigurierbaren Reifegradmodells zur zielgerichteten Gestaltung. Ph.D. thesis. Institute of Information Management. St. Gallen, University of St. Gallen (2010)
21. Hevner, A.R., et al.: Design science in information systems research. MIS Q. **28**(1), 75–105 (2004)
22. Conwell, C.L., Enright, R., Stutzman, M.A.: Capability maturity models support of modeling and simulation verification, validation, and accreditation. Papper presented at the 2000 Winter Simulation Conference, San Diego, USA, pp. 819–828 (2000)
23. Mettler, T.: Maturity assessment models: a design science research approach. Int. J. Soc. Syst. Sci. **3**(1–2), 81–98 (2011)
24. Von Wangenheim, C.G., et al.: Creating software process capability/maturity models. IEEE Computer Society, Bd. Voice of Evidence, pp. 92–94, July/August 2010

Selected Legislative Aspects of Cybernetic Security in the Slovak Republic

Tomáš Peráček$^{(\boxtimes)}$, Boris Mucha, and Patrícia Brestovanská

Faculty of Management, Comenius University in Bratislava, Bratislava
Slovak Republic
{tomas.peracek, boris.mucha,
patricia.brestovanska}@fm.uniba.sk

Abstract. Networks and information systems play a crucial role in free movement and are often interconnected and connected to the Internet as a global tool. Their disturbance in one Member State therefore affects other Member States and the whole European Union. Resilience of networks and stability of the information system is therefore a basic prerequisite for a smooth and undistorted functioning of the European Union's internal market and a prerequisite for international cooperation. The National Security Authority, as the central authority for cyber security, has prepared a bill on Cyber Security on the basis of several approved documents. By its approval and effect from 1 March 2018, the content of Directive 2016/1148 of the European Parliament and of the Council from 6 July 2016 measures to ensure a high common level of network security and information systems in the Union and is also transposed into the law of the Slovak Republic. The aim of this paper is to investigate selected legal institutes of the law using a number of scientific methods and to point out their crucial importance for the cyber security of the Slovak Republic. The subject matter of the survey is the field of law of information and communication technology, military and administrative law. However, this is a multidisciplinary examination of this problem, which interferes with security management.

1 Introduction

Networks and information systems play a crucial role in the free movement of information and are often interconnected and connected to the Internet as a global tool. Their disturbance in one Member State therefore affects other Member States and the whole European Union. According to Maisner et al. Resilience of networks and stability of the information system is therefore a basic prerequisite for a smooth and undistorted functioning of the European Union's internal market and a prerequisite for credible international cooperation [7]. The basic European legislative framework governing cyber security is Directive Nr. 2016/1148 of the European Parliament and of the Council of 6 July 2016 to ensure a high common level of the Union's security of information systems in the European Union (hereinafter referred to as "the Directive") [4].

The Directive is the first pan-European legislation on cyber security, aimed at strengthening the competences of the relevant national authorities, enhancing their co-ordination and introducing safety conditions for key sectors. Its main objective is to

© Springer Nature Switzerland AG 2019
F. Xhafa et al. (Eds.): INCoS 2018, LNDECT 23, pp. 273–282, 2019.
https://doi.org/10.1007/978-3-319-98557-2_25

guarantee a common network security and information systems within the European Union. Achieving it is possible in particular by increasing the security of the Internet and private networks and information systems. These are largely based on economic and social interests. The National Security Authority, as the Central State Office for Cyber Security, prepared a draft of the Cyber Security Law, which transposes the content of this Directive into the legal order of the Slovak Republic, and, for the first time, legislatively regulates this issue.

As the issue of cyber "security research is at the edge of the interest of legal theory, the authors are interested in filling this gap. We want to achieve the stated goal using the knowledge of the specialized literature on information systems, information and communication technology law as well as the application of related legislation. Using selected research methods, including analysis, description, deduction, doctrinal and scientific interpretation, we set ourselves as the main goal of reviewing selected institutes of the Cyber Security Act. Another goal is to find out and answer the research question whether there is a need to extend some of its regulation [2].

2   Problem Definition

The priority of cyber security is highlighted in several conceptual and strategic documents of the European Union and the North Atlantic Alliance, as well as in other major organizations' documents around the world. Generally speaking, the disruption of cyberspace is seen as one of the key threats of the present. For this reason, progressive governments are taking steps to implement effective cybercrime and reinforcement measures to prevent, detect, prevent and recover from cyber-attacks. The Member States' obligations under the Directive are set at the lowest acceptable level necessary to achieve the required preparedness and to ensure cross-border confidence-based cooperation. It allows the Member States, within the framework of the measures adopted, to take into account their national specificities, with the Member States transposing it only in the real risks in their company.

The aim of the law is therefore to create a functional legislative framework necessary for the effective implementation of key national cyber security measures. The law transposes the priorities and requirements created at European level and adopted by the general consensus through the Directive. Its main areas of adaptation following the directive include these areas:

- organizations and powers of the public authorities in the field of cyber security,
- a national cyber security strategy,
- a single cyber-security information system,
- the status and obligations of the Basic Services Operator (hereinafter referred to as the "BSO") and the Supplementary Service Provider (hereinafter referred to as "SSP"),
- the organization and operation of accreditation of cyber security incidents (hereinafter referred to as "CSIRT"),
- a cyber-security system and minimum requirements for cyber security,
- control and audit.

In addition to the above, the Act addresses some other requirements of the Directive, such as the definition of international cooperation in the field of cyber security, the fulfillment of notification obligations, reporting of cyber security incidents. We also include adjusting the voluntary reporting of cyber security incidents or supporting cyber-security research and education. Due to its uniqueness in individual articles, it also amends related legislation, the Law on Military Intelligence, the Defense Act of the Slovak Republic or the Critical Infrastructure Act. The law further addresses in a comprehensive manner employee remuneration on the state side so that the state is able to employ cyber security professionals and compete with private employers [8].

3   Subject of the Act

The Cyber Security Act no. 69/2018 was adopted on 30 January 2018 and was published in the Slovak Republic on 9 March 2018. In view of its importance, it became effective on 1 April 2018 [2]. Following the Directive, the subject matter corresponds to the requirements of its transposition and constitutes an unavoidable set of cyber-security. However, it priorities the rights and obligations of persons as well as the powers of the public authorities. This includes in particular the organization, scope and responsibilities of public authorities in the area of cyber security, the national cyber security strategy, an unified information security system for cyber security, the organization and scope of units for cyber security incidents and their accreditation. The legislator also included regulating the status and responsibilities of the primary service provider and digital service provider, security measures, cyber security and control over compliance with the law, including audit, into a broad scope of activity.

However, the law also defines its substantive scope in a negative way. This means that it does not apply to the requirements for network security and information systems under the General Rules on the Protection of Classified Information. The provisions of this Act are not related to the activities of the Slovak Information Service and the Ministry of Defense of the Slovak Republic in activities and threats in cyberspace if they threaten state security (cyber defense). This means that it can not be applied even in the context of the role and authority of the state body in the protection of cyberspace under the Military Intelligence Act, or the provisions of the Criminal Liability Act of the Investigation, Detection and Prosecution of Criminal Offenses.

We see the key importance in codifying and trying to unify the issue. We did not have a formally established terminology in the field of cyber security. For this reason, the law is a breakthrough, as it defines the basic concepts. By way of example, we are referring to the term "cybernetic", which does not appear in any generally binding legal regulation. Cyber-security means a collection of resources designed to protect the cyberspace. These means may of course be of a different nature, but for the purposes of the Cyber Security Act the term "cybernetic security" is otherwise required [6]. This is a state in which networks and information systems are able to withstand at a certain level of reliability any conduct that threatens the availability, authenticity, integrity or confidentiality of the data stored, transmitted or processed, or related services provided or accessible through these networks and information systems.

4 Cybernetic Space and Its Protection

The main goal of the law is therefore the protection and functionality of cyberspace. It is a global dynamic open system of networks and information systems that make up activated elements of cyberspace, people who perform activities in this system, and relationships and interactions between them. It is particularly important to define the definition of a cyber security incident. According to the legislator, any event that has a negative impact on cyber security due to a disruption of network and information system security or a violation of a security policy or a binding methodology [10].

As it is clear from §3 (b) of the Act, cybernetic space is limited by the use of electronic devices and electronic spectrum for the creation, storage, modification, exchange and use of data through interdependent and interconnected networks. As Gregušová and Halaszová cyber security states, it is a system provides a secure, protected, open, cybernetic space the environment of the state's socio-economic structure. It guarantees the security of the electronic, information, communication and controls systems in this area and the stored, processed and transmitted data as well as these systems of provided services [5]. Legal regulation applies to BSOs and SSPs providing their services in designated regulated sectors of the economy, which are specified in the law. In order to ensure a consistent approach across the European Union, the definition of BSO and SSP is based on the criteria set out in the Directive.

Paragraph 4 of the Act addresses the issue of the competence of the public authorities to which the Cyber Security Law has been imposed. These include the National Security Authority, several ministries, the Slovak Intelligence Service, the military intelligence or the Prosecutor General's Office of the Slovak Republic. The law is further clearly identified by the public authorities in charge of cyber security. In accordance with the Directive, the Slovak Republic also designated the National Security Authority as a national single point of contact for network and information security. This body also performs a liaison role in order to ensure cross-border cooperation between the competent authorities in other Member States, the cooperation group and the network of CSIRT units. The Office, as the central state administration body in the field of cyber security, manages and coordinates the performance of state administration. The legislator gave him the opportunity to conclude cooperation agreements to solve problems with well-recognized security experts and information systems whose state can not employ because of limited wages. It is a kind of agreement concerning rights and obligations arising from employment or other employment relationships where the contractor is required to execute orders and perform tasks not from his employer but from the National Security Authority. Paragraph 6 of the Act took into account the wording of the Directive in that each Member State had to designate at least one CSIRT unit. In our conditions, it is the Office which has imposed a law in this respect on the obligation to ensure a high level of availability of its communication services through preventive measures. Communication channels are clearly defined and stakeholders and cooperating partners are well informed about them. The focus of their protection lies in their placement in secure places, the introduction of a suitable management system and adequate staffing to ensure the continued availability of their services.

According to Veselý, Greguš and Beňová the law is also subject to the State's obligation to adopt a national strategy on network security and information systems. It sets out strategic objectives and appropriate policy and regulatory measures to achieve and maintain a high level of network security and information systems [11]. This includes the designation of research and development plans related to the national strategy in the area of network security and information systems or a risk assessment plan for the identification of risks.

In order to promote an effective exchange of information and best practices, it is essential to create an adequate and secure information system based on the trust of stakeholders. In the near future, the Office will establish and maintain a Unified Cyber Security Information System ("UCSIS"), which will be available continuously and automated through its own web site via the "www.slovensko.sk" portal. Its services should be used to efficiently manage, coordinate, record and control public cyber security and UCSIS units as well as collect, process and evaluate information. It is also supposed to provide trusted information not only at the time of peace during crisis planning but also in crisis situations.

5 Government Unit for Solving Cyber Security Incidents

In order to ensure legal certainty and build mutual trust, this law establishes the CSIRT government unit. However, it is exclusively intended for the sub-sector information systems of the public administration. From the point of view of the structure, it is set up within the competence of the Office of the Government of the Slovak Republic, since the functional e-government belongs exclusively to its portfolio of activities. However, the analysis of the Act further indicates that due to the considerable scope as well as the respect of sensitive and confidential information contained in these information systems, the tasks of the CSIRT are not performed for other subsectors [5].

As explained in the explanatory report to the CSIRT Draft on Cyber Security, we understand that it is a team that detects security incidents to co-ordinate and co-operate in their solutions and to prevent further threats. CSIRT is an expert body that provides its services to a specific component or group of components on which protection is targeted. Under the component, in this case, the client is understood. A client group for which the service is provided by CSIRT and which builds its services on the basis of the requirements of each component. The component is the baseline part of the cyber environment in which CSIRT operates, and the role of CSIRT in this system is determined on the basis of the relationships created between the components. As a general rule, the CSIRT for the Components fulfills the role of the monitoring body, and in the event of an incident, it coordinates the activity of the individual components and proposes solutions for the elimination of cyber incidents, but the self-solving of the cyber incident itself is done by the client itself, in which the CSIRT operates. Subsequently, CSIRT may issue a binding or recommendatory opinion to prevent the recurrence of a cyber incident, to act preventively. However, CSIRT can not be generalized as a senior monitoring or tracking, as it can, in correlation with the size and complexity of the environment in which it operates, also perform tasks related to direct cyber incident, forensic analysis, research and development of security technologies, and similar things.

In this context, the law defines the services that the CSIRT unit performs. These tasks - services are categorized under the Act into two main service areas, namely preventive services and reactive services. Preventive services provided by the single CSIRT focus on prevention of cyber security incidents. Reactive services directly respond to the resulting or threatened security situation.

In addition to the government unit that is set up by law, other "private" units are also created. However, they must meet demanding technical, technological and personal criteria, as well as complete a complex accreditation process. However, the accredited unit must also meet the accreditation conditions continuously throughout its operational life and fulfill statutory tasks. Otherwise, the Office will disqualify it from the list of accredited units. As further described by Davidekova and Greguš within its scope of responsibility, it is responsible for cyber security incidents and performs preventive services, monitoring and recording of cyber security incidents and reactive services, draft measures to prevent the continued continuation, spread and re-occurrence of cyber-security incidents [3].

Within the meaning of the Directive, the responsibility for ensuring network security and information systems is the responsibility of the BSO. The risk management culture, including risk assessment and the implementation of safeguards that are commensurate with existing risks, should be promoted and developed through appropriate regulatory requirements and voluntary sectoral practices. Therefore, it is necessary for the BSO to adopt technical and organizational measures to manage the risks related to network security and the information systems they use in their operations. In the latest technological developments, these measures provide for a level of cyber-security that corresponds to the level of risk involved. BSOs must take measures to prevent and minimize the impact of incidents affecting the security of the networks and information systems used to provide these basic services in order to ensure their continuity.

According to Peráček, Mucha and Kočišová under the Directive, existing capabilities are not sufficient to guarantee a high level of network security and information systems in the Union. There is a different level of preparedness in the Member States, leading to fragmentation of approaches in the Union. This results in a different level of consumer and business protection which undermines the overall level of network and information security within the Union [9]. The absence of common requirements for operators of basic services and digital service providers makes it impossible for a Union to establish a global and effective cooperation mechanism. Effective response to network and information security challenges therefore requires a comprehensive Union-level approach to common minimum requirements for capacity building and planning, information exchange, cooperation and common safety requirements for essential service providers and providers of digital services.

As mentioned earlier, the provisions of the Security Act act to the requirement of the Directive and introduce minimum security "standards" in relation to network and information systems security. In this construction, the Office builds on generally accepted and available safety standards in cyber security. We emphasize that the BSO and the SSP do not prevent the law from applying more stringent security measures than the provisions of the law. Furthermore, the law directly defines the security measures necessary to ensure cyber security through BSO and characterizes them as roles, processes and technologies in the organizational, personal and technical field, which are implemented depending on the categorization of networks and information systems within the scope of the BSO.

The special part of the Act regulates the areas for which security measures are taken. As can be seen from the analysis of the provisions of Paragraph 20, Sects. 3 and 4 of the Act, they are generally defined and their fulfillment can be achieved by various technological means. Entities that are required to apply these security measures may choose their own specific way of implementing these security measures. This premise corresponds to the principle of technical neutrality of the Cyber Security Act. However, security measures are taken in a specific form. It is a document - security documentation that must be up to date and in line with the real situation.

6 Reporting Cyber Attacks

The Paragraph 24 of the Act regulates the serious issue of the obligation to report cyber- security incidents and classify cyber security incidents on the basis of exceeding the thresholds set out in the implementing regulation issued by the Office. A first, second and third cyber security incident is distinguished. When determining it is based on Article 6 of the Directive and the following factors are taken into account when determining the importance:

- the number of users using the service provided by the entity,
- the dependence of other industries on the service provided by the entity,
- the impact that incidents could have on the scale and duration of economic and social activities or public security,
- the market share of the entity,
- the geographical spread in terms of the area the incident could have affected,
- the importance of the entity in terms of maintaining a sufficient level of service, taking into account the availability of alternative ways of providing the service.

As a clear positive, we can evaluate the Paragraph 26 of the Act, which allows voluntary reporting of a cyber security incident by natural persons or legal entities via UCSIS. However, in relation to the voluntary reporting of cyber-security incidents, it is no longer essential that a cyber-security incident exceeds the threshold criteria for each category of cyber-security incident.

The extensive Paragraph 27 is devoted to addressing cyber-security incidents by the Office if the response to a serious or ongoing serious cyber-security incident is inadequate. Cyber-security incident resolution is all procedures to support the detection, analysis and mitigation of cyber-security incident and its response. In practice, the procedures are:

- in procuring alerts and warnings declared in UCSIS or by mass media, if this is in urgent interest of society and the state. This is a precaution against the occurrence of a cyber-security incident in order to prevent it from occurring or at the time of the incident for the purpose of its resolution and to prevent or reduce the harmful consequences,
- imposing an obligation to address a cyber security incident. The obligation to deal with a cyber-security incident shall be imposed by the Office's decision on the CSIRT, SSP and BSO,
- the obligation to take a reactive action as a specific targeted response to a cyber-security incident. The obligation to execute a reactive measure is imposed by the Office by decision of the SSP in times of crisis situation and BSO,
- the obligation to design and implement a protection measure. The protective measure has the obligation to design and execute the BSO.

7 Fines for Breaking the Law

The complexity of the law is also examined in the Paragraphs 30 and 31 governing the correct punishment for breaching of the law. In general, the term "offense" is understood to mean an act that violates or threatens the interest of the company and is expressly identified in the law as a violation. As can be seen from the law, a cyber-crime offense is committed by a natural person who violates the obligation of confidentiality, has provided false information, does not observe minimum security measures, violates the obligation to provide information, does not accept security documentation that must be current and must correspond to the real state, technical, organizational or personal measures adopted by the basic service operator. As Vrabko et al. describes this act, whether intentional or negligent, may be imposed on the Office by a fine of between EUR 100 and EUR 5 000 [12]. From a procedural point of view, it should be emphasized that the law does not contain any specific procedure and therefore the legislator refers to the subsidiary scope of Act No. 372/1990 Coll. about offenses. The authority to deal with offenses and to impose fines that become revenues to the state budget is also entrusted to the Office [1].

Administrative offenses, as well as offenses, constitute unlawful conduct, the features of which are laid down in the law for which the administrative authority (the Office) always only by law imposes sanctions on natural persons for entrepreneurs and legal entities. According to Vrabko et al. the Office will impose a fine of between EUR 300 and EUR 30 000 on a basic service operator who will commit an administrative offense by breaching the obligation, report a cyber-attack, change through a unified cyber security information system, keep the security documentation up to date and realistic [13].

The Office will impose a fine of EUR 300 to a maximum of 1% of the total annual turnover for the preceding accounting year, but not more than EUR 300 000, the basic service operator committing an administrative offense by breaching the obligation to notify the Office that the identification criteria of the service, if it does not accept and disregard general safety precautions. The same penalty shall imposed to the person who fails to report a serious cyber-security incident or has sent an incomplete report. In this group of errors, the legislator also included a non-resolution of a cyber security incident on the basis of a decision of the Office, failure to take reactive measures on the basis of its decision, or failure to communicate and demonstrate the implementation of a reactive measure and its outcome.

By a maximum penalty of up to EUR 300,000 may be penalized a digital service provider who violates selected legal obligations, if he fails to notify the law, does not report a cyber-attack, or does not take a reactive action at the discretion of the Office.

When imposing a fine for an administrative offense, the Office shall apply the principle of administrative discretion and, in determining the amount of the fine, shall take into account the seriousness of the administrative offense, in particular the manner in which it was committed, the duration, the consequences and the circumstances in which it was committed. It is possible to consider the Paragraph 31, Sect. 7 of the Act as a matter of concern to the Slovak Republic for its own cybernetic security. If, within one year of the effective date of the decision to impose a fine, a breach of the obligations for which the fine has been imposed has been committed, the Office shall impose a double fine [8].

As a negative we can consider the fact that in dealing with administrative offenses, the Office is time-limited by subjective time because the fine may be imposed within two years from the date of the finding of an infringement. However, the decision to impose the penalty must be enforceable no later than four years from the date on which the breach occurred. Probably in order to avoid speculation on the part of the law-abusers, the issue of the fine for the administrative offense was strictly regulated, for a maximum of 30 days from the date on which the decision on its imposition came into force. The actual proceeds from the fines are revenues to the state budget.

8 Conclusion

The content of the contribution shows that the law is based on the current legal situation in the Slovak Republic, where cyber security issues are governed by a number of legal regulations. Due to their mutual inconsistency, the level of protection is diverse and incompatible, and as a result, it does not reach the required level of the advanced European Union Member States. As a result, there is no adequate level of cyber security against existing threats. This state of affairs can result in irreparable loss and disruption of the credibility of organizations and the Slovak Republic. Therefore, the primary objective of cyber security for each state is to minimize the potential for such threats. In the event of the consequences, the secondary objective is to minimize their impact on public administration as well as on the private sphere.

In general, however, we anticipate that in the future, the directive will be changed into line with the coordination of member statest' action so as to ensure pan-European cyber-security. This change will also have to be met by national legislation. In our view, in the case of a serious cyber-attack on a Member State, another non-EU country can proceed according to existing European and international treaties guaranteeing collective security.

By examining the law, we have found that its task is to create, in particular, uniform legislative conditions to ensure adequate protection of the cyberspace of the Slovak Republic against potential threats. Their existence could create irreparable damage and our international credibility would be compromised. We expect statutory measures to fully implement the prerequisites for a coherent, coordinated and effective system for protecting the cyberspace of the Slovak Republic. At present, therefore, we consider the content of the law to be sufficient. However, we believe that serious forms of cyber-attacks must be punished by the state, not by standards of administrative law but by criminal law. For this reason, in our opinion, the legislator also had to introduce into a separate section of the Criminal Code a specific section dealing with cyber security crimes, which would be sanctioned by unconditional punishments of deprivation of liberty.

References

1. Act No. 372/1990 Coll. about offenses as amended
2. Act No. 69/2018 Coll. Cyber Security Act as amended
3. Davideková, M., Gregus, M.: Concept Proposal for Integration of Virtual Team Collaboration in a University Study Subject, vol. 20, pp. 206–215. Mendel University, Brno (2017)
4. Directive Nr. 2016/1148 of the European Parliament and of the Council of 6 July 2016
5. Gregušová, D., Halásová, Z.: Elektronická podoba výkonu pôsobnosti orgánov verejnej moci, 1st edn. Vysoká škola Danubius, Sládkovičovo (2015)
6. Jansa, L., Otevřel, P.: Softwarové právo, 1st edn. Computer Press, Brno (2014)
7. Maisner, M., Zeman, J., Loebl, Z., Donat, J., Mičinský, Ľ., Menkeová, L., Beňa, J.: Základy práva informačných technológii, 1st edn. Iura Edition, Bratislava (2014)
8. Maisner, M., Černý, J., Donát, J., Loebl, Z., Nielsen, T., Polčák, R., Zeman, J.: Základy softwarového práva, 1st edn. Wolters Kluwer ČR, Praha (2013)
9. Peráček, T., Mucha, B., Kočišová, L.: Importance of the e-government act and its impact on the management and economy of the enterprise in the Slovakia. Management and economics in manufacturing. Technická Univerzita vo Zvolene, Zvolen (2017)
10. Štědroň, B., Ludvík, M.: Právo v informačních technologiích, 1st edn. Computer Media, Praha (2008)
11. Veselý, P., Greguš, M., Beňová, E.: Current Approaches to Increased Protection against Trojan Horses in Cloud Server Solutions, vol. 5, pp. 1092–1095. Central Bohemia University, Prague (2017)
12. Vrabko, M., Pekár, B., Srebalová, M., Škrobák, J., Vačok, J.: Správne právo hmotné: všeobecná časť, 1st edn. C. H. BECK, Bratislava (2012)
13. Vrabko, M., Pekár, B., Srebalová, M., Škrobák, J., Vačok, J.: Správne právo procesné: všeobecná časť, 1st edn. C. H. BECK, Bratislava (2013)

A Readiness Model for Measuring the Maturity of Cyber Security Incident Management

David Rieger(✉) and Simon Tjoa

Institute of IT Security Research, Josef Ressel Centre TARGET,
St. Pölten University of Applied Science, Matthias-Corvinus-Strasse 15,
3100 St. Pölten, Austria
{is171837, simon.tjoa}@fhstp.ac.at

Abstract. Hardly a week goes by without headlines about new cyber-attacks. As the sophistication of cyber-attacks constantly increases, organizations have to consider to be affected by attacks. In order to effectively and efficiently react to an incident, professional and well-organized incident management has to be in place. The major goal of this paper is to support organizations to develop and improve their cyber-security incident management. Therefore, in this work, a readiness model, covering nearly 80 topics and 500 requirements in the domain of incident management, is introduced.

Keywords: Cyber-security · Security incidents · Incident readiness
Readiness model · Readiness assessment · Cyber-security capabilities
Incident response · Incident management · Management system
Incident management capabilities · Incident response technologies
Incident response tools

1 Introduction

One of the most demanding challenges for companies today is the strong dependence on information systems. Incidents of the past impressively demonstrated that the information infrastructure is a vulnerable spot across all sectors (e.g. energy, water, finance or healthcare).

Driven by this development, cybercrime activities are rapidly increasing leading to serious operational and financial impacts (e.g. Carbanak bank robbery, ransomware campaign such as WannaCry) [1–4, 6]. Additionally, new legislation (e.g. NIS directive [18] or GDPR [17]) and regulations demand adequate protection against cyber-attacks and IT-failures. As a result, it is indispensable for all businesses to maintain an appropriate security and resilience level of the technical infrastructure.

However, due to the fast digitalization in the recent past, the complex nature of modern information systems and the increasing sophistication of attackers, it is impossible to solely rely on countermeasures trying to prevent security breaches.

In addition, it is necessary to develop adequate capabilities for reacting during and after a cyber-security incident in order to ensure cyber resilience [5]. Depending on the

© Springer Nature Switzerland AG 2019
F. Xhafa et al. (Eds.): INCoS 2018, LNDECT 23, pp. 283–293, 2019.
https://doi.org/10.1007/978-3-319-98557-2_26

organization, this can lead to forming cyber defense centers, security operation centers or computer security incident response teams.

In this paper the significant role of cyber-security incident management capabilities is analyzed. The major contribution is the introduction of a readiness model supporting the measurement of the maturity and readiness level of cyber security incident management capabilities.

The remainder of this paper is as follows: In Sect. 2, important publications in the domain of incident management are presented. In Sect. 3 the major contribution (i.e. the readiness model) is introduced. In Sect. 4 we briefly present how we implemented it in order to improve usability, before we conclude in Sect. 5 the paper with our key findings and open future research areas.

2   Related Work

Structured cyber-security incident management has been an important factor in critical infrastructures for many years. Thus, comprehensive literature from many international standardization bodies and research organizations exist.

In this section, we briefly outline the most important representatives [7–15] in the area of cyber-security incident management, which built the foundation of our readiness assessment model.

Each of these documents contains recommendations, requirements, best practices, standardized controls, or methods for designing, implementing, performing or maintaining an efficient and effective cyber-security incident management.

The ISO/IEC 27035:2016 [7] is one of the most commonly and widely used standards in the field of information security incident management and describes a general, well-structured and flexible model for an organizations incident management framework including all processes, roles and rules required for successfully handling cybersecurity incidents.

The ISO/IEC 27002:2013 [8] provides a list of general information security controls and therefore also covers relevant points in the context of cybersecurity incident management.

The NIST SP 800-61 [9] serves as an excellent base for defining and concepting an appropriate cybersecurity incident management within an organization as it provides many recommendations, concepts and requirements together with a highly sophisticated life-cycle-model.

The SANS Institute's "Incident Handler's Handbook" [10] offers hands-on oriented insight in the development, establishment and operation of cybersecurity incident response controls and activities. It is highly appreciated by many CSIRT members and especially relevant when searching for tested procedures. Another relevant source published by the SANS [11] establishes basic recommendation and important considerations for establishing and operating a CSIRT. The major goal is to provide a good overview about important requirements regarding the CSIRT within an organization which builds an essential part of successful cybersecurity incident management.

The German Federal Office BSI outlines in its IT-Grundschutz [12] a very detailed standard covering cybersecurity incident management. A main advantage of IT-Grundschutz is the deep level of granularity.

CREST "Cyber Security Incident Response Guide" [13] includes requirements and recommendations for organizations establishing a cybersecurity incident response process. It focuses especially on criteria for the successful long-term operation of cybersecurity incident management activities.

The European Network and Information Security Agency [14] summarizes recommendations and guidelines for European organizations on how to ensure the appropriate management of cybersecurity incidents. It provides also useful references to many additional sources which are relevant in this context.

CMU's Software Engineering Institute [15] provides one of the most well-known guides for the establishment and operation of a CSIRT. Beside requirements and recommendations about a variety of incident management aspects it provides detailed insights and explanations about the underlying ideas behind them.

3 Cyber-Security Incident Management Readiness Model

Based on the literature review, we structured the readiness model into three layers, which represent essential incident management capabilities. In order to structure the three layers and build a readiness model, we came up with a tree-based structure consisting of the three key-elements Domain, Topic and Requirement. Figure 1 schematically outline the overall structure of the proposed readiness model.

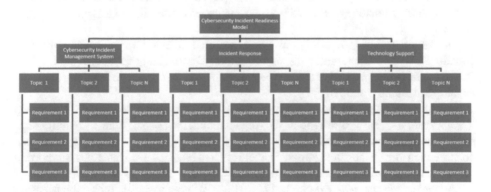

Fig. 1. Schematic overview of the cyber-security incident management readiness model

Domain elements describe a general field of cyber-security incident management and equals the layer model described before. The first domain *Incident Management System* represents all processes, policies, roles, workflows, strategies and other relevant aspects for the organization of incident management. The second domain *Incident Response* category represents all topics that are relevant for the actual handling of incidents. This includes for example specific reactions and measures to respond to a certain type of cyber-security incident. The last domain *Technology Support* includes all technical and tool-related aspects of incident management. This can exemplarily be proper support of incident response activities or detection of cyber-security incidents through appropriate technical tools.

Topic elements further refine **Domain** elements and describe a bundle of recommendations and requirements. The structure for topics of the domain *Incident Management System* was derived from the ISO 27035:2016 [7] as it represents the most commonly adopted publication in this context. For the *Incident Response* domain, the basic categorization of requirements followed the NIST incident response lifecycle [9]. The *Technology Support* domain applies a slightly modified version of the OODA Loop to categorize respective requirements [16]. The adaption eliminates the "Decision" phase since there is currently no option for organizations to implement controls for supporting this phase (except for some experimental artificial intelligence solutions).

Requirement elements specify concrete capabilities in the context of a **Topic**. Due to space restrictions, in the following subsections we focus on topic-level and only exemplarily outline requirements.

3.1   The Incident Management System Domain

Similar to other management systems, the cyber-security incident management system represents the organizational model including processes, policies, roles required to perform incident management capabilities.

Due to its touchpoints to the information security and business continuity domains, most organizations integrate the incident management systems as part of an information security management system (ISMS) or business continuity management system (BCMS).

In the following, we briefly highlight selected relevant topics within the first domain and provide examples for requirements. Due to length restrictions, the presented topics and requirements are not exhaustive but indicate examples of the developed readiness model.

In order to plan and prepare for incidents it is important to address *Roles and Responsibilities* [7–9] throughout the organization. Amongst others, requirements include management commitment, dedicated roles for detecting, assessing, handling, documenting and reporting computer security incidents as well as roles supporting incident management such as dedicated interfaces to legal affairs or finance.

Incident Management Processes [7, 8, 11, 12] have to be defined which ensure continual improvement and provide the framework for successfully performing incident management. Examples for requirements in this phase include the definition of clear escalation procedures and escalation triggers or communication during and following an incident.

The *Incident Management Policy and related statements* [7, 8, 12] clearly highlight the importance of incident management and provide general organizational rules and conditions. Requirements within this topic include, inter alia, scope and objectives of an organization's incident management applying rules and constraints.

Interfaces to Other Domains [7–9, 11–13], such as crisis management, are vital to ensure efficient and effective incident management. Requirements in this topic range include interaction processes and structures between the incident management team and risk management or corporate crisis management teams.

The *Organizational Structure* [7, 8, 11, 12], of an incident response team is paramount for its performance. Basic requirements exemplarily consist of defined roles and responsibilities of the CSIRT and other interested parties, such as supplier, authorities or external CERT teams.

Support Activities [7, 8, 11, 12], cover activities which are not directly dedicated to the management of incidents. Examples include travel management or procurement of hardware and software.

Awareness and Training [8, 9, 11, 12, 14] within the organization and among its partner highlight another important topic of the planning and preparation phase. Requirements in this area are ranging from managing awareness campaigns to specialized training for incident response teams.

For the detection and reporting of incidents, *Organization of Assessment and Reporting of Incidents* [7, 8, 11, 12] is a key area which defines clear rules and processes on managing detection sources, assessing observed events and reporting them corresponding to the organizational framework.

Log Management [7–9, 12, 13] covers all incident management related aspects on the logging and is a crucial part throughout all incident management phases. Examples include the exact definition of log events that have to be captured, definition of appropriate storage of logs or mechanisms ensuring consistency, integrity and availability of logs.

The consideration of *Forensic Processes* [7, 9, 11–13] is crucial to ensure evidence is acquired in a forensically sound manner and insights gained during investigations can be used in legal cases.

For promoting continual improvement and to maintain incident management capabilities it is crucial to have formal *Lessons Learned* [7–9, 11, 12] activities in place.

Compliance [7–9, 11, 12, 17] is a further key aspect to mention in this context. It increasingly becomes important for organizations to demonstrate compliance to laws, contracts and regulations through a proper incident management. Through recent laws, such as the EU-GDPR [17], increased attention has been paid to business continuity and incident management. Requirements in this category include trainings, skills and knowledge in the context of data protection laws and handling of personal data on the staff level.

3.2 The Incident Response Domain

The incident response domain covers required response capabilities for reacting to occurring cybersecurity incidents. In order to categorize the topics and requirements, the phase of the NIST incident response lifecycle [9] were slightly extended. In the following we briefly outline the topics of this domain structured by the above-mentioned categorization.

Within the *CSIRT* [9–11, 13, 15] category following topics have been identified:

- General security and incident related competencies skills, and knowledge: this topic covers the importance of skills and competencies which are required to identify, assess, respond to incidents. The requirements of this topic (i.e. skills and competencies) have been grouped by role and consist of overall competencies (e.g. analytical skills, stress resistance), technical competencies (e.g. digital forensics knowledge, such as evidence acquisition, knowledge on cryptographic technologies) and other competencies (such as experience in IT security.)
- Training and education: As adversaries constantly improve the sophistication of attacks and at the same time information systems evolve, it is crucial for incident response to train and educate response personnel to stay effective.
- Experience: The proper reaction to an advanced attack or any major incident requires good problem-solving and fast decision-making skills. An important factor to make good decisions in a stressful situation and under uncertainty is experience. Therefore, requirements for determining the readiness of this topic exemplarily include the experience of the incident response leader or personnel.
- Situation Awareness: The acquisition and combination of information during an incident is important to have a clear understanding about the big picture during an analysis. Therefore, this topic is dedicated to requirements, such as the technical and organizational implementation of the information gathering process or procedures to analyze information to enable well-informed decisions.

The next category *Prevention* [9–11, 13, 15] is dedicated to the topics technical controls (e.g. system hardening, implementation of an adequate access control system) and organizational controls (e.g. establishing and maintaining an ISMS).

The category *Incident Handling* [9–11, 13–15] can be further divided into sub-categories *Preparation, Detection and Analysis, Containment, Eradication and Recovery and Post-Incident Activities*.

Within the sub-category *Preparation for Incident Handling* [9–11, 13–15] topics like incident response plans, standard operation procedures, exercising, documentation or incident handling facilities are evaluated to determine the readiness. A very popular example for a requirement within this sub-category would be the establishment of incident response plans, which specify detailed steps for specific incident types (e.g. ransomware) including essential processes such as initial triage, classification procedure, immediate actions as well as long term resolution and follow-up strategies.

Detection and Analysis [9–11, 13–15] focuses on the topics detection mechanisms (e.g. processes for detecting and reporting incidents, implementation of tools gathering and distributing security events), analysis of incidents and their impact (e.g. categorization of incidents, procedures for conducting business impact analysis for occurred incidents).

The *Containment, Eradication and Recovery* [9–11, 13–15] category is the heart of incident response. Activities in this category counteract against incidents in order to keep the damage as low as possible. Topics within the proposed readiness model include: structural approach and containment strategies (e.g. definition of containment strategies for certain scenarios), digital forensics (e.g. evidence acquisition and handling), eradication plans and mechanisms (e.g. implementation of eradication mechanisms), recovery to operational state (e.g. establishment of recovery policies and plans).

Within the last sub-category *Post-Incident Activity* [9–11, 13–15] continual improvement is analyzed, taking into account the topics knowledge increase, improvement of CSIRT performance and analysis of weak spots.

The readiness of the category *Escalation and Interactions* [9–11, 15] is assessed by taking a deep look into the topics interfaces to other management domains and escalation of incidents and/or activities.

To evaluate the state of *Incident Response Services* topics such as triage and analysis services, incident handling and coordination of involved parties are examined.

The last category covers the large part of *Communication*. This includes topics on internal communication (e.g. CSIRT communications or In-house communications of the organization) as well as external communication (communications with external parties, or public relation activities).

All of these previous listed areas of the "Incident Response Domain" represent a basic structure for successfully handling cybersecurity incident. An organization should especially ensure that their "Requirements", which fall under the different "Topic"-categories of "Incident Handling" and the related sub-categories, are properly developed since they represent the general workflow. This is required for successfully resolve incident on the long term [9, 10, 15]. The other categories listed in this section are however also important since they represent basic enablers for the successful operation of this core workflow [9–12].

3.3 The Technology Support Domain

The technology support layer uses a slightly modified version of the widely-known OODA Loop [16] created by US Air Force military strategist John Byod. As briefly outlined before, the adaption consisted mainly in the elimination of the "Decision" phase as there is currently no option for organizations to implement controls for supporting this phase (except some experimental artificial intelligence solutions) and adding a new phase representing the preparation for cyber-security incident handling. Analog to the other domains above, we grouped the topics in different phases (see Fig. 2) to improve the applicability and usability of the approach.

All of the topics which were identified in this layer represent certain abstract requirements which are implied by the goal of the operation of a successful incident management within an organization [10, 14]. An example would be the requirement topic of "incident handling systems" which includes requirements for an appropriate technical solution to actively manage, document and assign new reported incident within an organization as well as the requirements for such a solution should be considered.

4   Implementation of the Readiness Models

In order to enhance the usability and feasibility in real-world environments, it was necessary to further list, analyze and explain the detailed "Requirement"-elements of each "Topic"-element identified before. This step was performed within the thesis upon which this paper is based on.

Incident Management Field	Phase	Technology Category
Incident Management and Preparation	Prepare	Incident handling systems
		Information management tools
		Incident management tools
		Implementing CSIRT procedures
	Observe	Network monitoring
		Detection (IDS and Host-Monitoring)
		SIEM
		Timeline tools
Incident Response	Orient	Log and Memory Analysis
		Supporting tools for handling evidences
		Network monitoring
		Gathering evidence
		Investigating evidence
		Implementing CSIRT procedures
		Information management tools
		Log and Memory Analysis
		Sandboxing/reversing tools
		Forensic Tools
	Act	Implementing CSIRT procedures
		Multipurpose-Tools (e.g. IR/Forensic-Workstations, Distributions,...)
		Secure remote access
		Forensic Tools
		Recovering the system
		Incident management tools

Fig. 2. Overview of "Topic"-elements for the "Technology Support"-domain

For each "Topic"-element six to eight "Requirement"-elements have been identified and described in order to build a comprehensive readiness model resulting in 79 "Topic"-elements over 500 "Requirement"-elements.

All topics and requirements have been combined in a spreadsheet. The spreadsheet has been discussed with various security experts to evaluate the feasibility of the approach. First evaluations have been very promising and a tailored version is currently evaluated in a real-world setting. The reason for tailoring the readiness model lies in the workload which would be caused by thoroughly assessing around 500 requirements.

Req.-Nr.	Requirement	Detailed Requirements and Explanation	Maturity
1	Plan and Prepare of incidents		
1.1	All roles and responsibilities for the organizations incident management are defined, assigned and communicated to all members of the organization.	For the successful operation and establishment of an incident management system and the organization of incidents overall it is essential that all roles and responsibilities for detecting, reporting, assessing, handling and analyzing security events and incidents are defined, documented, communicated and implemented throughout the whole organization. This especially includes the roles and responsibilities of the organizations CSIRT, the responsibilities of the top management, support roles of multiple departments and basic responsibilities of all employees in the scope of the organization's activities related to event and incident management. The basic requirements for roles and responsibilities in a sophisticated incident management system include: • Responsibilities and commitment of the top-level management are defined, assigned and communicated • Roles of the organizations CSIRT are defined according to a logical structured role schema. • Support roles and responsible. This means that all departments of an organization which could play a potential role in the incident management processes need to define responsible persons for supporting and enabling a successful incident management. Examples of such departments are especially HR, legal affairs, PR, marketing, finance, resource management and logistics. • Responsibilities of all employees related to the reporting of incidents and suspicious cybersecurity events. • Roles and responsibilities during an incident covering all incident response aspects. This also includes the assignment of a single point of contact (SPoC) for the incident management. • Responsible management roles for the incident management system itself. • Ensuring that all roles/responsibilities are assigned to employees with appropriate skills and competences	

Fig. 3. Screenshot of the practical implementation of the model

Figure 3 shows an example screenshot of the implementation in Excel. The actual "Requirement" elements can be seen in the column "Detailed Description" but it is important to bear in mind that the figure shall only serve as illustrative example in this context. Figure 4 shows how such a practical implementation might be used for assessing the actual situation of an organizations incident management by assigning these requirements with certain maturity levels.

Req.-Nr.	Requirement	Maturity
1	Plan and Prepare of incidents	
1.1	All roles and responsiblities for the organizations incident management are defined, assigned and communicated to all members of the organization.	Managed
1.2	All processes for the management, handling and further processing of security incidents are defined, established and actively managed.	Defined
1.3	Proper incident management policies with statements suited to the organizations needs are established.	Repeatable
1.4	Interfaces for general interaction between the organizations security incident management and its crisis, culnerability and risk management are defined and implemented.	Minimal
1.5	The organization defined a clear and logical structure for a CSIRT (Cybersecurity Incident Response Team) and implemented It accordingly.	Non-existent

Fig. 4. Screenshot of the "Incident Management System" first "Topic" elements assigned with different maturity levels.

5 Conclusions

Due to high significance of information systems and the dependence of nearly all business functions, incident management is evolving to one of the key areas to ensure continuous business operations.

Within this work, we presented a readiness model with over 500 requirements which can be used to evaluate an organization's state of incident management. Beside the readiness model, this paper also makes visible that the complexity of building a mature cyber-security incident management team is quite high.

Another contribution of this work is the introduction of a structure which can be used to aggregate available information and sources in the context of cyber-security management.

The last conclusion which can be drawn out of this work is the fact, that the topic of cyber-security incident management is often misunderstood in its complexity, content and necessity. Often the entire field is reduced to some tasks regarding components of an organizations IT-infrastructure or some certain roles and processes which need to be defined.

During the process of developing the readiness model huge potential for future work was identified. Especially a more detailed investigation and analysis of the requirements within the "Technology Support"-domain, defined within this work would provide a good starting point for future work.

Another research area for future work could be the formalization and extension of the model developed within this work to provide a more sophisticated source for organizations to assess their incident management capabilities. This could help to specify live sources (e.g. connection to monitoring, assessing training states through interfaces to HR tools) in order to get a live/realtime risk view on the topic.

Furthermore, a structured approach to tailor the herein presented readiness model to individual organizations could improve usability as the workload for assessing all 500 requirements could impede its usage especially for smaller organizations when trying to assess their overall cyber-security incident readiness.

Acknowledgements. The financial support by the Austrian Federal Ministry for Digital and Economic Affairs and the National Foundation for Research, Technology and Development is gratefully acknowledged.

References

1. CERT.at "CERT.at Statistiken". https://www.cert.at/services/statistics/statistics.html. Accessed 20 June 2018
2. Trevor White, D.L., Anderson, M., Team, S.: Global economic crime survey 2016 - adjusting the lens on economic crime preparation brings opportunity back into focus. pricewaterhousecoopers, Technical report (2016). http://www.pwc.com/gx/en/economic-crime-survey/pdf/GlobalEconomicCrimeSurvey2016.pdf
3. Wood, P., et al.: Internet security threat report. Symantec, Technical report, April 2016
4. Center for Strategic and International Studies, "Net losses: Estimating the global cost of cybercrime," June 2014. https://www.sbs.ox.ac.uk/cybersecurity-capacity/system/files/McAfee%20and%20CSIS%20-%20Econ%20Cybercrime.pdf. Accessed 18 June 2018
5. Deloitte, Cyber crisis management: Readiness, response and recovery (2016). https://www2.deloitte.com/content/dam/Deloitte/global/Documents/Risk/gx-cm-cyber-pov.pdf. Accessed 20 June 2018
6. Bromiley, M., Lee, R.: Incident response capabilities in 2016: the 2016 sans incident response survey. SANS Institute InfoSec Reading Room, June 2016. https://www.sans.org/reading-room/whitepapers/incident/incident-responsecapabilities-2016-2016-incident-response-survey-37047. 20 June 2018
7. ISO/IEC 27035 - Information technology - Security techniques - Information security incident management, ISO/IEC Std.
8. ISO/IEC 27002:2014 - Information technology - Security techniques - Code of practice for information security controls (ISO/IEC 27002:2013 + Cor 1:2014), ISO/IEC Std. 27002:2014, Rev. cor. 1:2014 (2014)
9. Computer Security Incident Handling Guide, NIST Std. 800-61, Rev. 2, August 2012. https://nvlpubs.nist.gov/nistpubs/specialpublications/nist.sp.800-61r2.pdf. Accessed 20 June 2018

10. Kral, P.: Incident handler's handbook. SANS Institute InfoSec Reading Room, December 2011. https://www.sans.org/reading-room/whitepapers/incident/incident-handlers-handbook-33901. 20 June 2018

11. Proffitt, T.: Creating and managing an incident response team for a large company. SANS Institiute InfoSec Reading Room (2007). https://www.sans.org/reading-room/whitepapers/incident/creating-managing-incidentresponse-team-large-company-1821. 20 June 2018

12. IT-Grundschutz B 1.8 Behandlung von Sicherheitsvorfällen, Bundesamt für Sicherheit in der Informationstechnik Std., Rev. 11. EL Stand (2009). https://www.bsi.bund.de/DE/Themen/ITGrundschutz/ITGrundschutzKataloge/Inhalt/_content/baust/b01/b01008.html

13. Creasey, J.: Cyber security incident response guide. CREST (2013). https://www.crest-approved.org/wp-content/uploads/2014/11/CSIR-Procurement-Guide.pdf. Accessed 20 June 2018

14. ENISA, "Good practice guide for incident management," December 2010. https://www.enisa.europa.eu/publications/good-practice-guide-for-incident-management. Accessed 20 June 2018

15. West-Brown, M.J., et al.: Handbook for computer security incident response teams (csirts), April 2003. http://resources.sei.cmu.edu/asset_files/Handbook/2003_002_001_14102.pdf. Accessed 20 June 2018

16. Bazin, A.A.: Boyds OODA Loop and the Infantry Company. Infantery Magazin, January–February 2005. https://www.academia.edu/attachments/34552740/download_file?st=MTQxNzczOTU2MSwxMDguMjYuMTIzLjE2MQ%3D%3D&s=popover. Accessed 18 June 2018

17. Council of European Union: Regulation (eu) 2016/679 of the european parliament and of the council of 27 April 2016 on the protection of natural persons with regard to the processing of personal data and on the free movement of such data, and repealing directive 95/46/ec (general data protection regulation), April 2016. https://eur-lex.europa.eu/legal-content/EN/TXT/?uri=uriserv:OJ.L_.2016.119.01.0001.01.ENG. Accessed 20 June 2018

18. Councile of European Union: Directive (EU) 2016/1148 of the European Parliament and of the Council of 6 July 2016 concerning measures for a high common level of security of network and information systems across the Union, July 2016. https://eur-lex.europa.eu/legal-content/EN/TXT/?uri=uriserv:OJ.L_.2016.194.01.0001.01.ENG. Accessed 20 June 2018

Model for Generation of Social Network Considering Human Mobility

Naoto Fukae[1], Akihiro Fujihara[2], and Hiroyoshi Miwa[1(✉)]

[1] Graduate School of Science and Technology, Kwansei Gakuin University,
2-1 Gakuen, Sanda-shi, Hyogo, Japan
{fukae,miwa}@kwansei.ac.jp
[2] Faculty of Engineering, Chiba Institute of Technology,
2-17-1 Tsudanuma, Narashino, Chiba, Japan
akihiro.fujihara@p.chibakoudai.jp

Abstract. It is well known that many actual networks have the scale-free property that the degree distribution follows the power law. This property is found in many actual networks in the real world such as the Internet, WWW, a food chain network, an airline network, and a human relations network. As for a generation mechanism of a human relations network, we should consider the human mobility that, in general, a person moves around, meets another person, and makes friend relations stochastically. However, there are few models considering human mobility so far. In this paper, we propose a new model that generates a human relations network considering human mobility, and we show that this model has the scale-free property by numerical experiments.

1 Introduction

It is well known that some actual networks have scale-free property. This means that the degree distribution follows the power law, that is, $p(k) = O(k^{-\gamma})$ where $p(k)$ is the ratio of the number of vertices whose degree is k and γ is a constant positive integer. This property is found in many networks of the real world such as the network structure of the Internet, the hyperlink structure of the WWW, a food chain network, an airline network, and a human relations network.

In this paper, we deal with a model that generates a human relations network considering human mobility. In general, while a person is moving in space, the person meets another person and becomes friends each other. However, few previous models explicitly consider human movement and encounter. Moreover, some models considering them have no scale-free property. For example, TF (Travel and Friend) model [1], after a person v moves to the place near a friend with the probability of P_v, or after v moves to a new place with the probability of $1 - P_v$, v becomes new friends with people within a certain range around v with the probability of P_c. However, the human relations network generated by this model does not have the scale-free property. Indeed, the degree distribution attenuates exponentially.

© Springer Nature Switzerland AG 2019
F. Xhafa et al. (Eds.): INCoS 2018, LNDECT 23, pp. 294–305, 2019.
https://doi.org/10.1007/978-3-319-98557-2_27

In this paper, we propose a human relations network generation model that takes into consideration human mobility. The basic idea of this model is the combination of a mobility model such as Homesick Lévy Walk (HLW) and the model making human relation, and so on. The numerical experiments indicate that the proposed model generates a human relations network having the scale-free property.

2   Related Works

The scale-free property is often found in various real networks. For example, many networks such as a gene network, a food chain network, an E-mail network, etc. have the scale-free property.

Many network generation models including scale-free networks have been proposed. Some are based on the two principles "growth" where vertices are added as time passes and "preferential attachment" where newly added vertices are connected to vertices with probability according to the degrees (Ex. [2]). The Barabási and Albert model (called the BA model) is analyzed in detail, and it is known that $p(k) \propto k^{-3}$ [3,4], and it is proved that the diameter is $O(\log n / \log \log n)$ [3,5].

As a network generation model not based on the principle of "growth" and "preferential attachment," there are some models that edges are stochastically added based on the value of a function of the weights of the end vertices [6–11]. In particular, the threshold model is the model that, when the sum or product of vertices v_i and v_j exceeds a threshold, an edge is added between v_i and v_j. If the distribution of the vertex weight follows the exponential distribution or the Pareto distribution, it is known that the generated graph has the scale-free property [11,12]. Furthermore, the limit theorem is also obtained [13]. The extended threshold model considering the spatial structure has also been proposed [14].

On the other hand, several studies on the properties of human mobility have been conducted. For example, the observation that human mobility follows the Lévy Walk (LW) is known. In particular, in the case of the spatial dimension $d = 2$, the probability of a linearly moving distance l to the next destination follows the power law $P_{LW}(l) = \mathcal{O}(l^{-(1+\beta)})$, where β $(0 < \beta \leq 2)$ is an power exponent in the moving distance.

It is also known that the distribution of frequency that two persons passing each other follows the power law [15], and, as the mobility model which is consistent with this property, the Homesick Lévy Walk (HLW) model that a person moves based on Lévy Walk and it returns to a point (base) with a certain probability (return rate) has been proposed.

TF (Travel and Friend) model [1] is a human relations network generation model considering human mobility. After a person v moves to the place near a friend with the probability of P_v, or after v moves to a new place with the probability of $1 - P_v$, v becomes new friends with people within a certain range around v with the probability of P_c. The move in TF model is "jump"(or "flight") that a node moves from the current place to the destination in one step (Fig. 1), which is different from "walk" that a node moves from the current place to the destination at a constant speed (Fig. 2)

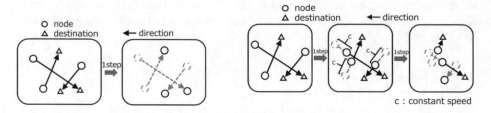

Fig. 1. Jump (flight)   **Fig. 2.** Walk

When a human relations network is generated by the TF model, the degree distribution attenuates exponentially, so that it has no scale-free property.

3 Human Relations Network Generation Model Considering Mobility

We describe a human relations network generation model considering human mobility in this section.

The basic idea is that nodes (persons etc.) move according to the mobility model, and when two nodes approach within a certain range, they become probabilistically friends (linking human relations) (Fig. 3).

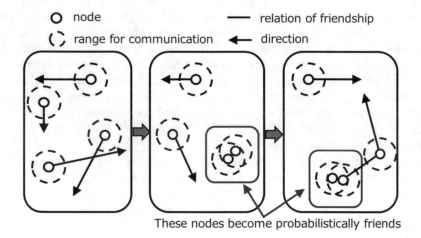

Fig. 3. Human relations network generation model considering mobility

A specific human relations network generation model is obtained by combining mobility models and stochastic relations models. We investigate the following combinations.

- HUC Model (HLW/Uniformly distributed bases/Constant probability for human relations)
  The nodes move according to Homesick Lévy Walk (HLW). The bases in HLW are uniformly distributed in the area. Each node is preliminarily given the probability P_f. When two nodes approach within range of d, the nodes make the relations with the probability of P_f.
- HUT Model (HLW/Uniformly distributed bases/Threshold model for human relations)
  The nodes move according to HLW. The bases in HLW are uniformly distributed in the area. Each node is given the intrinsic value according to the Pareto distribution. When two nodes approach within range of d, the nodes make the relations, if the value of the sum of the intrinsic values is more than a threshold of uniform random number.
- HGT Model (HLW/Gaussian distributed bases/Threshold model for human relations)
  The nodes move according to HLW. The position of the bases in HLW are determined as follows: the distance of a base from the center of the area is determined according to Gaussian distribution and the direction of the base is determined according to the uniform distribution. Each node is given the intrinsic value according to the Pareto distribution. When two nodes approach within range of d, the nodes make the relations, if the value of the sum of the intrinsic values is more than a threshold of uniform random number.
- HUI Model (HLW/Uniformly distributed bases/Increasing threshold model for human relations)
  The nodes move according to HLW. The bases in HLW are uniformly distributed in the area. Each node is given the intrinsic value according to the Pareto distribution. When the nodes make relations, the intrinsic values are increased by a constant. The upper limit of the intrinsic value is given, and the updated intrinsic value does not exceed the upper limit. When two nodes approach within range of d, the nodes makes the relations, if the value of the sum of the intrinsic values is more than a threshold of uniform random number.

A human relations network is generated by an above model.

We investigate the complementary cumulative distribution (CCDF) of the degree distribution of a network generated by an above model by numerical experiments.

The shape of the area is the rectangular area of $L \times L$. We show the values of the parameters in Table 1.

3.1 HUC Model

Figure 4 shows the degree distribution of a human relations network generated by the HUC Model, and Fig. 5 shows the complementary distribution of the degree distribution (CCDF) by double logarithmic chart.

Table 1. Values of parameters

Number of nodes	$n = 10000$
Size of area	$L = 5000$
Moving distance of 1step	$s = 20$
Range that two nodes can communicate	$d = 20$
Return rate of HLW	$\alpha = 0.3$
Exponent of HLW	$\beta = 1.0$
Probability of making relation	$P_f = 0.7$

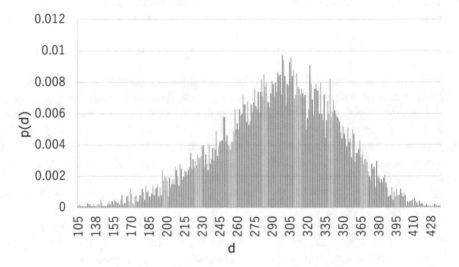

Fig. 4. Degree distribution of a network generated by HUC Model

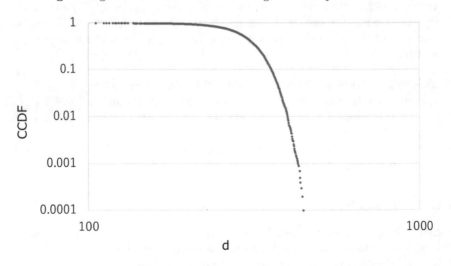

Fig. 5. CCDF (HUC Model)

In Fig. 5, the attenuation is not linear. Moreover, since the degree distribution is a bell curve, it has an average value to reflect the scale (Fig. 4). Therefore, the HUC model does not have the scale-free property. This is because, since the probability that a node is connected to another node is uniform, the degrees of some specified nodes do not increase in this model.

3.2 HUT Model

The HUT model reflects that the probability that a node is connected to another node is different among individuals. A node given a large intrinsic value according to the Pareto distribution tends to have a large degree. Figure 6 shows the degree distribution of a human relations network, and Fig. 7 shows the complementary distribution of the degree distribution by double logarithmic chart.

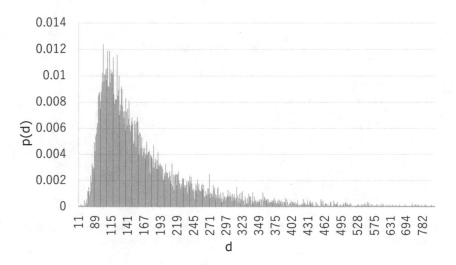

Fig. 6. Degree distribution (HUT Model)

In Fig. 6, the probability that the degree of a node is decreases gradually. However, in Fig. 7, the attenuation is not linear and there is a cut off; therefore, this model does not the scale-free property.

3.3 HGT Model

The HGT model reflects the bias of the density of the positions of the bases. A node given a large intrinsic value according to the Pareto distribution tends to have a large degree. Figure 8 shows the randomly located bases and Fig. 9 shows the bases located according to the Gaussian distribution.

Figure 10 shows the degree distribution of a human relations network, and Fig. 11 shows the complementary distribution of the degree distribution by double logarithmic chart.

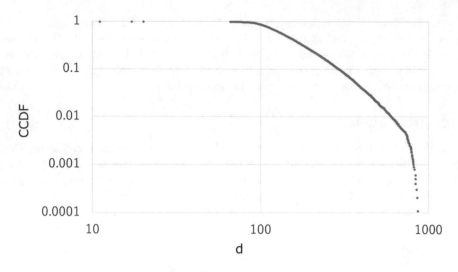

Fig. 7. CCDF (HUT Model)

Fig. 8. Bases located according to uniform distribution

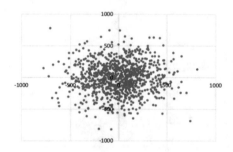

Fig. 9. Bases located according to Gaussian distribution

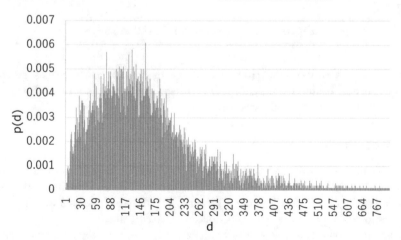

Fig. 10. Degree distribution (HGT Model)

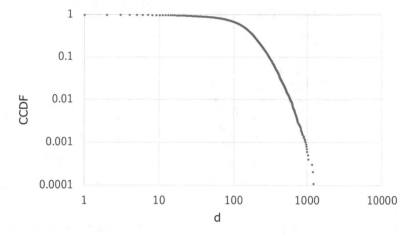

Fig. 11. CCDF (HGT Model)

In Fig. 11, the attenuation is not linear and there is a cut off; therefore, this model does not also the scale-free property.

3.4   HUI Model

In the HUI model, when the nodes make relations, the intrinsic values are increased by a constant (**EXT**). This reflects the tendency that a node having more friends becomes easier to make more friends (Fig. 12).

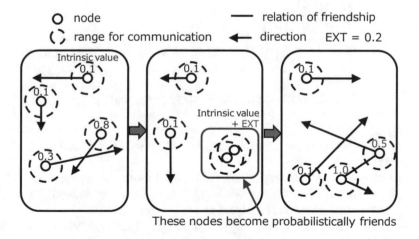

Fig. 12. HUI Model

We show the values of the parameters in Table. 2.

Table 2. Values of parameters for HUI model

Number of nodes	$n = 5000$
Size of area	$L = 2000$
Moving distance of 1step	$s = 10$
Range that two nodes can communicate	$d = 5$
Return rate of HLW	$\alpha = 0.3$
Exponent of HLW	$\beta = 1.0$
EXT	0.2
Exponent of Pareto distribution	$\gamma = 2$

Figure 13 shows the degree distribution of a human relations network, and Fig. 14 shows the complementary distribution of the degree distribution by double logarithmic chart.

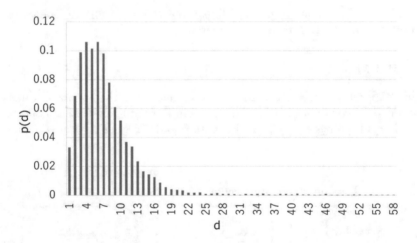

Fig. 13. Degree distribution (HUI Model)

Figure 14 indicates that a network generated by the HUI model has the scale-free property.

To investigate the property of the HUI model more, we replace the mobility model of the HUI model to Lévy Walk (LW) and Random Walk (RW), and we investigate whether these models have the scale-free property.

Figures 15 and 16 show the complementary distribution of the degree distribution of networks generated by the models replaced to LW and RW, respectively.

Both of Figs. 15 and 16 indicate that the replaced models do not have the scale-free property.

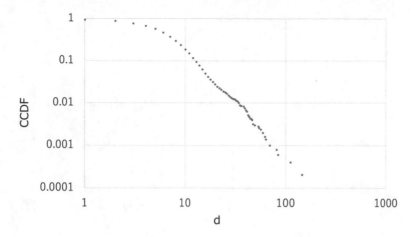

Fig. 14. CCDF (HUI Model)

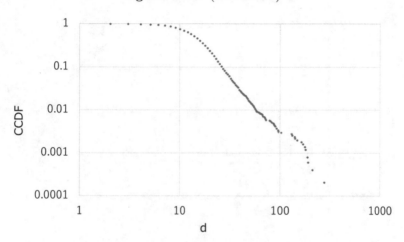

Fig. 15. CCDF (Levy Walk)

Next, we replace the Pareto distribution to the exponential distribution. Figure 17 shows the complementary distribution of the degree distribution by logarithmic chart.

Figure 17 indicates that this replaced model also does not have the scale-free property.

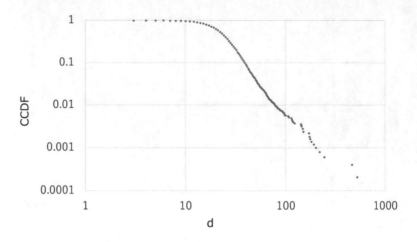

Fig. 16. CCDF (Random Walk)

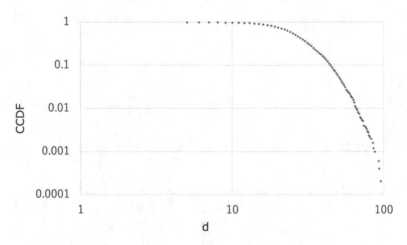

Fig. 17. CCDF (Exponential distribution)

4   Conclusions

In this paper, we proposed a new model that generates human relations networks considering human mobility. The model is described as follows: the nodes move according to Homesick Lévy Walk (HLW); the bases in HLW are uniformly distributed in the area; each node is given the intrinsic value according to the Pareto distribution; when the nodes make relations, the intrinsic values are increased by a constant; the upper limit of the intrinsic value is given and the updated intrinsic value does not exceed the upper limit; when two nodes approach within a range, the nodes makes the relations, if the value of the sum of the intrinsic values is more than a threshold of uniform random number.

The numerical experiments indicated that a network generated by the proposed model has the scale-free property.

Furthermore, we investigated some variations of this model, but no models can generate a network with the scale-free property.

Consequently, we found that HLW, the bias of initial intrinsic values and the increase of the value are essential for the scale-free property.

For the future work, we show that the proposed model has the scale-free property theoretically.

Acknowledgements. This work was partially supported by the Japan Society for the Promotion of Science through Grants-in-Aid for Scientific Research (B) (17H01742), JST CREST JP-MJCR1402, and Grants-in-Aid for Scientific Research (C) (17K00141).

References

1. Przemyslaw, A.G., Jose, J.R., Bruno, G., Victor, M.E.: Entangling mobility and interactions in social media. PLoS ONE **9**(3), 1–19 (2014)
2. Barabasi, A., Reka, A.: Emergence of scaling in random networks. Science **286**, 509–512 (1999)
3. Reuven, C., Shlomo, H.: Scale-free networks are ultrasmall. Phys. Rev. Lett **90**(5), 058701 (2003)
4. Bollobas, B., Oliver, R., Joel, S., Gabor, T.: The degree sequence of a scale-free random graph process. Random Struct Algorithms **18**, 279–290 (2001)
5. Bollobas, B., Oliver, R.: The diameter of a scale-free random graph. Combinatorica **24**(1), 5–34 (2004)
6. Caldarelli, G., Capocci, A., De Los Rios, P., Muñoz, M.A.: Scale-free networks from varying vertex intrinsic fitness. Phys. Rev. Lett. **89**(25), 258–702 (2002)
7. Vito, D.P.S., Guido, C., Paolo, B.: Vertex intrinsic fitness: how to produce arbitrary scale-free networks. Phys. Rev. E **70**, 056126 (2004)
8. Bo, S.: General formalism for inhomogeneous random graphs. Phys. Rev. E **66**, 066121 (2002)
9. Marian, B., Romualdo, P.S.: Class of correlated random networks with hidden variables. Phys. Rev. E **68**, 036112 (2003)
10. Goh, K.I., Kahng, B., Kim, D.: Universal behavior of load distribution in scale-free networks. Phys. Rev. Lett. **87**(27), 278701 (2001)
11. Masuda, N., Miwa, H., Konno, N.: Analysis of scale-free networks based on a threshold graph with intrinsic vertex weights. Phys. Rev. E **70**, 036124 (2004)
12. Fujihara, A., Uchida, M., Miwa, H.: Universal power laws in the threshold network model: a theoretical analysis based on extreme value theory. Phys. A **389**(5), 1124–1130 (2010)
13. Konno, N., Masuda, N., Rahul, R., Anish, S.: Rigorous results on the threshold network model. J. Phys. A Math. Gen **38**, 6277–6291 (2005)
14. Masuda, N., Miwa, H., Konno, N.: Geographical threshold graphs with small-world and scale-free properties. Phys. Rev. E **71**, 036108 (2005)
15. Fujihara, A., Miwa, H.: Homesick Levy walk: a mobility model having Ichi-go Ichi-e and scale-free properties of human encounters. In: Proceedings 2014 IEEE 38th Annual Computer Software and Applications Conference (COMPSAC2014), Vasteras, Sweden, 21–25 September 2014

Crowdfunding – An Empirical Study on the Entrepreneurial Viewpoint

Valerie Busse[✉]

University of Vienna, Oskar-Morgenstern-Platz 1, 1090 Vienna, Austria
valerie.busse@infinanz.de

Abstract. Crowdfunding has gained increased attention as an alternative type of investment opportunity. However, the rationale and the drivers of why this form of funding has grown quickly is partially unexplored. Engaging the crowd and capturing the wisdom of many is a growing phenomenon, which benefits entrepreneurs opening an alternative funding route as well as a path to knowledge acquisition. The entrepreneurial perspective when selecting a crowdfunding platform and the resulting decision parameters is the main topic of this research. By using the method of semi-structured interviews the research *(i)* analyzes the question what makes the entrepreneur trust in the online platform, and *(ii)* provides a deeper understanding of which kind of risk the entrepreneur has to face.

The findings suggest, that both experience factors such as design, relevant content, easiness of navigation, security and customer feedback, as well as sensitivity factors including reliability, intention, integrity, competence, openness and perceived concern are major trust drivers which lead to a higher use of crowdfunding web platform.

The components of perceived risk – in particular the risk of losing intellectual property (IP), the threat of a cyber-attack, and the risk of loss of autonomy by being dependent on a third party – were further observed results of analyzing the entrepreneurial perspective.

1 Introduction

The importance of crowdfunding has significantly increased since the financial crisis in 2008, when platforms such as *IndieGoGo* and *Kickstarter* were launched. The concept of crowdfunding is utilised by entrepreneurs as controversial start-up funding method besides venture capital, financial support by friends and family, and the fund raising through a business angel [2, 14].

Hence, crowdfunding is a new funding method, which allows founders of cultural, social and for-profit projects to gain funding from different individuals, usually in return for equity or potential products [6, 7, 37].

Projects funded through the crowd may range from projects requiring only smaller amounts to large projects of entrepreneurs "seeking hundreds of thousands of dollars" [31]. Therefore, crowdfunding offers entrepreneurs an alternative strategy to boost their early stage level into the next one and has therefore emerged as "the next big thing in entrepreneurial financing" [23, 26, 28].

© Springer Nature Switzerland AG 2019
F. Xhafa et al. (Eds.): INCoS 2018, LNDECT 23, pp. 306–318, 2019.
https://doi.org/10.1007/978-3-319-98557-2_28

Crowdfunding via intermediary online platforms has at least doubled its worldwide volume in recent years and represents a rapidly and further increasing concept of funding new ventures [12, 15]. In 2011, $99 million were raised on a single crowdfunding platform by *Kickstarter.com* [10, 13]. In 2014 the online crowdfunding market counted as $85.74 billion in the market of North America and $10.54 billion market around Asia and Europe [15].

The estimated funding volume as well as the annual growth rates show the immense increase of global crowdfunding [17]. As all of the components in the triadic relationship – the crowd, the intermediary, and the entrepreneur – have to face a high level of mutual trust, this research focuses on the trust components between the entrepreneur and the intermediary represented as a crowdfunding platform and web portal. The remainder of this paper is structured as follows: Sect. 2 provides the theoretical background. Section 3 explains the used research method, the findings and the contribution towards the theory. Section 4 summarizes the major findings and provides implications for future research.

2 Theoretical Background

Trust in intermediary online platforms is an essential factor in the growth of online networks therefore protection against fraudulent projects is a crucial issue.

Trust in intermediaries and their web-portals is associated with appropriate outcomes such as satisfaction, the overall firm performance, competitive advantages and several other positive economic outcomes [33].

Because of strong need and lack of trust-related research in the context of crowdfunding this research investigates the complex and hard to measure trust relationships. Existing literature from McKnight et al. [27], Mayer et al. [25], and Mühl [24] indicates first investigations in trust in context of web-based platforms.

McKnight et al. [27] suggest in their web-trust model how trust can be generated towards the online intermediary. The authors propose that trust arises from personal innovativeness combined with personal attitude towards trust and results in trust-related behaviours while focusing on three out of the seven dimensions of trust.

McKnight et al. [27] propose that aspects such as security and confidence are essential factors in terms of characterising trust in web-sites. Therefore, trust is a crucial supporting factor when insecurity and opportunism during web-based interactions occur [27]. Even if this model includes several elements in more detail than similar models, which contributes to a more detailed analysis when applying, it lacks of an important component as it does not consider risk in web-trust relations.

Similarly, to McKnight's et al. web-trust model, Mayer et al. [25] investigated the "proposed model of trust". Different to McKnight's et al. [27] investigations, Mayer et al. also take ability into account. The authors define this term by explaining "*the domain of ability is specific because the trustee may have little aptitude, training or experience in another area, for instance in interpersonal communication*" [25]. By introducing "Trustor's propensity" in their model, the authors aim to take the cultural aspects, personality backgrounds and experiences into account by including Hofstede's

[17] approaches. Hence, propensity measures how much the trustor trusts a trustee before realising the risky trust-relationship [25].

Furthermore, the authors implement perceived risk in their model by stating that the trustor has to achieve a higher level of trust than the level of perceived risk in order to engage in a trust-related relationship, which is not included in McKnight's et al. [27] model [27]. Compared to McKnight's et al. [27] model Mayer et al. take several components into account, however, not as detailed as McKnight et al. [27] which might lead to a lack of detail when applying the model [25, 27]. Therefore, by analysing trust in crowdfunding, the strengths from both models, speaking of risk as well as detailed components will be applied to gain a most appropriate analysis.

As already mentioned and approved by several authors measuring trust is a complex and still under research process [27]. Even if, trust evaluating has been examined in several fields, it is questionable to which extend these measurements are tangible or can be applied and formed [24]. McKnight and Chervany [27] point out, that "researcher's purposes may be better served if they focus on specific components of trust rather than the generalised case [27]. Hence, Mühl [24] listed several operational trust dimensions, which will be now examined in further detail [24].

Beginning with *benevolence*, which is a dimension defined as the willingness to support the trustor in general [25]. Black [1] states that this term is often indicated as altruism or love. Kant [19] underlines Black's [1] statement by pointing out that benevolence is an essential indicator for good relationships.

Mühl [24] investigated that benevolence is often provided in organisations if political environment favours the stakeholders in a trust relationship "on the trustee or trustor side to increase the wellbeing of the other person" [13].

The second trust dimension is called *integrity*. According to Levin et al. [22] integrity can be seen as a degree of fairness and is described by the authors as "rules that are applied equally to individuals" [22]. Investigations show that other authors use integrity similar to the other two trust dimensions honesty and openness In order to separate these terms, this paper uses "fairness" as synonym for integrity.

Being concerned is described by Mühl [24] as the "degree of how much the trust relationship concerns the trustee from his individual opportunistic standpoint" [24]. Other authors such as Weber rather see the term as how much value oneself has according to the vulnerability of the trustor [36].

The fourth dimension of trust is *competence* and is explained by Mayer et al. [25] as having a high and adequate level of qualification combined with the capability to perform in a specific role. Other investigators describe the term more as a combination of skills, knowledge, education and social behaviour, which is used to improve performance [22]. The author also suggests that in conditions where a high level of competence is crucial, a trustor might be less willing to trust the trustee if the competent factor is not given, hence the trustee behaves unqualified [22].

Openness the fifth dimension of trust. According to Mühl [24] this dimension indicates some similarities to empathy, integrity and being concerned. Mayer et al. [25] describe this term rather as the ability changes the point of view to another perspective [25].

Reliability is expressed as the degree of reliability of the trustee [24]. Levin et al. [22] rather describe the term as "being consistent with word and deed" [22].

Investigators such as Kramer and Cook [21] came up with similar ideas about relia-bility by proposing that trust increases though learned, interactive experiences between two parties [35]. The last dimension of trust is called *intention* and is the level of intended performance of action of the trustee against the trustors favour [25].

Rousseau et al. [32] investigated that positive beliefs of the intention of another person will directly lead to a higher level of trust. Hence, Mühl [24] suggests that in the case of a comparable intention of trustor and trustee it will directly lead to a higher trust- level of the two parties [24, 32].

As already mentioned above, several models where developed in order to make trust-measurement more tangible, however, few of them, including Mayer et al. [25] consider the risk factor which is directly correlated to a trust–relationship.

Josang and Stephane [18] give evidence to this fact by stating that both factors, trust and risk are directly correlated to a high level of uncertainty and unpredictability what leads to a high complexity in their modelling.

According to Rousseau et al. [32] risk is defined by the supposed probability of harm or loss. Current, investigations show that it is still not clarified, whether appro-priate models of risk and trust can be designed [16].

However, Josang and Stephane [18] propose that risk in a trust-relationship arises *"when the value of the stake in a transaction is high"* [18]. Castelfranchi and Tan [9] argue that having a high level of trust in a certain person does not necessary lead to the willingness of entering the trust-relationship. As the authors state [8, 9]:

> *"For example it is possible that the value of the damage per se (in case of failure) is too high to choose a given decision branch, and this independently either from the probability of the failure (even if it is very low) or from the possible payoff (even if it is very high). In other words, that danger might seem to the agent an intolerable risk."* [8, 9].

Additionally, Povey [29] implements risk in McKnight's and Chesbrough's web-trust model and states that risk is visible by trusting behaviour and has a high influence on the overall decision to trust [11, 29].

Other approaches analysed the trust in terms of risk in electronic commerce [21]. The authors developed a theoretical framework examining the trust-based- decision making process via a web-survey. Results show that the customers' trust and risk have highly impacts the purchasing decisions.

By not only paying attention to the trust components this research also indicates the risk factors which are directly related to the trust dimensions.

3 Empirical Study

3.1 Research Method

Data was collected through in-depth interviews with five entrepreneurs from different sectors, who used crowdfunding platforms to start their businesses.

The interviews were done either face to face. The interviewees were selected and qualified members of a broader network of The "Bride Red Triangle" association from Napier University. The interviews were voice-recorded with the permission of the participators and were scheduled for approximately forty minutes per interview.

However, there was no time limit given which enabled the participators to go deeper into details of specific answers. The questions were based on the theoretical background of crowdfunding platforms, transactional costs, trust dimensions, and customer loyalty.

The interviews were semi-structured and differentiate themselves from structured interviews by not following a certain organized and standardized structure of questions [15]. The form of semi-structure interviews enables in-depth interviews, which follow a relatively loose level of standardization. Furthermore, semi-structured interviews are less formal and are more an open conversation about a certain topic [4]. Consequently, semi-structured interviews follow a certain 'road-map', where the interviewer covers the same topics in every interview and the interviewee has the possibility to answer these and conduct new questions within the interview [4].

The interview questions were divided into three main types [33]: open questions, closed questions, and probing questions.

Open questions allow the participator to describe specific scenarios in detail and give personal statements, experiences and backgrounds to the questioned topic. Closed questions enable the interviewer to gain specific numbers or dates. Probing questions allow the interviewer to have a deeper understanding of certain topics and discuss further topics in detail [33].

A purposive sampling method was used in order to allow the researcher to choose a sample, which indicates a suitable amount of professionals in order to get the research questions answered properly [4]. Especially, by researching this topic it seems necessary to interview people who have certain experiences in crowdfunding to gain useful responses.

The sample size of five participators was considered as sufficient for this research as a limited number of participators allows the researcher to go into more depth [3].

In order to generate a deeper insight in the importance of trust-components from an entrepreneurial point of view the approach of qualitative interview was applied. This part of the analysis was gained through an interview, which was conducted with five entrepreneurs, who have currently used a crowdfunding-platform in order to start their business. All entrepreneurs funded their idea through different crowdfunding webpages such as Kickstarter.com, Indiegogo.com and Crowdfunder.com. To broaden the information, entrepreneurs with completely different project areas were asked. To be specific, project topics covered theatre, finance, education, healthcare and fashion ventures. Whereas the male Participator 1 (P1) had 7 years' experience in the start-up business and focused on study platforms, the female Participator 1 (P2) had 3 years' experience and set a fashion project for her crowdfunding campaign. Participator 3 (P3), also female and 3 years' experience, however, focused in theatre. Participator 4 (P4) was female as well with a 4 years expertise and with focus on finance. The last participator of the interview was male with 5 years' experience participated in a crowdfunding project focusing on healthcare.

All participators became entrepreneurs due to independency, which gave them *"a certain flexibility to achieve their set goals and previous stated vision"*.

A further aspect declared by the participators was that they were additionally to monetary benefits driven by social ones. Hence, they want the idea, which is provided for the society *"also be funded by the society"* as stated by P5. Just one out of five, to be specific, P4, became an entrepreneur just because of monetary benefit, the other four gave more weight to emotional and social drivers which satisfy the needs of the society. The entrepreneurs were asked the following four questions with sample answers in case to guide the participator:

(i) Why did you choose a crowdfunding approach rather than a business angel or venture capital? (e.g., emotional or social benefits, benefits for society, not just monetary benefits, establishing relationships, receiving validation, expanding awareness of work through social media, feedback)

(ii) Why did you choose your specific webpage rather than another one? (e.g., did you consider reputation, feedback, loyalty, and emotional components)

(iii) When you chose the web-page, how important is it to feel trust in this web-page? (when we are talking about trust, what does trust mean for you, e.g., security, safety, reputation, or appealing design? Did you consider risk? In which terms, e.g. identity loss, cyber-attack, or being dependent on a third party?)

(iv) What do you think is the future outlook for crowdfunding web-pages? Do you think there will be a dominant design? Do you think there will be more specialised platforms?

3.2 Findings

By analysing the interview, four different main topics including financial methods, trust, risk and the future outlook for crowdfunding platforms could be examined, which will now be evaluated in more detail.

By asking why participators chose the crowdfunding approach rather than a business angle or venture capital participators pointed different reasons. While P1 stated that the main reason was the lack of a suitable network, which would be necessary in order to take advantage of a business angel others mentioned the high dependency on one specific person. Most of them also proposed that they did not want to write a business plan and according to P4 *"make the decisions on the way instead of having something set and stoned from the beginning and have to go with it even if you maybe want to change later on"*. Furthermore, participators mentioned that *"other companies will influence a business angel and investors to its favour"*.

In terms of venture capital and depth and equity, participators summarized similar aspects. Additionally, to the aspect of the dependency, P4 also mentioned the *"high amount of interest rate whether the business becomes successful or not"* Whereas, by *"using crowdfunding you are most likely to give a reward if you are actually successful and making the money so you are much more flexible"*.

Subsequently, all participators stated the high advantage of getting *"a first impression what other people are thinking"* about their initial idea. Hence, they indicated that the continuous feedback is an essential factor while deciding for one specific funding method *"especially, when they are from different sectors"*.

Some also stated that *"it helped to market the business before the actual business exists"*. Therefore, a high advantage of building relationships and a huge network of future customers *"helps the business to get more successful"* according to P2. Moreover, it *"gives the opportunity to build new relationships that can be beneficial, both for personal and professional benefits"*.

Consequently, entrepreneurs already have a network, which will further expand through *"the word–of mouth"* as concluded by P4. Summarizing, according to P3, *"it was a much faster route to the market"* than choosing one of the other financing methods.

Additionally, all interviewees claimed the high advantage of crowdfunding of not being depended on one single party as *"you should not put all eggs in one basket, you need to spread the risk because if you drop the basket all the eggs will be destroyed"*, as P1 mentioned. Hence, most of the participators chose crowdfunding because of the major advantage of a "spread risk".

Every participator confirmed the importance of feeling *"a high trust-level in the web-page, especially when dealing with sensitive topics such as money and intellectual property"*. By talking about trust three main indicators were identified.

- Security and reliability
- Design and easiness to navigate
- Good reputation of the page

First, interviewees confirmed that security and reliability in the webpage are considered as the most important aspects in terms of trust. P2 underlined this fact by stating that *"trust means security and reliability with the page, so you want to know if it's actually legit"*. Another participator, P1, broadened this importance by stating:

"Trust is very important especially when dealing with crowdfunding, you can compare it with Facebook for example, there you have all your private and sensitive data and you want to be secure as well and we have seen so many outbreaks that Facebook uses our information and I want to avoid this because I think the risk of leaking an idea that could be worth millions is very important that the trust level on the webpages is super high especially when dealing with money and intellectual property".

Secondly, the design and easiness of navigation were identified as a major factor which can be seen by the following statements.

P3: *"I also considered easiness and accessibility for everyone so if the page is structured and you know where everything is on the page it is more reliable for my point of view."*

P1: *"Nice and easy design, no pop-up windows and cookies, hence the design has also something to do with the amount of trust I feel for a webpage"*.

P2: *"When you take the example of amazon, where everything is saved and easy to navigate as well as a good set up makes me trust in the webpage"*.

One interviewee, P4, summarized these statements as *"the more user-friendly of the platform, a good web-page design combined with an easy navigation is crucial for success."*

Thirdly, a good reputation and customer feedback for the webpages was an essential trust criterion for all participants. One Interviewee, P4, claimed that:

"You want to know that people don't just take advantage of you and you don't get anything out of it and with the idea of word of mouth when people tell you something about the webpage then

it is to be trusted from my side, it is always important to talk to people who have already used the web page".

Furthermore, another interviewee, P2, added: *"and also the reputation of course, when people have a good and trustful experience with the platform encourages me to use it because if others had success why shouldn't I have it as well?"*

Hence, all of them informed themselves in feedback platforms, compared different platforms and asked for professional advice.

It was interesting to see that all of the participators considered big and known crowdfunding platforms such as Kickstarter.com, Indiegogo.com and Crowdfunder.com as more trustful and reliable. All mentioned similar reasons for choosing these platforms including a high successful rate in past projects as well as feedback of successful entrepreneurs and "relied on professional advice". Additionally, participators wanted to "reach a lot of different people" and supposed that *"the audience we wanted to have would trust Indiegogo more than a smaller niche one"*. However, one participator, P3, argued that they considered to use a smaller niche one *"which covers the specific topic and gets the right target group"*. Nevertheless, in the end they decided to go for the bigger one because *"the more people who go to the webpage, the more people will see the idea"*. One also claimed that they used the webpage because of its funding type as they preferred the "keep it all" rather than the "all or nothing approach".

There could be three main risks identified from the interviews:

- Risk of losing intellectual property IP
- Risk of cyber-attacks
- Risk of being dependent on a third party

Most participators claimed the risk of losing their intellectual property, summarised by P4:

> *"leaking the idea out to someone else that could take it and use it as their own is the highest risk"* and *"that your idea gets stolen with some kind of identity theft that you don't know if someone else at the end of the world thinks that's a good idea and you are left without anything pretty much".*

However, most of the interviewees confirmed that this risk was quite low as most of them went to a *"quite competitive market that already exists"* so *"the ideas are not really new"* and therefore hard to copy.

Another participator, P5, claimed that he did not take the risk of losing intellectual property into account by stating *"especially in my case this risk was really restricted as my business model is only with a lot of effort copyable"*. Others mentioned that they just put the basic idea on the platform and leak the "design and details". Hence, a possible copy of the idea is very unlikely.

Concerning the risk of a cyber-attack, all interviewees took this risk into account. However, P4 of them mentioned that *"especially in the high sensitive financial start-up"* a cyber-attack seems very likely but another proposed that "I *didn't really think of a cyber-attack because I think that this can always happen, I didn't really take this into consideration".*

By asking of considering the risk of being dependent on a third party most of them stated that *"you are always depended on a third party either the third party is a bank or a business angel or someone else"* so *"whatever form of financial form you use you are always dependent on a third party, so when you do equity you be dependent on shareholder's, by doing a loan you be depended on the bank, you just have to figure out what is most efficient for your model"*. P1 stated that *"you need to be depended on a third party if you want to grow of course it has been a consideration but we didn't really have another choice as we depend on a monetary injection in order to grow"*. Hence according to P5, *"you can't completely be on your own because you don't have your own resources"*.

Concerning the future outlook participators agreed on the point that *"the idea on crowdfunding will grow in the future"*. Even if, currently the *"traditional resources of finance"* are more preferred, they proposed if *"the word crowdfunding is more spread it becomes more popular and when people start to give it a chance it will develop, as its actual and efficient way"*.

Participators insinuate a general increase in the Start-up Scene due to *"the highly risk adverse generation Y"* which is characterized by their *"willingness of independency and freedom"*.

Most of the participators do not think that there will be a single dominant design within the different platforms. P5 mentioned that *"there is often a benchmark made with Facebook, however, I think that the phenomenon Facebook cannot be reflected to crowdfunding pages because of its participants and all of these individual ideas"*.

Another one, P3, stated that there will not be a dominant design but *"different dominant ones combined with more niche ones along the way"*.

However, all of the participators agreed on "the separation and deviation in different sub-categories" on the pages. Some even broaden this suggestion by P3, stating that there might be different specialized platforms with categories such as *"clothing, finance and stuff like that"*, in order to *"pair the people who have the* knowledge *in this specific sectors with the people that gets the ideas concerning this sector"* and get more industry specialists for the networking and expertise and be with people who come from the same industry so the ideas you share and the feedback you gain will be more helpful"*.

Another one suggested that there will be a separation in *"products and services"* within the pages.

One financial participator, P4, argued that *"in certain parts such as the financial industry, I believe it could be possible that there will be separations to its own part, as the whole sector is in a change right now which can be seen with all these FinTechs. However, in other sectors such as health care and social concerns I can hardly imagine that they will have separate pages"*.

Another point of view could be observed by the argumentation of P1, *"I could imagine that some small platforms get purchased by bigger platforms to gain a bigger user basis and try to get out the competition as it been seen in Facebook and Twitter or other internet based platforms"*. However, the participator also claimed the *"uncertainty of different financial regulations in different countries"*.

3.3   Contribution Towards Theory

By contributing the interviews to theory the following outcomes can be identified:

Including the risk in the web-trust model of Mayer et al. [25] is according to the interview justified to some extent. Participators claim that they considered risk in the first place but did not necessarily conclude it in their decision by using a web-based model as financial support [25].

Concerning the Intellectual Property rights by Khalique et al. [20] it can be identified concerns of losing the three major Intellectual property rights such as human capital, social capital and technological capital [20]. However, participators eliminated this concern though stating that their idea is either hard to copy or the market already exists and therefore the risk of losing their intellectual property is quite low.

The risk of being dependent on a third party as analysed by Peisl et al. [31] and Williamson [37] was considered by the participators however, not as a barrier of going through an intermediary, as they mention that an intermediary is a main as requirement of being dependent on a third party [31, 37].

Concerning the trust dimensions which were first illustrated in McKnight's et al. [27] web-based model and later extended by Mühl [24], analysis of the Interview shows that the major factors of trust-importance was in security, reliability and competence of the webpages [24]. Additionally, being concerned could be observed as an essential factor by analysing the risk components. Integrity and openness was minor mentioned and benevolence was not taken into consideration.

As previously stated by Pavlou and Ba [30] the high importance of customer feedback in order to trust a webpage was highly confirmed by every participator [30].

The diffent financial opportunities including their disadvantages and advantages anaysed by McKaskill [26] and Busse [5] were also confirmed by the interviewees including new observations of crowdfunding advantages such as spreading the risk, not being influenced by other parties the high independeny of the entrepreneur.

Furthermore, the not emotional trust drivers defined by Peisl et al. [31] as *"Experience Drivers"* such as the design of the webpage, easiness of navigaton and the content were confirmed by the participators as essential trust-components [31].

Participators mentioned the likelihood of a dominant design combined with more specialized web-pages. However, there are several discrepancies between the answers and different opinions how crowdfunding might develop. Nevertheless, most assumed more specialized pages in the future without having one specific dominant design.

Summarizing, from the entrepreneurial point of view, experience factors such as easiness to navigate, content and design of the webpage as well as sensitivity factors such as security, reliability and competence were considered as highly important by analysing trust contributors. However, perceived risk is also termed as one of the major barriers of choosing a platform. In particular, the high risk of losing intellectual property, the risk of being dependent on a third party and the risk of a cyber-attack. Additionally, most participators mentioned the need for more specialized platforms without a precise dominant design as mentioned beforehand.

4 Conclusion

This research has focused on the entrepreneur-intermediary interactions in the triadic relationship of entrepreneur-intermediary-crowd. The major outputs of the research show that the entrepreneur is driven by design, relevant content, easiness of navigation, security and customer feedback as well as Mühl's [24] trust drivers including intention, integrity, competence openness and perceived concern. However also risks could be identified such as the risk of losing IP, risk of being dependent on a third party as well as the threat of a cyber-attack. The research on hand helps companies to improve the trustworthiness of their web-pages.

Further research is needed in understanding trust factors between the entrepreneur and the crowd as well as the perspective from the intermediary towards the crowd in order to cover all different perspectives of the triadic relationship. Additionally, the risk factors of the crowd are not yet fully investigated. Furthermore, there is a lack of research in the influences of the law as well as in the financial regulations within the crowdfunding platforms. Besides, further research has to be done on the value of social media connecting to crowdfunding platforms. Subsequently, there is a research limitation of how different topics of crowdfunding sectors, e.g. fashion, finance, and theatre are advertised. Furthermore, future research could also take behavioural models or decision making concepts into consideration.

References

1. Black, S.: Trust and commitment: reciprocal and multidimesional concepts in distribution relationships. A&M Univ.-Corpus Christi J., 46–53 (2008)
2. Brettel, M.: Business Angles in Germany: A Research Note, Venture Capital, vol. 3. Routledge, Germany (2010)
3. Brinkmann, S.: Qualitative interviewing. Oxford, New York (2013)
4. Bryman, A., Bell, E.: Business Research Methods. Oxford, New York (2011)
5. Busse, F.-J.: Grundlagen der Betriebswirtschaftslehre. Oldenbourg, München (2003)
6. Buerger, B., Mladenow, A., Strauss C.: Equity crowdfunding market: assets and drawbacks. In: International Conference on Information Systems (ICIS), SIGBD, (ICIS 2017), Proceedings, vol. 6, pp. 1–5. Seoul, South Korea, 10th–13th December 2017
7. Buerger, B., Mladenow, A., Novak, N.M., Strauss, C.: Equity Crowdfunding: quality signals for online-platform projects and supporters' motivations. In: CONFENIS 2018 - International Conference on Research and Practical Issues of Enterprise Information Systems, Poznan, Poland. Springer Lecture Notes in Business Information Processing (LNBIP), 18th–19th September 2018, forthcoming
8. Castelfranchi, C., Tan, Y.H.: Trust and Deception in Virtual Societies. Springer, Rome (2001)
9. Castelfranchi, C., Falcone, R.: Principles of Trust, pp. 55–99. Kluwer (2001)
10. Catalini, C., Fazio, C., Murray, F.: Can equity Crowdfunding democratize access to capital and investment opportunities? Innovation, Science Report, pp. 1–16 (2016)
11. Chesbrough, H.: Open Innovation: The New Imperative for Creating and Profiting from Technology. Havard Business Press, Boston (2006)

12. Cumming, D., Leboeuf, G., Schwienbacher, A.: Crowdfunding Models: Keep-It-All vs. All-Or-Nothing. York University-Schulisch School of Business, York (2015)
13. Deutsch, M.: The Resolution of Conflict: Constructive Process. Yale University Press, New Haven (1973)
14. Duoqi, X., Mingyu, G.: Equity-based Crowdfuning in China: beginning with the first Crowdfunding financing case. Asian J. Law Soc., 81–107 (2017)
15. Fisher, C.: Reaserching and Writing a Dissertation: A Guidebook for Business Students. Prentice Hall, Upper Saddle River (2007)
16. Grandison, T., Solman, M.: A survey of trust in internet applications. IEEE Commun. Surv. Forth Q. **3** (2000)
17. Hofstede, G.: Culture's Consequences: iNternational Differneces in Work-Related Values. Sage Publications, Beverly Hills (1980)
18. Josang, A., Stephane, P.: Analsying the Relationship between Risk and Trust. Springer, pp. 1–377 (2004)
19. Kant, I.: Grundlegung zur Methaphysik der Sitten. In: Bibliothek, P. (ed.), vol. 3, Verlag der Dürrchen Buchhandlung, Leipzig (1785)
20. Khalique, M., Shaari, A., Isa, H.A.: Intellectual capital and its major components. Int. J. Curr. Res. **3**(6), 343–347 (2011)
21. Kramer, R., Cook, K.: Trust and Distrust in Organizations: Dilemmas and Approaches. Russel Sage Foundation, New York (2004)
22. Levin, D., Cross, R., Abrams, L., Lesser, E.: Nuturing inerpersonal trust in knowledge-sharing networks. Acad. Manag. Rev. **17**(4), 64–76 (2003)
23. Li, Y., Rakesh, R.: Project success predictin in Crowdfunding environments. ACM Guide Comput. Lit. (2016)
24. Mühl, J.K.: Organizational Trust: Measurement, Impact, and the Role of Management Accountants. Springer, London (2014)
25. Mayer, R., Davis, J., Schormann, D.: An integrative model of organizational trust. Acad. Manag. Rev. **3**, 709–734 (1995)
26. McKaskill, T.: Raising Angel & Venture Capital Finance. Breakthrough Publications, Melbourne (2009)
27. McKnight, H., Chervany, N.: Trust in Cyber societies: Trust and Distrust Definitions: One Bite at a Time. Springer, Heidelberg (2001)
28. Mollick, E.: The dynamics of crowdfunding: an exploratory study. J. Bus. Ventur. **29**, 1–16 (2014)
29. Pavlou, P.A., Ba, S.: Does online reputation matter? An empirical investigation of reputation and trust in online auction markets. Association for Information Systems, pp. 948–950 (2000)
30. Pavlou, P., Gefen, D.: Building effective online martplaces with institution-based trust. Inf. Syst. Res. **1**, 37–59 (2004)
31. Peisl, T., Raeside, R., Busse, V.: Predictive crowding: the role of trust in crowd selection. In: 3E Conference Irland, pp. 1–19 (2017)
32. Rousseau, D., Sikins, S., Burt, R., Cramerer, C.: Not so different after all: a cross-discipline view of trust. Acad. Manage. Rev. **23**, 393–404 (1998)
33. Saunders, M., Lewis, P., Thornhill, A.: Research Methods for Business Students. Pearson (2009)
34. Schwienbacher, A., Laaralde, B.: Crowdfuning of Small Entrepreneurial Ventures: Handbook of Entrepreneurial Finance. Oxford Univeristy Press, Luxembourg (2010)
35. Tomkins, C.: Interpendencies, trust and information in relationships, alliances and networks. Acc. Organ. Soc. **34**, 161–191 (2001)

36. Weber, J.: Das Advances-Controlling Handbuch, Richtungsweisende Konzepte, Steuerungssysteme und Instrumente. Wiley, Stuttgart (2008)
37. Williamson, O.: The economics of organization: the transaction cost approach. Am. J. Sociol. **87**(3), 548–577 (1981)

The 10th International Workshop on Information Network Design (WIND-2018)

Reliable Network Design Problem Considering Cost to Improve Link Reliability

Keyaki Uji and Hiroyoshi Miwa[✉]

Graduate School of Science and Technology, Kwansei Gakuin University,
2-1 Gakuen, Sanda-shi, Hyogo, Japan
{keyaki-uji,miwa}@kwansei.ac.jp

Abstract. It is important to design robust networks that are resistant to network failures, because high reliability is required for communication networks as important infrastructure. For that purpose, there is an approach of link protection to decrease link failure probability by the backup mechanism and the fast recovery mechanism; however, enormous cost is required, if the failure probability of all links must be decreased. It is realistic to preferentially protect only some highly required links so that the reliability of the entire network is improved. In this paper, we assume that the failure probability of each link can be reduced according to cost for protection. Since the failure probability of each link is decreased by the cost allocated to the link, the network reliability defined as the probability that the entire network is connected, is increased. We define the network design problem that determines the cost allocated for each link, under the constraint that the sum of the cost allocated for each link must be less than or equal to a given threshold, so that the network reliability is maximized. We show the NP-hardness of this optimization problem and design a polynomial-time heuristic algorithm. Moreover, we evaluate the performance by using the topology of some actual communication networks, and we show that the proposed algorithm works well.

1 Introduction

In order to realize a highly reliable communication network, it is important to design and operate a robust network that is resistant to network failures. As the Internet is an important social infrastructure, the reliability of a communication network is getting more important. Disruption of communication has a large influence on services on the Internet; however, it is difficult to avoid all failures by large-scale disasters such as earthquake, especially in a large network. Although it is desirable that a communication path exists between nodes even at the event of a failure, it is not realistic to completely prevent a link or a node from failing at the event of a disaster, and it is assumed that a failure occurs stochastically.

© Springer Nature Switzerland AG 2019
F. Xhafa et al. (Eds.): INCoS 2018, LNDECT 23, pp. 321–331, 2019.
https://doi.org/10.1007/978-3-319-98557-2_29

There is an approach of link protection to reduce the link failure probability by the backup mechanism and the fast recovery mechanism. We can make the failure probability of a link in an IP layer small by fast switches and backup resource reserved in advance in a lower layer, if it is carefully avoided that backup resources share network facility such as fibers and routers in the lower layer. If such a recovery system is provided in the lower layer and the backup resource is sufficiently prepared in comparison to the magnitude of the failure, a failure of the link is rapidly recovered and the failure cannot be detected in the IP layer.

The link protection improves the network reliability defined as the probability that the entire network is connected. However, enormous cost is required, if all links must be protected so that the failure probability is sufficiently decreased. Therefore, since it is not practical to protect all links, it is realistic to preferentially protect only some highly required links so that the reliability of the entire network is improved.

In this paper, we assume that the failure probability of each link can be reduced according to cost for protection. Since the failure probability of each link is decreased by the cost allocated to the link, the network reliability is increased. We define the network design problem that determines the cost allocated for each link, under the constraint that the sum of the cost allocated for each link must be less than or equal to a threshold, so that the network reliability is maximized.

First, we formulate the network design problem and show the NP-hardness of this optimization problem. Next, we design a polynomial-time heuristic algorithm. Moreover, we evaluate the performance by using the topology of some actual communication networks, and we show that the proposed algorithm works well.

2   Related Works

The connectivity of a graph is often used as a measure for evaluating the reliability of the graph. There is a method in which the connectivity is increased to improve reliability by augmenting edges of the graph. This is known as the connectivity augmentation problem. The purpose is to determine augmented edges so that the graph does not become disconnected against any removal of edges. Some theoretical results related to this problem are known; a set of the minimum number of edges for increasing the edge-connectivity of the graph to an arbitrary value can be determined in polynomial time; on the other hand, for the vertex-connectivity, the problem is generally NP-hard [1–3].

Another measure for evaluating the reliability of a network is the number of connected components when the network becomes disconnected due to a failure [4]. The reference [4] deals with an optimization problem for maximizing the number of connected components disconnected due to the failure of nodes. There is also some results on link protection [5–8]. The reference [5] deals with the problem of determining a set of protected links so that the diameter of a graph becomes a certain value or less. As the smaller diameter of a network leads to the smaller communication latency, an algorithm for the problem gives a network design method. In addition, the reference [6,7] deals with the problem of determining a set of protected links that keep the reachability to a specific node, and the reference [7] proposes an approximation algorithm to this problem. The reference [8] deals with the problem of determining protected links so that the reachability is kept within a fixed distance to multiple mirror servers even at the event of a failure. It is known that this problem is NP-hard and that the polynomial-time algorithm for a single-link failure is proposed. The reference [9] deals with the problem of determining protected links so that the master server and all edge servers are connected and that the capacity restrictions are satisfied. The reference [10] treats the problem of determining protected links so that all server nodes, regardless of whether they are master or edge-servers, are connected and so that the capacity restrictions are satisfied; the reference [11] deals with the problem of determining protected links so that the master server and all edge servers are connected and that the capacity restrictions are satisfied and that the increase ratio of the distance in a network to the distance in a failed network does not exceed a threshold. The reference [12] deals with the problem of determining server placement and link protection simultaneously.

When each edge has the probability that the edge is removed, the network reliability is defined so that the entire network is connected. The network design problems that maximizes the reliability or minimizes cost under reliability constraint are extensively studied (ex. [13,14]).

In this paper, we deal with the network reliability defined as the probability that the entire network is connected. Since calculating the network reliability of a given network is generally #P-complete problem, the main purpose of the previous studies is the design of a fast approximation algorithm (ex. [15]).

As mentioned above, there are many studies on link protection, link failure probability and connected probability of the entire network, but there are few combinations of them. As a problem similar to this problem, the problem of determining the protected links to maximize the network reliability under the constraint that the number of protected links is restricted was defined in [16] by the authors. In the reference [16], it was assumed that the link failure probability can be set to zero by link protection in order to simplify the problem. However, it is not easy to set the link failure probability zero in an actual network. It is realistic that the failure probability of each link can be reduced according to cost for protection. We assume such a condition in this paper.

3 Link Protection Using Cost Problem Failure Probability

Let $G = (V, E)$, where V and E are the vertex set and the edge set of G, respectively, be the graph representing a network structure and let the cost function be $h : E \times \mathbb{R}_+ \to \mathbb{R}_+$. When assigning cost c to edge e, the probability that e is not removed, in other words, the reliability of e, increases $h(e, 0)$ to

$h(e, c)$. Let $R(i)$ be a set of edge subset $E_K(\subseteq E)$ $(|E_K| = i)$ such that $G_r = (V, E \setminus E_K)$ is connected. For the network $N = (G, h)$ and the cost $c(e)$ allocated for edge $e(\in E)$, the network reliability $Rel(N, c)$ is defined as follows.

$$\sum_{i=1}^{m} \sum_{E_K \in R(i)} \prod_{\forall e \in E_K} (1 - h(e, c(e))) \prod_{\forall e' \notin E_K} (h(e', c(e')))$$

This is the probability that the network is connected when the probability that edge e is not removed increases $h(e, 0)$ to $h(e, c(e))$.

When a vertex pair set Q is given, the network probability $Rel(N, Q, c)$ that all vertex pairs included in Q are connected is similarly defined.

Problem PPCP
Input: *Network $N=(G=(V,E),h)$ Set of vertex pairs Q, a positive integer B.*
Constraint: $\sum_{e \in E} c(e) \leq B$
Objective: *$Rel(N, Q, c)$ (maximization)*
Output: *cost $c(e)$ allocated for each edge e.*

The solution of PPCP corresponds to the cost allocated for each link to increase the probability that the link does not fail so that the network reliability is maximized.

We show an example as follows (see Figs. 1, 2, and 3). For the sake of simplicity, let $Rel(N, c)$, the probability that all vertex pairs are connected, be the objective function. Let $Rel(N)$ be the network reliability of the original network. The reliability of the edges are given in Fig. 1(a). In Fig. 1(b), all the events that all vertex pairs are connected and their probability are shown. The sum of the probabilities means the network reliability. The network probability of the original network, $Rel(N)$, is 0.6132.

For the sake of simplicity, we assume that $h(e, c) = c$, that is, the reliability of edge e increases by c when the cost c is allocated for e. We assume that $B = 0.4$. In Fig. 2(a); $c(t_1, t_2) = 0.3$; $c(t_2, t_4) = 0.1$; the network reliability is 0.918. On the other hand, in Fig. 3(a), $c(e) = 0.1(e \in E)$; the network reliability is 0.776. In both cases, the total cost is the same; however, the network reliabilities are different according to the cost allocated for edges.

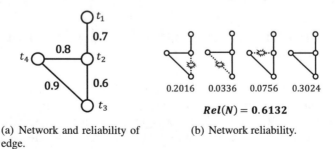

(a) Network and reliability of edge.

(b) Network reliability.

Fig. 1. Example of network and reliability.

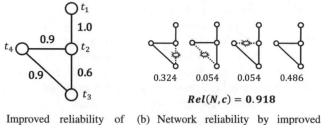

(a) Improved reliability of edge (1)

(b) Network reliability by improved edge reliability (1)

Fig. 2. Network reliability by cost allocated for edges (1)

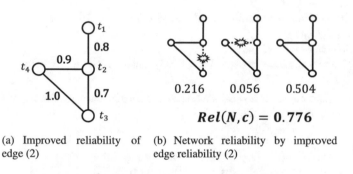

(a) Improved reliability of edge (2)

(b) Network reliability by improved edge reliability (2)

Fig. 3. Network reliability by cost allocated for edges (2)

Theorem 1. *PPCP is generally NP-hard.*

Proof. When the cost function is restricted so that the reliability of an edge is not continuously but discretely increased to one, the restricted problem is equivalent to the problem in [16] which is NP-hard. Therefore, PPCP is also NP-hard. □

4 Heuristic Algorithm for PPCP

Since the problem PPCP is generally NP-hard as shown in the previous section, we design a polynomial-time heuristic algorithm.

For the sake of simplicity, let $Rel(N, c)$ be the objective function, which is the probability that all vertex pairs are connected. Furthermore, we assume that $h(e, c) = h(e, 0) + c$ (if $h(e, 0) + c \geq 1, h(e, c) = 1$), that is, the reliability of edge e increases by c when the cost c is allocated for e; $h(e, 0)$ is abbreviated as $h(e)$.

We describe the basic idea of the proposed algorithm. First, when $N = (G = (V, E), h)$, we make the network $N^{log} = (G, log(h))$ where the cost of edge e is $log(h(e))$ for all edges. Let the maximum spanning tree of N^{log} be $T_{max} = (V, E_{Tmax})$. T_{max} consists of the edges with "large" reliability. We find the locally edge-connectivity between both end vertices of each edge in E_{Tmax}.

Next, we allocate the cost to the edges in E_{Tmax} in the ascending order of the edge-connectivity. This is because the reliability of the edge such that the edge-connectivity is small between the end vertices of the edge, must be preferentially increased. The cost is allocated so that the minimum of the increased reliability of the edges is maximized under the constraint of the total cost. Since this optimization problem is a linear programing problem, it can be solved in polynomial time. We show the algorithm in Algorithm 1.

Algorithm 1. Algorithm PPCP

Input: Network $N = (G = (V, E), h)$, $(|V| = n, |E| = m)$, a positive integer B
Output: $c(e)(e \in E)$

1 $c(e)(e \in E) \leftarrow 0$, $b' \leftarrow B$
2 Find the maximum spanning tree $T_{max} = (V, E_{Tmax})$ of network $N^{log} = (G, log(h))$
3 **for** *all vertices e in* E_{Tmax} **do**
4 $\quad$ Calculate the locally edge-connectivity between the end vertices of e

5 **for** $1 \leq i \leq n$ **do**
6 $\quad$ $E_c \leftarrow$ the edges in E_{Tmax} such that the locally edge-connectivity between the end vertices of the edge is i
7 $\quad$ **for** $1 \leq j \leq |E_c|$ **do**
8 $\quad\quad$ $e^* \leftarrow \emptyset$, $e^{**} \leftarrow \emptyset$
9 $\quad\quad$ Sort $h(e, c(e))$ $(e \in E_c)$ in the ascending order
10 $\quad\quad$ $e^* \leftarrow$ the smallest edge e of $h(e, c(e))$ $(e \in E_c)$
11 $\quad\quad$ $h_{min} \leftarrow h(e^*, c(e^*))$
12 $\quad\quad$ Add j edges from the smallest of $h(e, c(e))(e \in E_c)$ to E_P
13 $\quad\quad$ **if** $j < |E_c|$ **then**
14 $\quad\quad\quad$ The smallest edge in $e^{**} \leftarrow h(e, c(e))(e \in E_c \setminus E_P)$
15 $\quad\quad\quad$ $b_{diff} \leftarrow h(e^{**}, c(e^{**})) - h_{min}$

16 $\quad\quad$ **else**
17 $\quad\quad\quad$ $b_{diff} \leftarrow 1 - h_{min}$

18 $\quad\quad$ **if** $b' > b_{diff} \cdot j$ **then**
19 $\quad\quad\quad$ $c(e)(e \in E_c) \leftarrow c(e) + b_{diff}$
20 $\quad\quad\quad$ $b' \leftarrow b' - b_{diff} \cdot j$

21 $\quad\quad$ **else**
22 $\quad\quad\quad$ $c(e)(e \in E_c) \leftarrow c(e) + b'/j$
23 $\quad\quad\quad$ $b' \leftarrow 0$

24 **return** $c(e)(e \in E)$

---

5   Performance Evaluation

In this section, we evaluate the performance of the proposed algorithm in Sect. 4 by using the graph structures of actual ISP backbone networks provided by CAIDA (Center for Applied Internet Data Analysis) [17].

First, we describe the algorithm for the comparison. The algorithm PPCP-beta1 omits the procedure to find a spanning tree in the proposed algorithm PPCP. The algorithm PPCP-beta2 omits the procedure of the locally edge-connectivity in the algorithm PPCP. The algorithm PPCP-beta3 randomly assigns the cost 10000 times and adopts the assignment with the highest network reliability among them.

In the following experiments, the initial reliability of each edge is randomly determined at 0.5 or more.

First, we investigate the change in network reliability according to the change of the total cost. The result is shown in Table 1. Let $Rel(N)$ be the network reliability of the original network. n is the number of vertices; m is the number of edges, and B is the total cost.

Table 1. Comparison of network reliability

n	m	B	$Rel(N)$	$Rel(N, c)$			
				PPCP	beta1	beta2	beta3
46	55	0.5	0.0247	0.0420	0.0417	0.0342	0.0332
		1.0		0.0613	0.0586	0.0490	0.0439
		1.5		0.0876	0.0814	0.0695	0.0566
37	44	0.5	0.0086	0.0170	0.0170	0.0146	0.0125
		1.0		0.0291	0.0278	0.0218	0.0177
		1.5		0.0452	0.0411	0.0313	0.0247
35	50	0.5	0.0756	0.1209	0.1154	0.0975	0.0932
		1.0		0.1645	0.1399	0.1296	0.1137
		1.5		0.2193	0.1757	0.1711	0.1373
8	24	0.2	0.9481	0.9787	0.9705	0.9787	0.9516
		0.4		0.9995	0.9854	0.9941	0.9550
		0.6		1.0000	0.9939	1.0000	0.9582

The proposed algorithm PPCP can achieve the higher network reliability than the algorithms PPCP-beta1, the algorithm PPCP-beta2, and the algorithm PPCP-beta3.

Figures 5, 6, 7, and 8 show the results by applying the algorithms to the network of Fig. 4, when the total cost B is restricted to 1.5. In these figures, the value written at the side of each edge is the reliability of the edge.

In Figs. 5 and 6, the reliabilities of the edges incident to vertices with degree of one are preferentially increased to one, which rapidly increases the network reliability. The network reliability improved by PPCP is better than one by PPCP-beta1. This is because the procedure to find a spanning tree in the algorithm PPCP is effective.

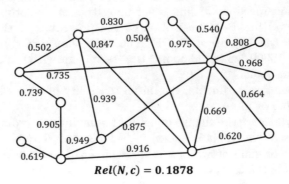

Fig. 4. Example of actual network topology.

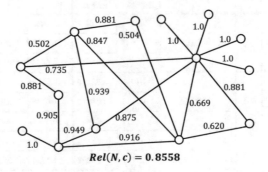

Fig. 5. Network improved by algorithm PPCP.

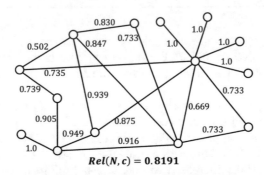

Fig. 6. Network improved by algorithm PPCP-beta1.

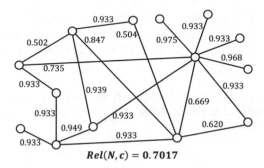

$$Rel(N, c) = 0.7017$$

Fig. 7. Network improved by algorithm PPCP-beta2.

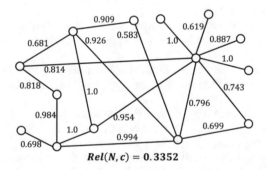

$$Rel(N, c) = 0.3352$$

Fig. 8. Network improved by algorithm PPCP-beta3.

6 Conclusions

In this paper, we proposed the network design method based on the link protection. The reliability of each link is increased more by spending more cost for link protection. The network reliability defined as the probability that the entire network is connected, is also increased by increasing the reliability of edges. However, since the link protection requires the enormous cost, it is necessary to preferentially protect only some highly required links so that the reliability of the entire network is improved. We defined the network design problem that determines the cost allocated for each link, under the constraint that the sum of the cost allocated for each link is less than or equal to a given threshold, so that the network reliability is maximized.

We proved the NP-hardness of this optimization problem and designed the polynomial-time heuristic algorithm. Moreover, we evaluated the performance by using the topology of some actual communication networks. The results show that the proposed algorithm works better than the other algorithms.

In this paper, the reliability is assumed to linearly increase as the cost increases. However, this relationship does not always hold in an actual network in general. For the future work, it remains to design an effective algorithm even if the cost function is not linear.

Acknowledgements. This work was partially supported by the Japan Society for the Promotion of Science through Grants-in-Aid for Scientific Research (B) (17H01742) and JST CREST JPMJCR1402.

References

1. Frank, A.: Augmenting graphs to meet edge-connectivity requirements. In: Proceedings of 31st Annual Symposium on Foundations of Computer Science, St. Louis, 22–24 October 1990
2. Ishii, T., Hagiwara, M.: Minimum augmentation of local edge-connectivity between vertices and vertex subsets in undirected graphs. Discrete Appl. Math. **154**(16), 2307–2329 (2006)
3. Kortsarz, G., Krauthgamer, R., Lee, J.R.: Hardness of Approximation for Vertex-Connectivity Network Design Problems, vol. 2462, pp. 185–199. Springer, Heidelberg (2002)
4. Arulselvan, A., Commander, C.W., Elefteriadou, L., Pardalos, P.M.: Detecting critical nodes in sparse graph. Proc. Comput. Oper. Res. **36**(7), 2193–2200 (2009)
5. Imagawa, K., Fujimura, T., Miwa, H.: Approximation algorithms for finding protected links to keep small diameter against link failures. In: Proceedings of INCoS2011, Fukuoka, Japan, 30 November 2011–December 1–2 2011
6. Imagawa, K., Miwa, H.: Detecting protected links to keep reachability to server against failures. In: Proceedings of ICOIN2013, Bangkok, 28–30 January 2013
7. Imagawa, K., Miwa, H.: Approximation algorithm for finding protected links to keep small diameter against link failures. In: Proceedings of INCoS2011, Fukuoka, Japan, pp. 575–580, 30 November 2011–2 December 2011
8. Maeda, N., Miwa, H.: Detecting critical links for keeping shortest distance from clients to servers during failures. In: Proceedings of HEUNET2012/SAINT2012, Turkey, 16–20 July 2012
9. Irie, D., Kurimoto, S., Miwa, H.: Detecting critical protected links to keep connectivity to servers against link failures. In: Proceedings of NTMS 2015, Paris, pp. 1–5, 27–29 July 2015
10. Fujimura, T., Miwa, H.: Critical links detection to maintain small diameter against link failures. In: Proceedings of INCoS2010, Thessaloniki, pp. 339–343, 16–24 November 2011
11. Yamasaki, T., Anan, M., Miwa, H.: Network design method based on link protection taking account of the connectivity and distance between sites. In: Proceedings of INCoS2016, Ostrava, pp. 339–343, 7–9 September 2016
12. Irie, D., Anan, M., Miwa, H.: Network design method by finding server placement and protected links to keep connectivity to servers against link failures. In: Proceedings of INCoS2016, Ostrava, pp. 439–344, 7–9 September 2016
13. Aggarwal, K.K., Chopra, Y.C., Bajwa, J.S.: Topological layout of links for optimising the overall reliability in a computer communication system. Microelectron. Reliab. **22**, 347–351 (1982)

14. Jan, R.-H., Hwang, F.-J., Chen, S.-T.: Topological optimization of a communication network subject to a reliability constraint. IEEE Trans. Reliab. **42**, 63–70 (1993)
15. Koide, T., Shinmori, S., Ishii, H.: The evaluations on lower bounds of all-terminal reliability by arc-packings for general networks. IEICE Trans. Fundam. Electron. Commun. Comput. Sci. **E82A**(5), 784–791 (1990)
16. Uji, K., Miwa, H.: Method for finding protected links to keep robust network against link failure considering failure probability. In: Proceedings of International Conference on Intelligent Networking and Collaborative System (INCoS2017), Toronto, Canada, 24–26 August 2017
17. CAIDA. http://www.caida.org/

Two-Factor Blockchain for Traceability Cacao Supply Chain

Andi Arniaty Arsyad[(⊠)], Sajjad Dadkhah, and Mario Köppen

Graduate School on Creative Informatics, Kyushu Institute of Technology,
Kitakyushu, Japan
{arsyad.andi-arniaty825,dadkhah.sajjad197}@mail.kyutech.jp,
mkoeppen@ci.kyutech.ac.jp

Abstract. The primary aim of food traceability is to increase food safety but traceability systems can also bring other benefits to production systems and supply chains. Technology has helped supply chain distribution to be successful and more competitive in the global market by improving their customers satisfaction level and providing more reasonable prices compared to their competitors. However, nowadays supply chains are long and complex and include many different actors, beginning with the farmers, followed by collectors, traders, manufacturers. At the end, the processed products are difficult to trace back to their origins in a trusted way. Best way so far is the continuous medial or legal documentation of the chain of transactions, available only to a limited circle of persons. The recent efforts toward blockchain-based trust management open a new perspective here. Here, we propose a solution to link legal documentation and blockchain technology within a traceability system, with the specific application case of cacao and chocolate production in mind. For improved data security, the common blockchain concept is extended to a two-factor blockchain, where both blockchains are connected via digital watermarking of the documentation media, one blockchain traces the documentation steps, the other the watermarking embedding. By a unique reading-verification approach, the integrity of their relatedness can be proven, thus improving also reliability of the whole product tracing.

Keywords: Cacao · Traceability · Supply chain
Digital watermarking · Blockchain

1 Introduction

One of the challenges to realize sustainable cacao sector is the low level of productivity and quality of cacao crops, especially those produced by farmers. Low inputs which then result in low output (quantity and quality), chronically inhibit the growth of cacao commodities. This condition is worsened by climate and weather factors, pests and plant diseases, and other maintenance factors that

---

The original version of this chapter was revised: Typographical error in second author name has been corrected. The correction to this chapter is available at https://doi.org/10.1007/978-3-319-98557-2_50

© Springer Nature Switzerland AG 2019
F. Xhafa et al. (Eds.): INCoS 2018, LNDECT 23, pp. 332–339, 2019.
https://doi.org/10.1007/978-3-319-98557-2_30

affect the quality of production from cacao beans to processed products such as chocolate bars and so on. Nowadays supply chains are long and complex and include many different actors, beginning with the farmers, followed by collectors, traders, manufacturers. At the end, the processed products are difficult to trace back to their origins in a trusted way. Traceability (tracking and tracing) has become a major issue in the food chain [1–6], and quality and supply concerns are merging with traceability issues [7]. Recently, the researchers are paying more attention to the need for investigating different factors that can contaminate the products which belong to global food supply chains. These type of researchers can result in faster, cheaper, real-time, more accurate and ratify testing method for food safety and quality warranty [8].

In recent years, a series of serious cacao accidents have occurred. There was Mars chocolate manufacturers recalls chocolate bars in 55 countries after plastic was found in their products. The recall, affecting 55 countries, could end up costing the company tens of millions of dollars [9]. An investigation that reported in some news organizations about Ivory Coast shows some cases of "dirty cocoa". Much of the world's cocoa grows in Ivory Coast, the examination result showed that a significant amount of cocoa that used in famous companies such as Mars, Nestle, Hershey's, Godiva, and other major chocolate companies was grown illegally in national parks and other protected areas in Ivory Coast and Ghana [10]. These illegal products can be mixed in with "clean" beans in the supply chain, which make it extremely difficult to know what products are safe and which one is affected. The importance of these cases can pose the following question: how does the industry make sure that illegal or "dirty" cocoa beans do not end up in the world's most famous and well-loved chocolate brand?

With the establishment of information technology, the food traceability system which can degrade individuals' disquiet about food safety by serving precise information about the safety and quality of the entire process, from producers to consumers—has been amply disseminated in the food industries [11]. The primary purpose of this paper is to link legal documentation and blockchain technology within a traceability system, with the specific application case of cacao and chocolate production in mind. For improved data security, the common blockchain concept is extended to a two-factor blockchain, where both blockchains are connected via digital watermarking of the documentation media, one blockchain traces the documentation steps, the other the watermarking embedding. By a unique reading-verification approach, the integrity of their relatedness can be proven, thus improving also reliability of the whole product tracing. This paper is organized as follows: Sect. 2 introduces the proposed concept of a 2-factor blockchain and followed Sect. 3 discussed about the advantages of the proposed algorithm.

2 Concept and Method

2.1 Concept

The cacao supply chain traceability system which we propose in this paper mainly relies on existing media-based documentation practice, i.e. we assume that the documentation, enriched by media like photos or videos, is made such that it provides sufficient evidence and cues to retrace the complete process. When the watermark information is not visible to naked eye, it is referred to as Invisible watermarking. The data from the raw material from the farmer will be embedded into the image of the cacao generated by the farmer or manufacturer. The recent efforts toward blockchain-based trust management open a new way. This difficult and challenging process of traceability can be automated, simplified and accelerated by efficient use of Blockchain technology. While one blockchain would be sufficient to ensure the integrity of the sequence of media files generated during the documentation process, it would not allow to ensure the integrity of the media itself. For this reason, we propose the use of digital watermarking technology. However, the need automatically arises to ensure the integrity of the watermarking process itself, and a second blockchain is needed. Assume blockchain to be used in default way, i.e. including automatic hashing and hash control, community validation and proof of work. Linking documentation and blockchain by digital watermarking done by using two blockchains where both blockchains are connected via digital watermarking of the documentation media, one blockchain traces the documentation steps, the other the watermarking embedding. Digital watermarking is the process of embedding information in digital multimedia content such that the data (which is called the watermark) can later be extracted for a variety of purposes including tamper prevention and authentication. As mentioned, the whole watermarking algorithm can have different purposes such as copyright violation prevention, tamper localization, hiding data or authentication. The watermarking processes can be either fragile, robust or semi-fragile. The fragile watermarking techniques are perfect for tamper detection and localization, on the other hand, robust watermarking are useful for authentication and bypassing different operation such as digital noises and compressions [12,13]. Linking is done by embedding textual information in the media data and logging the embedding in the second blockchain. A unique reading-verification process ensures the correct linking of both blockchains thus improving also reliability of the whole product tracing.

The Two-factor blockchain for traceability cacao supply chain procedure is depicted in Fig. 1 followed by a detailed explanation.

1. The farmer and manufacturer provide textual information and an image (photograph). A token system can be used here to control the textual information.
2. The image is watermarked with the text information, giving DWMImage.
3. The hash value of DWMImage together with text and (not-watermarked) image is stored in a Block of blockchain 1(BC 1). Proof-of-work is done to add Block to BC 1.
4. DWMI will be stored in a block of blockchain 2 (BC 2) as documentation.
5. Consumers will see the documentation results in BC 2.
6. For the tracing process then the validation process is done between the existing hash blockchain 1 with the watermark documentation in blockchain 2.

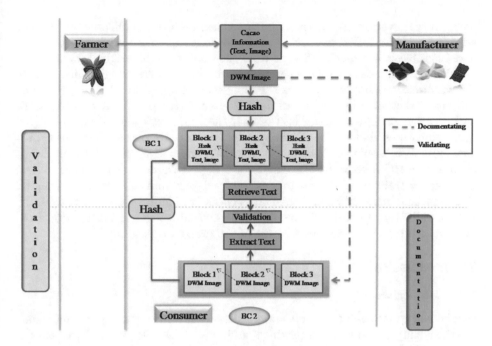

Fig. 1. Concept of Two-factor blockchain for traceability cacao supply chain.

2.2 Method

Texts containing cocoa information are transformed to binary form, as well as images to be transformed into binary form. For the embedding process frequency domain Discrete Wavelet Transform (DWT) is used. The basic idea of DWT transformation in image processing is to decompose an image into 4 parts of frequency: The low frequency part that called LL, and another high-frequency parts are HL, LH, and HH. The output of the sub-level frequency district will be found when DWT is transformed by low-frequency district. The differences from sub-images are as shown below:

1. LL: A bearish estimation to the original image conceived the entire information about the whole image and it applied of the low-pass filter on both x and y coordinates.
2. HL and LH: The high-pass filter on one coordinate and the low-pass filter on the other coordinate is gained of HL and LH.
3. HH: The high-pass filter on both x and y coordinates found from the high-frequency component of image in the diagonal direction [14].

The wavelet transform allows high compression ratios with reconstructed images to be of good quality. The wavelet transform can provide time and frequency information simultaneously, thus providing a time-frequency representation of the signal [15]. Discrete Wavelet Transform has advantages in identifying parts in the cover image, where watermarks can be inserted effectively [16]. The advantage of the DWT method is that the image quality that is the place for text insertion is not much different from the quality of the original media.

The watermarked than goes through the hashing process stage which aims to get watermarking hashed to check the integrity of image files. This way it can be validated whether a byte series of data is altered (tampered). A cryptographic hash is a sort of 'signature' for text or data files. SHA-256 generates 256-bit (32-byte) signatures that are nearly unique to text and are one of the most robust hash functions available. Hashing is used as a method to store data in BC 1. As addition BC 2 contains the results of a digital watermarking image. When a consumer trails a validation process, extract text and images from a digital watermarking image BC 2 and then cross-check (validate) by retrieving from BC 1. The validation process will result in the authentication of information from farmers and manufacturers as a result of the process traceability.

3   Discussion

3.1   Objectives

Blockchain system can overcome complexities occurred in the supply chain such as traceability process of the products and lack of security which measures the whole procedure. In case that a farmer wants to send images to the manufacturer or vice versa, authorization is required from each party. Furthermore, blockchain is known as a peer to peer structured system which can solve different securities problems such as lack of confidentiality in data or information received by consumers and other parties of the process. However, the security aspect of the blockchain process can be covered by applying different hashing function and utilizing different watermarking methods. Using a good fragile watermarking algorithm for blockchain data makes it almost impossible for attackers to intrude.

3.2   Potential for Manipulation

As mentioned already, the proposed watermarking and blockchain algorithm in this paper can provide particular transparency for both the consumer and manufacturer from beginning to end. However, one of the fundamental problems with the whole system is that there is no way to guarantee the correctness of the photographs before the digital image is in our supply chain. Another problem is how to track the growth of cocoa and ensure that the pictures of cocoa at different times of growth are consistent with the same physical objects. However, this class of problems has to be delegated to the documentation process itself, i.e. it is expected that it has to follow common standards discipline. For example, the documentation process already involves accounting means, back confirmations, witnesses, or provides sufficient number of cues. The essential add-on in trust provided by using blockchain is to prevent manipulation from third parties, as well as a simplifying the identification of an injection point of a rogue intermediate producer in the whole processing chain.

3.3   Access Control

In terms of access control, even if blockchain is considered to be public, still information can be controlled by user because the proposed blockchains algorithm in this paper is secured by cryptographic verification using mechanisms such as proof of work.

3.4   Advantages of Proposed Algorithm

The following can be considered as advantages of the proposed algorithm:

1. Level of security: As mentioned before with the addition of different hashing and watermarking algorithm, the originality of the pictures is verified every step of the way.
2. Effort in general: Blockchain systems as general keeps data in textual format while in this system digital images or even other media are utilized to keep different types of information such the size of fruit, chemical content, the time when the picture is taken etc. So in this way less effort and storage is used to gather all the information for a particular bean or a fruit.
3. Convenience: Usually, like bitcoin, blockchain entries are rather hard to read, only long numbers. In this paper, we used digital images which make it more to understand for the user.

3.5   Energy Usage

Energy usage is known to be too costly on the long run. Some of the alternative developments, for example the stakeholder-based proof of work, make such problems obsolete. In other cases, a peer-to-peer energy trading model to solve the energy usage problem is already envisaged.

3.6   Concept of Implementation

The security of the proposed concept relies on the fact that it is not possible to manipulate a watermarked image such that the hash value of the manipulated image is equal to the hash of the original image that is stored in the second blockchain. Neither the stored hash value can be manipulated, per definitionem of a blockchain, nor the image manipulated such to have a different image with same hash value.

Figure 2 shows how the concept of security occurs within the blockchain system. The image before and after digital watermarking appear to be the same but the hashing process shows the difference, once the modified watermarking image can be known by the difference through the hashing process.

3.7   Traceability

The 2-factor blockchain systems can be used to improve the security and traceability on farmers', manufacturers' and consumers' level. BC 1 is used to store

Fig. 2. Hash values and watermarking of a cacao fruit: upper left the original image and the SHA256 hash values; upper right, same image after watermarking, perceptually not changed, but a different hash value; lower image, an attempt to remove the digital watermark by Gaussian blurring, but even then, the original hash value can't be restored.

hashing of the watermarked data and information derived from both farmers and manufacturers. BC 2 is used to store the watermarked document. When traceability occurs, the system will generate hashing from BC 2 and compare with hashing stored in BC 1. The hashing validation process will ensure the integrity and authenticity of the data. BC 2 is open to the public, which means that it can be accessed in the process traceability. Moreover, the proof of Work is run by Farmers and manufacturers when making new blocks. A new block is added by each manufacturer who's issue a new production which is different from the previous production while the farmer level can create new blocks if they have different cacao content and quality. Here, we assume 4–5 peers per each farmer and manufacturer.

4   Conclusion

In this paper, a blockchain-based traceability system for the cacao production and supply chain is proposed. Information about cacao is stored safely through watermarking techniques at all user levels, so it is hard to manipulate. The watermarking utilized in the proposed algorithm make sure the information at each stage are authenticated and in case there is change or manipulation of data of any kind it can be detected. The cacao traceability process becomes robust through a two-factor blockchain to perform documentation and data validation. For the future work, we are going to set up a practice with a chocolate factory which is located in Japan, to find all the pros and cons of the proposed traceable blockchain algorithm in the real-life case.

References

1. Dabenne, F., Gay, P.: Food traceability systems: performance evaluation and optimization. Comput. Electron. Agric. **75**, 139–146 (2011)
2. Gellynck, X., Verbeke, W., Vermeire, B.: Pathways to increase consumer trust in meat as a safe and wholesome food. Meat Science. **74**, 161–171 (2006)
3. Jacquet, J.L., Pauly, D.: Trade secrets: renaming and mislabeling of seafood. Marine Policy **32**, 309–318 (2008). K. Elissa, Title of paper if known, unpublished
4. Monteiro, D.M.S., Caswell, J.A.: Traceability adoption at the farm level: an empirical analysis of the portuguese pear industry. Food Policy **34**, 94–100 (2009)
5. Opara, L.: Engineering and technological outlook on traceability of agricultural production and products. Agric. Eng. Int.: CGIR J. Sci. Res. Develop. **4**, 1–13 (2002)
6. Regattieri, A., Gamberi, M., Manzini, R.: Traceability of food products : general framework and experimental evidence. J. Food Eng. **81**, 347–356 (2007)
7. Kapllinsky, R.: Competitions policy and the global coffee and cocoa value chains. Paper for United Nations Conference for Trade and Development (UNCTAD). www.unctad.org (2004)
8. van der Vorst, J.G.A.J.: Product traceability in food supply chains. Accred Qual. Assur. **11**, 33–37 (2006)
9. The Guardian Magazine, International edition. https://www.theguardian.com/lifeandstyle2016/feb/23/mars-chocolate-product-recalls-snickers-milky-way-celebrations-germany-netherlands
10. The Guardian Magazine, International Edition. https://www.theguardian.com/environment/2017/sep/13/chocolate-industry-drives-rainforest-disaster-in-ivory-coast
11. Badia-Melis, R., Mishra, P., Ruiz-García L.: Food traceability: New trends and recent advances. Food Control **57**, 393–401 (2015). Elsevier Science
12. Dadkhah, S., Abd Manaf, A., Hori, Y., Ella Hassanien, A., Sadeghi, S.: An effective SVD-based image tampering detection and self-recovery using active watermarking. Signal Process. Image Commun. **29**, 1197–1210 (2014)
13. Jabade, V.S., Gengaje, S.R.: Literature review of wavelet based digital image watermarking techniques. Int. J. Comput. Appl. **31**(1), 28–30 (2011)
14. Wang, K.: Region-based three-dimensional wavelet transform coding, thesis submitted to the faculty of graduate studies and research in partial fulfillment of the requirements for the degree of Master of Applied Science. Carleton University, Ottawa, Canada, May 2005
15. Bedi, S.S., Kumar, A., Kapoor, P.: Robust secure SVD based DCT ñ DWT oriented watermarking technique for image authentication. In: International Conference on IT to Celebrate S. Charmonman's 72nd Birthday, pp. 1–7 (2009)
16. Al-Haj, A.: Combined DWT-DCT digital image watermarking. J. Comput. Sci. **3**, 740–746 (2007)

On the Characteristics of TCP/NC Tunneling in Heterogeneous Environments

Nguyen Viet Ha[✉] and Masato Tsuru

Kyushu Institute of Technology, Kitakyushu, Japan
nguyen.viet-ha503@mail.kyutech.jp, tsuru@cse.kyutech.ac.jp

Abstract. Transmission Control Protocol (TCP) with a loss-based congestion control is still dominantly used for reliable end-to-end data transfer over diverse types of network although it is ineffective when traversing lossy networks. We previously proposed an IP tunneling system across lossy networks using the TCP with Network Coding (TCP/NC tunnel) and showed its potential to significantly mitigate the goodput degradation of end-to-end TCP sessions without any change of end-device's communications protocol stack, but it was shown only in homogeneous conditions. On the other hand, reliable end-to-end data transfer in diverse and heterogeneous IoT environments in a cost-efficient manner is an emerging challenge. Therefore, in this paper, we investigate the characteristics of the TCP/NC tunnel on heterogeneous networks with/without network congestions, to assess the applicability of the TCP/NC tunnel-based intelligent gateway system to IoT environments where end-devices are connected to a gateway with different link bandwidths or connected to different gateways in terms of network topology. The simulation results suggest the TCP/NC tunnel can efficiently utilize the bottleneck bandwidth in such heterogeneous situations even with congestion and achieve a significantly high goodput of end-to-end TCP sessions in a wide range of link loss degree especially when the tunnel link bandwidth is sufficient.

1 Introduction

Transmission Control Protocol (TCP), a transport layer communication protocol between end-nodes with a long history, is still dominantly used for reliable end-to-end (E2E) data transfer with network congestion control. However, TCP suffers a low data transfer goodput on lossy networks especially with a long Round-Trip Time (RTT) because its loss-based congestion control reduces the sending rate mistakenly when non-congestion origin packet losses occur due to lossy links [1]. Actually, widely used TCP variants such as NewReno, CUBIC, and Compound-TCP mainly adopt a loss-based congestion control.

To mitigate the goodput degradation of TCP on lossy networks, a variety of solutions have been studied. In E2E solutions, one approach is to improve the congestion detection and the sending rate control in TCP adaptive to wireless

© Springer Nature Switzerland AG 2019
F. Xhafa et al. (Eds.): INCoS 2018, LNDECT 23, pp. 340–349, 2019.
https://doi.org/10.1007/978-3-319-98557-2_31

networks such as TCP Westwood+ [2], and another approach is to mask packet
losses possibly caused by lossy links based on proactively sending redundant
packets such as TCP with Network Coding (TCP/NC) [3]. Furthermore, new
types of E2E transport protocol on the top of UDP, e.g., QUIC, are also being
developed and already in use [4]. Among them, the TCP/NC, on which the
authors focused, enables a proactive recovery of lost packets with the redundant
transmission of coded packets in cooperation with a reactive recovery of lost
packets with retransmission based on the standard TCP-ACK mechanism, so
as to maintain the goodput properly even in lossy networks. Among succeeding
variants of TCP/NC, our developed one significantly improves the performance
by an efficient retransmission and an adaptively optimized redundancy of coded
packets based on online packet loss observation [5]. We showed the fundamental benefits of the improved TCP/NC over a wide range of loss rates and loss
burstiness degrees with dynamic but slow changes. However, in general, any E2E
solution requires a change on all involved end-nodes. In addition, TCP/NC is
inefficient when a session traverses multiple heterogeneous networks (i.e., with
different link loss rates). In middle-box solutions, the Performance Enhanced
Proxy (PEP) approach, such as TRL-PEP [6] and D-Proxy [7], has been widely
developed. While they do not require any change of end-nodes, complicated and
costly per-TCP session management should be performed on the proxy (gateway) nodes.

On the other hand, the demand for Internet of Things (IoT) applications is
growing rapidly with penetration of IoT devices in wild network environments,
which are often characterized by a high loss rate and a long RTT. Especially for
tiny end-devices, e.g., with less memory and power, it is an emerging challenge to
realize E2E data transfer in diverse and heterogeneous IoT environments with no
or few changes on end-device in cooperation with a scenario-specific application
[8]. Please note that the use of packet-level coding for loss recovery in a variety
of data transfer scenarios including IoT has attracted attention [9].

By considering the above issues, we proposed the TCP/NC tunnel to convey
end-to-end TCP (E2E-TCP) sessions over lossy networks on a single TCP/NC
session between two gateways [10] or on cascaded TCP/NC sessions involving
more than two gateways [11] as shown in Fig. 1, without any change in TCP on
each end-device. The TCP/NC tunnel is a kind of middle-box solution. However,
in contrast to the PEP approach, the "tunneling" approach does not require a

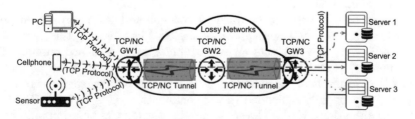

Fig. 1. Example of TCP/NC tunnel with three TCP/NC gateways

complicated per-session management on each gateway. On the other hand, the encapsulation overhead (e.g., header space and processing time) is introduced in general.

In our previous papers [10,11], we developed the TCP/NC tunnel system on Network Simulator 3 (ns-3) [12] and showed its potential to significantly mitigate the goodput degradation of E2E-TCP sessions in a wide range of link loss rates, but only in homogeneous conditions. Therefore, the purpose of this paper is to investigate the characteristics of TCP/NC tunnel on heterogeneous networks in terms of the link bandwidths and link positions of end-devices with/without network congestions. Such heterogeneous conditions are essential to IoT networking. Please note that we assume a stationary lossy network such as a large-scale Power Line Communications network or a multi-hop fixed long-distance wireless network. The TCP/NC tunnel for mobile ad-hoc networks remains as future work.

The remainder of this paper is organized as follows. In Sect. 2, the TCP/NC and its tunnel are briefly explained. Simulation evaluation is presented in Sect. 3 and a conclusion is given in Sect. 4.

2   Overview of TCP/NC Tunnel

2.1   TCP/NC

TCP/NC protocol introduces an idea of adding NC layer between TCP and IP layer. In every receiving n TCP segments (called original packets) from TCP layer, NC layer combines them to produce m combination packets (called combination) with $m \geq n$. The sink is expected to recover all the original packets without retransmission using the received combinations even some combinations are lost over a lossy network. TCP layer does not detect any loss events; thus, it maintains the CWND appropriately to stable the goodput performance. The processes of creating m combinations and regenerating n original packets are called encoding and decoding, respectively. Encoding and decoding processes are separated in every n original packets; hence, n also corresponds to Coding Window (CW) size. Redundancy factor $R = \frac{m}{n}$ and the recovery capacity $k = m - n$ in one CW are two parameters to express the recovery ability of the system and must be chosen carefully. Too large R incurs an unnecessary redundancy causing a small goodput and too large k affects to the decoding delay and the hardware limitations.

NC layer uses the degree of freedom concept and the seen/unseen definition [3] for returning ACK process to avoid TCP layer seeing the losses which can be recovered. The sink sets an ACK number as a sequence number of the oldest "unseen" packet that it needs more information for the decoding process. Figure 2 is an example of the encoding, decoding and acknowledgment processes. The packets p_1 to p_4 are encoded to the combinations $C[1]$ to $C[6]$. Due to the two lost combinations, the NC layer cannot decode any combination until receiving $C[6]$. For each received combinations, NC layer returns an ACK packet whose ACK number corresponds to the smallest sequence number of the unseen packet.

During the process, the TCP layer is unaware of any loss events; thus, the CWND keeps increasing to stable the performance. In this example, R equals to $\frac{6}{4}$, k equals to 2. The packets from p_1 to p_4 are in the same CW. If a new packet comes e.g., p_5, it will be in the next CW.

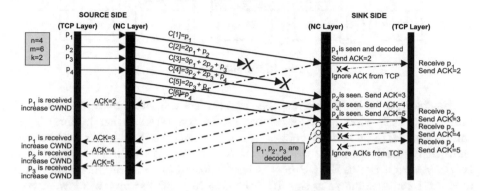

Fig. 2. Network coding process

In our previous work, we did a significant enhancement on TCP/NC in terms of a retransmission mechanism (retransmitting more than one lost packet quickly and efficiently, allowing encoding the retransmitted packets for reducing the repeated losses, and handling the dependent combination packets for avoiding the decoding failure), and an adaptation mechanism for encoding parameters (k and R) to bursty and time-varying losses (estimating both packet loss rate and burstiness by observing transmitted packets, computing appropriate parameters based on a mathematical model of packet losses, and updating the parameters of the current CW promptly) [5]. In this paper, our enhanced version of TCP/NC is used as "TCP/NC".

2.2 TCP/NC Tunnel

We proposed an IP tunneling system using TCP/NC in which two or more TCP/NC gateways in the middle of the lossy network, called TCP/NC tunnel [10,11]. Without any change of end-device's communication protocol stack, the TCP/NC tunnel has brought the benefits of TCP/NC that maintains TCP goodput properly even in lossy networks. The protocol stack and structure of TCP/NC tunneling are illustrated in Fig. 3.

TCP/NC gateway includes two interface types. The first is the "internal" interface which is connected to the local network. The second is the "external" interface which is connected to other networks. If a data packet comes from the internal interface and goes to the external interface, it will join TCP/NC flow. Otherwise (e.g., local delivery), TCP/NC gateway works as a normal router.

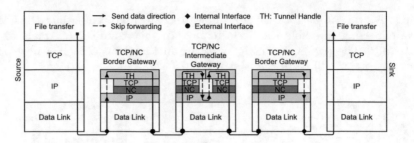

Fig. 3. Tunnel handler

Basically, a TCP/NC gateway has two buffers to store a data including link buffer and TCP sending buffer which can be congested. Other buffers (TCP-small-queue, Traffic control queue) are not covered on this paper and are not enabled or available in the simulation. At the sending TCP/NC gateway, an IP packet from the E2E-TCP flow is firstly stored in TCP sending buffer until it is ACKed. Assume that this IP packet is in TCP sending window, a copy of this packet will be sent to the link and stored in link buffer. Consequently, there are two congestion cases of two buffers. If the receiving data rate at TCP sending buffer is larger than the ACKing rate, the congestion will happen at TCP sending buffer. If the sending rate of the higher layer is larger than that of link buffer, the congestion will happen at link buffer.

Packet dropping in network congestion is necessary to inform for TCP layer to decrease the sending rate. But dropping the packet at TCP/NC flow causes goodput degradation because of the lack of function to distinguish the congestion loss, resulting in the NC-parameters estimation and adaptation work inefficiently. The link buffer size should be chosen to limit the packet loss by network congestion. The maximum packets storing in link buffer are $R_{max} \times CWND_{max}$ where R_{max} is the maximum estimated value of the redundancy factor at a maximum link loss rate and $CWND_{max}$ is the maximum value of CWND. In the simulation, we consider the random loss channel with the maximum link loss rate of 0.2 and TCP window scaling option is disabled due to the small bandwidth connection; hence, R_{max} is about 1.5 and $CWND_{max}$ is 63 packets (a TCP segment size at TCP/NC gateway is 1000 bytes plus IP and TCP header). Therefore, the link buffer size in this paper is set to 100 packets. The network congestion packet dropping behavior now moves to TCP sending buffer which is set to 64 packets plus the link buffer size (100 packets).

3   Simulation Evaluation

3.1   Simulation Settings

The performance of TCP/NC tunnel in heterogeneous environments has been evaluated on ns-3 with two topologies shown in Figs. 4 and 5. Both topologies have three routers or TCP/NC gateways (called GW) dependent on the simulation case and accommodate three E2E TCP sessions from source j to sink j

$(j = 1, 2, 3)$. The delay of the links among GWs and the delay between the sinks and GW3 were set to 10 ms and 1 ms, respectively. The links connected to the sources were set to 1.5 ms to avoid the bias based on an artificial synchronization of packet arrival times.

Topology 1 shows the mix of different edge link bandwidths connecting three sources to GW1. Sources 1 and 2 are connected with 1 Mbps while source 3 with 3 Mbps. GW1 and GW2 are connected with 3, 5, and 7 Mbps in different cases. GW2 and GW3 are connected with 5 Mbps. Topology 2 is asymmetric where sources 1 and 2 are connected to GW1 while source 3 is connected to GW2. All sources and sinks are connected to GWs with 1 Mbps, respectively. GW1 and GW2 are connected with 2 Mbps. GW2 and GW3 are connected with 2, 3, or 4 Mbps in different cases.

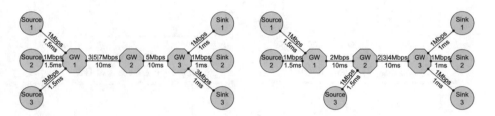

Fig. 4. Simulation topology 1   **Fig. 5.** Simulation topology 2

On both topologies, we evaluate the goodput of each of three E2E-TCP sessions with TCP/NC tunnel (called TCP/NC Tunnel option) and without TCP/NC tunnel (called E2E-TCP option). The TCP type is NewReno which the default settings are used except the Window Scaling, Time-stamp are disabled. One of default settings is Delayed-ACK which its delay is set to 200 ms and the number of packets to wait before sending an ACK is set to two packets. The TCP segment size is 1000 bytes. The link buffer and TCP sending buffer size of each TCP/NC gateway is set to 100 and 163 packets. The random lossy channel in the transferred data direction is enabled at the intermediate links (between GWs) with link loss rate per link ranging from 0 to 0.2% (r_0); the total link loss rate from GW1 to GW3 equals $r_0 + r_0 \times (1 - r_0)$. Note that, the link loss rate parameter in the x-axis of the result figures is the link loss rate per link. All the simulations are run 20 times to get the average results (goodput).

3.2 Performance with "Mix Edge Device Links" (Topology 1)

The bandwidth of the link between GW1 and GW2 is set to 3 Mbps (Case 1), 5 Mbps (Case 2), and 7 Mbps (Case 3). On this topology, session 3 has a higher bandwidth edge link compared with sessions 1 and 2. With lossy links, unacceptable goodput of all sessions is seen in E2E-TCP option as predicted.

In Case 1, the congestion happens on GW1 in both options. While the packet loss happens at link buffer in E2E-TCP option, the packet loss happens at TCP

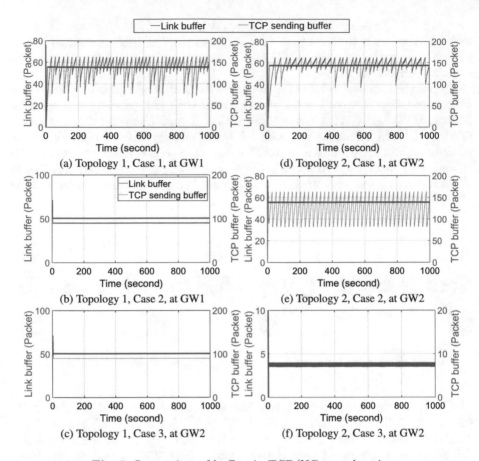

Fig. 6. Queue sizes of buffers in TCP/NC tunnel option

sending buffer in TCP/NC Tunnel option shown in Fig. 6(a). Without lossy links ($r_0 = 0$), regarding session 3's goodput, E2E-TCP option outperforms TCP/NC Tunnel while TCP/NC Tunnel outperforms E2E-TCP in sessions 1 and 2 shown in Fig. 7(a). Actually the total goodput of three sessions is 2.87 Mbps in E2E-TCP option and 2.77 Mbps in TCP/NC Tunnel option shown in Fig. 8(a), indicating a comparably effective utilization of the bottleneck bandwidth of 3 Mbps with a small difference resulting from the TCP/NC tunneling overhead (about 4%). As the link loss rate increases with lossy links, TCP/NC Tunnel option can keep the goodput of each session high and stable in a relatively fair manner.

In Case 2, the network congestion does not happen in E2E-TCP option, but it lightly happens in TCP/NC Tunnel option on GW1 due to the TCP/NC tunneling overhead shown in Fig. 6(b). The congestion in TCP/NC option still happens on GW2 in Case 3 because the bandwidth of the link between GW2 and GW3 remains of 5 Mbps shown in Fig. 6(c). However, even without lossy links, the goodput in TCP/NC Tunnel option is comparable to E2E-TCP option; the difference is negligible as shown in Fig. 7(b,c) and Fig. 8(a). In both cases, as the link loss rate increases, in TCP/NC Tunnel option, the goodput of session 3 decreases gradually while those of sessions 1 and 2 keep 1 Mbps because the advantage of 3 Mbps edge link of session 3 becomes less.

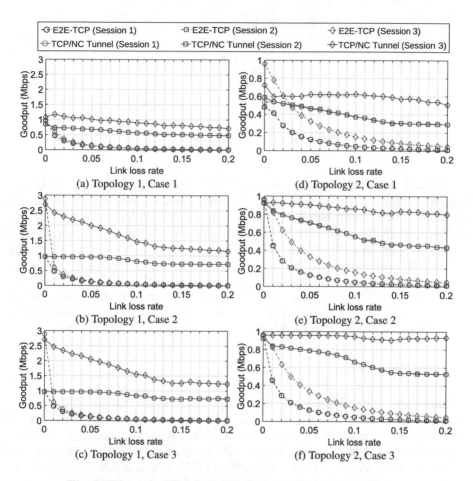

Fig. 7. Per-session Goodputs in three cases on two topologies

3.3   Performance on Asymmetric Topology (Topology 2)

The bandwidth of the link between GW2 and GW3 is set to 2 Mbps (Case 1), 3 Mbps (Case 2), and 4 Mbps (Case 3). On this topology, the network congestion happens at GW2 in both of two options in Case 1 (Fig. 6(d)), while it happens only in TCP/NC Tunnel option in Case 2 due to the TCP/NC tunneling overhead (Fig. 6(e)). But there is no network congestion in Case 3 (Fig. 6(f)). Besides, session 3 traverses a fewer number of lossy links compared with sessions 1 and 2. Therefore, the goodput of session 3 is always greater than that of sessions 1 and 2 in all cases.

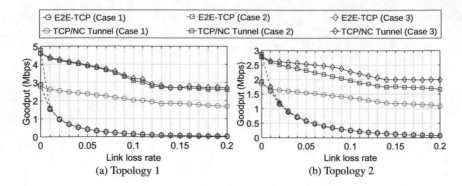

Fig. 8. Total Goodputs in three cases on two topologies

In Case 1, Fig. 7(d) shows the goodput of session 3 is greater than that of sessions 1 and 2 in both options at network congestion because of the shorter RTT. In Case 2 (less congestion with no packet dropped) and Case 3 (no congestion), as the link loss rate increases, an E2E-TCP option does not perform suddenly again. In contrast, in TCP/NC Tunnel option, the goodput of sessions 1 and 2 decreases gradually while session 3 keeps a high goodput because the advantage of passing only one lossy links becomes greater.

In all cases, TCP/NC Tunnel option achieves a fairness among sessions like E2E-TCP option. But it has a higher bandwidth utilization compared to E2E-TCP option as shown in Fig. 8(b).

4   Conclusions

We have investigated the characteristics of the TCP/NC tunnel on heterogeneous networks eventually aiming at the TCP/NC tunnel-based "intelligent gateway" system distributed over IoT environments. We have shown that the TCP/NC tunnel can efficiently utilize the bottleneck bandwidth even with congestion, and achieve a significantly high goodput of E2E-TCP sessions in a wide range of link loss degree especially when the tunnel link bandwidth is sufficient.

In future work, the issues to address include (i) network congestion detection by utilizing additional cross-layer information; (ii) modest redundancy control on links with a narrow bandwidth; and (iii) prediction and fast adaptation to rapid changes of network conditions. To solve those issues, more precise environment-dependent models on packet loss and its burstiness are required. An adaptive gateway selection and an integration of possible multiple network paths among gateways are also essential.

Acknowledgements. The research is supported by JSPS Grant-in-Aid for Scientific Research 16K00130 and "Resilient Edge Cloud Designed Network", the Commissioned Research of NICT, Japan.

References

1. Leung, K.-C., Li, V.O.K.: Transmission Control Protocol (TCP) in wireless networks: issues, approaches, and challenges. IEEE Commun. Surv. Tutorials **8**(4), 64–79 (2006)
2. Grieco, L.A., Mascolo, S.: Performance evaluation and comparison of Westwood+, New Reno and Vegas TCP congestion control. ACM Comput. Commun. Rev. **34**(2), 25–38 (2004)
3. Sundararajan, J.K., Shah, D., Medard, M., Mitzenmacher, M., Barros, J.: Network coding meets TCP. In: Proceedings of the IEEE International Conference on Computer Communication (INFOCOM), pp. 280–288 (2009)
4. Langley, A., Riddoch, A., Wilk, A., et al.: The QUIC transport protocol: design and internet-scale deployment. In: Proceedings of the ACM SIGCOMM, pp. 183–196 (2017)
5. Ha, N.V., Kumazoe, K., Tsuru, M.: TCP network coding with adapting parameters for bursty and time-varying loss. IEICE Trans. Commun. **E101–B**(2), 476–488 (2018)
6. Ivanovich, M., Bickerdike, P., Li, J.: On TCP performance enhancing proxies in a wireless environment. IEEE Commun. Mag. **46**(9), 76–83 (2008)
7. Murray, D., Koziniec, T., Dixon, M.: D-Proxy: reliability in wireless networks. In: Proceedings of 16th Asia-Pacific Conference on Communications (APCC), pp. 129–134 (2010)
8. Gomez, C., Arcia-Moret, A., Crowcroft, J.: TCP in the internet of things: from ostracism to prominence. IEEE Int. Comput. **22**(1), 29–41 (2018)
9. Sandell, M., Raza, U.: Application layer coding for IoT: benefits, limitations, and implementation aspects. IEEE Syst. J. 8 pages (2018, early access)
10. Ha, N.V., Kumazoe, K., Tsukamoto, K., Tsuru, M.: Masking lossy networks by TCP tunnel with network coding. In: Proceedings of 22nd IEEE Symposium on Computers and Communications (ISCC), pp. 1292–1297 (2017)
11. Ha, N.V., Kumazoe, K., Tsukamoto, K., Tsuru, M.: Benefits of multiply-cascaded TCP tunnel with network coding over lossy networks. In: Proceedings of 15th International Conference on Wired/Wireless Internet Communications (WWIC), pp. 247–258 (2017)
12. Network simulator (ns-3). https://www.nsnam.org/. Accessed 1 Mar 2018

Method for Determining Recovery Order for Successive Node Failures

Tsuyoshi Yamasaki and Hiroyoshi Miwa[✉]

Graduate School of Science and Technology, Kwansei Gakuin University,
2-1 Gakuen, Sanda-shi, Hyogo, Japan
{tsuyo,miwa}@kwansei.ac.jp

Abstract. It is difficult to completely prevent failures of links and nodes in a communication network. Therefore, it is necessary to quickly recover the failed nodes. Since resources are generally limited at the time of a disaster, it is impossible to recover all failure nodes at the same time, and it must be recovered sequentially. However, as is often seen in recent large-scale earthquakes, the aftershocks often cause failures, give large damage again and again, and prevent the recovery. In such a situation, since a new failure may occur at any time during sequentially recovering many failures, the order of recovery significantly affects the communication quality of the network. Therefore, it is important to determine an appropriate recovery order so that the communication quality of a network keeps even in a situation that failures occur successively. In this paper, we propose a method of determining the node recovery order that as many nodes as possible keep to be connected even if successive node failures occur. First, we formulate this problem as an optimization problem, prove that the problem is NP-hard, and design a heuristic algorithm. We also evaluate the performance of the proposed algorithm by using the topology of various actual networks.

1   Introduction

Reliabile communication networks is absolutely imperative, because the Internet is a social infrastructure today. Especially, it needs that a communication path between each pair of nodes exists even at the time of a network failure. From this background, there is a lot of studies on network design methods to keep high reliability.

On the other hand, recovery scheduling had not considered successive network failures; that is, if a failure occurs, it was assumed that the failure parts are recovered immediately. However, network failures are caused again and again by aftershocks. For example, in the Great East Japan Earthquake in 2011 and the Kumamoto earthquake in 2016, communication networks and road networks were damaged by not only the main shock but also the aftershocks. Moreover, in general, since the interval between successive aftershocks is short, it is not easy to recover the failed parts immediately. Indeed, at the time of a disaster, material resource and human resource are limited, so that it is impossible to recover

© Springer Nature Switzerland AG 2019
F. Xhafa et al. (Eds.): INCoS 2018, LNDECT 23, pp. 350–359, 2019.
https://doi.org/10.1007/978-3-319-98557-2_32

all failed parts at the same time; therefore, it is necessary to recover failed parts sequentially. Since a new failure may occur at any time during sequentially recovering many failures, the order of recovery significantly affects the communication quality of the network. Consequently, we must deal with reiterated network failures by a main shock and successive aftershocks. Even in such a situation, in order to keep network connectivity as much as possible, a recovery scheduling is important.

We show an example that the difference of the recovery orders causes the difference of connectivity in case of successive node failure in Fig. 1. In Fig. 1(a), it is assumed that five nodes fail at the main shock and that three nodes recovered until the next aftershock occurs, then one node fails at the time of the next aftershock. The recovery order must be determined so that as many nodes as possible can keep connectivity even in the successive node failures. If the failed nodes are recovered in the order in Fig. 1(c), the network after the next aftershock occurs can keep the connectivity among seven nodes. However, if the failed nodes are recovered in the order in Fig. 1(d), the network after the next aftershock occurs cannot keep the connectivity among seven nodes but among only five nodes. We can observe that the connectivity at any time during sequentially recovering failures depends on a recovery order.

In this paper, we propose a method to determine a recovery order such that as many nodes as possible can keep connectivity even in the successive node failures. First, we formulate the optimization problem of determining an appropriate recovery order of failed nodes against succesive node failures, and prove the NP-hardness of this problem. Furthremore, we design a heuristic algorithm, and we evaluate the performance of the algorithm by applying the proposed algorithm to the topology of an actual networks.

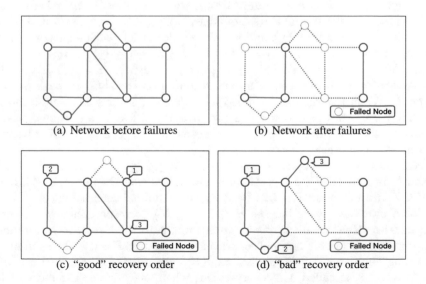

(a) Network before failures   (b) Network after failures

(c) "good" recovery order   (d) "bad" recovery order

Fig. 1. Difference of connectivity caused by different recovery order.

2   Related Works

There are many studies on network reliability from the viewpoint of protection of nodes and links; the studies on the problem of determining the protected links so that the master server and the edge server keep connectivity even if link failure [1–4]; the reference [2] deals with the problem of determining protected links to keep small the diameter; the reference [3,4] deals with the problem of determining a set of protected links to keep connectivity to a specified node; the reference [3] proposes the approximation algorithm; the reference [4] deals with the problem of determining the protected links to avoid congestion in case of failure.

The recovery order are studied in the research area of road networks [5–7], which are the problem of determining the recovery segments of a road network and the main measure for the evaluation of the performance is traveling time of cars. In these previous studies, it is implicitly assumed that a failure occurs only once, and hence the successive failures are not considered. Therefore, when the recovery orders is determined by a previous method, the connectivity is often lost by successive failures.

Regarding the problem of determining the recovery order, the group of the authors has investigated so far [8]. The reference [8] deals with the problem of determining the recovery order to keep connectivity between a master-server and edge-servers against link failures. Since this study focuses on link failure, we cannot apply the results to our problem dealing with node failures.

3   Problem for Determining Vertex Recovery Order

The graph representing a communication network is $G = (V, E)$, where V and E are the set of vertices and the set of edges of G, respectively. Each node corresponds to a router, a building, and a mobile base station; each edge corresponds to a link between nodes. Let $Q = \{q_1 = \{s_1, t_1\}, q_2 = \{s_2, t_2\}, \ldots, q_r = \{s_r, t_r\}\}$ be the demand set for each pair $q_i = \{s_i, t_i\}$ of vertices ($i = 1, 2, \ldots, r$). The communication path between s_i and t_i is routed on the shortest path between s_i and t_i. When the vertices included each demand set are connected and the length of the shortest path between the vertices is b or less for all demand sets, we define that the network is feasible. This means that the communication between the important nodes survives.

When graph $G = (V, E)$, demand set $Q = \{q_1, q_2, \cdots, q_r\}$, and a positive integer b are given, if $V'(\subseteq V)$ is the minimal subset that the network G' induced by $V \setminus V'$ from G is not feasible, V' is called a minimal separating set.

For a minimal separating set S and $F(\subseteq V)$ corresponding to the set of failure nodes, $|S \setminus F|$ is called the remaining number in S against F. The minimum remaining number of all minimal separating sets against F is called the minimum remaining number and it is referred as $a(F)$. An order of all the vertices in F, $\{f_1, f_2, \ldots, f_k\}$, is called a recovery order, when the vertices are recovered in this order; the sequence $\{a(F), a(F \setminus \{f_1\}), a(F \setminus \{f_1, f_2\}), \ldots, a(\emptyset)\}$ is called the remaining number sequence of the recovery order $\{f_1, f_2, \ldots, f_k\}$. Note that the difference of the adjacent values in a remaining number sequence is at most one.

We define the order relation by the lexicographic order relation in all the recovery orders of F. Let the first recovery order be $\{r_1^*, r_2^*, \ldots, r_m^*\}$ in the order relation. For a recovery order $seq = \{r_1, r_2, \ldots, r_m\}$, $\sum_{i=1}^{m}(r_i - r_i^*)$ is called the durability of seq. We define the maximal recovery order as the optimal recovery order. We explain the basic idea of the optimality of the recovery order as follows. When all the vertices in a minimal separating set are removed, the network becomes infeasible. This implies that, if the number of the vertices included in a minimal separating set is small, the probability that the network after the next removal of vertices becomes infeasible is large. Therefore, the removed vertices included in a minimal separating set whose remaining number is small should be recovered preferentially. When removed vertices are recovered, the remaining number sequence becomes large in the lexicographic order. The larger the durability is, the less likely network is to be infeasible. Therefore, a recovery order that the remaining number increases rapidly, in other words, a recovery order of remaining number sequence with large durability, is desirable. Thus, it is reasonable that the maximum durability is the optimum.

We define the problem of determining the robust "good" recovery order as an optimization problem for obtaining the recovery order with the maximum durability as follows.

Problem for Determining Vertex Recover Order(VROP)

INPUT: *Undirected connected graph* $G = (V, E)$, *demand set* $Q = \{q_1, q_2, \cdots, q_r\}$, $F(\subseteq V)$, *a posotive integer* b.
OBJECTIVE: *Durability of recovery order (maximize)*

Theorem 1. *VROP is generally NP-hard.*

Proof. The decision problem, which asks whether there is the recovery order whose durability is more than or equal to d, is also called VROP.

We reduce the vertex cover problem (VC) known as NP-hard to VROP in polynomial time.

Vertex cover problem is the problem asking whether there is a subset $V' \subseteq V(|V'| \le z)$ that includes at least one of the end vertices of each edge of G for all edges.

We construct the instance of VROP, $(G^s, Q, F, b, d = n - z - 1)$ from vertex cover problem(G, z) as follows (Fig. 2). We make an edge $(p_{v'}, v')$ and $(v', q_{v'})$ for each vertex v in G, vertices s_{xy}, t_{xy} for all edges (x, y) in G, and connect s_{xy} to vertex $p_{x'}$ and vertex $p_{y'}$, and connect t_{xy} to vertex $q_{x'}$ and vertex $q_{y'}$ (Fig. 2). Let $Q = \{(s_{xy}, t_{xy})|(x, y) \in E\}$, $F = \{v' \in V(G^s)|v \in V\}$, and durability $d = n - z - 1$. Let $b = 4$. This transformation can be executed in polynomial time.

First, we show that a solution of VROP, $(G^s, Q, F, b = 4, d = n - z - 1)$ can be constructed from a solution of VC, (G, z). Let the minimal separating set be $\{\{v', w'\}|v', w' \in V(G^s), (v, w) \in E\}$. We make the recovery order by lining up the vertices v' in the graph G^s corresponding to vertices v in $\tilde{V}(|\tilde{V}| = \tilde{z}(\leq z))$ which is the solution of (G, z) and then succesively lining up the other vertices arbitrarily. The remaining number sequence of the recovery order is $\{0, 0, \ldots, 0, 1, 1, \ldots, 1, 2\}$ where the number of zero is $\tilde{z}$, the number of one is $n - \tilde{z}$, and the last value is two. Since the first remaining number sequence is $\{0, 0, \ldots, 0, 1, 2\}$ whose recovery order is made by lining up all the other vertices except a minimal separating set and then lining up the the vertices of the minimal separating set, the durability is $(n - \tilde{z}) - 1 = n - \tilde{z} - 1 \geq n - z - 1 = d$; therefore, the recovery order is the solution of VROP.

Conversely, we show that a solution of (G, z) can be constructed from a solution of VROP, $(G^s, Q, F, b = 4, d = n - z - 1)$. The remaining number sequence whose durability is more than or equal to $n - z - 1$ is $\{0, 0, \ldots, 0, 1, 1, \ldots, 1, 2\}$ where the number of zero is z or less is placed first, the value of one is followed, and the last value is two. This is because, the values is lined up in the ascending order and the difference of the adjacent values in a remaining number sequence is at most one, and the last value is two. The vertex set of G corresponding to the vertex set of G^s including the first one in the remaining number sequence corresponds to a vertex cover with the size of z or less. This is because, when first one appears in the minimum remaining sequence, each minimal separating set includes a recovered vertex at least one. Furthermore, a vertex in each minimal separating set corresponds to a vertex in G. Thus, this vertex cover is the solution of (G, z).

Consequently, since VC is reduced to VROP in polynomial time, VROP is NP-hard.   □

Tsuyoshi Yamasaki and Hiroyoshi Miwa

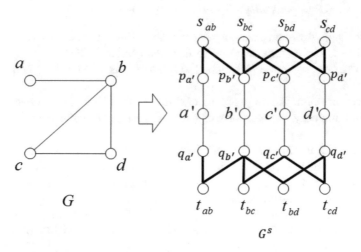

Fig. 2. G^s constructed from G.

4   Algorithm Determining Vertex Recovery Order

Since VROP is generally NP-hard, we cannot expect a polynomial-time algorithm for VROP. Therefore, we design the heuristic algorithm. We assume that the demand set consist of all the vertex pairs and $b = \infty$ in the rest of the paper. In this case, a graph is conneced, if and only if the graph is feasible.

We describe the basic idea of the proposed algorithm in Algorithms 1 and 2. This algorithm aims to keep the high ratio of the size of a maximum connected component to the number of all vertices. The ratio is referred to as MCCR. First, we enumerate all minimal separating sets that MCCR of the resulting graph after the removal of the minimal separating set is less than or equal to a threshold and that include at least a vertex of F. If there is a minimal separating sets whose remaining number is zero, a vertex in such a set is chosen and recovered. As long as there are the minimal separating sets whose remaining number is zero, the network is not connected and MCCR is small; therefore, the vertices included in such sets must be recovered as soon as possible.

When there are two or more minimal separating sets whose remaining number is zero, a vertex included in the minimal separating set whose size is one is preferentially chosen and recovered. This is because the recovery of such vertices makes the network connected rapidly. After these vertices are recovered, a vertex which is included in a minimal separating set whose remaining number is zero and which is included in as many minimal separating sets as possible, is preferentially chosen and recovered. This is because the recovery of such vertices increases the remaining number of as many minimal separating sets as possible.

The computational complexity of the algorithm MDS is an exponential time in general. Since the number of the minimal separating sets whose size is large is small in an actual communication network, it is sufficient to enumerate only minimal separating sets whose size is not more than a threshold.

Algorithm 1. Algorithm VROP-general

Input: Graph $G = (V, E)$, $(|V| = n, |E| = m)$, $F(\subseteq V)$
Output: Recovery order list D
1 $D \leftarrow \emptyset$
2 Enumerate all the minimal separating set in G by the algorithm MDS.
3 Calculate the remaining number in each minimal separating set against F.
4 Calculate the number of the minimal separating sets including vertex v for all vertices v.
5 **while** *there is a vertex of F whose recovery order is not determined* **do**
6   **if** *there are the minimal separating sets $\{C_1, C_2, \ldots, C_{k_1}\}$ whose remaining number is zero* **then**
7   **if** $k_1 = 1$ **then**
8   ⌊ Choose a vertex v from C_1 and add v to the tail of D.
9   **else**
10   ⌊ Choose a vertex v that the number of the minimal separating sets including v
                is the maximum, and add v to the tail of D.
11   **else**
12   ⌊ Choose a vertex v that the number of the minimal separating sets including v is the
            maximum, and add v to the tail of D.
13 **return** D

Algorithm 2. Algorithm determining minimal separating sets, MDS

Input: Graph $G = (V, E)$, a real number R.
Output: The family of minimal separating sets, C.

1 $C \leftarrow \emptyset$
2 **for** $1 \leq k \leq n$ **do**
3 $\quad\lfloor$ Determine C_k, the set of the vertex subsets whose number of vertices is k.

4 **for** $1 \leq k \leq n$ **do**
5 $\quad$ **while** *there is an unscanned vertex subset W in C_k* **do**
6 $\quad\quad$ **if** *the size of the maximum connected component of the induced subgraph G' induced by $V \setminus W$ is less than R* **then**
7 $\quad\quad\quad$ **if** *C has no vertex subset included in W* **then**
8 $\quad\quad\quad\quad\lfloor$ $C \leftarrow W$

9 **return** C.

---

5   Performance Evaluation

In this section, we evaluate the performance of the proposed algorithm by using the topology of the actual networks of CAID (Cooperative Association for Internet Data Analysis) [9]. We compare the performance of the proposed algorithm (Prop) with the algorithm (Simp) of determining the recovery order in the order of occurrence of failures.

We consider the situation of the successive aftershocks in the following numerical experiments. Let an event be the removal of nodes; let a round be the time period from an event to the next event. The events continues successively. Some nodes fail at an event and are recovered according to a recovery order during a round. The recovery order is determined at the time of an event by an algorithm. If all nodes are not recovered during a round, the failed nodes are carried forward to the next round. The time length of a round is not constant. In the following experiments, we determine the time length of a round by the number of the recovered nodes in a round, because we can consider that the time length for the recovery is almost constant in comparison with the time length of a round.

We show the network topology used for evaluation in Figs. 3 and 4.

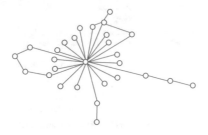

Fig. 3. Network(1)

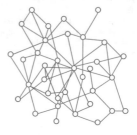

Fig. 4. Network(2)

We show the results in Tables 1, 2, 3, and 4. In these tables, n is the number of nodes and m is the number of links; R is the threshold of MCCR which is used

Table 1. Result of Network(1) for large interval between aftershocks

n = 28, m = 30

R = 0.75

The number of rounds	The number of connected nodes	
	Prop	Simp
1	26.2	26.2
2	24.4	21.8
3	24.8	24.6
4	24.3	23.7
5	24.7	24.4
6	24.9	24.5
7	23.9	23.6
8	25.0	21.6
9	24.0	23.8
10	24.0	23.7

Table 2. Result of Network(1) for small interval between aftershocks

n = 28, m = 30

R = 0.75

The number of rounds	The number of connected nodes	
	Prop	Simp
1	15.1	13.1
2	13.3	8.9
3	13.4	7.1
4	12.9	7.1
5	12.7	6.9
6	12.5	5.9
7	11.9	5.2
8	11.6	6.6
9	10.8	6.7
10	11.1	5.7

Table 3. Result of Network(2) for large interval between aftershocks

n = 35, m = 72

R = 0.75

The number of rounds	The number of connected nodes	
	Prop	Simp
1	29.9	30.1
2	30.5	30.1
3	29.9	30.0
4	31.5	31.4
5	29.2	28.7
6	29.9	29.6
7	30.5	30.2
8	30.8	30.2
9	29.4	29.1
10	30.3	30.0

Table 4. Result of Network(2) for small interval between aftershocks

n = 35, m = 72

R = 0.75

The number of rounds	The number of connected nodes	
	Prop	Simp
1	22.8	22.3
2	18.3	16.6
3	15.0	13.4
4	13.0	9.8
5	13.6	11.1
6	13.7	10.1
7	16.1	13.0
8	17.6	15.5
9	15.2	12.3
10	13.4	9.6

in the algorithm MDS; the number of the rounds is ten. We assume that the failed nodes are randomly chosen; the number of the recovery nodes in a round is randomly chosen in the range from $0.1|F|$ to $0.5|F|$ (this means that the interval between aftershocks is small), and the range from $0.6|F|$ to $1.0|F|$ (this means that the interval between aftershocks is large), where $|F|$ is the number of failed nodes. In addition, we assume that the recovered nodes can fail again.

In Tables 1, 2, 3 and 4, the number of the connected nodes by the algorithm Prop is larger than that by the algorithm Simp. When the interval between aftershocks is large, the difference of the performance between the algorithms is small, because both algorithms can recover almost all failed nodes in a round. Especially, when the network has the large connectivity such as the network in Fig. 4, the performance between the algorithms is almost same, because the probability that many nodes are disconnected is small and many nodes are always connected even after a failure.

6   Conclusions

In this paper, we proposed a method to determine a recovery order such that as many nodes as possible can keep connectivity even in the successive node failures. First, we formulated the optimization problem of determining an appropriate recovery order of failed nodes against succesive node failures, and proved the NP-hardness of this problem. Furthermore, we designed a heuristic algorithm, and we evaluated the performance of the algorithm by applying the proposed algorithm to the topology of the actual networks. The results show that the proposed algorithm can determine a recovery order to keep connectivity even in the successive node failures.

Acknowledgements. This work was partially supported by the Japan Society for the Promotion of Science through Grants-in-Aid for Scientific Research (B) (17H01742) and JST CREST JPMJCR1402.

References

1. Yamasaki, T., Anan, M., Miwa, H.: Network design method based on link protection taking account of the connectivity and distance between sites. In: Proceedings of International Conference on Intelligent Networking and Collaborative System (INCoS2016), pp. 445–450, Ostrava, Czech Republic, 7–9 September 2016
2. Imagawa, K., Fujimura, T., Miwa, H.: Approximation algorithms for finding protected links to keep small diameter against link failures. In: Proceedings of International Conference on Intelligent Networking and Collaborative System (INCoS2011), pp. 575–580, Fukuoka, Japan, 30 November 2011–2 December 2011
3. Imagawa, K., Miwa, H.: Detecting protected links to keep reachability to server against failures. In: Proceedings of The International Conference on Information Networking (ICOIN2013), pp. 30–35, Bangkok, Thailand, 28–30 January 2013

4. Noguchi, A., Fujimura, T., Miwa, H.: Network design method by link protection for network load alleviation against failures. In: Proceedings of International Conference on Intelligent Networking and Collaborative System (INCoS2011), pp. 581–586, Fukuoka, Japan, 30 November–2 December 2011
5. Furuta, H., Ishibashi, K., Nakatsu, K., Hotta, S.: Optimal restoration scheduling of damaged networks under uncertain environment by using improved genetic algorithm. Tsinghua Science and Technology, vol. 13, pp. 400–405 (2008)
6. Dielk, T.A., Ozdamar, O.: A mathematical model for post-disaster road restoration: Enabling accessibility and evacuation, vol. 66, pp. 56–67 (2014)
7. Lance, F.: An algorithm to prioritize road network restoration after a regional event, Technologies for Homeland Security (HST), pp. 19–25, 12–14 November 2013
8. Yamasaki, T., Miwa,H.: Method for determining recovery order against intermittent link failure. In: Proceedings of International Conference on Intelligent Networking and Collaborative System (INCoS2017), pp. 403–412, Toronto, Canada, 24–26 August 2017
9. CAIDA. http://www.caida.org/

The 8th International Workshop on Adaptive Learning via Interactive, Collaborative and Emotional approaches (ALICE-2018)

Using an Intelligent Tutoring System with Plagiarism Detection to Enhance e-Assessment

David Bañeres[(✉)], Ingrid Noguera, M. Elena Rodríguez,
and Ana Guerrero-Roldán

Open University of Catalonia, Barcelona, Spain
{dbaneres, inoguerafr, mrodriguezgo, aguerreror}@uoc.edu

Abstract. Nowadays, online learning has become a promising solution to personalize and increase flexibility in the learning-teaching process. However, e-assessment is still questioned in terms of authorship and identity checking. Some virtual learning environments are introducing technological solutions, such as plagiarism detection tools, to increase the security when submitting assessment activities. However, this is a partial solution. When the activities are performed on third-party tools, as it is the case of intelligent tutoring systems, the identity and authorship checking can fail. This paper introduces a modular plagiarism detection tool that combines different input data sources in order to verify the authorship. A case study is presented to show the potential of the tool.

Keywords: Plagiarism · Image comparison · Authorship · e·Assessment

1 Introduction

Assessment of students in online education is one of the most relevant ongoing challenges [1–3]. Educational institutions are working hard to support fully online learning and avoiding relying on traditional forms of assessment such as on-site exams or face-to-face meetings. Society at large and accrediting quality agencies are being reluctant to give the social recognition or credibility that online education deserves [4]. However, lifelong learners are creating a paradigm shift. New approaches are needed to fulfill the requirements of these students (e.g., they cannot continuously attend on-campus classrooms) [5–7].

The TeSLA project [8] proposes a promising solution to this challenge. The primary objective of the project is to develop an e-assessment system, which provides an unambiguous proof of students' academic progression during the learning process while avoiding the time and physical limitations imposed by face-to-face education. The TeSLA project aims to support any purpose of assessment (diagnostic, formative or summative) covering the teaching-learning process as well as ethical, legal and technological aspects. To do so, it is being developed an e-assessment system within the project where multiple instruments and pedagogical resources are available. The instruments are deployed in the e-assessment activities to capture students' data to ensure their authentication and authorship. Such instruments are transparently

F. Xhafa et al. (Eds.): INCoS 2018, LNDECT 23, pp. 363–372, 2019.
https://doi.org/10.1007/978-3-319-98557-2_33

integrated into the assessment activities and according to pedagogical criteria to avoid interfering in the learning process of the students.

The TeSLA system is a cloud-based platform that provides the services related to identity and authorship checking and information related to potential illegitimate behaviors. Additionally, all the information related to students is anonymously stored by a hash ID that only the institution where the student is enrolled can access. Currently, TeSLA is only supporting two Virtual Learning Environments (VLE), Moodle[1] and Blackboard[2], limiting the learning spaces where can be deployed.

Examples of learning spaces where currently cannot be used TeSLA are Intelligent Tutoring Systems (ITS) or custom VLE. This limitation reduces the interoperability among learning systems and such plagiarism checking platform cannot be reused within the ITS. Moreover, the ITS is constrained to be used as a tool to practice and limits its use for graded assessment activities. Note that, the possibility to enhance the security when performing the assessment activities by checking authorship and identity might increase the trustworthiness of the system. The aim of this paper is to analyze the potentiality of a plagiarism checker system to verify the authorship for enforcing the security of the e-assessment activities. Thus, a particular modular plagiarism checker system has been designed for the ITS VerilUOC [9]. Although, it is a particular solution, it can be easily extended to any learning tool.

The paper is organized as follows. Section 2 introduces the plagiarism checker system. Sections 3 and 4 describe the internal modules. Finally, Sect. 5 summarizes the experimental results and Sect. 6 outlines the conclusions and future work.

2   Plagiarism Checker System

In this section, the plagiarism checker system is described. The objective of the system is twofold. First, the system should automate the process of plagiarism checking. The students generate different outputs during their learning process such as text documents, images, videos, code, circuit schemas, among others. The system should be capable of applying the plagiarism checking process to these types of outputs. Second, these outputs will produce different sources of plagiarism information. The system should be capable of combining these sources of information to aid instructors to detect illegitimate behaviors. Note that, we will use indistinctively the terms plagiarism, fraud and illegitimate behaviors in this paper.

Figure 1 illustrates the design of the system to achieve both objectives focused on the ITS VerilUOC. The inputs of the system are: the circuits designed within the tool, timestamping information generated when students submit the circuits for validation, and textual submissions of the assessment activities in the VLE. The combination of these three sources makes possible the VerilUOC platform to deal with plagiarism.

The system is composed of two modules: The *Checker Module* and the *Merger Module*. The first one is responsible for running the plagiarism detection algorithms.

---

[1] https://moodle.org/.

[2] http://www.blackboard.com/index.html.

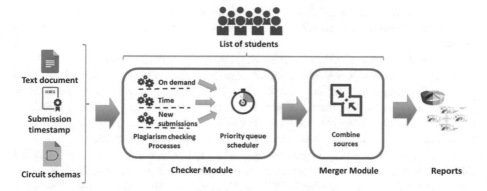

Fig. 1. Plagiarism checker system

The latter combines the information produced by the plagiarism tools aggregated by student to show the potential cheaters and offer a range of visualizations to analyze the output of the system for the teachers. The modules are further detailed in the next subsections.

3 The Checker Module

Aforementioned, the Checker module is responsible for running the antiplagiarism tools. In the case of VerilUOC, we identified three types of potential plagiarism:

- Circuit schemas: Students may cheat when sharing the designed circuits. A plagiarism detection tool based on components placement and routing of wires can help to identify parts of the circuit copied (i.e., identical components in the same place and wires with identical length and position).
- Images: Students perform cheating outside VerilUOC. The textual document performed for the assessment activity is submitted in the VLE. This document contains multiple images related to circuits, finite states machines and time diagrams. These images might be shared among students. Note that, we discarded text plagiarism detection. The submitted documents contain tables, small pieces of text and mostly present technical descriptions and formulas. Thus, they do not provide meaningful information for text plagiarism detection.
- Timestamping information: All circuits submitted to VerilUOC for validation are recorded with the submission time and all students who have ever manipulated the file containing the circuit (i.e., this information is stored within the file). Using this timestamping information, potential cheaters can be checked within the administrative section of VerilUOC platform as we will observe in Sect. 5.

Note that, this list could be extended in the future by adding new antiplagiarism tools such as [10, 11] or by adding external features such as the TeSLA reports.

We defined three possible trigger actions to run the tools: on demand, by time, and by new submission. The first one is useful when the current snapshot of the results is needed, the second one schedules the comparison at fixed intervals, and the third one is the most responsive by running the comparison when new submission files are detected.

Although there are currently three possible antiplagiarism tools, there could be many comparisons at once. Different courses may exist within VerilUOC, some of them with a large number of students (i.e., MOOC [12]) and different assessment activities may overlap in time. Aiming to avoid overloading the system, a scheduler based on priority queues has been implemented. The queues are prioritized based on the defined trigger actions: on demand, by time and by new submissions.

Note that, the circuit plagiarism tool is currently under development and the timestamping generator can be reviewed on [13]. Next subsection describes the image plagiarism tool designed for VerilUOC to analyze images within the textual documents.

3.1   Image Comparison Tool

As an example of plagiarism detection tool, we describe the image plagiarism checker that has been developed within the system. For performing the analysis, the input data are the submission documents in text format performed for an assessment activity. Additionally, the list of students is also provided to assign plagiarism evidence to the owner of the submissions. We next describe each phase.

First, the textual documents are processed. The standard text document formats (i.e., Word Office, LibreOffice/OpenOffice and PDF) are supported and all images are extracted. These images are linked to her owner (the student) to easily identify the cheater in case of detection.

Next, two filters are applied: (1) Small images in size are removed. After experimentation, we detected that images smaller than 2 KB can be removed since they tend to be thumbnail images on headers or meaningless images within the document. (2) Images from the document or template with the statement for the assessment activity. These images could produce false positives since many students reuse these documents to develop their answers. Thus, these images are excluded from any comparison.

After the cleansing process, the image comparison process starts. Images from different students are compared in pairs and the result is stored in a database. There are many well-known methods to compare images. Here, the perceptual hashing [14] algorithm[3] has been used due to the trade-off between speed and quality of the results. This technique follows the next steps:

1. The image is shrunk to 8×8 pixels and it is decolored to greyscale.
2. A 64-bit sequence is constructed representing the pixels in x-axis direction order. The value of a bit is one whether the intensity of the color is increasing from the current pixel with respect to the next one in the order.
3. Another 64-bit sequence is constructed representing the pixels in y-axis direction order. The bit values are computed the same way as the previous.

---

[3] https://github.com/Jetsetter/dhash.

4. Finally, both sequences are concatenated to build a 128-bit sequence representing the image.
5. Two images are compared by computing the Hamming distance between hashes. Values near 0 means that images are similar. Note that, the similarity percentage can be computed based on the total length of the hash (i.e., Hamming distance/Number of bits of the sequence).

Note that, this process is exhaustive and time-consuming due to the cross-comparison among all images. It is essential to perform a smart comparison and three implementation decisions accomplish this: (1) An image of one student is only compared with images of other students. There is no need to compare images from the same student; (2) The comparison is commutative. Thus, if an image of one student is compared with another one, there is no need to perform the reverse comparison; (3) In case of adding new students or submissions, only new images are compared with the rest of the already processed images. These decisions are trivial and intuitive. However, their application reduces the computation time to halve.

Finally, the checking threshold process remains. The threshold is a configurable value in the interval [0,100] that the instructor defines based on the similarity percentage. This value aids instructors to filter potential plagiarism evidence and showing only students that have submitted documents with similar images. Currently, the threshold is set to 10%. Thus, the tool shows all similar images on the interval [0,10].

4   The Merger Module

Aforesaid, this module stands for combining the data from different sources produced by the Checker module. Currently only timestamping information and the image plagiarism tool are available. The three available reports are described next:

- *Plagiarism detection graph based on timestamping*: The graph presented in [15] has been improved by gathering information over different semesters of the same course. VerilUOC is operating for eight years and many students have enrolled on the associated courses (more than 3000 students). Thus, many students have a set of solved activities that could be easily shared with the new students. The previous version of this graph obviated users from earlier editions of the course. Thus, plagiarism among different semesters was not detected. Figure 2 shows a real example of a circuit design activity. Student's names have been anonymized and some students have been filtered (user6 to user9) to simplify the illustration. We can observe that top users have the suffix *prev*. These users were students enrolled in previous semesters. The suffix means that the semester was not possible to be recovered since the activity was submitted on the VerilUOC version described in [15] where the semester was not stored in the timestamping information. On new editions of the course, this suffix *prev* is replaced with the semester where the activity is submitted in the ITS. As we can observe, all submissions related with the original submission are linked and the students who submitted the same (or partially modified) activity are timestamped on the activity submission jointly with the semester and the submission date.

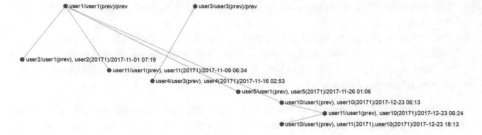

Fig. 2. Plagiarism detection graph based on timestamping

- *Plagiarism detection matrix based on image comparison*: To check the similarity percentage between each image, the system provides a matrix visualization. Figure 3 illustrates a partial view of the matrix where different comparisons can be seen with the respective similarity percentage and a thumbnail of the images. Note that, the name of the students have also been anonymized. The matrix allows to filter for a threshold value and only images below this value are shown. In this case, the threshold has been set to 100% to demonstrate multiple instances of comparison. The example shows potential plagiarism. A similarity percentage of 4.68% has been detected and, as we can observe, the images are identical.
- *Combined plagiarism report*: Finally, the system offers the combination of both sources. The results are shown for each student of the course who potentially commits fraud. Figure 4 shows an example for one of the assessment activities. We can observe two cases. In the first one, the student *user2* has used VerilUOC and the timestamping information shows that the activity *20171_PR_1C* has been shared with the *user7*. In this case, the image comparison tool also finds evidence on the textual document with a similarity percentage of 4.68%. The second case is related to a student *user3* that has not used VerilUOC to perform the assessment activity since no submission information is reported. However, the image detector tool is capable of finding within the submitted text document a circuit performed with VerilUOC which is identical to the circuit within the text document submitted by *user6*.

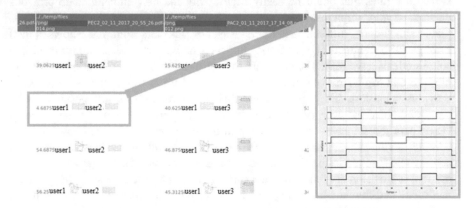

Fig. 3. Plagiarism detection matrix based on image comparison

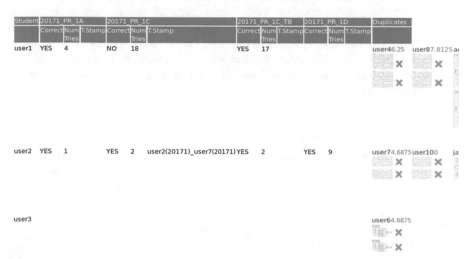

Fig. 4. Combined plagiarism report

5 Experimental Results

The objective of this section is to show validated plagiarism cases on a course where the ITS is currently used based on information reported by the plagiarism detection system.

The experiment has been performed within the course of Computer Fundamentals on the Degree of Computer Science at the Universitat Oberta de Catalunya. This course is a first-year course where students learn the principles for designing digital circuits and the basis of Computer Architecture. The course has a significant number of enrollments every semester (from 300 to 500) and it is the first contact with higher online education for many students. These two factors (first-year and newbies in higher online education) affects the dropout of the course significantly, which it is near to the 50% of the enrolled students in some semesters. The experiment was performed during the 2017 Fall semester where 567 students were enrolled.

The course has three assessment activities (CAA) and one final project (FP). The ITS VerilUOC is used from the second assessment activity until the final project. Also, VerilUOC offers different non-assessment activities to practice and this set of 150 activities are optional to be performed within the course. Although the non-assessment activities do not impact directly in the assessment activities, they can increase the mark of the final project up to 1 point. With the aim to perform the experiment, the ITS has been used as it is. Students have used the tool and the system has internally gathered and monitored all the submissions and the timestamping information as previous semesters to check possible plagiarism. This semester the image plagiarism detection tool has been enabled with the manual trigger action to check the results of the tool. Finally, results of both tools have been analyzed to confirm plagiarism cases.

Table 1 summarizes the plagiarism cases found for the Computer Fundamentals course by both plagiarism detection tools. The number of students who submitted the activity and the different plagiarism cases are reported by each assessment activity and divided by the tool which detected the fraud. Cases are also counted by the total unique students who performed plagiarism in the activity. Note that, some students have been detected by both tools. Absolute and percentage values are computed based on the submitted activities.

Table 1. Confirmed cases of plagiarism within the Computer Fundamentals course

	No. students who submitted	Time stamping		Image comparison		Total (unique)	
		No.	%	No.	%	No.	%
CAA2	339	5	1.47	19	5.60	23	6.78
CAA3	277	8	2.89	25	9.03	33	11.91
FP	258	9	3.49	13	5.04	16	6.20

It is worth noting that students have been informed on the teaching plan about what is plagiarism and which academic consequences are applied in the case that any fraud is detected in the assessment activities. Additionally, they are also informed in the dashboard of the classroom about the plagiarism detection tools deployed in the course (i.e., within the ITS and plagiarism detection on submitted textual documents). Thus, the students are aware that they are monitored concerning plagiarism detection.

Although students knew that they are surveilled, interesting findings were found. The timestamping tool found several cases of plagiarism on the assessment activities (1.47% on CAA1, 2.89% on CAA2 and 3.49% on FP). During the past semesters, instructors detected a decreasing trend in plagiarism cases on assessment activities within the ITS. The timestamping technique was initially deployed in 2014 and instructors assumed that students stopped to share solutions related to the assessment activities due to the corrective measures applied to classmates in previous semesters. However, the degree of plagiarism in non-assessment activities has not actually decreased. Students continue submitting non-assessment activities performed by other students. Also, instructors have detected that cheaters are mostly using non-assessment activities from students of past editions of the course. However, sharing these activities does not have any corrective measure since the activities stand for practicing the contents of the course. The only measure is that the extra point in the final project is not added in case of plagiarism.

However, this is an evidence that some students are trying to cheat and this is confirmed by the image comparison tool. The number of plagiarism cases increased significantly to 5.60% for CAA2, 9.03% for CAA3, and 5.04% for FP. Students are mostly cheating within the submitted textual document and particularly the students who are not using the ITS for practicing. We observe a common pattern that those students who plagiarized any image in the submitted document obtained an image from other students who used the ITS and validated the correct solution of the assessment activities.

The Table 2 reports total unique students who performed any illegitimate behavior and students who plagiarized any non-assessment activity in the ITS based on the timestamping information. When unique cases are analyzed, we observe that the percentage of plagiarism is greater than the rate detected on previous semesters when it was only used the timestamping tool. Thus, we confirm that the combination of different information sources enhances the detection of plagiarism cases.

Table 2. Unique confirmed cases of plagiarism within the Computer Fundamentals course

	No. students	Time stamping		Image comparison		Total (unique)	
		No.	%	No.	%	No.	%
Total ass. activities (unique)	567	18	3.17	54	9.52	63	11.11
Total non-ass. activities (unique)	567	41	7.23	–	–	–	–

6 Conclusions and Future Work

This paper has described a plagiarism detection system for ITS. The system can combine information from different plagiarism tools to enhance e-assessment by detecting illegitimate behaviors. Note that, enforcing the security on third-party tools can extend their use on assessable activities. Experimental results show an increment of the detected cases when combining two sources: timestamping and image comparison.

A prototype of image comparison tool has been presented based on the perceptual hashing technique. This simple technique was capable to find multiple cases. However, authors are aware of the limitations of the tool. Cropped, distorted and rotated images are not currently detected. As future work, we plan to explore various image comparison algorithms to enhance the plagiarism tool. Also, we plan to add new plagiarism information sources such as the TeSLA information and the circuit plagiarism tool.

Additionally, the prototype has been built based on running on a single server by using the priority scheduler. Note that, this design can be a bottleneck on massive courses such as MOOCs. We will study how the architecture described in Fig. 1 can be adapted to a MapReduce model to leverage workloads on the system.

Finally, the tool only displays the information and instructors should manually analyze each case. We are interested in adding more visualizations and a recommender system to help instructors to detect plagiarism cases.

Acknowledgments. This work is supported by H2020-ICT-2015 T project "An Adaptive Trust-based e-assessment System for Learning", Number 688520.

References

1. Herr, N., Rivas, M., Chang, T., Tippens, M., Vandergon, V., d'Alessio, M., Nguyen-Graff, D.: Continuous formative assessment (CFA) during blended and online instruction using cloud-based collaborative documents. In: Koç, S., Wachira, P., Liu, X. (Eds.) Assessment in Online and Blended Learning Environments (2013)
2. Kearns, L.R.: Student assessment in online learning: challenges and effective practices. J. Online Learn. Teach. **8**(3), 198 (2012)
3. Callan, V.J., Johnston, M.A., Clayton, B., Poulsen, A.L.: E-assessment: challenges to the legitimacy of VET practitioners and auditors. J. Vocat. Educ. Training **68**(4), 416–435 (2016). https://doi.org/10.1080/13636820.2016.1231214
4. Kaczmarczyk, L.C.: Accreditation and student assessment in distance education: why we all need to pay attention. SIGCSE Bull. **33**(3), 113–116 (2001)
5. Walker, R., Handley, Z.: Designing for learner engagement with e-assessment practices: the LEe-AP framework. In: 22nd Annual Conference of the Association for Learning Technology, University of Manchester, UK (2015)
6. Ivanova, M., Rozeva, A., Durcheva, M.: Towards e-Assessment models in engineering education: problems and solutions. In: International Conference on Web-Based Learning, pp. 178–181. Springer, Heidelberg (2016)
7. Baneres, D., Rodríguez, M.E., Guerrero-Roldán, A.E., Baró, X.: Towards an adaptive e-Assessment system based on trustworthiness. In: Caballé, S., Clarisó, R. (Eds.) Formative Assessment, Learning Data Analytics and Gamification in ICT Education, pp. 25–47. Elsevier (2016)
8. The TeSLA project website. http://tesla-project.eu/. Accessed 15 May 2018
9. Baneres, D., Clariso, R., Jorba, J., Serra, M.: Experiences in digital circuit design courses: a self-study platform for learning support. IEEE Trans. Learn. Technol. (IEEE TTL) **7**(3), 1–15 (2014)
10. Liu, C., Chen, C., Han, J., Yu, P.S.: GPLAG: detection of software plagiarism by program dependence graph analysis. In: Proceedings of the 12th ACM SIGKDD International Conference on Knowledge Discovery and Data Mining, pp. 872–881. ACM (2006)
11. Potthast, M., Stein, B., Barrón-Cedeño, A., Rosso, P.: An evaluation framework for plagiarism detection. In: Proceedings of the 23rd International Conference on Computational Linguistics: Posters, pp. 997–1005. Association for Computational Linguistics (2010)
12. Baneres, D., Saíz, J.: Intelligent tutoring system for learning digital systems on MOOC environments. In: The 10th International Conference on Complex, Intelligent and Software Intensive Systems (CISIS 2016), pp. 95–102 (2016)
13. Baneres, D.: Principles for an effort-aware system. In: Advances on P2P, Parallel, Grid, Cloud and Internet Computing: Proceedings of the 11th International Conference on P2P, Parallel, Grid, Cloud and Internet Computing (3PGCIC–2017), pp. 576–585 (2017)
14. Niu, X.M., Jiao, Y.H.: An overview of perceptual hashing. Acta Electronica Sinica **36**(7), 1405–1411 (2008)
15. Baneres, D.: Towards an analytical framework to enhance teaching support in digital systems design course. In: The 9th International Conference on Complex, Intelligent and Software Intensive Systems (CISIS 2015). pp. 148–155 (2015)

Multi-criteria Fuzzy Ordinal Peer Assessment for MOOCs

Nicola Capuano[1](✉) and Santi Caballé[2]

[1] Department of Information Engineering, Electric Engineering and Applied Mathematics, University of Salerno, Via Giovanni Paolo II 132, 84084 Fisciano, SA, Italy
ncapuano@unisa.it
[2] Faculty of Computer Science, Multimedia and Telecommunications, Universitat Oberta de Catalunya, Rambla Poblenou, 156, 08018 Barcelona, Spain
scaballe@uoc.edu

Abstract. Due to the high number of students enrolled and the relatively small number of available tutors, the assessment of complex assignments is deemed as one of the most critical tasks in Massive Open On-line Courses (MOOCs). Peer assessment is becoming an increasingly popular tool to face this problem and many approaches have been proposed so far to make its outcomes more reliable. A promising approach is FOPA (Fuzzy Ordinal Peer Assessment) that adopts and integrates models coming from Fuzzy Set Theory and Group Decision Making. In this paper we propose a FOPA extension supporting multi-criteria assessment based on rubrics. Students are asked to rank a small number of peer submissions against specified criteria, provided rankings are then transformed in fuzzy preference relations, expanded to obtain missing values and aggregated to establish a global ranking between students' works with respect to each criterion and globally. The absolute grades of all submissions are then calculated.

1 Introduction and Related Work

Peer assessment is an educational practice requiring students to grade a small number of their peers' assignments as part of their own assignment. Peer assessment has been used for many years to improve learning outcomes. In fact, the literature reports on many learning benefits for peer-assessors, such as the development of self-learning abilities, the exposure to different approaches, the enhancement of critical thinking, etc. [1]. As a matter of fact, with the growing diffusion of MOOCs and the related difficulties in students' evaluation, peer assessment is being used more and more often as a tool for formative and summative student's evaluation.

The positive aspect of peer assessment, when applied to this context, is its capability of easily scale to any size: the number of assessors naturally grows with the number of students. Conversely, peer assessment may be seen as unprofessional and unreliable given that it is based on grades assigned by students lacking the needed expertise, both didactical and on the specific subject to be assessed [2]. To mitigate this issue, several corrected methods have been identified so far by different researchers, such as *Calibrated Peer Review* [3], *PeerRank* [4, 5], etc.

© Springer Nature Switzerland AG 2019
F. Xhafa et al. (Eds.): INCoS 2018, LNDECT 23, pp. 373–383, 2019.
https://doi.org/10.1007/978-3-319-98557-2_34

In [6] authors have demonstrated that, asking students to provide ordinal feedback (e.g. "the report x is better than the report y"), allows for obtaining better results with respect to asking them to provide cardinal feedback (e.g. "the grade of report x is a B"). Ordinal feedback is easier to provide, more reliable and overcomes the so called *bias problem* occurring when students evaluate their peers on different scales. Elaborating on these assumptions, in [7] a new ordinal peer assessment model named FOPA (Fuzzy Ordinal Peer Assessment), based on GDM (Group Decision Making) and *fuzzy sets*, has been defined. Experimental results with synthetic and real data have shown that FOPA, in most cases, outperform other ordinal and cardinal models both in the ranking of submission and in the estimation of students' grades [8, 9].

To make FOPA even more reliable, in this paper we introduce a model extension aimed at supporting *multi-criteria evaluation* based on assessment rubrics. A rubric [10] makes explicit a range of assessment criteria and expected performance standards. In this way, students may evaluate peers' performance against such criteria rather than assigning a single subjective score. While the use of rubrics in cardinal peer assessment is quite common, its adoption within an ordinal model is rather new. Moreover, while some experimental peer assessment models based of *fuzzy sets* have been proposed so far, to the best of our knowledge, very few of them adopt a multi-criteria approach (similar to [11]) while none of them supports ordinal feedback.

The paper is organized as follows: after having described the ordinal peer assessment problem in Sect. 2, the existing FOPA model is summarized in Sect. 3 while Sect. 4 describes the proposed improvements based on multi-criteria GDM techniques. Then, Sect. 5 summarizes conclusions and outlines on-going work.

2   Ordinal Peer Assessment

In peer assessment, an *assignment* is given to n students from a set $S = \{s_1, \ldots, s_n\}$. Once students have elaborated their solutions to the assignment, they generate a set of *submissions* that must be evaluated by peers. Given that each student is expected to evaluate m peer submissions (with $m \leq n$), several algorithms with different levels of sophistication have been defined so far for the assignment of submissions to assessor students [12]. Such algorithms define an *assessment grid*: a Boolean $n \times n$ matrix $A = (a_{ik})$ where $a_{ik} = 1$ if the student s_k is asked to grade the submission of s_i and $a_{ik} = 0$ otherwise. The following properties must be verified:

- the sum of the elements in each row and column of A is equal to m (i.e. each student grades and is graded by m other students);
- the sum of the elements in the main diagonal is equal to 0 (i.e. nobody evaluates himself).

The assessment grid can be seen as the adjacency matrix of an m-regular directed graph, where each node represents a student and each arc represents an assessment to be performed (see Fig. 1). Once the grid is defined, in *ordinal peer assessment* [6, 13], each student $s_k \in S$ is asked to define an *ordinal ranking* $\succ_k$ over the subset of her assessees $S_k = \{s_i \in S | a_{ik} = 1\}$. Being s_i^k the generic element of S_k, such that $i \in \{1, \ldots, m\}$, an ordinal ranking takes the following form:

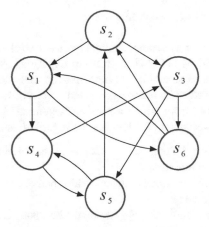

Fig. 1. Graph interpretation of a sample assessment grid with $n = 6, m = 2$: nodes represent students while edges represent assigned evaluations and connect assessors with assesses.

$$s^k_{p(1)} \succ_j s^k_{p(2)} \succ_j \ldots \succ_j s^k_{p(m)} \tag{1}$$

where $p : \{1, \ldots, m\} \to \{1, \ldots, m\}$ is a permutation function. Equation (1) means that, according to s_k, the submission of the student $s^k_{p(1)}$ is better than that of $s^k_{p(2)}$, etc.

The partial rankings defined by all students can be collected in a $n \times n$ *ranking matrix* $R = (r_{ik})$ whose generic element r_{ik} is the position of s_i in the ranking $\succ_k$ if $s_i \in S_k$, 0 otherwise. Starting from a ranking matrix, an *aggregation rule* is needed to compute a complete ranking over the whole set of submissions.

Several aggregation rules have been proposed so far by different researchers. A simple and effective rule is the classical *Borda count* [13, 14] where the partial ranking provided by each assessor is interpreted as follows: m points are given to the submission ranked first, $m - 1$ points to the one ranked second, etc. Based on the assessment grid A and the ranking matrix R, the Borda score of any $s_i \in S$ can be calculated as:

$$\phi_B(s_i) = \sum_{k=1}^{n} a_{ik} \cdot (m - r_{ik} + 1). \tag{2}$$

The global ranking is then computed by ordering all the submissions in decreasing order of their Borda scores.

3 Fuzzy Ordinal Peer Assessment

In [7, 8], an alternative peer assessment model named *FOPA* has been defined in light
of a GDM process. In GDM [15–17], a group of experts is asked to evaluate a set of
alternatives with the aim of selecting the best one. To this end, each expert expresses
preferences on alternatives, which are aggregated to arrive at a collective preference
degree on each alternative from which a ranking is generated.

Therefore, a peer assessment problem can be regarded as a GDM problem where:

- experts and alternatives belong to the same set (i.e. students evaluate the submissions made by other students);
- each expert only ranks a small subset of alternatives (i.e. few submissions are evaluated by each student);
- experts' opinion is not fully reliable (it should be taken into account that students are far to be perfect assessors).

These properties (in particular the last two) suggest referring to GDM approaches
able to deal with the uncertainty resulting from inaccuracy and lack of knowledge in
experts' evaluations, like those based on the *fuzzy set theory* [18]. In this section, after
having introduced basic GDM concepts, the standard FOPA model is introduced and
relevant literature is referenced. In the next section, an improved version of FOPA,
supporting multiple assessment criteria and rubrics is proposed and discussed.

3.1 Group Decision Making Concepts

A GDM problem is characterized by a group of experts $E = \{e_1, \ldots, e_m\}$ that express
their preferences on a finite set of alternatives $X = \{x_1, \ldots, x_n\}$. Experts' preferences
are commonly modelled with *Fuzzy Preference Relations* (FPRs) that specify the extent
at which each alternative $x_i \in X$ is at least as good as the others [19]. FPR is a fuzzy set
on $X \times X$ with a membership function $\mu_P : X \times X \to [0, 1]$ such that:

$$\mu_P(x_i, x_j) = \begin{cases} 1 & \text{if } x_i \text{ is definitely preferred to } x_j, \\ a \in (0.5, 1) & \text{if } x_i \text{ is slightly preferred to } x_j, \\ 0.5 & \text{if } x_i \text{ and } x_j \text{ are equally preferred}, \\ b \in (0, 0.5) & \text{if } x_j \text{ is slightly preferred to } x_i, \\ 0 & \text{if } x_j \text{ is definitely preferred to } x_i. \end{cases} \quad (3)$$

A FPR can be conveniently represented as a $n \times n$ matrix $P = (p_{ij})$ such that
$p_{ij} = \mu_P(x_i, x_j)$. Once experts have expressed preferences over the set of alternatives,
m individual FPRs $P_k = (p_{ij}^k)$ for $k \in \{1, \ldots, m\}$ are available. Individual FPRs are
then merged into a single collective one using an aggregation operator. Among the
existing ones, the *Ordered Weighted Average* (OWA) [20] is one of the most widely
used for this purpose. It is defined as:

$$OWA(p_{ij}^1, \ldots, p_{ij}^m) = \sum_{k=1}^{m} w_k p_{ij}^{\sigma(k)} \tag{4}$$

where $w_1, \ldots, w_m \in [0, 1]$ are weights such that $\sum_{k=1}^{m} w_k = 1$ while $\sigma : \{1, \ldots m\} \rightarrow \{1, \ldots m\}$ is a permutation function such that $p_{ij}^{\sigma(k)} \geq p_{ij}^{\sigma(k+1)}$ for $k \in \{1, \ldots, m-1\}$.

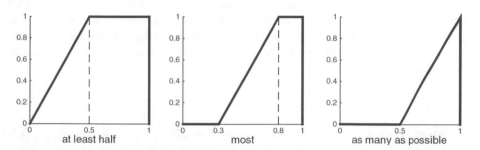

Fig. 2. Membership functions of sample increasing proportional linguistic quantifiers

The behaviour of OWA strictly depends on selected weights. In [21], the authors propose to initialize weights starting from a *non-decreasing proportional linguistic quantifier* carefully chosen to reflect the fusion strategy that the decision makers would apply i.e. the ratio of experts that are expected to be satisfied with the aggregated value. According to [7] feasible quantifiers include *at least half*, *most* and *as many as possible* whose membership functions are shown in Fig. 2. Once the appropriate quantifier Q is selected, OWA weights can be obtained from as follows:

$$w_k = \mu_Q\left(\frac{k}{m}\right) - \mu_Q\left(\frac{k-1}{m}\right); k \in \{1, \ldots, m\}. \tag{5}$$

Once the collective FPR P is obtained, the alternatives are rated by associating a degree of preference $\phi(x_i)$ to each alternative $x_i \in X$ based on P. Also, in this case, several measures are possible. Among others, in [19] the degree of preference of each alternative is calculated in terms of *Net Flow* as follows:

$$\phi_{NF}(x_i) = \sum_{j=1,j\neq i}^{n} p_{ij} - \sum_{j=1,j\neq i}^{n} p_{ji} \tag{6}$$

where the first summation is the *leaving flow* (i.e. the total degree of preference of x_i over the other alternatives) while the last one is the *entering flow* i.e. the total degree of preference of the other alternatives over x_i. After having rated available alternatives, the one with the highest preference degree is the solution of the GDM problem.

3.2 Application to Peer Assessment

As described in Sect. 2, in ordinal peer assessment each student of a set S ranks the submissions coming from m other students. By setting $E = X = S$ and assigning to each $s_k \in S$ a subset $S_k = \{s_i \in S | a_{ik} = 1\}$ of submissions to be evaluated, FOPA obtains a GDM problem corresponding to ordinal peer assessment [7, 8].

As in ordinal peer assessment each student s_k is asked to define a partial ranking over S_k by leveraging on GDM, preferences can be expressed in terms of a FPR P_k. Unfortunately, the definition of FPR may result too complex and time-consuming for students with the risk of introducing errors and inconsistencies impacting in assessment performances. To overcome this issue, FOPA adopts a simpler intermediate preference model based on fuzzy rankings. According to [22], a *fuzzy ranking* on the set S_k can be defined as the sequence:

$$s^k_{p(1)} \sigma_1 s^k_{p(2)} \cdots s^k_{p(m-1)} \sigma_{m-1} s^k_{p(m)} \tag{7}$$

Terms in odd positions represent elements of S_k while $p : \{1, \ldots m\} \rightarrow \{1, \ldots m\}$ is a permutation function. Terms in even positions belong to the set $\{\gg, >, \geq, \approx\}$ and define a degree of preference between subsequent terms in the sequence (with $\gg$ meaning *much better than*, $>$ *better than*, $\geq$ *a little better than* and $\approx$ *similar to*). Each submission appears at most once in the ranking.

For example, let suppose that the student s_1 has to evaluate the subset of students $S_1 = \{s_2, s_4, s_5, s_6\}$. By providing the ranking $R_1 = (s_4 \gg s_5 \approx s_2 > s_6)$ she states that, according to her opinion, the submission of s_4 is much better than that of s_5 that, in turn, is similar to the submission of s_2 that, in turn, is better than the submission of s_6.

Each fuzzy ranking is then converted in a FPR P_k by applying one of the methods defined in [22]. When individual FPRs are obtained, an aggregation step, based on the OWA operator introduced in Sect. 3.1, is applied to obtain the collective FPR $P = (p_{ij})$. Given n students and m assignments per student, the conversion step produces individual FPRs where only a fraction of m^2/n^2 elements are defined, so the collective FPR often also presents undefined values. In fact, when none of the assessors expressed a preference between the i-th and j-th submissions for some $i, j \in \{1, \ldots, n\}$, then the corresponding values p_{ij} and p_{ji} of the collective FPR can't be calculated. Missing values are then estimated according to methods described in [23, 24].

Once all values of P are defined, it is possible to calculate the degree of preference $\phi_{NF}(s_i)$ for each $s_i \in S$ according to *Net Flow* as seen in Sect. 3.1. The global ranking between alternatives is then computed by ordering all the submission decreasingly on their preference degree. The cardinal grade of each submission can be also calculated, provided that a cardinal assessment is made by a reliable expert (e.g. the teacher) to the best and the worst submissions (i.e. the first and the last in the final ranking). Let g_{min} and g_{max} be the grades assigned to the best and the worst submissions, the estimated grade g_i for every $s_i \in S$ can be obtained as follows:

$$g_i = \frac{(\phi(s_i) - \phi_{min}) \cdot (g_{max} - g_{min})}{(\phi_{max} - \phi_{min})} + g_{min} \qquad (8)$$

where ϕ_{min} and ϕ_{max} are the preference degrees associated to the best and the worst submissions.

4 Introducing Multiple Criteria

To guide students during the evaluation process, an *assessment rubric* can be provided [10, 11]. A rubric is a coherent set of *criteria* $C = \{c_1, \ldots, c_q\}$, usually defined by the teacher, that makes explicit the factors that must be considered when evaluating a submission as well as the expected levels of performance for each of them. Using a rubric, assessors evaluate submissions against all the defined criteria rather than by assigning a single score. Several studies show that assessment rubrics make evaluations more reliable and objective, even in peer assessment contexts [25].

In cardinal peer assessment, rubrics are often presented in table format where rows refer to submissions to be evaluated while columns to criteria to be considered. Thus, each cell represents the evaluation given to a specific submission with respect to a given criteria. The application of rubrics can be extended to *ordinal peer assessment* asking each student $s_k \in S$ to define several *ordinal rankings* $\succ_{kl}$ (one for each criterion $c_l \in C$) on the subset $S_k \subset S$ of her assessees rather than just a summative one.

Partial rankings defined for the l-th criteria can be then collected in a *ranking matrix* $R_l = (r_{ik}^l)$ whose generic element r_{ik}^l represents the position of s_k in the ranking $\succ_{kl}$ if $s_i \in S_k$, 0 otherwise. Applying Eq. (2) on each r_{ik}^l for $l \in \{1, \ldots, q\}$ rather than on r_{ik}, the overall score $\phi_B^l(s_i)$ of the i-th submission against the l-th criteria can be obtained. To aggregate them in a summative score for each submission $s_i \in S$, a weighted arithmetic mean may be used as follows:

$$\phi(s_i) = \sum_{l=1}^{q} w_l \cdot \phi_l(s_i). \qquad (9)$$

where w_l, for $l \in \{1, \ldots, q\}$, is a weight stating the importance of each assessment criteria such that $w_l \in [0, 1]$ and $\sum_{l=1}^{q} w_l = 1$.

The standard FOPA model does not provide any support to assessment rubrics. To fill this lack, in this section, after having introduced useful concepts of multi-criteria GDM, we propose a FOPA extension that enables evaluation based on rubrics.

4.1 Extension of FOPA to Multi-criteria GDM

A *Multiple Criteria Group Decision Making* (MCGDM) problem is characterized by a group of experts $E = \{e_1, \ldots, e_m\}$ that express their preferences on a set of alternatives $X = \{x_1, \ldots, x_n\}$ according to a set of criteria of factors $F = \{f_1, \ldots, f_q\}$ of quantitative or qualitative nature [15]. The literature contains several strategies for extending GDM models to MCGDM in order to obtain solutions that reflect the collective vision of a problem. In [26] such strategies have been classified according to the different points of the decision-making process in which the preference aggregation is performed.

Notably, the *aggregation of preferences per criterion* strategy requires experts to evaluate alternatives against each criterion with a feasible preference model (like *fuzzy rankings* defined in Sect. 3.2). Then $m \times q$ FPRs are generated i.e. one for each expert and criterion. Individual FPRs are aggregated along each criterion with standard operators like OWA so resulting in q collective FPRs that are subsequently aggregated to obtain a single FPR used to establish a ranking of the alternatives.

Being $P^1, \ldots, P^q$ the q collective FPRs resulting from the aggregation per criterion, the last aggregation step can be done by using the OWA operator again. Nevertheless, when the importance of each criterion is important, specific operators are needed. The Importance Induced OWA (I-IOWA) [27] considers the importance of each FPR while being guided by a linguistic quantifier. Let $(p_{ij}^1, \ldots, p_{ij}^q)$ with $i, j \in \{1, \ldots, n\}$ be a list of preferences to aggregate, let $u_l \in [0, 1]$ be the importance degree of each $f_l \in F$ and Q a non-decreasing proportional quantifier, I-IOWA is defined as follows:

$$I - IOWA\left(\left(p_{ij}^1, u_1\right), \ldots, \left(p_{ij}^q, u_q\right)\right) = \sum_{l=1}^{q} w_l p_{ij}^{\sigma(l)} \tag{10}$$

were $\sigma : \{1, \ldots q\} \to \{1, \ldots q\}$ is a permutation function so that $u_{\sigma(l)} \geq u_{\sigma(l+1)}$ for each $l \in \{1, \ldots, q-1\}$ and the l-th weight w_l is obtained as follows:

$$w_l = \mu_Q\left(\frac{S(l)}{S(q)}\right) - \mu_Q\left(\frac{S(l-1)}{S(q)}\right); l \in \{1, \ldots, q\}. \tag{11}$$

where $S(l) = \sum_{k=1}^{l} u_{\sigma(l)}$.

4.2   Adoption of Assessment Rubrics

As seen in Sect. 3.2, a GDM problem corresponding to ordinal peer assessment can be obtained by setting $E = X = S$ (i.e. by matching the set of experts and alternatives of the GDM problem with the set of students of a peer assessment). Introducing multiple criteria needs to match, in addition, the set of factors of a MCGDM problem with the assessment criteria defined in the peer assessment rubric by setting $F = C$. Then, each student $s_k \in S$ is required to define q partial rankings $R_k^1, \ldots, R_k^q$ over S_k, one against each criterion $c_l \in C$.

Defined fuzzy rankings are then converted into $n \times q$ individual FPRs P_k^l with $k \in \{1, \ldots, n\}$ and $l \in \{1, \ldots, q\}$ and aggregated along each criterion through the OWA operator so obtaining q collective FPRs $P^1, \ldots, P^q$ such that:

$$P^l = OWA\left(P_1^l, \ldots, P_n^l\right) \text{with } l \in \{1, \ldots, q\}. \tag{12}$$

As for the standard FOPA model, given the sparsity of individual FPRs, collective ones may present several missing values that should be estimated according to methods described in [23, 24]. Once all values of the collective FPRs have been defined, it is possible to merge them through the I-IOWA operator defined in Sect. 4.1 as follows:

$$P = I - IOWA\big((P^1, w_1), \ldots, (P^q, w_q)\big) \tag{13}$$

where w_l, for $l \in \{1, \ldots, q\}$, is a weight stating the importance of each assessment criteria $c_l \in C$ such that $w_l \in [0, 1]$. Once the aggregated collective FPR P is obtained, the degree of preference $\phi_{NF}(s_i)$ of each $s_i \in S$ is calculated according to Eq. (6) and the cardinal grade of each submission is obtained according to Eq. (8).

5 Final Remarks

In this paper we have proposed an extension of an existing fuzzy GDM-based ordinal peer assessment model aimed at introducing the support for assessment rubrics through the application of MCGDM techniques. Rubrics are quite spread in cardinal assessment contexts because of their positive aim to improve the reliability and objectivity of student evaluations. Nevertheless, their adoption within an ordinal model is rather new as well as the application of peer assessment models based of *fuzzy sets*.

The proposed extension also allows teachers to associate different weights to defined criteria and to take them into account during the aggregation process. A future extension of such model could be the involvement of students not only to rank peer submissions against the defined criteria but also to rank criteria themselves based on their perceived importance. Thus, the weights associated to assessment criteria could be generated through the aggregation of such additional information.

A thorough experimentation of the proposed model with synthetic datasets as well as with real users is also planned as future work to measure its performance both with respect to the standard FOPA model as well as against other existing cardinal and ordinal peer assessment models.

Acknowledgement. This work has been supported by the project *colMOOC* "Integrating Conversational Agents and Learning Analytics in MOOCs", co-funded by the European Commission within the Erasmus + Knowledge Alliances program (ref. 588438-EPP-1-2017-1-EL-EPPKA2-KA).

References

1. Glance, D.G., Forsey, M., Riley, M.: The pedagogical foundations of massive open online courses. First Monday **18**(5) (2013)
2. Bouzidi, L., Jaillet, A.: Can online peer assessment be trusted? Educ. Technol. Soc. **12**(4), 257–268 (2009)
3. Carlson, P.A., Berry, F.C.: Calibrated peer review™ and assessing learning outcomes. In: Proceedings of the 33rd International Conference Frontiers in Education (2003)

4. Walsh, T.: The peer rank method for peer assessment. In: Proceedings of the 21st European Conference on Artificial Intelligence (2014)
5. Capuano, N., Caballé, S.: Towards adaptive peer assessment for MOOCs. In: Proceedings of the 10th International Conference on P2P, Parallel, GRID, Cloud and Internet Computing (3PGCIC 2015), Krakow, Poland (2015)
6. Raman, K., Joachims, T.: Methods for ordinal peer grading. In: Proceedings of the 20th SIGKDD International Conference on Knowledge Discovery and Data Mining (2014)
7. Capuano, N., Loia, V., Orciuoli, F.: A fuzzy group decision making model for ordinal peer assessment. IEEE Trans. Learn. Technol. **10**(2), 247–259 (2017)
8. Capuano, N., Orciuoli, F.: Application of fuzzy ordinal peer assessment in formative evaluation. In: Proceedings of the 12th International Conference on P2P, Parallel, Grid, Cloud and Internet Computing (3PGCIC 2017), Barcelona, Spain (2017)
9. Albano, G., Capuano, N., Pierri, A.: Adaptive peer grading and formative assessment. J. e-Learn. Knowl. Soc. **13**(1), 147–161 (2017)
10. Reddy, Y., Andrade, H.: A review of rubric use in higher education. Assess. Eval. High. Educ. **35**(4), 435–448 (2009)
11. Lan, C.H., Graf, S., Lai, K.R., Kinshuk, K.: Enrichment of peer assessment with agent negotiation. IEEE Trans. Learn. Technol. **4**(1), 35–46 (2011)
12. Capuano, N., Caballé, S., Miguel, J.: Improving peer grading reliability with graph mining techniques. Int. J. Emerg. Technol. Learn. **11**(7), 24–33 (2016)
13. Caragiannis, I., Krimpas, A., Voudouris, A.A.: Aggregating partial rankings with applications to peer grading in massive online open courses. In: Proceedings of the International Conference on Autonomous Agents and Multiagent Systems, Istanbul (2015)
14. Borda, J.C.: Memoire sur les elections au scrutin, Histoire de l'Académie Royale des Sciences (1781)
15. Pedrycz, W., Ekel, P., Parreiras, R.: Fuzzy Multicriteria Decision-Making: Models, Methods and Applications. Wiley, Hoboken (2010)
16. Lu, J., Zhang, G., Ruan, D., Wu, F.: Multi-objective group decision making, methods, software and applications with fuzzy set techniques, World Scientific (2007)
17. Capuano, N., Chiclana, F., Fujita, H., Herrera-Viedma, E., Loia, V.: Fuzzy group decision making with incomplete information guided by social influence. In: IEEE Transactions of Fuzzy Systems. In Press
18. Zadeh, L.A.: A computational approach to fuzzy quantifiers in natural languages. Comput. Math Appl. **9**, 149–184 (1983)
19. Wang, Y.M., Fan, Z.P.: Fuzzy preference relations: aggregation and weight determination. Comput. Ind. Eng. **53**(1), 163–172 (2007)
20. Yager, R.R.: Families of OWA operators. Fuzzy Sets Syst. **59**(2), 125–148 (1993)
21. Chiclana, F., Herrera, F., Herrera-Viedma, E.: Integrating three representation models in fuzzy multipurpose decision making based on fuzzy preference relations. Fuzzy Sets Syst. **97** (1), 33–48 (1998)
22. Capuano, N., Chiclana, F., Fujita, H., Herrera-Viedma, E., Loia, V.: Fuzzy rankings for preferences modeling in group decision making. Int. J. Intell. Syst. **33**(7), 1555–1570 (2018)
23. Alonso, S., Chiclana, F., Herrera, F., Herrera-Viedma, E., Alcala-Fdez, J., Porcel, C.: A consistency-based procedure to estimate missing pairwise preference values. Int. J. Intell. Syst. **23**(1), 155–175 (2008)
24. Alonso, S., Herrera-Viedma, E., Chiclana, F., Herrera, F.: Individual and social strategies to deal with ignorance situations in multi-person decision making. Int. J. Inf. Technol. Decis. Making **8**(2), 313–333 (2009)
25. Jonsson, A., Svingby, G.: The use of scoring rubrics: reliability, validity and educational consequences. Educ. Res. Rev. **2**(2), 130–144 (2007)

26. Ekel, P., Queiroz, J., Parreiras, R., Palhares, R.: Fuzzy set based models and methods of multicriteria group decision-making. Nonlinear Anal.: Theor. Methods Appl. **71**(12), 409–419 (2009)
27. Chiclana, F., Herrera-Viedma, E., Herrera, F., Alonso, S.: Some induced ordered weighted averaging operators and their use for solving group decision-making problems based on fuzzy preference relations. Eur. J. Oper. Res. **182**(1), 383–399 (2007)

Conversational Agents in Support for Collaborative Learning in MOOCs: An Analytical Review

Santi Caballé[✉] and Jordi Conesa

Faculty of Computer Science, Multimedia and Telecommunications,
Universitat Oberta de Catalunya, Rambla Poblenou, 156, 08018 Barcelona, Spain
{scaballe,jconesac}@uoc.edu

Abstract. Massive Open Online Courses (MOOCs) arose as a way of transcending formal higher education by realizing technology-enhanced formats of learning and instruction and by granting access to a wide audience way beyond students enrolled in any one Higher Education Institution. However, while MOOCs have been reported as an efficient and important educational tool, yet there is a number of issues and problems related to their educational impact. More specifically, there is an important number of drop outs during a course, little participation, and lack of students' motivation and engagement overall. To overcome these limitations, Conversational pedagogical agents have arisen to guide and support student dialogue using natural language both in individual and collaborative settings. Conversational agents have been produced to meet a wide variety of applications and studies exploring the usage of such agents have led to positive results. Integrating this type of artificial agents into MOOCs is expected to trigger productive peer interaction in discussion groups. In this paper, we present a state-of-the-art study of the use of conversational agents to support collaborative learning in the context of MOOCs. The ultimate goal of this study is to analyze the potential of conversational agents to considerably increase the engagement and the commitment of MOOC students, reducing consequently, the overall MOOCs dropout rate. The research reported in this paper is currently undertaken within the research project colMOOC funded by the European Commission.

1 Introduction

MOOCs have been praised for their potential to democratize education and help ensure individuals access to quality education regardless of their geographic location, financial means, schedule, language or background [17]. Due to their high enrolment numbers, MOOCs offer a middle ground for teaching and learning between the highly organized and structured classroom environment and the chaotic open web of fragmented information [27]. The first reaction to MOOCs tended to be one of fearful respect for the disruption that they seemed likely to cause in the higher education scene. It seems though that MOOCs, universities, and learners will all likely continue to adapt to new methods and mediums as with 23 million new MOOC users in 2017 and major venture capital funding and investments of universities and companies alike it seems unlikely

© Springer Nature Switzerland AG 2019
F. Xhafa et al. (Eds.): INCoS 2018, LNDECT 23, pp. 384–394, 2019.
https://doi.org/10.1007/978-3-319-98557-2_35

that they will be disappearing anytime soon [26]. However, as MOOCs have evolved, and educators and academics began to caution against extreme optimism and instead pointed to the challenges of MOOCs. In past few years MOOCs have gathered a substantial amount of criticism for, among other things, lacking pedagogical innovation, having high drop-out rates, and failing to reach exactly those people they claim to target [4]. Therefore, despite their undeniable potential to deliver high quality learning experience to large audiences, MOOC also pose additional issues and challenges for the teacher, the learning designer, and the underlying technology [8, 19, 25].

Among the most relevant pedagogical innovations, conversational pedagogical agents have arisen to guide and support student dialogue using natural language both in individual and collaborative settings. According to MIT [21], agents are "computer systems to which one can delegate tasks. Software agents differ from conventional software in that they are long-lived, semi-autonomous, proactive, and adaptive". Hence software agents can use artificial intelligence methods in order to analyze information and react to it. Integrating this type of artificial agents into MOOCs is expected to trigger productive peer interaction in discussion groups, thus increasing the engagement and the commitment of MOOC students, reducing consequently, the overall MOOCs dropout rate [9].

Indeed, the opportunities for using agents in e-learning courses and MOOCs in particular are enormous. Agent characteristics like autonomy, abilities to perceive, reason and act in specialized domains, as well as their capability to cooperate with other agents makes them ideal for e-learning applications [24]. The potential use of several agents in e-learning environments has been researched many years before the advent of MOOCs, defining some potential roles for them like for example Pedagogical Agents (tutor, mentor, assistant), Web Agents (working with Internet applications and social networking tools), Learner's agents and mixed agents which could teach and learn [16]. In MOOC environments, agents can be used to analyze data produced by the MOOC platform, systems and participants, and use it intelligently or mechanically to improve design, delivery and assessment [8]. Indeed, some applications of agents to MOOCs design and delivery could help identify potential problems, gaps and limitations of the initial course design, for example improper planning, improper distribution of course constituents, inadequate time assignment to the course different issues, errors in tests and evaluations, etc. [15].

The goal of this work is to provide a throughout review and discussion on the application of software agents to support educational environments, and specifically alleviate the main problems that limit the enormous potential of MOOCs. To this end, following a systematic review methodology, Sect. 2 reports on existing approaches of software agents in support for MOOCs while Sect. 3 discusses on the opportunities and challenges to enhance MOOC environments further, and in particular the innovative use of conversational agents to increase engagement in MOOCs by enhancing peer collaboration. Finally, Sect. 4 outlines the main conclusions and provides on-going and future directions of research.

2 Pedagogical Agents for MOOCs

The chosen methodology for this review is an adaptation of the process for systematically reviewing literature explained in [22]. In this section we present the steps and tasks of our methodology, which are applied sequentially but may overlap in time or even repeat iteratively as needed (Fig. 1).

Fig. 1. Steps and tasks in the chosen study methodology.

Based on the experiences gained through analysis of pedagogical agents in MOOC settings, some systems have been developed and successfully deployed that support effective collaboration and learning in those settings [12] developed two systems built on pedagogical agents designed to provide opportunities for discussion-based learning in MOOCs.

One of the pedagogical agents, called QuickHelper [12] (Fig. 1), focused on help exchange while the other, called Bazzar, focused on collaborative reflection. Both agents were deployed in a real MOOC that was offered on the edX MOOC platform [11, 12]. QuickHelper was motivated by the fact that while virtually all MOOCs offer threaded discussion affordances where students can post help requests, some students are reticent to ask for help, and even when students do post help requests, many of these requests go unanswered. Therefore, this pedagogical agent was designed to support help seeking as well as increase the probability that help requests were met with a satisfactory response. Using this help request, a social recommendation algorithm selects three potential help providers from the pool of student peers. The student is then given the option to invite one or more of these potential helpers to their thread, as shown in Fig. 2. Once selected, an email with a link to the help request thread is then automatically sent to the selected helpers inviting them to participate in the thread.

A second pedagogical agent, named Bazaar Collaborative Reflection (Fig. 3), makes synchronous collaboration opportunities available to students in a MOOC context. This approach follows research outcomes in Computer-Supported

These students are good matches for answering you question. Would you like to invite any of these potential helpers to your discussion thread via private message?

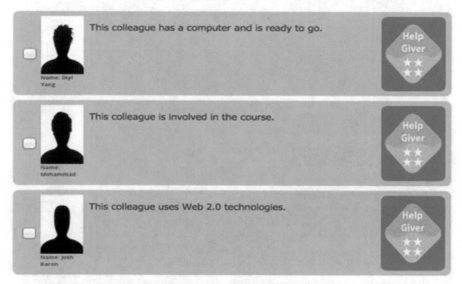

IF YOU SELECT NONE, YOUR HELP REQUEST WILL BE POSTED TO THE COURSE DISCUSSION BOARD WITHOUT SENDING A PRIVATE MESSAGE TO ANY OF THESE POTENTIAL HELPERS.

Fig. 2. Screenshot of quick helper showing the helper selection [12].

Collaborative Learning, which has demonstrated that conversational computer agents can serve as effective automated facilitators of synchronous collaborative learning [10]. However, typical MOOC providers do not offer students opportunities for synchronous collaboration, and therefore have not so far benefitted from this technology. To overcome this limitation, Bazaar is implemented as a collaborative chat environment with interactive agent support with the aim to facilitate the formation of ad-hoc study groups for the chat activity, where students are matched and provided with a link to a chat room where they can work with their partner students on a synchronous collaboration activity supported by a conversational computer agent [12]. This work is built on research findings [1, 10] where conversational agents appear as regular users in the chat. An important lesson learned from this approach was that whereas providing the opportunity for synchronous chat was positive for students for whom it was possible to be matched for a chat easily, this positive effect was balanced with a negative effect in the case where the lack of critical mass despite the total enrollment of 20,000 students from their MOOC was not sufficient to enable a quick match [13].

A different approach is to integrate Intelligent Tutoring Systems (ITS) with MOOCs, considering that the pedagogical approaches used in ITS and MOOCs are

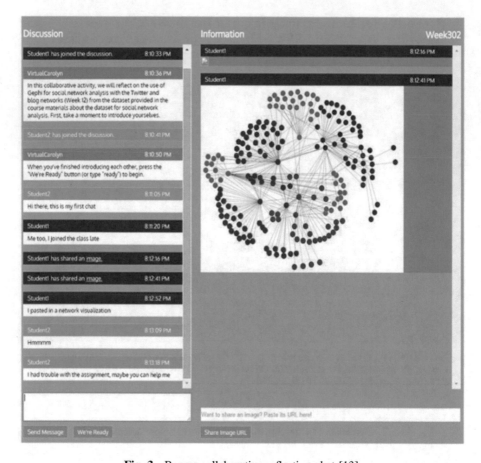

Fig. 3. Bazaar collaborative reflection chat [13].

largely complementary [2]. Indeed, MOOCs support many forms of instruction, including video lectures, reading with conceptual questions, discussion boards, and various forms of learning-by-doing with automated or peer feedback. Although these solutions have been quite successful at scale, they have their drawbacks. For instance, peer discussion and feedback are not always timely; peers may not know the right answer or may disagree, and many learners may be reluctant to post questions and concerns to a large audience [5].

In addition, MOOCs sometimes have limited capabilities to support individual learning or personalizing instruction [8]. ITSs can address some of these limitations by, for example, providing step-by-step guidance during complex problem solving, tracking learners' skill growth and select problems on an individual basis, and adaptively responding to student strategies and errors [23]. The purpose is to integrate ITS-style learning-by-doing into MOOCs. To this end, a system named Cognitive Tutor Authoring Tools (CTAT) was developed to build and embed software tutors from ITSs into MOOCs and in particular in the edX MOOC platform [11] so that an author can build a tutor interface with CTAT-enabled interface components using ordinary HTML coding techniques. Several projects and studies [2] have demonstrated the feasibility of the MOOC/ITS integration between edX and CTAT HTML tutor interfaces, including several prototypes of tutors used in online statistics courses.

In line with the previous approaches, some projects and developments based on software agents have the potential to stimulate the collaborative learning interaction by agents with the aim to overcome relevant issues arisen in these contexts, in particular in massive MOOC-like courses, such as (i) the discussion is based on a long list of messages, which is hard to follow by students and tedious to monitor by tutors and moderators; (ii) after the collaborative activity is over the discussion is not available anymore and the collaborative knowledge produced is lost; (iii) usual text-formatted posts are far from real-life discussions and physical participation, thus chances for social benefits from actual collaboration are not available. All these deficiencies lead to rudimentary collaborative learning practices, little attractive and lack of interest, thus having a negative effect on learners' self-motivation and engagement in their learning process [6].

In order to overcome these and other related limitations and deficiencies, a system called Virtualized Collaborative Session (VCS) [14] that enables the virtualization and registration of live collaborative sessions in order to produce interactive and attractive learning resources to be experienced and played by learners. Learners can observe how students represented by agents discuss and collaborate, how discussion threads grow, and how knowledge is constructed, refined and consolidated (Fig. 4). The registered agent-based discussions are eventually packed and stored for further reuse, enriching live sessions of collaborative learning with balanced levels of interaction, challenge and empowerment [7].

3   Discussion

From the above review, we summarize and discuss next on the main opportunities and challenges of using software agents to support MOOCs, and most importantly to support peer collaboration, in order to increase student engagement and commitment.

3.1   Agent Application to MOOCs

In MOOC environments, agents can be used to analyze data produced by the MOOC platform, systems and participants, and use it intelligently or mechanically to improve design, delivery and assessment [4, 8]. Indeed, some applications of agents to MOOCs design and delivery could help identify potential problems, gaps and limitations of the initial course design, for example improper planning, improper distribution of course constituents, inadequate time assignment to the course different issues, errors in tests and evaluations, etc.

Moreover, MOOC managers can receive information collected by agents that allow them to analyze the cost/effectiveness ratio of the courses, measure the quality of the learning offering provided, predict success/failure and drop-out rate of their learners and adjust their learning offer accordingly. In addition, software agents could be used

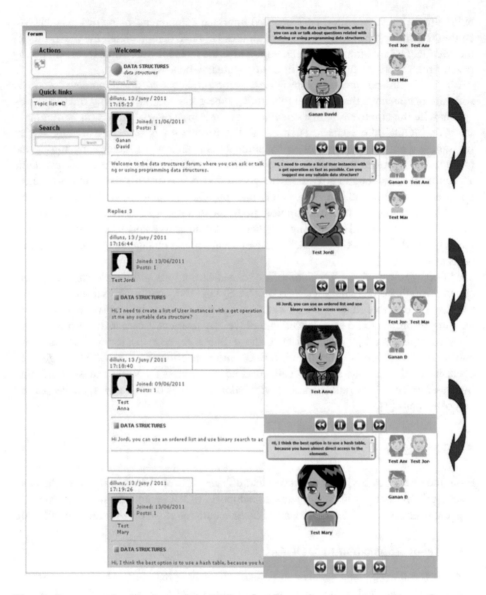

Fig. 4. Sequence of screenshots of the VCS embedding software agents guiding a discussion evolving over time, produced by the virtualization of a live collaborative session performed in a web forum [7].

for real time analysis of content accessed by the MOOC participants, in order to detect potential problematic areas, solve some predefined detected issues automatically, enhance participants support and optimize the use of human support time by making them focus on the most critical issues.

Agents can also improve MOOC assessment further [9, 20]. Agents can help define evaluation parameters that are personalized according, for example, to the participant's educational level, previous performance, etc. as well as design evaluation indicators that consider other parameters like peer-evaluation, social participation, creative thinking, problem solving, application of knowledge to a local reality, etc. In addition, agents can be used to improve testing methods in order to measure other achievement indicators in MOOCs, besides delivering automated tests.

Finally, by analyzing usage parameters and content, agents can be used to detect potential cheating, such as plagiarism and alert tutors. Software agents can also be used to validate the identity of a person (e.g. by typing patterns and face recognition).

3.2 Conversational Agents to Support Collaboration in MOOCs

Despite the inclusion of discussion forums in virtually all MOOCs, the typical experience of participation in a MOOC is solitary. In particular, MOOC contexts allow for asynchronous interaction and possibly even collaboration. However, the asynchronous nature leads to some unsavory experiences. Sometimes MOOC participants spend time posting a thoughtful post but get only a cursory reply or have to wait days or even weeks to get a response to a question. This lack of immediacy implies that information can be out of date by the time someone views it, or worse, that the student needing help or feedback gave up and dropped out before the response was posted. These types of issues can hinder motivation. In this context, scaffolding can be provided through the facilitation moves of an intelligent conversational agent taking a teacher role. The goal is for the agent to aid the students in integrating their respective understandings of the concepts they previously encountered individually in. On the other hand, it also provides an opportunity to develop their collaborative skills. However, the use of agents in this context have been, to the best of our knowledge, scarcely explored [30].

Based on the above considerations and expectations, a newly research project called "Integrating Conversational Agents and Learning Analytics in MOOCs (col-MOOC)" participated by the authors of the present study [9] leverages the potential of Conversational Agents (CA) to address the MOOC challenges described above. The main priority tackled of the colMOOC project is to identify that there is a need to make MOOCs more collaborative by the use of agents in order to automatically support teachers and engage and motivate students. To this end, the type of learners' interventions is modelled according to the "Academically Productive Talk (APT)" model [28, 29], which emphasizes the orchestration of teacher-students talks and highlights a set of useful discussion practices that can lead to reasoned participation by all students, thus increasing the probability of productive peer interactions to occur [18]. Therefore, the project will design and implement APT-based agents able to be integrated in forum and chat tools (Fig. 5). This type of automated form of CA support can enhance students' explicit reasoning on domain concepts and improve individual and group learning in the context of a collaborative learning activity in higher education settings and MOOCs. However, all available research so far has been conducted in controlled experimental settings only. The colMOOC project aims to expand and explore this approach in actual MOOC settings at large scale [9].

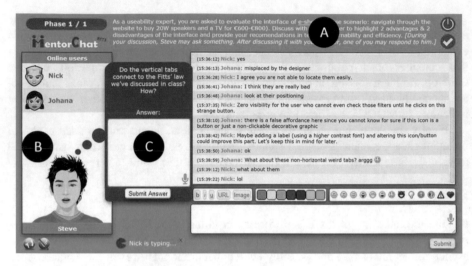

Fig. 5. A conversational agent's intervention in a collaborative activity to stimulate participation [29].

4   Conclusions

In this paper, a systematic review study was conducted to analyze the potential of software pedagogical agents to increase engagement and commitment of MOOC students while reducing dropout rates. Based on this review, we identified and discussed the main benefits and challenges of integrating a specific version of agents named Conversational Agents into collaborative tools, such as forums and chats, so as to support peer collaboration in MOOCs. From the main issues identified in this study we have proposed some direction to work, which are currently being addressed by the European research project "colMOOC" with the aim to integrate an agent software component, which attempts to trigger learners' discussion through appropriate interventions in dialogue-based collaborative activities.

Current and future directions of research, in line with the colMOOC project, is the provision of assessment in MOOCs as a critical component to monitor individual achievement and achieve constant engagement. To this end, the power of Learning Analytics to support MOOCs [3] will be investigated in order to achieve more meaningful assessment, thus allowing educators and trainers to apply learning processes where learners participate actively and are engaged throughout the course Following our review methodology (see Sect. 2), these goals will in turn conduct the next iteration of this review study.

Acknowledgements.   This research was funded by the European Commission through the project "colMOOC: Integrating Conversational Agents and Learning Analytics in MOOCs" (588438-EPP-1-2017-1-EL-EPPKA2-KA).

References

1. Adamson, D., Dyke, G., Jang, H.J., Rosé, C.P.: Towards an agile approach to adapting dynamic collaboration support to student needs. Int. J. AI Educ. **24**(1), 91–121 (2014)
2. Aleven, V., Sewall, J., Popescu, O., Ringenberg, M., van Velsen, M., Demi, S.: Embedding intelligent tutoring systems in MOOCs and e-learning platforms. In: 13th International Conference on Intelligent Tutoring Systems (ITS), pp. 409–415. Springer (2016)
3. Bañeres, D., Caballé, S., Clarisó, R.: Towards a learning analytics support for intelligent tutoring systems on MOOC platforms. In: 10th International Conference on Complex, Intelligent, and Software Intensive Systems (CISIS), pp. 87–94. IEEE (2016)
4. Bassi, R., Daradoumis, T., Xhafa, F., Caballé, S., Sula, A.: Massive open online courses: Innovation in education? In: McGreal, R. et al. (eds.) Open Educational Resources: Innovation, Research and Practice, Chap. 1, pp. 5–15. Athabasca University Press, Athabasca (2014). Software, Siemens, G.A. (2013)
5. Baxter, J.A., Haycock, J.: Roles and student identities in online large course forums: implications for practice. In: The International Review of Research in Open and Distance Learning, vol. 15 (2014)
6. Caballé, S., Daradoumis, T., Xhafa, F., Juan, A.: Providing effective feedback, monitoring and evaluation to on-line collaborative learning discussions. Comput. Hum. Behav. **27**(4), 1372–1381 (2011)
7. Caballé, S., Mora, N., Feidakis, M., Gañán, D., Conesa, J., Daradoumis, T., Prieto, J.: CC-LR: providing interactive, challenging and attractive collaborative complex learning resources. J. Comput. Assist. Learn. **30**(1), 51–67 (2014)
8. Daradoumis, T., Bassi, R., Xhafa, F., Caballé, S.: A review on massive e-learning (MOOC) design, delivery and assessment. In: Eighth International Conference on P2P, Parallel, Grid, Cloud and Internet Computing (3PGCIC), Track on eLearning and Groupware Systems, pp. 208–213. IEEE (2013)
9. Demetriadis, S., Karakostas, A., Tsiatsos, T., Caballé, S., Dimitriadis, Y., Weinberger, A., Papadopoulos, P.M., Palaigeorgiou, G., Tsimpanis, C., Hodges, M.: Towards integrating conversational agents and learning analytics in MOOCs. In: 6th International Conference on Emerging Intelligent Data and Web Technologies. (EIDWT), pp. 1061–1072. Springer (2018)
10. Dyke, G., Howley, I., Adamson, D., Kumar, R., Rosé, C.P.: Towards academically productive talk supported by conversational agents. In: Productive Multivocality in the Analysis of Group Interactions, pp. 459–476. Springer (2013)
11. edX: MOOC platform (2018). https://www.edx.org/. Accessed 21 Mar 2018
12. Ferschke, O., Howley, I., Tomar, G., Yang, D.: Fostering discussion across communication media in massive open online courses. In: Proceedings of the 11th International Conference on Computer Supported Collaborative Learning (CSCL), pp. 459–466 (2015)
13. Ferschke, O., Yang, D., Tomar, G., Rosé, C.P.: Positive impact of collaborative chat participation in an edx mooc. In: 17th International. Conference on Artificial Intelligence in Education (AIED), pp. 115–124. Springer (2015)
14. Gañán, D., Caballé, S., Conesa, J., Xhafa, F.: An application framework to systematically develop complex learning resources based on collaborative knowledge engineering. Int. J. Appl. Math. Comput. Sci. **25**(2), 361–375 (2015)
15. Guàrdia, L., Maina, M.F., Sangrà, A.: MOOC design principles. A pedagogical approach from the learner's perspective. eLearning papers, vol. 33, pp. 1–6 (2013)

16. Jaques, P., Andrade, A., Jung, J., Bordini, R., Vicari, R.: Using pedagogical agents to support collaborative distance learning. In: Proceedings of the Conference on CSCL: Foundations for a CSCL Community, pp. 546–547 (2002)
17. Kucirkova, N., Littleton. K.: Digital learning hubs: theoretical and practical ideas for innovating massive open online courses. Learn. Media Technol. (2015). https://doi.org/10.1080/17439884.2015.1054835. Advanced online publication
18. Kumar, R., Rosé, C.P.: Architecture for building conversational agents that support collaborative learning. IEEE Trans. Learn. Technol. **4**(1), 21–34 (2011)
19. Mackness, J., Mak, S., Williams, R.: The ideals and reality of participating in a MOOC. In: Proceedings of the 7th International Conference on Networked Learning, pp. 266–275 (2010)
20. Miguel, J., Caballé, S., Prieto, J.: Providing information security to MOOC: towards effective student authentication. In: Fifth IEEE International Conference on Intelligent Networking and Collaborative Systems, pp. 289–292 (2013)
21. MIT software agents group (2013). http://agents.media.mit.edu/. Accessed 15 Mar 2018
22. Oates, B.J.: Researching Information Systems and Computing. Sage Publications Ltd., London (2006)
23. Pane, J.F., Griffin, B.A., McCaffrey, D.F., Karam, R.: Effectiveness of cognitive tutor algebra I at scale. Educ. Eval. Policy Anal. **36**(2), 127–144 (2013)
24. Papazoglou, M.P.: Agent-oriented technology in support of e-business. Commun. ACM **44**(4), 35–41 (2001)
25. Razmerita, L., Kirchner, K., Hockerts, K., Tan, C.: Towards a model of collaborative intention: an empirical investigation of a massive online open course. In: 51st Hawaii International Conference on System Sciences, pp. 727–736 (2018)
26. Shah, D.: By the numbers: MOOCS in 2017, 18 Jan 2018. https://www.class-central.com/report/mooc-stats-2017/. Accessed 14 Mar 2018
27. Siemens, G.: Massive open online courses: Innovation in education? In: McGreal, R., et al. (eds.) Open Educational Resources: Innovation, Research and Practice (Chap. 1), pp. 5–15. Athabasca University Press, Athabasca (2013)
28. Tegos, S., Demetriadis, S.: Conversational agents improve peer learning through building on prior knowledge. J. Educ. Technol. Soc. **20**(1), 99–111 (2017)
29. Tegos, S., Demetriadis, S., Karakostas, A.: Promoting academically productive talk with conversational agent interventions in collaborative learning settings. Comput. Educ. **87**, 309–325 (2015)
30. Tomar, G.S., Sankaranarayanan, S., Wang, X., Rosé, C.P.: Coordinating collaborative chat in massive open online courses. In: Proceedings of International Conference of the Learning Sciences, ICLS, vol. 1, pp. 607–614 (2017)

A Teaching Application to Improve Access and Management of Web-Based Academic Materials

Antonio Sarasa Cabezuelo[1(✉)] and Jordi Conesa Caralt[2]

[1] Universidad Complutense de Madrid, C/Profesor García Santesmases, 9,
28040 Madrid, Spain
asarasa@fdi.ucm.es
[2] Universitat Oberta de Catalunya, Rambla del Poblenou, 156,
08018 Barcelona, Spain
jconesac@uoc.edu

Abstract. The seminars are a learning mechanism that serve to complement regular education with more specialized content. Different learning materials are generated from the seminars activity, such as documentation, presentations, videos, exercises… These materials can also be used for teaching purposes in other contexts. The management and distribution of these materials is normally done by hand or stored in a digital repository. The problem with these tools is that they facilitate storage management but are not focused to support their formative application. This article describes the experience of creating a web tool that allows to take profit of the materials generated in a seminar called Zaragoza Linguistics for later teaching use. The functionality has been implemented using web scraping techniques and the YouTube web services API.

1 Introduction

Zaragoza Linguistics is a monthly scientific seminar organized by General Linguistics professors at the Faculty of Philosophy and Letters of the University of Zaragoza. The main objective of the seminar is to disseminate topics of interest on general and applied linguistics [7]. To this end, prestigious national and international speakers are invited. Access to the seminar is completely free unless there is some kind of specific workshop that requires payment. With regard to the target audience, the seminar is mainly aimed at students of the Philology degrees and anyone who is interested in these topics. In this sense, there is no limitation to be able to access it.

As for the organization, the seminar is structured in the form of lectures with a discussion time at the end, or in the form of practical workshops where participants participate actively with exercises and guided practices. All the sessions of the seminar are recorded by a professional team of contents creation of the University. Thus, after an editing and revision process, all the resulting videos are uploaded to a YouTube thematic channel belonging to the Faculty of Philosophy and Letters [12] so that they are accessible to any interested person. In addition to the videos, the seminars generate

© Springer Nature Switzerland AG 2019
F. Xhafa et al. (Eds.): INCoS 2018, LNDECT 23, pp. 395–405, 2019.
https://doi.org/10.1007/978-3-319-98557-2_36

other types of materials such as documentation that the speaker can provide, presentations in electronic format and other resources.

On the other hand, the seminar has several teaching applications. First, the seminar can serve students of linguistics subjects of the Philology degrees to delve into some of the topics covered in class. In this sense, the seminar constitutes a second opportunity to revisit certain topics but with the additional motivation that they will be presented by highly prestigious researchers and professors (in some cases they are the authors of the textbooks that are recommended in the subjects).

Another teaching application of the seminar is the use of the materials generated in it as part of the content covered in the subjects. These materials present as an advantage the fact that the contents presented have been created by specialists of the subject treated, so that the quality of the contents cannot be better. For this reason, in many cases, teachers use in their classes the videos or documents that have been generated in a seminar talk to present the subjects of their subjects.

The last teaching application consists of using the seminar as part of the evaluation of the subjects [3]. In this case, the teachers offer the possibility of obtaining an extra grade in the subject by attending the seminar and doing some type of activity related to the subject taught in the seminar. The activities can be a summary of the contents seen, answer some reflection questions, and perform a work extending the explained or relating it to other topics.

The aspect dealt with in this article refers to the management of information generated in the seminar for educational purposes. Currently, the tool used for this purpose has been a WordPress blog [2] associated with the seminar [13]. The blog serves as an element of centralization and distribution of all information that is generated directly from the seminar such as videos or associated documentation, as well as indirectly generated information such as blog entries on topics that it will be discussed in the seminar or topics related to the talk given. This blog is managed directly by the teachers who organize the seminar, although anyone can participate freely by adding comments in the blog entries. A blog can be an interesting tool [11] to disseminate and manage information for a closed group of people, however when the number of participants and the number of materials to be managed grows too much, then the blog is no longer a suitable tool, and it has certain limitations. In this article it will be discussed the limitations that have been detected in the blog, and it will be described a web application that it has developed in order to improve the access to the materials generated from the seminar.

The structure of the paper is as follows. In Sect. 2, the limitations of the blog as a tool for disseminating the materials generated in the seminar will be discussed. Section 3 will present the specification of requirements of a web application that aims to improve access to information associated with the seminar. Next, in Sect. 4, a brief introduction to the web scraping techniques and the YouTube web services API will be made, which is the technology that has been used to create the web application. Then, in Sect. 5, the functionalities of the implemented web application are shown. Finally, in Sect. 6 a set of conclusions and lines of future work will be presented.

2   The Limitations of the Zaragoza Linguistics Blog

Zaragoza Linguistics is a blog that is hosted on WordPress. It is used by teachers of general linguistics who are part of the PSylex research group as a means of publicizing the activities carried out around the Zaragoza Linguistics seminar as well as other activities such as congress advertising, dissemination of articles, manuals and other type of documentation. Figure 1 shows the main page of the blog.

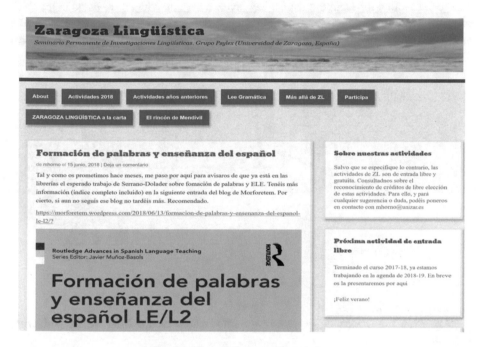

Fig. 1. Blog of Zaragoza linguistics.

Essentially, the blog is organized in several sections that show the activities of the seminar planned for the current year, the activities conducted in previous years, other activities carried out in which members of the research group participate, a section called "El Rincón de Mendivil" with entries on topics related to general linguistics written by one of the professors of the Psylex research group, and a section called "Zaragoza Linguistics on demand" that allows access to the links to the recorded videos of the seminar that are hosted on YouTube.

With respect to the activity of the seminar, the blog is used as follows [9]. At the beginning of each academic year the agenda of the seminar activities is published as a blog post. During the course, before and/or after each seminar activity, entries are made on the blog to promote each activity. In the previous entries to an activity an introduction is made about the talk that is going to be taught and about the researcher who gives it. There may also be a post-seminar entry in which the talk is discussed or used to disseminate material provided by the speaker or links to deepen. In all entries, anyone can participate by entering comments, marking the entries with a "like" or sharing the entry in a social network. All the activities of the seminar are recorded by a multimedia professional team. The resulting videos are uploaded to a YouTube channel belonging to the Faculty of Philosophy and Letters of the University of Zaragoza. Likewise, the links to the videos are published on the blog. To do this, there is a section in the blog called "Zaragoza Linguistics on demand" where the links to all the recorded videos of the seminar that are hosted on YouTube appear. The videos are presented categorized into 4 different thematic areas. In each area the videos are listed in the form of a link that directs to a page with information about the video: link to the YouTube video, summary of the talk, associated links or any additional documentation. An example is shown in Fig. 2.

Fig. 2. Example of link to video.

The main limitations for accessing the information on the blog are the following [6]:

- There is no specialized search system on the content of the videos. The only way to search is to use the generic search engine that has the blog that searches about the keywords of the pages that make up the blog. Thus, the results of the searches are imprecise and in many cases are not significant with respect to the contents sought.
- From the user's point of view, the possibilities of interaction with the videos is reduced since it is only possible to visualize the content. For example, it is not possible to add comments or download materials.
- The page in which the videos are shown, although it is responsive, does not allow a good user experience.
- From the manager's point of view, the possibilities of change and maintenance are expensive. For example, a change of the categories or of the videos that make up each category would make it necessary to manually transfer the links from one category to another.

- In general, the possibilities of exploiting the material generated are very limited, since any modification of the behavior of the blog can only be done using the specific features offered by the WordPress system.

3 Application Requirements

In order to improve the situation described, the development of a web application that improves the user experience, facilitates greater possibilities of interaction with the contents and improves the management of contents has been proposed. A web application presents the same advantages as a blog in terms of universal access through the web, but without its functional limitations. In this sense, the objectives to reach with this development were:

- Create a layer of value-added services oriented towards the user.
- Facilitate the association of complementary resources to the videos.
- Develop a search engine on the contents of the videos.
- Improve the visual accessibility of the materials generated from the seminars.
- Facilitate information management tasks for seminar administrators.

In order to satisfy these requirements, a set of functionalities aimed at two types of actors was defined: the general users and the administrator of the application. In this sense, Figs. 3a and b show the functionalities proposed for a user and an administrator respectively.

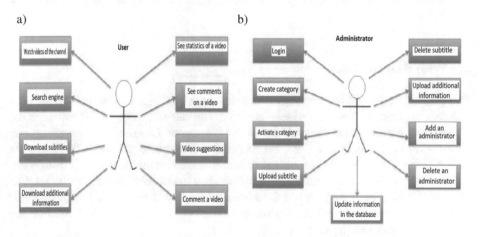

Fig. 3. (a) Functions defined for a user, (b) Functions defined for an administrator.

4   The YouTube Service API and the Web Scraping Techniques

The main functionality of this application is based on two technologies:

- Web scraping techniques. These are programming frameworks that allow writing programs that behave like a web browser and recover web pages [5]. In addition, the programmer offers a set of methods for processing with different purposes [4]: (a) Browse freely on the data of a page, (b) Search information on a page using search patterns based on regular expressions, (c) Automate recognition of certain information structures to carry out their recovery; (d) Transformations of the structure of the information retrieved to be stored in a database and thus be able to be exploited; or (e) Recovery of information fields specific to a document. Likewise, an advantage of these frameworks is their ability to recover from coding errors that often appear on web pages [10] such as labels that are not closed, incorrect attributes...
- YouTube service API [1]. It is a set of web services [8] that allow a user to perform certain operations on videos and YouTube channels. To use them it is necessary to have an account and be authenticated in the system. The operations can be of consultation such as access to various fields of information about the videos (statistics, number of downloads, comments made on a video...) or modification such as adding a video to a channel, adding comments to a video, modify the access configuration... The API can be used directly using the web service, or being embedded within a program. In this sense, the API is compatible with the most used programming languages. In general, the results of a query are returned in JSON format.

5   Web Application

The general design of the application is summarized in Fig. 4. The web application retrieves the links to YouTube videos as well as the categorization of the videos of the "Zaragoza Linguistics on demand" tab of the "Zaragoza Linguistics" blog using web scraping. From the recovered links, the application connects to YouTube, and using the web services API retrieves certain fields of information about the videos from You-Tube. With the information retrieved, a website is dynamically generated with a set of services aimed at both the user and the administrator. These services allow queries and certain update operations on the videos, as well as perform management tasks on the information shown to the user.

Regarding the persistence of the system information, a relational database consisting of 5 tables has been used:

- Table of Administrators: it stores information about the administrators of the application.
- Table of Videos: it stores the information collected via web scraping of the videos found in the blog of Zaragoza Linguistics.

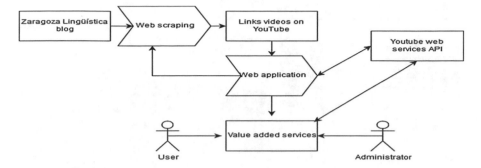

Fig. 4. General scheme of operation of the web application.

- Table of Categories: it stores the categories used to classify the videos in the Zaragoza Linguistics blog extracted by the web scraping technique.
- Table of Subtitles: it stores subtitle files associated with videos.
- Table of Additional Information: it stores the associated files as extra information related to a video.

In order to describe the implemented functionality, we will distinguish between the public interface and the administrator-oriented interface.

(a) Public interface. In the main page of this interface there are two zones well differentiated. In the upper part there is information about "Zaragoza Linguistics on demand " and a "Start Session" button for administrators; and in the lower part the grouping categories of the videos are shown next to a search button, that is connected to a search engine that allows searching the contents of the videos. When you click on a category, a list of all the associated videos associated is displayed, together with relevant information of each video (name of the video, number of views, number of comments, number of likes, duration and quality of the video, subtitles or extra information in case it exists). Figure 5 shows the categories and a selected category.

When a specific video in the list is clicked, a new page is created to see the video and access to its related information. This page shows all the video data that already appeared in the list, a brief description of the video, a set of descriptive labels, and a set of services for downloading subtitles by choosing the language and for downloading a document with additional information of the video (it can be an image, a pdf, a txt, etc.). Figure 6 shows the page of a selected video.

Finally, the user has a search engine about the videos. To use it, a set of words is entered in the search engine, and two search processes are launched, one about the subtitles of the videos (if they exist) and another that is done directly on YouTube. As a result, a consolidated list of links to videos is generated from the results of both searches. In the case of the search about subtitles, the results generated are links to the instants within the contents of a video where the searched terms have been mentioned exactly. Figure 7 shows an example of searching for the term "history".

Fig. 5. List of categories and a specific category.

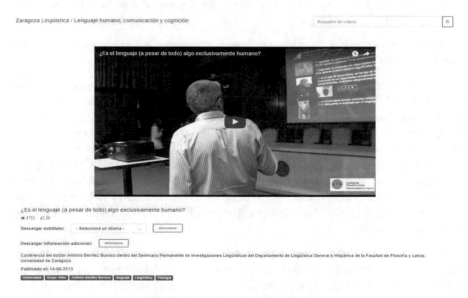

Fig. 6. Example of recovered video

(b) Administrator interface. The page is formed by a side menu with links to the functions assigned to the administrator and a central area that contains a set of icons that represent links to the same functions. The administrator has 4 functions defined: "Category management", "Add/Remove administrator", "Upload/Delete subtitle" and "Upload additional information". When the "Category Management" function is selected, the categories currently defined are displayed, and an option to create new categories is displayed. For this case there is a form where the administrator can enter the name of the new category as well as select the videos

Resultados de búsqueda: "historia"

Buscador de vídeos   🔍

Vídeo	Título
	Lenguaje y cognición
	👁 4115 visualizaciones 💬 3 comentarios
	👍 20 ⏱ 1h 12m 🗒 Subtítulo 🔗 Info extra
	Este término se ha pronunciado en el vídeo en los siguientes momentos: 💬
	Minuto: 00:05:18 - "interés a un poco de historia hace (...)" Minuto: 00:08:39 - "en la historia de las ideas en el ataque (...)" Minuto: 00:13:21 - "conocer esta historia no los colobos (...)" Minuto: 00:15:55 - "estructurado en la historia de esas (...)" Minuto: 00:20:32 - "cuando pedrosa y muchísima historia (...)" Minuto: 00:47:08 - "historia inah (...)"
	¿Pero qué es una lengua? Biología, historia y cultura en el lenguaje humano
	👁 3172 visualizaciones 💬 2 comentarios
	👍 20 ⏱ 1h 45m 8s 📺 HD
	XIV Reunión Científica de la Fundación Española de Historia Moderna
	👁 341 visualizaciones 💬 1 comentarios
	👍 2 ⏱ 1h 23m 3s 📺 HD

Fig. 7. Results of the search for the term "history".

that will be part of this category. The function "Add/Remove administrator" allows you to delete an existing administrator or add a new one by entering the data of the same. The "Upload/Delete Subtitle" function allows you to delete subtitle files associated with a video or add them. In the latter case, before making the association, the system checks that it is a subtitle file. Finally, the function "Upload additional information" allows associating information to a video. To do this, from a form, the administrator must choose the video and the file that you want to associate. If more than one file should be associated, they can be provided in a compressed file (in a .zip format) (Fig. 8).

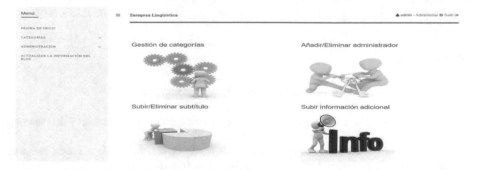

Fig. 8. Administrator interface.

6   Conclusions and Future Work

The article presents a web application that improves the access and management to the materials generated in a scientific diffusion seminar called "Zaragoza Linguistics". The key elements have been the creation of a set of value-added services based on the use of the YouTube web services API, and the extraction of the starting information from the original hosting in a WordPress blog. Regarding the teaching utility, the developed tool provides improvements in the following aspects:

(a) Encourages access to information. In this sense, it offers an intuitive and friendly access to the videos and all the relevant information about them extracted directly from YouTube. And on the other hand, it allows a precise search based on You-Tube's own search engine and on the possibility of searching on subtitles (provided they have been associated). In this last case, the search will be even more precise, since it will allow to search on the own content.
(b) Allows associating complementary documentation to the videos. Thus, this option could be used to associate exercises based on the contents of the videos such as reflections, arguments, true/false questions…

The lines of future work are two. On the one hand we want to enrich the information about a video using linked data and SPARQL queries, and on the other hand we want to add a whole layer of services oriented to the use of materials for evaluation purposes. For this purpose, we intend to add to the application the teacher actor, who will be able to create different types of questions or tests based on the contents of the videos, and the student actor, who will be able to access the questions or tests, keep information about the tests carried out and the results obtained.

Acknowledgments. I would like to thank the student Jhersy Andrés Nolasco Arévalo, who carried out the coding of the application described in this paper. This work has been partially supported by the European Commission through the project "colMOOC: Integrating Conversational Agents and Learning Analytics in MOOCs" (588438-EPP-1-2017-1-EL-EPPKA2-KA) and the projects Santander-UCM GR3/14 (group 962022) and eLITE-CM S2015/HUM-3426.

References

1. API YouTube. https://www.youtube.com/yt/dev/es/api-resources.html
2. Brazell, A.: WordPress Bible, Vol. 726. Wiley (2011)
3. Cheng, X., Dale, C., Liu, J.: Understanding the characteristics of internet short video sharing: YouTube as a case study. arXiv preprint arXiv:0707.3670 (2007)
4. Villamor, J.I.F., Garcia, J.B., Fernandez, C.A.I., Ayestaran, M.G.: A semantic scraping model for web resources-Applying linked data to web page screen scraping (2011)
5. Glez-Peña, D., Lourenço, A., López-Fernández, H., Reboiro-Jato, M., Fdez-Riverola, F.: Web scraping technologies in an API world. Briefings Bioinform. **15**(5), 788–797 (2013)
6. Heuristics of Nielsen. https://es.wikipedia.org/wiki/Heurísticas_de_Nielsen
7. Chéliz, Mª.C.H., Ibarretxe-Antuñano, I., Giró, J.L.M.: Panorama actual de la ciencia del lenguaje. Primer sexenio de Zaragoza Lingüística. KLC 5. Zaragoza: PUZ (2016)

8. Masse, M.: REST API Design Rulebook: Designing Consistent RESTful Web Service Interfaces. O'Reilly Media, Inc. (2011)
9. Sarasa-Cabezuelo, A., Horno-Chéliz, M.C., Ibarretxe-Antuñano, I.: Zaragoza Lingüística o cómo divulgar la ciencia del lenguaje a través de la innovación docente. Actas de la Jornada Las TIC en la Enseñanza: Experiencias en la UCM, pp. 119–124 (2017)
10. Vargiu, E., Urru, M.: Exploiting web scraping in a collaborative filtering-based approach to web advertising. Artif. Intell. Res. 2(1), 44 (2012)
11. Williams, J.B., Jacobs, J.S.: Exploring the use of blogs as learning spaces in the higher education sector. Australas. J. Educ. Technol. 20(2), 232–247 (2004)
12. YouTube channel Faculty of Philosophy and Letters. https://www.youtube.com/channel/UCG_fXqOLea9FSTLoWZieaqw
13. Zaragoza Linguistics. https://zaragozalinguistica.wordpress.com/

Providing Timely Support to Students in Educational Virtual Worlds

Anupam Makhija[1], Deborah Richards[1(✉)], Santi Caballé[2], and Jordi Conesa[3]

[1] Macquarie University, Sydney, Australia
anupam.makhija@students.mq.edu.au,
deborah.richards@mq.edu.au
[2] Faculty of Computer Science, Multimedia and Telecommunications,
Universitat Oberta de Catalunya, Rambla Poblenou, 156, 08018 Barcelona, Spain
scaballe@uoc.edu
[3] eHealth Center, Universitat Oberta de Catalunya,
Rambla Poblenou, 156, 08018 Barcelona, Spain
jconesac@uoc.edu

Abstract. Educational Virtual Worlds (EVWs) provide an interesting and engaging educational platform to promote student learning. Virtual classrooms should be designed to be responsive and adaptive enough to judge the emotional level of the learner and should be capable of providing required support and feed-back to maintain the engagement of the learner to achieve their learning goals. Virtual agents with affective capabilities have potential to improve the effectiveness and efficiency of the e-learning and virtual learning platforms. This research study will explore understanding of learners' emotional states that evolve during the learning process within a EVW by drawing on theories related to epistemic emotions and develop approaches to recognise and intervene in a timely manner based on the individual's feelings being experienced at that moment. To derive that emotional state, this paper also presents a multi-data and multimodal approach to judge the underlying emotional state of the learner within EVW and the most appropriate intervention for that learner.

1 Introduction

The 21st century classroom has not significantly transformed despite promises of a digital education revolution. We need intelligent systems in our classrooms that understand the individual learner and exploit the potential benefits of advanced educational technologies such as Education Virtual Worlds (EVWs). In history classes, students could witness recreations of historical sites and events to recognise that different perspectives and explanations exist. Science students could use EVWs to go on virtual field trips, observe phenomena, collect data, conduct experiments and learn science inquiry skills by practising doing what scientists do. We also need intelligent systems that detect the learning needs of users, including their current knowledge, attitudes and motivations, and provide support as needed. The project presented in this paper involves use of an EVW for science inquiry together with methods from artificial

© Springer Nature Switzerland AG 2019
F. Xhafa et al. (Eds.): INCoS 2018, LNDECT 23, pp. 406–419, 2019.
https://doi.org/10.1007/978-3-319-98557-2_37

intelligence, (data mining, language processing, and agent-technology) to identify when students need support and how best to deliver that support based on the EVW's deep understanding of the student.

There is a strong connection between affect and learning [1–3]. In the journey towards facilitating deeper learning and better understanding of new concepts, learners go through various positive and negative states of the mind. Previous studies indicate that it is crucial to manage negative emotions effectively to maintain the learning process at that time otherwise students might get disengaged and give up on the learning process [4]. Pedagogical virtual agents can act as peers or tutors to guide the learner through the learning material or environment (see the comprehensive review by [5]). According to the persona effect [6], learners attribute social ability to these lifelike agents and this has a positive influence on the learning experience within the EVW. We aim to use affective animated pedagogical agents within an EVW to help students to overcome their negative emotions and motivate them to stay on the optimal path of learning by providing right support at right time.

We present an overview of the related literature in the next section. Section 3 presents our research questions and our proposed framework for emotion detection. Section 4 describes evaluation of the framework. Conclusions are given in Sect. 5.

2 Literature Review

This section reviews literature to understand the emotions related to learning and engagement, what affective states may need intervention, when to intervene and how.

2.1 Defining Epistemic Emotions and Engagement

There are many psychology-based theories regarding categorisation of human emotions. Emotions can be divided into two main types namely basic and non-basic emotions. Anger, sadness, surprise, happiness, fear and disgust are the examples of basic emotions while confusion, boredom, engagement, curiosity and frustration fall under the umbrella of non-basic emotions [4, 7]. While learning new a concept, non-basic emotions are more frequently experienced. Emotions that are generated during the learning process are referred to as epistemic emotions. These emotions have characteristic of being dynamic and are associated with cognitive activities [8].

Human-computer interfaces should have potential to support learners to overcome their negative affective state and continue with their learning [9, 10]. These negative emotional states should be detected in the early stages to avoid frustration and positively influence the learning goals. During the learning process, certain patterns are observed in terms of emotional transition from one state to another [11]. This transition pattern allows prediction of the next state and may help to guide how to respond to the learner's current state to move or keep them in the desired emotional state. During learning activities, as complexity of the task increases, a learner can get into a state called stuck and go through a phase of non-optimal experience. The feeling of being 'stuck' can be defined as a feeling of being out of control along with lack of focus, mental fatigue and distress. In this stage the likelihood of loss of motivation to learn increases [12].

Figure 1 illustrates that while working to achieve the superordinate goal of completing a task, a learner is more engaged initially. In the next phase is the state of confusion that arises because of some misconceptions and unexpected feedback, and learner reaches an impasse (stuck). Subsequently the learner enters into the state of frustration that emerges because the impasse is not resolved. Finally, boredom emotion is experienced because of continued failure and eventually the learner gets disengaged from the task [4].

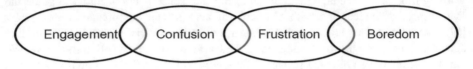

Fig. 1. Emotional transition pattern during learning D'Mello, Lehman [11]

Student engagement is a significant factor in the learning process. Disengagement is considered to be a key factor influencing student dropout rates [13]. The academic performance of students can be predicted based on four types of engagement: academic, psychological, behavioural and cognitive. Appleton, Christenson [13] have suggested that cognitive (e.g. goal setting) and psychological (e.g. belonging) types are equally important to determine the academic performance of students as compared to academic and behavioural engagement factors but there is insufficient research evidence concerning these engagement categories.

2.2   Challenging Students and Providing Help at the Right Times

Confusion is one of the significant epistemic emotions that is associated with complex learning and influences the learning process in a positive manner by improving learners' engagement [11]. This research idea aligns with the theory suggested by VanLehn, Siler [14] that highlights the positive effects of impasse driven learning where understanding of concepts is more obvious when the learners reach an impasse and engage themselves in effortful cognitive activities to come out of that state. Recent research claims that confusion allows deeper learning if it is managed within the zone

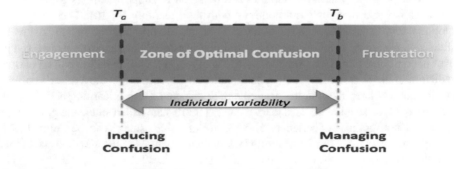

Fig. 2. Zone of optimal confusion [15]

of optimal confusion as illustrated in Fig. 2. This means that induced confusion should be resolved within certain thresholds in terms of discrepancy and duration otherwise it might lead to frustration, disengagement and boredom and can influence learning in negative direction [11, 15].

VanLehn, Siler [14] emphasised the need to analyse the activities performed during human tutoring that help students to learn successfully. From their study, it was concluded that it is most beneficial for students to reach an impasse before they are able to make learning gains. Students should be allowed to come up with their own explanations as compared to the explanations received from tutor. If needed, tutors can provide content free prompts to encourage problem solving and meta-cognitive reasoning. Where tutor explanations are required, they should be very short. Consistent with Van Lehn's notion of impasse learning, Kapur and Bielaczyc [16] proposed a theory of productive failure that suggests that long term learning will be greater if the student is allowed to work with problem solving activities without initial instructional support from the teacher. Once the student has engaged with the problem, even if they fail to find a solution, they can revise their theories through a consolidation phase led by the teacher. Through this process the student learns more deeply. The past empirical studies have demonstrated the efficacy of the theory of productive failure and impasse driven learning that motivates a student to take responsibility for their own learning [14].

Arguel and Lane [15] demonstrated that induced confusion can encourage cognitive disequilibrium in digital environment and this confusion should be managed for deep learning. One of the ways to incite cognitive disequilibrium can be to compare prior and current knowledge or by providing some inconsistent information. To avoid the negative effects of confusion during complex learning and to help students come out of an impasse or cognitive disequilibrium, various strategies have been proposed [17]. Related research studies claim that curiosity is another significant intrinsic emotion that can prove to be a motivating factor to promote learning by enhancing critical thinking and observational skills of learners and it should be simulated to achieve positive outcomes in the learning [18, 19].

2.3 Identifying When the Learner is Stuck or Confused and Providing Support

The crucial aspect of providing appropriate help in the form of encouragement and support is to correctly recognise the learners' epistemic emotions otherwise it can negatively affect the learning. In the case of an EVW, or other virtual learning environment, it is a complex process to judge when a learner needs help to move forward. Such environments also lack the social element that is a critical factor of successful learning [20].

In order to provide appropriate help and support in EVW, it is essential to identify the engagement level of the learner. While many research studies take this direction, much work remains to correctly identify the satisfaction level of students in this environment. Eliciting learner's emotions in Virtual Learning Environments is a challenging aspect [21]. Some of the affect detection techniques that have been implemented so far are the observation of learner's emotions by human observers [2], self-reporting [22], measuring physiological data [23] including facial expressions [1, 20, 24], gaze patterns

[25], use of machine learning techniques like Decision-Trees, Support Vector and Bayesian networks [2, 17], analysis of learning data generated during usage [26]. Another study demonstrated the use of combination of physiological signals, such as conversational cues, facial expression, skin conductance, heart beat and body temperature [27]. All of these methods to elicit the learners' affective state have their own pros and cons. Some of these techniques like measuring physiological signals need proper lab equipment to be set up and that makes it hard to use in all scenarios [17], particularly with school children. Analysis of student interaction data with an EVW provides the benefit of collecting data without interrupting student learning activities and also does not need any extra lab equipment. Tian, Gao [26] demonstrated this technique in their study where they used system interaction data such as temporal positions, location data along with physiological data to determine user's emotional state during usage of EVW.

Human beings experience the world through five senses that relates to the concept of multiple modality. This is an emerging research trend in computing domains including affect detection. In contrast to single modality input, multimodality approach allows to integrate input from a variety of channels (Audio, Video, Text etc.) using a variety of fusion algorithms for accurate analysis and prediction of outcome. Relying on more than one modality can help in overcoming some of the issues faced during single modality approach as information through different modalities can complement each other and there is more probability of accurate outcome [28, 29].

Once the need for support is detected, methods are needed to deliver timely and appropriate support. The techniques such as self-explanations [30], verbal or non-verbal feedback [31–33], providing motivation [12, 34], improving self-efficacy skills [35] and scaffolding [36, 37] can be implemented to regulate learners' emotions and provide appropriate help to them to maintain learning.

A study conducted by Kim, Baylor [38] demonstrated that empathic responses by animated agents can influence learning positively by changing their affective state. McQuiggan, Robison [39], investigated the effectiveness of two types of empathy named reactive and parallel empathy by proposing empathy modelling framework named 'CARE' in an enquiry-based learning environment called 'Crystal Island'. The parallel empathy refers to mimicking the affective state of user while in case of reactive empathy, virtual agent displays different emotion in order to help user to regulate their emotions during complex learning task. This study suggests that demonstrating both types of empathy can make positive impact on learning outcomes.

3   Research Framework, Design and Methodology

The main aim of this research study is to develop techniques that can positively aid the learning of students by providing appropriate affective and cognitive support in EVWs by recognising and responding to negative emotional states that can generate during the learning process. To align with this objective, we intend to explore the following research questions:

- *How can the system detect when a learner reaches an impasse (i.e. stuck or confused) in real time within an EVW?*

- *How can virtual agents provide timely help to a learner within an EVW based on their affective state?*

In answer to these research questions, this study aims to create a framework to develop a strategy to provide timely help to the learners within EVWs based on the data elicited through multiple channels. To evaluate our framework, we will use the design-based research methodology that involves development and use of an artefact, an EVW in our context.

This study will create learner profiles, using data such as the learner's engagement score calculated through the Student Engagement Instrument (SEI) [13] and analyse in real-time various forms of system interaction data including a variety of data types and modalities (e.g. free text, scores, graphical, keystrokes, navigation paths, etc.). As discussed in the Literature Review, data relating to the learner and understanding their possible need for assistance may include psychological data, verbal, non-verbal, audio and video data. As a fundamental individual difference, we are also interested to explore if personality has any relationship to student learning. To capture personality we will use the Big Five Factor model of personality and use instruments, such as IPIP-International Personality Item Pool [40], depending on the age of the learners. Another alternative would be to infer personality from behavioural data, similar to the approach by Capuano, D'Aniello [41] using social network data. Table 1 summarises the possible types of data and categories of multimodal channels (audio, video, text, physiology and interaction with virtual world) to be collected and the measurement variables [29].

Table 1. A framework to measure learner's affective state in virtual world.

Modality/Data source	Evaluation criteria
Interaction with learning Content (specific to each EVW)	Number of agents talked with, Number of inventory items collected, Workbook questions answered correctly, Workbook completed
Behavioural patterns (common to EVWs)	Navigation Pattern followed, Time spent in virtual world, Number of areas visited/ available number of areas
Video	Body postures, Eye Gaze, Facial Expressions
Physiology	EEG (Electroencephalography- Brain Signals), ECG (Electrocardiography- Heart Response), EMG (Electromyography- Muscle Response), EDR (Electrodermal- Skin Conductivity), Respiration
Audio acoustic-prosodic features	Pauses, Intonation, Speech rate, Volume, Intensity
Text linguistic/semantic features	Keyword spotting, Sentence parsing, Sentiment analysis
Learner Engagement Profile	Cognitive, Emotional/Psychological, Academic
Learner Personality	Personality Traits

These modalities need to be fused to determine the affective state of learner, their learning performance in the virtual world and the most appropriate time and method to provide assistance. Past empirical studies suggest that the fusion of more than one modality for emotion detection allows to classify emotions with higher precision of accuracy yielding better performance [42]. Baltrušaitis, Ahuja [28] discussed two approaches to integrate multiple modalities namely early fusion (feature-based and late fusion (decision-based) (see Fig. 3). The first approach refers to combining individual modalities in early stage before classification is done for individual modalities; while late fusion approach combines the data after classification is done by individual modalities. The proposed model (Fig. 4) suggests the early fusion approach and shows how these different sources of data and modalities can be fused together to provide a picture of the emotional state of the user. Depending on the emotional state and the profile of the learner, a decision can be made by the system (or the pedagogical agent) about the appropriate action.

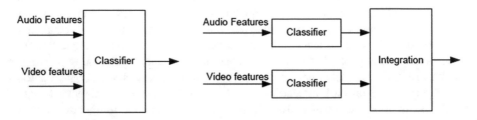

Fig. 3. Early and late fusion approaches [42]

To exemplify how the framework presented will be used to identify the learner's emotional state, we consider the framework in the context of the EVW to be used in our experimental studies. A multimodal channel network (see Fig. 5) for emotion detection and provision of affective feedback will be used with four different data channels, namely Psychological; Bio-physiological; Motor-behavioural and Learning Analytics [46–49]:

- *Psychological:* The representation of the learner's affective profile will be exploited, instantiated by using data coming from non-verbal input mechanisms from pictorial scales for learners to explicitly express their affective state anytime and its evolution during the class (see Fig. 6-left).
- *Bio-physiological:* The data from various physiological sources, such as EEG for Brain Signals [43], ECG for Heart Response [44], EMG for Muscle Response, EDA (Electro-dermal Activity) for Skin Conductivity and Respiration [45] can be mapped in to the emotions experienced by the learner (see Fig. 6-right).
- *Motor-behavioural:* Body movements will be considered to elicit learner's emotion to track facial movements and body gestures of learners by using specific equipment, such as body sensors (see Fig. 7) and webcams in order to extract data pertaining to learners' physical features (head, arms, cheeks, lips, eyes and eyebrows, etc.) [7, 48].
- *Learning analytics:* Textual responses to questions can be (semi-)automatically scored to determine progress and correct understanding. Sentiment analysis of posts in an online chat forum with peers may reveal the student's emotional state and attitude towards the activity [49]. In addition, the prosodic features and lexical contents such as pitch, speed etc. can be used as their acoustic cues for emotion recognition [46].

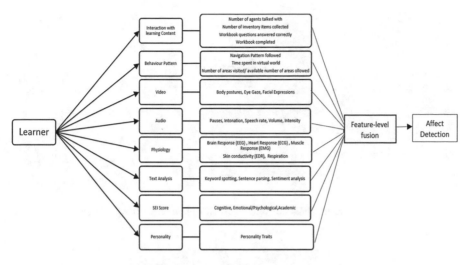

Fig. 4. Fusion model for detecting when and how to provide assistance

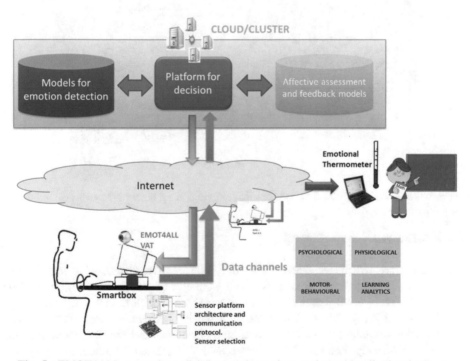

Fig. 5. EMOT4ALL network to elicit learners' emotions and provide affective feedback

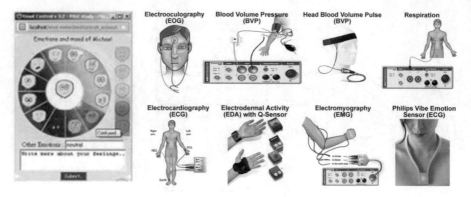

Fig. 6. Non-verbal mechanism (left) [46]; Bio-physiological signals (right) [47]

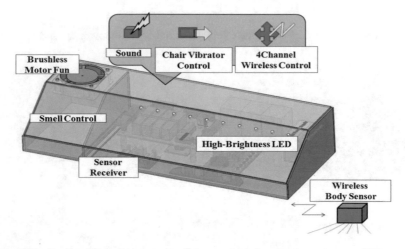

Fig. 7. SmartBox functions to elicit emotions from body movement [48]

4 Framework Evaluation

This framework will be evaluated by a study that involves detection in real time of when a participant feels stuck or gets into the state of cognitive disequilibrium while using an EVW. We will conduct an experimental study with human participants and their use of an EVW. Some participants will receive help when detected. Some will receive help at regular intervals not based on their interactions/profiles. Some students will receive no help (control). We will use an EVW created for science inquiry that represents an ecosystem for a fictitious island. The EVW has been created using Unity 3D for teaching biology and science inquiry concepts to high school students. It has also been used in studies with undergraduate students. It consists of five main locations, that are the village, the research lab, the hunting ground, the animal counting station and the weather station and that can be visited by students by clicking on the map in the

EVW. There are virtual agents present in each of the main locations that can converse with students to provide help in the form of written answers to an already defined set of questions. The questions are based on the facts presented during the visit to the island covering scientific concepts about biological systems.

Participants will be allowed to navigate through the island following any path. They can talk to different characters during their journey in the island and gather relevant information that helps them to answer certain questions asked from them in the form of online workbooks. The workbooks will guide them about what learning activities they should do each day in the EVW. All the navigational information along with log books will get stored in the database on the server.

To answer the first research question, the data from navigation pattern, log files and work books will be used to determine whether student is stuck and needs help while visiting the EVW. At the beginning of the experiment, participants will be required to provide informed consent and complete some survey questions about themselves. The purpose of these survey questions would be to assess various aspects such as the participant's personality, technology preference, student engagement score etc. At the start of the interaction with the virtual environment, the participants will be given the instructions regarding how to use the system through dialog with the first character they meet. As learners use the EVW, their interaction data will be stored in the log file in regular intervals in terms of which place are they visiting, which character are they talking to, when are they idle and not using the system, the inventory items they picked on the way etc. Each of these data points will provide the starting point to determine the affective state of the learner at that particular time. Learners' answers to the workbook questions can be stored and answers can be automatically marked to be used to aid reflection of their performance. To answer the second research question, the existing agent dialogs will be adjusted to provide appropriate feedback to participants to maintain their learning in the EVW.

In this particular scenario, we can imagine that the student has talked to 2 agents. For the problem they are solving, there are 3 characters they need to talk to. Audio input processing indicates the fast speech rate with slow tempo and high intensity for this learner. The body posture is detected to be leaning forward and eyes are fixated on the screen. The SEI score of the learner indicates that their attitude towards learning is positive and student is engaged cognitively, psychologically and academically as is reflected by high score value. After deep understanding of all these elements through different modalities, the quantification of learner's behaviour (using the given condition) suggests that the student is confused and needs help to continue on the learning path. By taking the dimensions of engagement score and personality traits into account, it makes sense not to intervene straight away, rather monitor the progress of the learner for some time and provide appropriate tips (e.g. 'suggesting them to use alternative strategies' or 'speak to another character') to help them come back to the optimal path of learning.

We devise a formula to unify the features from dissimilar modalities along with engagement score and personality dimension for affect recognition in EVW.

If ('SEI score' is 'high' AND 'Speech rate' is 'fast' AND 'Personality' is 'extravert' AND 'Body Posture' is "leaning forward' AND 'Eye Gaze' is 'fixed'

AND' Navigation pattern' is 'teleporting' AND 'Agents talked to' is less than 3 AND 'workbook completed' is less than 50% AND 'QuestionsCompleted' is 6/10 AND 'Heart rate' is 'high' AND 'Respiration' is 'Fast')
Then learner is 'confused or stuck'

For comparison, we imagine an alternative scenario that represents the outcome for a student being 'stuck' by employing the same condition used for first scenario but his SEI score is very low. In that situation, it would be more productive to provide support earlier and offer a concrete suggestion of what to do to avoid the detrimental effect of confusion on the learner.

5   Conclusion

The motivation for this research is inspired by the fact that awareness of the learner's affective state during the learning process in an EVW can bring substantial improvement in their learning. We believe that in the near future, multimodal input can be used more effectively to better understand the learners' emotion during complex learning and suggest appropriate help strategies matching with the individual need. The suggested solution offers a flexible approach with potential to add new channels in the future to elicit the learner's emotions with more accuracy. Furthermore, this research study aims to determine an effective support strategy for the learners by taking individual engagement scores and personality traits into consideration and responding appropriately to engage the learner with the learning activity and bring them back to the optimal learning path.

Acknowledgments.   This research was funded by the European Commission through the project "colMOOC: Integrating Conversational Agents and Learning Analytics in MOOCs" (588438-EPP-1-2017-1-EL-EPPKA2-KA).

References

1. Kort, B., Reilly, R., Picard, R.W.: An affective model of interplay between emotions and learning: Reengineering educational pedagogy-building a learning companion. In: IEEE International Conference on Advanced Learning Technologies, Proceedings. IEEE (2001)
2. Woolf, B., et al.: Affect-aware tutors: recognising and responding to student affect. Int. J. Learn. Technol. 4(3–4), 129–164 (2009)
3. Wu, C.H., Huang, Y.M., Hwang, J.P.: Review of affective computing in education/learning: trends and challenges. Br. J. Educ. Technol. 47(6), 1304–1323 (2016)
4. D'Mello, S., Calvo, R.A.: Beyond the basic emotions: what should affective computing compute? In: CHI 2013 Extended Abstracts on Human Factors in Computing Systems. ACM (2013)
5. Johnson, W.L., Rickel, J.W., Lester, J.C.: Animated pedagogical agents: Face-to-face interaction in interactive learning environments. Int. J. Artif. Intell. Educ. 11(1), 47–78 (2000)

6. Lester, J.C., et al.: The persona effect: affective impact of animated pedagogical agents. In: Proceedings of the ACM SIGCHI Conference on Human Factors in computing systems. ACM (1997)
7. Ekman, P.: An argument for basic emotions. Cogn. Emotion **6**(3–4), 169–200 (1992)
8. Pekrun, R., Linnenbrink-Garcia, L.: Academic emotions and student engagement. In: Handbook of Research on Student Engagement, pp. 259–282. Springer (2012)
9. Klein, J., Moon, Y., Picard, R.W.: This computer responds to user frustration: Theory, design, and results. Interact. Comput. **14**(2), 119–140 (2002)
10. Mcquiggan, S.W., Lee, S., Lester, J.C.: Early prediction of student frustration. In: International Conference on Affective Computing and Intelligent Interaction. Springer (2007)
11. D'Mello, S., et al.: Confusion can be beneficial for learning. Learn. Instr. **29**, 153–170 (2014)
12. Burleson, W., Picard, R.W.: Affective agents: sustaining motivation to learn through failure and a state of stuck. In: Workshop on Social and Emotional Intelligence in Learning Environments (2004)
13. Appleton, J.J., et al.: Measuring cognitive and psychological engagement: validation of the student engagement instrument. J. School Psychol. **44**(5), 427–445 (2006)
14. VanLehn, K., et al.: Why do only some events cause learning during human tutoring? Cogn. Instr. **21**(3), 209–249 (2003)
15. Arguel, A., Lane, R.: Fostering deep understanding in geography by inducing and managing confusion: an online learning approach (2015)
16. Kapur, M., Bielaczyc, K.: Designing for productive failure. J. Learn. Sci. **21**(1), 45–83 (2012)
17. Arguel, A., et al.: Inside out: detecting learners' confusion to improve interactive digital learning environments. J. Educ. Comput. Res. **55**(4), 526–551 (2017)
18. Pluck, G., Johnson, H.: Stimulating curiosity to enhance learning. GESJ: Education Sciences and Psychology, vol. 2 (2011)
19. Wu, Q., Shen, Z., Miao, C.: Stimulating students' curiosity with a companion agent in virtual learning environments. In: EdMedia: World Conference on Educational Media and Technology. Association for the Advancement of Computing in Education (AACE) (2013)
20. Duo, S., Song, L.X.: An e-learning system based on affective computing. Phys. Procedia **24**, 1893–1898 (2012)
21. Leony, D., et al.: Provision of awareness of learners' emotions through visualizations in a computer interaction-based environment. Expert Syst. Appl. **40**(13), 5093–5100 (2013)
22. Sabourin, J., Mott, B., Lester, J.: Computational models of affect and empathy for pedagogical virtual agents. In: Standards in emotion modeling, Lorentz Center International Center for workshops in the Sciences (2011)
23. Prendinger, H., Ishizuka, M.: The empathic companion: a character-based interface that addresses users'affective states. Appl. Artif. Intell. **19**(3–4), 267–285 (2005)
24. Ammar, M.B., et al.: The affective tutoring system. Expert Syst. Appl. **37**(4), 3013–3023 (2010)
25. Levitski, A., Radun, J., Jokinen, K.: Visual interaction and conversational activity. In: Proceedings of the 4th Workshop on Eye Gaze in Intelligent Human Machine Interaction. ACM (2012)
26. Tian, F., et al.: Recognizing and regulating e-learners' emotions based on interactive Chinese texts in e-learning systems. Knowl. Based Syst. **55**, 148–164 (2014)

27. D'mello, S.K., Graesser, A.: Multimodal semi-automated affect detection from conversational cues, gross body language, and facial features. User Model. User-Adapt. Interact. **20** (2), 147–187 (2010)
28. Baltrušaitis, T., Ahuja, C., Morency, L.-P.: Multimodal machine learning: a survey and taxonomy. arXiv preprint arXiv:1705.09406 (2017)
29. D'mello, S.K., Kory, J.: A review and meta-analysis of multimodal affect detection systems. ACM Comput. Surv. (CSUR) **47**(3), 43 (2015)
30. Chi, M.T., et al.: Eliciting self-explanations improves understanding. Cogn. Sci. **18**(3), 439–477 (1994)
31. Hattie, J., Timperley, H.: The power of feedback. Rev. Educ. Res. **77**(1), 81–112 (2007)
32. Ranjbartabar, H., Richards, D.: Student designed virtual teacher feedback. In: Proceedings of the 9th International Conference on Computer and Automation Engineering. ACM (2017)
33. Lin, L., et al.: Animated agents and learning: does the type of verbal feedback they provide matter? Comput. Educ. **67**, 239–249 (2013)
34. van der Meij, H.: Motivating agents in software tutorials. Comput. Hum. Behav. **29**(3), 845–857 (2013)
35. Kim, Y.: Empathetic virtual peers enhanced learner interest and self-efficacy. In: Workshop on Motivation and Affect in Educational Software, in Conjunction with the 12th International Conference on Artificial Intelligence in Education (2005)
36. Wu, L., Looi, C.-K.: Agent prompts: scaffolding students for productive reflection in an intelligent learning environment. In: Intelligent Tutoring Systems. Springer (2010)
37. Villarica, R., Richards, D.: Educational scaffolding for students stuck in a virtual world. ACIS (2014)
38. Kim, Y., Baylor, A.L., Shen, E.: Pedagogical agents as learning companions: the impact of agent emotion and gender. J. Comput. Assist. Learn. **23**(3), 220–234 (2007)
39. McQuiggan, S.W., et al. Modeling parallel and reactive empathy in virtual agents: an inductive approach. In: Proceedings of the 7th International Joint Conference on Autonomous Agents and Multiagent Systems-Volume 1. International Foundation for Autonomous Agents and Multiagent Systems (2008)
40. Goldberg, L.R.: The development of markers for the Big-Five factor structure. Psychol. Assess. **4**(1), 26 (1992)
41. Capuano, N., et al.: A personality based adaptive approach for information systems. Comput. Hum. Behav., **44**(C), 156–165 (2015)
42. Busso, C., et al.: Analysis of emotion recognition using facial expressions, speech and multimodal information. In: Proceedings of the 6th International Conference on Multimodal Interfaces. ACM (2004)
43. Bos, D.O.: EEG-based emotion recognition. Influence Vis. Auditory Stimuli **56**(3), 1–17 (2006)
44. Agrafioti, F., Hatzinakos, D., Anderson, A.K.: ECG pattern analysis for emotion detection. IEEE Trans. Affect. Comput. **3**(1), 102–115 (2012)
45. Jerritta, S., et al.: Physiological signals based human emotion recognition: a review. In: 2011 IEEE 7th International Colloquium on Signal Processing and Its Applications (CSPA). IEEE (2011)
46. Cowie, R., et al.: Emotion recognition in human-computer interaction. IEEE Sign. Process. Mag. **18**(1), 32–80 (2001)

47. Feidakis, M., Daradoumis, Th., Caballé, S., Conesa, J.: Embedding emotion awareness into e-learning environments. Int. J. Emerg. Technol. Learn. **9**(7), 39–46 (2014)
48. Caballé, S., Barolli, L., Feidakis, M., Matsuo, K., Xhafa, F., Daradoumis, Th., Oda, T.: A study of using SmartBox to embed emotion awareness through stimulation into e-learning environments. In: 6th International Conference on Intelligent Networking and Collaborative Systems, pp. 469–474. IEEE (2014)
49. Pousada, M., Caballé, S., Conesa, J., Bertrán, A., Gómez-Zúñiga, B., Hernández, E., Armayones, M., Moré, J.: Towards a web-based teaching tool to measure and represent the emotional climate of virtual classrooms. In: 5th International Conference on Emerging Intelligent Data and Web Technologies, pp. 314–327. Springer (2017)

The 6th International Workshop on Frontiers in Intelligent Networking and Collaborative Systems (FINCoS-2018)

Integration of Cloud-Fog Based Environment with Smart Grid

Hanan Butt[1], Nadeem Javaid[1(✉)], Muhammad Bilal[2], Syed Aon Ali Naqvi[1], Talha Saif[1], and Komal Tehreem[1]

[1] COMSATS University, Islamabad 44000, Pakistan
nadeemjavaidqau@gmail.com
[2] School of Computing and IT, Taylor's University, Subang Jaya, Malaysia
http://www.njavaid.com

Abstract. Smart grid (SG) is an efficient electrical grid that provides opportunities to manage the energy load in a reliable and efficient way. Moreover, smart meters (SMs) are introduced, which play a vital role in communication between homes and SG. SMs manage and monitor the energy consumption of homes. SGs and SMs produce a big data, which is very hard to store and process even with cloud computing. For this purpose fog computing concept is introduced, which provides a good environment for computing and storing the data of SGs and SMs before transmitting them to cloud. The concept of fog computing, acts as a bridge between cloud and SGs. In this paper, a cloud-fog based architecture is integrated with SG for efficient energy management of buildings with SMs. To manage the energy requirements of consumers, micro grids (MGs) are available near to the buildings. Fogs are distributed over the world, overhaul the cloud via important features, including low latency and increased security for MGs. To balance the load on cloud and fogs, three load balancing algorithms are used. These algorithms are round robin (RR), throttled and greedy. Closest datacenter policy is used to compare their results. Greedy gives better results than RR and throttled.

Keywords: Micro grid · Smart grid · Cloud computing
Fog computing · Greedy algorithm · Load balancing · Virtual machine

1 Introduction

Both customers and utility companies are getting benefits from the services of smart grid (SG). In SG, smart meters (SM) monitor the electricity usage by consumers using the different internet of things (IoT) devices and appliances in a home. The SG has ability to control different parts of electrical system by using the benefits of information and communication technology (ICT) [1].

A micro grid (MG) comprises distributed generators, energy storage and load, which may operate autonomously [2]. SG is integrated with cloud computing and fog computing to overcome the scalability issue. Different scenarios have been

© Springer Nature Switzerland AG 2019
F. Xhafa et al. (Eds.): INCoS 2018, LNDECT 23, pp. 423–436, 2019.
https://doi.org/10.1007/978-3-319-98557-2_38

introduced to integrate cloud computing and fog computing with SG. Practically, this integration provides multiple facilities at consumers end.

Among many properties of cloud computing some are quite attractive to the field of power system, such as cloud computing can flexibly expand or reduce itself according to the specific demand of the users, which provide significant scalability [3]. Cloud computing provides shared storage, information and communication infrastructure to the public users, which can reduce the service cost [4].

Metering mechanism of cloud computing, allows users to only pay for their consumed resources. Which provides space for cost reduction by mean of optimal resource allocations [5]. Basically in cloud computing and fog computing three service models are used. These are service model of software, service model of platform and service model of infrastructure [6].

In this paper, integration of cloud computing and fog computing with SG is proposed. The proposed scenario is divided into four layers. The first layer is cloud layer, which stores complete information of other three layers in cloud information service (CIS). Fog layer is the second layer; each fog is registered with the cloud through CIS. Each fog has some hardware configurations (Ram = 500, Bandwidth = 1000 etc.). Different numbers of virtual machines (VMs) are installed on each fog. Basically main resources of fogs are VMs.

Further, fog layer is connected with MG layer and SM layer. MG layer fulfills the needs of consumers by supplying electricity. The generation and distribution of electricity is performed by MG layer. When MGs are unable to fulfill the needs, macro grid is used from the cloud layer. MGs are distributed into those regions which are closed to the end users. Fourth layer is SM layer, this layer comprises c number of brokers and c number of clusters and each cluster contain b number of buildings. Further each building comprises s number of SMs. SM performs one-way communication with MG and performs two-way communication with broker. From SM layer end users sent request to the broker, then broker communicates with cloud layer and gets the information of registered fogs through CIS. Broker selects the fog with the help of broker policies. Closest datacenter policy is used in this paper. The request of the end user is forward to the fog, one of the installed VM is allocated to the request. To allocate VM three load balancing algorithms are used in this paper. These are round robin (RR), throttled and greedy load balancing algorithms.

1.1   Motivation

Two tier cloud-fog based platform is presented in [7]. To provide better services fogs are installed in that regions which are closed to the end users. On the other hand, SG is a modernized electrical grid that contains several smart devices like SMs, smart appliances, etc.,[8]. In this paper, the author study the information usage of SG, how SG manages the information flow of different power networks. Further SG equipped with distributed storage and distributed generation units [9]. SG performs two-way communication with the end users through SMs[a]. By integrating SG with the cloud-fog based platform, scalability and resource utilization pattern will improve efficiently [10]. For good resource utilization several load balancing techniques are used in [11]. In these papers, Authors focus on the load balancing techniques for response time (RT) and processing time (PT) optimization.

1.2 Contributions

In this paper, to get benefits from the cloud-fog based architecture, SG is integrated' The proposed scenario covers the six continents of the world. Each continent comprises large number of consumers. The contributions of the work are described as below:

- To improve latency (RT, PT and request time), fog computing is integrated with cloud computing.
- To provide location awareness smart devices are used such as SM and home appliances.
- MGs are scattered geo-graphically, which provide better scalability support.
- In cloud-fog based architecture with SG, three load balancing algorithms are used RR, throttled and greedy. Greedy algorithm gives better results than RR and throttled. Average RT and average PT of greedy algorithm is less than RR and throttled.

2 Related Works

In the past few years, numerous algorithms are proposed to balance the resource allocations in SG and cloud computing. Further to deal with time and energy management, lots of scheduling models have been presented. In this paper, related works are classified into three classes.

2.1 SG Without Cloud Computing

In [12], the authors compare the pervious system of load management with the current system of load management. The clear difference of load management with SG and without SD is discussed in this paper. Utilization of ICT in demand side management (DSM) of the power system has been considered as one the main characteristics of SG [4]. Three technical domains generation, transmission and, distribution are discussed in [13]. Generation side generates electricity, which is distributed through transmission side. The number of techniques are used for resource allocations in SG. In [14], the authors compare three parallel processing techniques for resource allocations in SG.

2.2 Cloud Computing and Fog Computing

Three-layer architecture end devices, fog nodes and, a cloud server is discussed in [15]. According to [16], the cloud has all the information of registered fogs. Fogs are located near to the end users than cloud. Fogs allocate the resources

Table 1. State of the art work survey

Technique(s)	Feature(s)	Achievement(s)	Limitation(s)
RR [29]	Equal time slice is given to each request	Treats the entire server equally	If request is not completed in 1 slice, it has to wait for next slice
MMLBA [20]	Task are selected with maximum execution time and VM is selected with minimum load.	Efficiency of parallel execution is improved	That request which will take long time has to wait for VM with minimum load
OLBA [19]	Based on the framework of the system	Keeps every node busy	Sometimes node is busy but execution time is completed
GA [18]	Scheduling of home appliances	User Satisfaction with reduced bill	Execution time increased

of VMs to the requests of end users with the help of different algorithms of load balancing [17]. In [18], Genetic algorithm (GA) is used to manage the resources of cloud. According to this algorithm, a fitness value is given to each VM. After the process of crossover and mutation, VMs are allocated. Opportunistic load balancing algorithm (OLBA) is used in [19], this algorithm works according to the framework of the system. OLBA does not consider the present state of VMs. It allocates the VMs to the requests without consideration the state of VMs. In [20], max-min load balancing algorithm (MMLBA) is used for resource allocations, tasks are selected with maximum execution time and VM is selected with minimum load of tasks.

2.3   SG with Cloud and Fog Computing

In [21], cloud computing is integrated into SG, electricity supply stations are used as MGs, from where electric vehicles are being charged or discharged. Two priority algorithms are used here for the allocation of resources. Comparison of two load balancing algorithms throttled and RR is discussed in [6]. In this paper, authors used different broker policies for fog selection. End users send requests to the controller and then the controller selects the fog with help of these broker policies. In [21], two priority algorithms are used, which deal with electrical vehicles (EVs). Supply stations are used as MGs, from where these EVs can be charged or discharged. These algorithms transfer the load explicitly from over process to under load process. Many researchers are working to overcome these issues [22, 23].

3   System Model

This paper discusses the communication flow between the SG, fog and, the cloud as shown in Fig. 1. To reduce the scalability issue, the fogs are introduced [25]. Computer information system company (Cisco) proposed the concept of fog computing in 2014. This proposed model comprises four layers cloud layer, fog layer, MG layer and SM layer. In the first layer, the cloud communicates with fogs and macro grid. The cloud has a kind of registry cloud information (CIS), which store the relevant information of fogs. In second layer eighteen fogs are located in different regions of the world. The World is divided into 6 regions. Three fogs are located in each region. Each fog has h number of hosts. Because the cloud environment works on virtualization, each host is virtualized into v number of VMs. Each fog is registered with CIS. The broker also has the information of each fog; it gets resource information from CIS. Broker decides which fog is best for a specific cloudlet. Broker sends the cloudlet to the one of the VM, which are installed in the hosts of fog. To balance the load of cloudlets of end users, three load balancing algorithms are used RR, throttled and greedy. The third layer is MG layer, which actually fulfills the need of the end users. M numbers of MGs are distributed in six regions. When a user sends the request for energy to fog through the broker, fog checks the nearest MG which can fulfill the need of user. If fog is found then locate the MG, otherwise request the cloud to fulfill the need of the user. The fourth layer comprises b number of brokers. Three brokers for each region, each broker further comprises b number of buildings. Each building contains s number of SMs. SMs are used for two-way communication between homes and broker. Further, SM performs one-way communication between homes and MG. Sometimes SM performs one-way communication between homes and macro grid. The important feature of cloud computing is the computational load profile characteristics of the computing applications. Different load balancing techniques are used for efficient computational load profile management.

In SG, it is as similar to electricity load profile concept. So, by integrating SG with cloud-fog based environment computational load profile of all SGs will be efficiently managed. In this work, different algorithms are used for effective load balancing.

3.1   Problem Formulation

Every system is stabled with some performance parameters. For this purpose, system is virtualized according to the requirements. In this paper, cloud-fog architecture with SG is considered. This architecture is virtualized into v number of VMs, to balance the load of user requests (URs). The total VMs can be calculated as:

$$Total_{VMs} = \sum_{i=1}^{v}(VM_i) \tag{1}$$

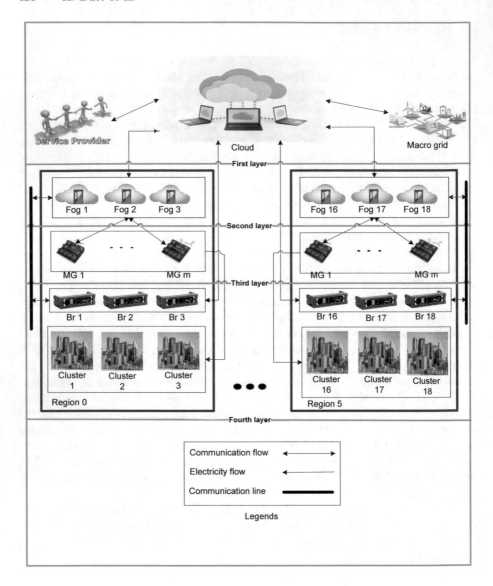

Fig. 1. A system model.

Let $Total_{UR}$ is the set of URs in Eq. 2.

$$Total_{URs} = \sum_{i=1}^{n} (UR_i) \tag{2}$$

Here user requests are varying from 0 to u. each UR processes on different VMs. Each VM is supervised by the manager of VM. Assignment of VMs to URs is dependent of RT and PT of VMs. Linear programing is used to formulate these parameters in [25].

3.1.1 Processing Time

PT is the time taken by the VMs of fog to process the requests of end users. Here PT taken by VMs is PT_{gh}, state of UR is defined in Eq. 3.

$$S_{gh} = \begin{cases} 1; & \text{If VM is assigned to UR} \\ 0; & \text{Other wise} \end{cases} \tag{3}$$

The PT of VM_h to process the UR_g is formally represented as:

$$PT_{gh} = \frac{\text{Size of } UR_g}{\text{Capacity of } VM_h} \tag{4}$$

Total PT can be estimated by:

$$Total_{PT} = \sum_{c=1}^{n} \sum_{k=1}^{m} (PT_{gh} * S_{gh}) \tag{5}$$

Different VMs have different efficiencies to process the same UR according to [27]. The main objective is to minimize the PT of UR on VM as in Eq. 6.

$$minimise \quad max_{UR} \left\{ PT_{ck} = \frac{WL_{UR}}{\sum_{k=1}^{m} r_{ck} EP_{ck} CP_{ck}} \right\} \tag{6}$$

PT_{ck} is PT of VM_k on UR_c. WL_{UR} indicates the working load of all URs. To show the relationship between VM_k and UR_c, r_{ck} is used in Eq. 6. EP_{ck} and CP_{ck} denote the efficiency of VM_k to processes the UR_c and computing power of VM_k respectively.

3.1.2 Response Time

The RT is time taken by the fog to receive and response the request from any cluster of building. Equation 7 represents the RT.

$$RT = D_{Time} + F_{Time} - A_{Time} \tag{7}$$

D_{time}, F_{time} and A_{time} indicate the delay time, finishing time and the arrival time of UR.

3.1.3 Costs

Cost of VMs is calculated in Eq. 8.

$$VM_{Cost} = \frac{T_V MT}{VM_{Costperhour} * MH} \tag{8}$$

Here T_VMT shows the total time taken by the VM as in Eq. 9.

$$T_{VMT} = Finish_{Time} - Start_{Time} \tag{9}$$

MH indicates the value of time converting from millisecond (ms) to hours in Eq. 8. Equation 10 represents the data transfer cost.

$$Data_{TC} = Data_{gigabytes} * Data_{\text{Cost per gigabytes}} \tag{10}$$

Total cost of entire system is calculated in Eq. 11.

$$Total_{Cost} = Data_{TC} + VM_{Cost} + MG_{Cost} \tag{11}$$

3.2   Load Balancing Algorithms

Load balancing is a process of managing the total load of the end users. The main purpose of load balancing is to improve utilization of resources and reduce computations. Cloud and fogs contain physical and virtual servers located at different locations to provide services to the end users efficiently as shown in Fig. 2. Different number of VMs are installed on the servers and allocated for load balancing. Cloud and fog require an algorithm to allocate VMs. The load balancer is responsible to allocate the VMs to the requests. Load balancing algorithms improve overall system performance with reasonable cost [28]. Three load balancing algorithms are used in this paper RR, throttled and greedy.

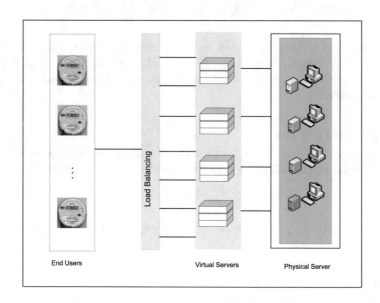

Fig. 2. Load balancing in cloud computing.

3.2.1   Round Robin

RR is a simple scheduling algorithm with least complexity [29], used for allocating resources to each host by defining equal time slice resource utilization. In this paper, this algorithm is used to allocate VMs to balance the load of requests. This algorithm, uses the concept of time quantum; equal time quantums are defined for resource allocations [30].

3.2.2   Throttled

In this algorithm, the load balancer maintains an index table of VMs [31]. Load balancer checks which VM is able to load the request of the respective client. On the other hand, consumers sent the requests to the broker for assigning suitable fog, which easily loads the requests and performs the respective operations.

3.2.3   Greedy

This algorithm selects that option which seems to give best solution. It always looks the local best solution for specific task which will lead to the global best solution. In this paper, this algorithm deals with VMs and distributes the tasks to VMs according to their work load. Greedy load balancer (GLB) Algorithm 1, maintains indexes of VMs with their current state for every VM. At the start, all VMs are in ready state. The Fog receives the user requests. Find a VM which has minimum allocations. The GLB allocates the VM and add one in its allocations. After processing the request, it notifies the GLB to minus by one in its allocations. GLB distributes the total load of cloudlets in an efficient way over VMs. This algorithm works on recent allocations.

4   Simulation Results

In this paper, Cloud Analyst is used as a simulation tool. The results of simulations have appeared in demonstrated figures. The world is divided into six regions. For each region, there are three fogs, three clusters, one broker and m numbers of MGs are used. The numbers of requests are sent to the broker, and then broker decides which fog is best to handle the specific request of end-user. In this paper, closest datacenter policy is used to manage the requests. On the other hand, each fog contains v number of VMs, memory, and storage etc. Each fog manages the number of requests by different load balancing algorithms. In this paper three load balancing algorithms RR, throttled and, greedy are used to compare the simulation results. For experimental purpose eighteen clusters, eighteen fogs and one cloud are considered. Further one fog and one cluster are considered to compare the performance parameters of PT and RT. Figure 3 shows the average RT of eighteen clusters of six regions with RR, throttled and greedy load balancing algorithms. There is no major difference of average RT between these algorithms still the performance of greedy is better than RR and throttled.

Algorithm 1. Greedy load balancing algorithm

1: Input: V_List (), V_table, V_id, Vids (), T_id .
2: Output: Selected V_id.
3: Initialize, C_table and VMs , V<– total VM, V_id <—1, Vids () =-1, i<– 0,
   Curr_count <– 0, Min_count <– Max_count, T_id <– -1;
4: Passes V_List() to Load_balancer:
5: **for** i = 1:v **do**
6:     | TV_id <–(V_List (i)).
7:     | V_id <– TV_id
8:     | **if** V_id exist in C_Table (V_id) **then**
9:     | | Curr_count <– C_Table (V_id)
10:     | **else**
11:     | | Curr_count <– 0
12:     | **end if**
13:     | Vids () <– (VM_id, currCount).
14: **end for**
15: TV_id <– -1
16: Curr_count <– 0
17: **for** i = 1:k **do**
18:     | TV_id <– i
19:     | Curr_count <–Vids (TV_id)
20:     | **if** Curr_count less than Min_count **then**
21:     | | Min_count = Curr_ count
22:     | | V_id <– TV_id
23:     | **end if**
24: **end for**

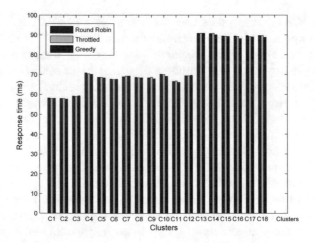

Fig. 3. Response time.

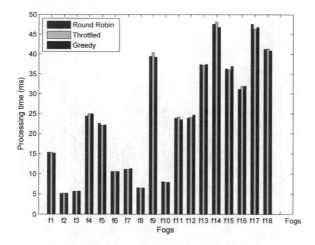

Fig. 4. Processing time.

Table 2. Overall RT and PT

Load balancing algorithm	RR	Throttled	Greedy
Avg RT (ms)	73.69	73.66	73.63
Min RT (ms)	40.11	39.89	40.09
Max RT (ms)	221.57	221.57	220.79
Avg PT (ms)	23.95	23.87	23.84
Min PT (ms)	0.26	0.26	0.26
Max PT (ms)	169.66	169.66	169.65

Overall RT of these three load balancing algorithms is displayed in Table 1. Average RT of RR, throttled and greedy are 73.69 ms, 73.66 ms and 73.63 ms respectively. Minimum RT of RR, throttled and greedy are 40.11 ms, 39.89 ms and 40.09 ms respectively. 221.57 ms, 221.57 ms and 220.79 ms are the maximum RT of the RR, throttled and greedy respectively. In Fig. 4 we can see the average PT of eighteen fogs with RR, throttled and greedy load balancing algorithms. Table 2 shows the average, minimum and maximum PT of RR, throttled and greedy load balancing algorithms. The average PT of greedy is 23.84 ms, which is less than the average PT of throttled and RR.

The comparison of hourly PT of fog number fifteen by these three load balancing algorithms is depicted in Fig. 5, here average PT of greedy is less than throttled and RR. Figure 6 shows the comparison of hourly RT of cluster number fifteen by these three load balancing algorithms. Greedy and throttled have less RT than RR From the given results, the performance of greedy is better than RR and throttled. Different load balancing algorithms are used in cloud computing, similarly for cloud-fog based SG architecture three load balancing algorithms are used in this paper.

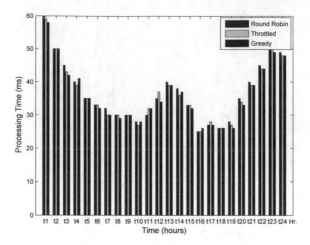

Fig. 5. Fog 15 hourly average processing time comparison.

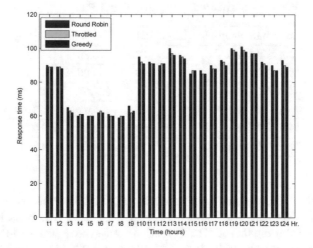

Fig. 6. Cluster15 hourly average response time.

5   Conclusion

Fog computing with SG integration provides numerous benefits and opportunities for information management. In this paper, an integrated cloud-fog based architecture is proposed to compute the energy consumption of residential buildings in all regions. Managing the energy requirement of buildings is the main purpose of this work. In cloud computing for efficient resource allocations, different load balancing algorithms are used, similarly here; three load balancing algorithms are used. Simulations are performed on JAVA platform in eclipse. Furthermore, we observe that the overall performances of greedy are better than RR and throttled. Average RT and average PT of greedy is less than RR and throttled.

References

1. Al Faruque, M.A., Vatanparvar, K.: Energy management-as-a-service over fog computing platform. IEEE Internet Things J. **3**(2), 161–169 (2016)
2. Nisar, A., Thomas, M.S.: Comprehensive control for microgrid autonomous operation with demand response. IEEE Trans. Smart Grid **8**(5), 2081–2089 (2017)
3. Lehrig, S.: CloudStoretowards scalability, elasticity, and efficiency benchmarking and analysis in cloud computing. Future Gener. Comput. Syst. **78**, 115–126 (2018)
4. Cao, Z.: Optimal cloud computing resource allocation for demand side management in smart grid. IEEE Trans. Smart Grid **8**(4), 1943–1955 (2017)
5. Narayan, A., et al.: Resource procurement, allocation, metering, and pricing in cloud computing. In: Research Advances in Cloud Computing, pp. 141–186. Springer, Singapore (2017)
6. Fatima, I., Javaid, N., Iqbal, M.N., Shafi, I., Anjum, A., Memon, U.: Integration of cloud and fog based environment for effective resource distribution in smart buildings. In: 14th IEEE International Wireless Communications and Mobile Computing Conference (IWCMC-2018)
7. Moghaddam, M.H.Y., Leon-Garcia, A., Moghaddassian, M.: On the performance of distributed and cloud-based demand response in smart grid. IEEE Trans. Smart Grid (2017)
8. Yaghmaee, M.H., Moghaddassian, M., Garcia, A.L.: Power consumption scheduling for future connected smart homes using bi-level cost-wise optimization approach. In: Smart City 360, pp. 326–338. Springer, Cham (2016)
9. Damavandi, Ghasemi, M., Marti, J.R., Krishnamurthy, V.: A methodology for optimal distributed storage planning in smart distribution grids. IEEE Trans. Sustain. Energy (2017)
10. Fang, B.: The contributions of cloud technologies to smart grid. Renew. Sustain. Energy Rev. **59**, 1326–1331 (2016)
11. Pham, N.M.N., Le, V.S.: Applying ant colony system algorithm in multi-objective resource allocation for virtual services. J. Inf. Telecommun. **1**(4), 319–333 (2017)
12. Pujara, A.: A novel approach for UI charge reduction using AMI based load prioritization in smart grid. J. Electr. Syst. Inf. Technol. **4**(2), 338–346 (2017)
13. Zhao, C.: Consensus-based energy management in smart grid with transmission losses and directed communication. IEEE Trans. Smart Grid **8**(5), 2049–2061 (2017)
14. Celik, B., et al.: A comparison of three parallel processing methods for a resource allocation problem in the smart grid. In: 2017 North American Power Symposium (NAPS). IEEE (2017)
15. Wang, H., Wang, Z., Domingo-Ferrer, J.: Anonymous and secure aggregation scheme in fog-based public cloud computing. Future Gener. Comput. Syst. **78**, 712–719 (2018)
16. Yolda, Y.: Enhancing smart grid with microgrids: challenges and opportunities. Renew. Sustain. Energy Rev. **72**, 205–214 (2017)
17. Lagwal, M., Bhardwaj, N.: A survey on load balancing methods and algorithms in cloud computing, pp. 46–51 (2017)
18. Nadeem, Z., Nadeem, J., Malik, A.W., Iqbal, S.: Scheduling appliances with GA, TLBO, FA, OSR and their hybrids using chance constrained optimization for smart homes. Energies **11**(4), 888 (2018) ISSN: 1996-1073. https://doi.org/10.3390/en11040888

19. Razzaghzadeh, S.: Probabilistic modeling to achieve load balancing in expert clouds. Ad Hoc Netw. **59**, 12–23 (2017)
20. Mangla, P., Goyal, S.K.: Heuristic vs meta-heuristic approaches for load balancing in cloud environment (2017)
21. Chekired, D.A., Khoukhi, L.: Smart grid solution for charging and discharging services based on cloud computing scheduling. IEEE Trans. Ind. Inf. **13**(6), 3312–3321 (2017)
22. Nikhit, P., Lilhore, U.M., Agrawal, N.: A hybrid ACHBDF load balancing method for optimum resource utilization. In: Cloud Computing (2017)
23. Javaid, N., Ahmed, A., Iqbal, S., Ashraf, M.: Day ahead real time pricing and critical peak pricing based power scheduling for smart homes with different duty cycles. Energies **11**(6), 1464 (2018). ISSN: 1996-1073. https://doi.org/10.3390/en11061464.
24. Gupta, H., et al.: iFogSim: a toolkit for modeling and simulation of resource management techniques in the Internet of Things, edge and fog computing environments. Softw. Pract. Exp. **47**(9), 1275–1296 (2017)
25. Wickremasinghe, B., Buyya, R.: CloudAnalyst: a cloudsim-based tool for modelling and analysis of large scale cloud computing environments. MEDC project report **22**(6), 433–659 (2009)
26. Shi, L., Zhang, Z., Robertazzi, T.: Energy-aware scheduling of embarrassingly parallel jobs and resource allocation in cloud. IEEE Trans. Parallel Distrib. Syst. **28**(6), 1607–1620 (2017)
27. Li, D., Wu, J.: Minimizing energy consumption for frame-based tasks on heterogeneous multiprocessor platforms. IEEE Trans. Parallel Distrib. Syst. **26**(3), 810–823 (2015)
28. Xu, M., Tian, W., Buyya, R.: A survey on load balancing algorithms for virtual machines placement in cloud computing. Concurrency Comput. Pract. Exp. **29**(12), e4123 (2017)
29. Thakur, P., Mahajan, M.: Different scheduling algorithm in cloud computing: a survey. Int. J. Modern Comput. Sci. **5**(1), (2017)
30. Tani, H.G., El Amrani, C.: Smarter round robin scheduling algorithm for cloud computing and big data. J. Data Mining Digit. Hum. (2018)
31. Parida, S., Panchal, B.: Review paper on throttled load balancing algorithm in cloud computing environment (2018)

Improved Functional Encryption Schemes by Using Novel Techniques

Xu An Wang[1(✉)], Jindan Zhang[2,3], Guangming Wu[1], and Chenghai Yu[4]

[1] Key Laboratory of Cryptology and Information Security,
Engineering University of CAPF, Xi'an, China
wangxazjd@163.com
[2] Xianyang Vocational Technical University, Xianyang, People's Republic of China
[3] State Key Laboratory of Integrated Service Networks, Xidian University,
Xi'an, People's Republic of China
[4] Zhejiang Sci-Tech University, Hangzhou, China

Abstract. Nowadays more and more people or enterprises prefer to outsource their data to the remote cloud servers. However how to efficiently and flexibly share the outsourced data storage secure is a very challenge problem, the cryptographic primitive of functional encryption is such a technique to solve this problem. In this paper, we show several new functional encryption schemes, which are more efficient than the original schemes, by using a novel technique. This technique maybe also has potential applications in other related schemes. Finally we conclude our paper with many interesting open problems.

1 Introduction

Functional encryption is now a very important cryptographic primitive these days. Functional encryption can support the paradigm of encryption as such: the encrypter encrypts the plaintext with attributes or access control policies f and gets the ciphertext $Enc(x)$, and later the decryptor with private key associated with access control policies f or attributes can decrypt $Enc(x)$ to get $f(x)$, instead of x. Functional encryption extends the concept of encryption to very large contents. Generally speaking, functional encryption includes identity based encryption, fuzzy identity based encryption, key-policy attribute based encryption, and ciphertext-policy attribute based encryption, inner product encryption and even proxy re-encryption etc. Until now, there are many good functional encryption schemes [3–8,13], some of them have wide applications in our practical life such as cloud storage [9–12]. However, in this paper, we show there exists a very simple novel technique which can improve the efficiency of most of the existing functional encryption schemes. We give an improved KP-ABE scheme and an improved IPE scheme to show our technique.

We organize our paper as following: in Sect. 2, we give our improved KP-ABE proposal based on RW's KP-ABE scheme. In Sect. 3, we first review Guo et al.'s IPE scheme and then give an improved IPE scheme. Finally we conclude our paper with interesting open problems.

© Springer Nature Switzerland AG 2019
F. Xhafa et al. (Eds.): INCoS 2018, LNDECT 23, pp. 437–444, 2019.
https://doi.org/10.1007/978-3-319-98557-2_39

2 Our Improved KP-ABE Proposal

Our scheme is based on RW's KP-ABE scheme [6], and it consists of following four algorithms:

1. Setup($1^\lambda \rightarrow (pp, msk)$): The setup algorithm takes as input a security parameter and a universe U of attributes. To cover the most general case, we let $U = \{0,1\}^*$. It then choose a bilinear group G of prime order p, generators $g, h, u, w \in G$ In addition, it choose random exponents $\alpha \in Z_p$. The authority sets $MSK = (\alpha, PK)$ as the master secret key. It publishes the public parameters as:

$$PK = (G, p, g, h, w, e(g,g)^\alpha)$$

   We assume that the universe of attributes can be encoded as elements in Z_p.

2. Extract($MSK, (M, \rho)) \rightarrow SK$: The extract algorithm take as input the master secret key MSK and an $LSSS$ access structure (M, ρ). Let M be an $l \times n$ matrix. The function ρ associates rows of M to attributes. The algorithm initially chooses random values $y_2, \cdots, y_n \in Z_p$. It then computes l shares of the master secret key as $(\lambda_1, \cdots, \lambda_l) := M \cdot (\alpha, y_2, \cdots, y_n)^T$ (where T denote the transpose). It then picks l random exponents $t_1, \cdots, t_l \in Z_p$ and also a random exponent $\boxed{T_1 \in Z_p}$. It first computes $K_0 = g^{T_1}$, and for $i = 1$ to l, it computes underline

$$K_{\tau,0} = g^{\lambda_\tau} w^{t_\tau}, K_{\tau,1} = (u^{\rho(\tau)} h)^{-t_\tau},$$

$$\boxed{K_{\tau,2} = (t_\tau - T_1) \bmod p}$$

   and the private key is $\underline{SK = ((M, \rho), K_0, \{K_{\tau,0}, K_{\tau,1}, K_{\tau,2}\}_{\tau \in [1,l]})}$

3. Encrypt($m, S = \{A_1, \cdots, A_k\}) \rightarrow CT$: Initially, the algorithm picks $k + 2$ random exponents $s, r_1, \cdots, r_k, \boxed{T_2} \in Z_p$. It computes $C = me(g,g)^{\alpha s}, C_0 = g^s, \boxed{C_1 = g^{T_2}}$, and for every $j \in [1,k]$, it computes

$$\boxed{C_{\tau_1,1} = (r_\tau - T_2) \bmod p}, C_{\tau,2} = (\mu^{A_\tau} h)^{r_\tau} w^{-s}$$

   The ciphertext is $\underline{CT = (S, C, C_0, C_1, \{C_{\tau,1}, C_{\tau,2}\}_{\tau \in [1,k]})}$

4. Decrypt($SK, CT) \rightarrow m$: The algorithm takes as input a ciphertext $CT = (S, C, C_0, Caa_1, \{C_{\tau,1}, C_{\tau,2}\}_{\tau \in [1,k]})$ for attribute set S and a private key $SK = ((M, \rho), K_0, \{K_{\tau,0}, K_{\tau,1}, K_{\tau,2}\}_{\tau \in [1,l]}$ for access structure (M, ρ). If S dose not satisfy this access structure, then the algorithm issues an error message. Otherwise, it set $I = (i : \rho(i) \in S)$ and computes constants $\{w_i \in Z_p\}_{i \in I}$ such that $\sum_{i \in I} w_i M_i = (1, 0, \cdots, 0)$, where M_i is the i-th row of the matrix M. Then it calculates

$$B = \prod_{i \in I} e((C_0, K_{i,0}) e(C_1 g^{C_{\tau,1}}, K_{i,1})$$

$$\cdot e(C_{\tau,2}, K_0 \cdot g^{K_{i,2}}))^{w_i} = e(g,g)^{\alpha s}$$

where j is the index of the attribute $\rho(i)$ in S(it depends on i).

Correctness: If the attribute set S of the ciphertext is authorized, we have that $\sum_{i \in I} w_i \lambda_i = \alpha$. Therefore,

$$
\begin{aligned}
B &= \prod_{i \in I} (e(C_0, K_{i,0}) e(C_1 g^{C_{\tau,1}}, K_{i,1}) e(C_{\tau,2}, K_0 \cdot g^{K_{i,2}}))^{w_i} \\
&= \prod_{i \in I} (e(C_0, K_{i,0}) e(g^{r_\tau}, K_{i,1}) e(C_{\tau,2}, g^{t_i}))^{w_i} \\
&= \prod_{i \in I} (e(g^s, g^{\lambda_i} w^{t_i}) e(g^{r_\tau}, (u^{\rho(i)} h)^{-t_i}) e((u^{A_\tau} h)^{r_\tau} w^{-s}, g^{t_i}))^{w_i} \\
&= \prod_{i \in I} (e(g,g)^{s \lambda_i} e(g,g)^{s t_i}) e(g, (uh)^{-\rho(i) r_\tau t_i}) \\
&\quad \cdot e((uh), g)^{\rho(i) r_\tau t_i} e(g, w)^{-s t_i})^{w_i} \\
&= \prod_{i \in I} e(g,g)^{s \lambda_i w_i} \\
&= e(g,g)^{s \alpha}
\end{aligned}
$$

But note here

$$
\begin{aligned}
B &= \prod_{i \in I} (e(C_0, K_{i,0}) e(C_1 g^{C_{\tau,1}}, K_{i,1}) e(C_{\tau,2}, K_0 \cdot g^{K_{i,2}}))^{w_i} \\
&= (e(C_0, \prod_{i \in I} K_{i,0}^{w_i}) e(C_1, \prod_{i \in I} K_{i,1}^{w_i}) e(g, \prod_{i \in I} K_{i,1}^{w_i C_{\tau,1}}) e(\prod_{i \in I} C_{\tau,2}^{w_i}, K_0) e(\prod_{i \in I} C_{\tau,2}^{K_{i,2}}, g))
\end{aligned}
$$

Remark 1. Compared with RW's KP-ABE scheme, this scheme has advantage in computation complexity. This scheme which needs 5 pairings instead of $3|I|$ pairings for the original scheme, and the original scheme needs $2|I| + 1$ modular exponentiation while this scheme needs $4|I| + 1$ modular exponentiation. For one pairing computation is much more costly than one modular exponentiation, thus our improved scheme is more efficient.

3   Our Improved Online/Offline IPE Scheme

3.1   Guo's Inner Product Encryption Scheme

In this subsection, we review of Guo's IPE scheme [1]:

1. IPE.Setup: The setup algorithm takes as input a security parameter λ and an integer n to denote the length of vector. It first chooses $PG = (\mathbb{G}, \mathbb{G}_T, g, p, e)$. The algorithm then chooses random α_i, β from Z_p for all $i = 1, 2, \cdots, n$. Finally, for all $i = 1, 2, \cdots, n$, it computes group elements $g_i = g^\alpha$ and $u = e(g,g)^\beta$. The master public key $IPE.mpk$ and the master secret key $IPE.msk$ are defined as follows.

$$
IPE.mpk = (G, G_T, g, p, e, g_i, u)
$$

$$
IPE.msk = (\alpha_1, \alpha_2, \cdots, \alpha_n, \beta)
$$

2. **IPE.Key:** The key generation algorithm takes as input an n-length vector $\vec{z} = (z_1, z_2, \cdots, z_n) \in Z_p^n$ and the master public/secret key pair $(IPE.mpk, IPE.msk)$. The algorithm randomly chooses $t \in Z_p$ and computes the private key $IPE.sk_{\vec{z}}$ as follows.

$$IPE.sk_{\vec{z}} = (g^{\beta + t \sum_{i=1}^{n} \alpha_i z_i}, g^t) \in \mathbb{G} \times \mathbb{G}$$

3. **IPE.Enc:** The encryption algorithm takes as input an n- length vector $\vec{w} = (w_1, w_2, \cdots, w_n) \in Z_p^n$, a message $M \in G_T$ and the master public key $IPE.mpk$. It chooses random r, s from Z_p and creates the ciphertext as follows.

$$IPE.CT = (u^r \cdot M, g^r, g_1^r g^{sw_1}, g_2^r g^{sw_2}, \cdots, g_n^r g^{sw_n})$$

4. **IPE.Dec:** Suppose that $IPE.CT = (C_m, C_0, C_1, \cdots, C_n)$ is a ciphertext encrypted with $\vec{z}$ and we have a private key $IPE.sk_{\vec{z}}$ for z satisfying $< \vec{w}, \vec{z} > = 0$. The decryption algorithm begins by computing

$$e_0 = e(g^t, \prod_{i=1}^{n} C_i^{z_i})$$

$$e_1 = e(g^{\beta + t \sum_{i=1}^{n} \alpha_i z_i}, C_0)$$

Then it decrypts message by computing

$$C_m \cdot e_1^{-1} e_0 = M$$

3.2   Their Online/Offline Variant

Guo et al. also give an online/offline variant. Concretely, without changing the algorithms of setup and key generation, the other two algorithms for online/offline IPE is below:

1. **IPE.Enc:** The encryption algorithm is split into the follow- ing two algorithms.
   a. **Offline:** Taking as input the master public key, the offline encryption algorithm randomly chooses $K_R \in \mathbb{G}_T$, $\boxed{r, s, s_1, s_2, \cdots, s_n \in \mathbb{Z}_p}$ and computes the offline ciphertext $IPE.CT.off$ as

$$IPE.CT.off = (u^r \cdot K_R, g^r, g_1^r g^{-s_1}, g_2^r g^{-s_2}, \cdots, g_n^r g^{-s_n})$$
$$= (C_R, C_0, C_1', C_2', \cdots, C_n')$$

   It stores K_R and $\boxed{(IPE.CT.off, s, s_1, s_2, \cdots, s_n)}$, for the online encryption after the presence of $\vec{w}$. Here, K_R is the symmetric key.
   b. **Online:** Taking as input offline parameters $\boxed{(IPE.CT.Off, s, s_1, s_2, \cdots, s_n)}$, an n-length vector $\vec{w} = (w_1, w_2, \cdots, w_n) \in Z_p^n$ and the master public key $IPE.mpk$, the online encryption algorithm computes

$$R_i = s_i + sw_i (\bmod p)$$

   It outputs encryption key K_R and its ciphertext $IPE.CT$ as

$$IPE.CT = (IPE.CT.off, R_1, R_2, \cdots, R_n)$$

2. IPE.Dec: Suppose that a ciphertext $IPE.CT$ is encrypted with $\overrightarrow{w}$ and we have a private key $IPE.sk_{\overrightarrow{z}}$ for $\overrightarrow{z}$ satisfying $<\overrightarrow{w}, \overrightarrow{z}> = 0$. Let the ciphertext $IPE.CT$ be

$$IPE.CT = (C_R, C_0, C_1', C_2', \cdots, C_n', R_1, R_2, \cdots, R_n)$$

Taking as input $(IPE.CT, sk_{\overrightarrow{z}}, \overrightarrow{z})$, the decryption algorithm begins by computing

$$e_0 = e(g^t, g^{\sum_{i=1}^{n} R_i z_i} \cdot \prod_{i=1}^{n} C_i'^{z_i})$$

$$e_1 = e(g^{\beta + t \sum_{i=1}^{n} \alpha_i z_i}, C_0)$$

Then, the symmetrickey K can be decrypted by

$$C_R \cdot e_1^{-1} e_0 = K_R$$

The decryption is correct the same as the proposed IPE scheme except the e_0 computation. We have

$$g^{\sum_{i=1}^{n} R_i z_i} \cdot \prod_{i=1}^{n} C_i'^{z_i} = \prod_{i=1}^{n} C_i^{z_i}$$

the same as the basic IPE scheme, and therefore the correctness holds.

3.3 Our Reusable Online/Offline IPE

In this subsection, we give our improved online/offline IPE. The two algorithms for online/offline IPE is described as follows, other algorithms are the same as the original scheme.

1. IPE.Enc: The encryption algorithm is split into the follow- ing two algorithms.
   a. Offline: Taking as input the master public key, the offline encryption algorithm randomly chooses $K_R \in \mathbb{G}_T$, $\boxed{r, s_1, s_2, \cdots, s_n \in Z_p}$ and computes the offline ciphertext $IPE.CT.off$ as

   $$IPE.CT.off = (u^r \cdot K_R, g^r, g_1^r g^{-s_1}, g_2^r g^{-s_2}, \cdots, g_n^r g^{-s_n})$$
   $$= (C_R, C_0, C_1', C_2', \cdots, C_n')$$

   It stores K_R and $(IPE.CT.off, s_1, s_2, \cdots, s_n)$, for the online encryption after the presence of $\overrightarrow{w}$. Here, K_R is the symmetric key.
   b. Online: Taking as input offline parameters $\boxed{(IPE.CT.Off, s_1, s_2, \cdots, s_n)}$, an n-length vector $\overrightarrow{w} = (w_1, w_2, \cdots, w_n) \in Z_p^n$ and the master public key $IPE.mpk$, the online encryption algorithm first chooses randomly $\boxed{s \in Z_p^*}$ and computes
   $$R_i = s_i + sw_i (\bmod p)$$
   It outputs encryption key K_R and its ciphertext $IPE.CT$ as
   $$IPE.CT = (IPE.CT.off, R_1, R_2, \cdots, R_n)$$

2. IPE.Dec: Suppose that a ciphertext $IPE.CT$ is encrypted with $\vec{w}$ and we have a private key $IPE.sk_{\vec{z}}$ for $\vec{z}$ satisfying $<\vec{w}, \vec{z}> = 0$. Let the ciphertext $IPE.CT$ be

$$IPE.CT = (C_R, C_0, C_1', C_2', \cdots, C_n', R_1, R_2, \cdots, R_n)$$

Taking as input $(IPE.CT, sk_{\vec{z}}, \vec{z})$, the decryption algorithm begins by computing

$$e_0 = e(g^t, g^{\sum_{i=1}^{n} R_i z_i} \cdot \prod_{i=1}^{n} C_i'^{z_i})$$

$$e_1 = e(g^{\beta + t \sum_{i=1}^{n} \alpha_i z_i}, C_0)$$

Then, the symmetrickey K can be decrypted by

$$C_R \cdot e_1^{-1} e_0 = K_R$$

The decryption is correct the same as the proposed IPE scheme except the e_0 computation. We have

$$g^{\sum_{i=1}^{n} R_i z_i} \cdot \prod_{i=1}^{n} C_i'^{z_i} = \prod_{i=1}^{n} C_i^{z_i}$$

the same as the basic IPE scheme, and therefore the correctness holds.

Remark 2. At first sight, this improved online/offline IPE is the same as the original one. But there exists crucial difference. In the original scheme, s is fixed in the offline phase, which means the offline part can not be reused. For example, if the offline part is reused, the adversary could get the following two equations

$$R_i = s_i + s w_i \pmod{p} \tag{1}$$
$$R_i' = s_i + s w_i' \pmod{p} \tag{2}$$

and the adversary can easily compute

$$s = \frac{R_i' - R_i}{w_i' - w_i}$$

While in our scheme, s is randomly chosen in the online phase, which means the offline part can be reused for many times. For example, although the adversary could get the following two equations

$$R_i = s_i + s w_i \pmod{p} \tag{3}$$
$$R_i' = s_i + s' w_i' \pmod{p} \tag{4}$$

but she can not easily compute s or s' again, and thus s_i. That means, s_i can be reused many times.

Remark 3. This improvement although seems little, but it has great impact for practical applications. For example, if we use mobile phone to implement the online/offline IPE encryption scheme. If the mobile phone does the offline part when charging, the online part can be done when the users outside the home, in this case, if the offline part can be reused many times, the mobile phone user can do many times of IPE encryption, which will largely increase this paradigm's advantage.

4 Conclusion

In this paper, we give two improved functional encryption schemes by using novel techniques. One is an KP-ABE scheme and the other one is an reusable online/offline IPE scheme. However, these results are very basic, there are many open problems need to be solved, such as how to prove the proposals' security formally, how to construct efficient reusable functional encryption by using multilinear map etc.

Acknowledgements. The first author and the fourth author are the corresponding authors. This work is supported by National Cryptography Development Fund of China Under Grants No. MMJJ20170112, National Natural Science Foundation of China (Grant Nos. 61772550, 61572521, U1636114, 61402531), National Key Research and Development Program of China Under Grants No. 2017YFB0802000, Natural Science Basic Research Plan in Shaanxi Province of china (Grant Nos. 2018JM6028, 2016JQ6037) and Guangxi Key Laboratory of Cryptography and Information Security (No. GCIS201610), Public Welfare Technology Application Projects of Zhejiang Province under Grant (2016C31072), Research Project of Educational Reform in Zhejiang (jg20160053).

References

1. Guo, F., Susilo, W., Mu, Y.: Distance based encryption: how to embed fuzziness in Biometric Based Encryption. IEEE Trans. Inf. Forensics Secur. **11**(2), 247–257 (2016). https://doi.org/10.1109/TIFS.2015.2489179
2. Goyal, V., Pandey, O., Sahai, A., Waters, B.: Attribute-based encryption for fine-grained access control of encrypted data. In: Juels, A., Wright, R.N., De Capitani di Vimercati, S. (eds.) ACM CCS 2006, pp. 89–98. ACM Press, October/November 2006. Available as Cryptology ePrint Archive Report 2006/309
3. Green, M., Hohenberger, S., Waters, B.: Outsourcing the decryption of abe ciphertexts. In: Proceedings of the USENIX Security Symposium, San Francisco, CA, USA (2013)
4. Lai, J., Deng, R., Guan, C., Weng, J.: Attribute-based encryption with verifiable outsourced decryption. IEEE Trans. Inf. Forensics Secur. **8**(8), 1343–1354 (2013)
5. Hohenberger, S., Waters, B.: Online/offline attribute-based encryption. In: Krawczyk, H. (ed.) PKC 2014, LNCS, vol. 8383, pp. 293–310. Springer, March 2014
6. Rouselakis, Y., Waters, B.: Practical constructions and new proof methods for large universe attribute-based encryption. In: Sadeghi, A.-R., Gligor, V.D., Yung, M., (eds.) ACM CCS 2013, pp. 463–474. ACM Press, November 2013

7. Li, J., Huang, X., Li, J., Chen, X., Xiang, Y.: Securely outsourcing attribute-based encryption with checkability. IEEE Trans. Parallel Distrib. Syst. (2013, in press). https://doi.org/10.1109/TPDS.2013.27
8. Qin, B., Deng, R.H., Liu, S., Ma, S.: Attribute-based encryption with efficient verifiable outsourced decryption. IEEE Trans. Inf. Forensics Secur. 10(7), 1384–1393 (2015)
9. Puzar, M., Plagemann, T.: Data sharing in mobile ad-hoc networks-a study of replication and performance in the MIDAS data space. Int. J. Space-Based Situated Comput. 1(2/3), 137–150 (2015)
10. Petrlic, R., Sekula, S., Sorge, C.: A privacy-friendly architecture for future cloud computing. Int. J. Grid Util. Comput. 4(4), 265–277 (2013)
11. Wang, Y., Du, J., Cheng, X., Liu, Z.: Degradation and encryption for outsourced PNG images in cloud storage. Int. J. Grid Util. Comput. 7(1), 22–28 (2016)
12. Ye, X., Khoussainov, B.: Fine-grained access control for cloud computing. Int. J. Grid Util. Comput. 4(2/3), 160–168 (2013)
13. Wang, X.A., Ma, J., Xhafa, F.: Outsourcing decryption of attribute based encryption with energy efficiency. In: Proceeding of the International Conference on P2P, Parallel, Grid, Cloud and Internet Computing 3PGCIC 2015, pp. 444–448. IEEE (2015)

Improved Provable Data Transfer from Provable Data Possession and Deletion in Cloud Storage

Yudong Liu[1,2], Xu An Wang[1,2(✉)], Yunfei Cao[3], Dianhua Tang[3], and Xiaoyuan Yang[1,2]

[1] Key Laboratory for Network and Information Security of the People's Armed Police, Engineering University of the People's Armed Police, Xi'an 710086, Shaanxi, China
wangxazjd@163.com
[2] School of Cryptographic Engineering, Engineering University of the People's Armed Police, Xi'an 710086, Shaanxi, China
[3] Science and Technology on Communication Security Laboratory (CETC 30), Chengdu, China

Abstract. Due to the limited computational resources of data owners and the developments in cloud computing, more and more data owners choose to store data on the cloud to reduce their own storage burden. When the data owner wants to re-select a new cloud service provider, the data needs to be transmitted from Cloud A to Cloud B. How to ensure that data has been safely transferred to Cloud B and Cloud A honestly implementing data deletion has become a hot topic for many scholars. To solve these problems, Liang et al. proposed a provable data transfer protocol based on provable data possession and deletion for secure cloud storage. However, we find a security flaw in their scheme. In this paper, we first review their scheme and then present our attack in detail. Finally, we propose an improved scheme which can resist the attack well.

1 Introduction

In this age of information, more and more data are generated by data owners, and the storage and processing of these data has put tremendous pressure on users. Cloud service providers can provide data outsourcing services for data owners, so data owners prefer to hand over data to cloud servers to ease their burden [2]. In this case, the data owner does not need to spend a lot of money to buy hardware or software, and only need to purchase the required services from the cloud service providers, which greatly reduces the economic burden on the data owner.

Cloud servers bring great convenience to data owners while there are also some security issues [3]. For reasons of economic interest, the cloud server may delete or modify the data stored by the data owner, which makes the integrity and reliability of data destroyed. To verify the integrity of outsourced data, many improved PDP protocols, such as [4–6] have been proposed to achieve more advantages then the original PDP scheme [7]. Sometimes the data owner needs to transfer partial data or whole data

© Springer Nature Switzerland AG 2019
F. Xhafa et al. (Eds.): INCoS 2018, LNDECT 23, pp. 445–452, 2019.
https://doi.org/10.1007/978-3-319-98557-2_40

from Cloud A to cloud B [8]. In order to reduce the communication overhead, the data owner usually does not retrieve the data from Cloud A. Instead, a transfer request is sent to Cloud A by the data owner, and then Cloud A transfers data to Cloud B directly. However, in order to save bandwidth or for other reasons, Cloud A may not transfer the data in accordance with the data owner's requirements. Secure data deletion is also a hot topic of privacy-preserving, a malicious cloud may not delete the data for hidden business value [9]. If the above problems are not solved well, cloud service providers will not obtain the trust of data owners, and the public acceptance of cloud servers will be greatly reduced.

In order to solve the above problems, Liang [1] et al. proposed a novel and provable data transfer scheme based on the Merkle Hash Tree with rank to achieve the following three goals: Provable data transfer, provable data possession, provable data deletion.

Unfortunately, we find a security flaw in their scheme. Specifically, for the modified block m_i, the corresponding polynomial-based authentication tag can be forged by the malicious cloud easily. In this paper, we point out this security flaw and propose an improved scheme.

Organization

The rest of the paper is organized as follows. In Sect. 2, we first review the system model of their scheme. The brief constructions of their scheme are given in Sect. 3. In Sect. 4, we present our concrete attack which indicates that their scheme is not secure. In Sect. 5, we give an improved scheme and make a security analysis for it. Finally, a summary of this paper is given in Sect. 6.

2 Their System Model

In order to reduce the storage burden, the data owner chooses Cloud A to store his data, he can check the integrity of data in Cloud A or Cloud B by sending a verification query to TPA. Besides, when the data owner wants to re-select Cloud B to store his data, a transfer request will be send by him to Cloud A. Next, Cloud A preforms the transfer request and deletes this part of data locally. TPA can not only verify the data transfer is successful or not, but also check whether Cloud A has already deleted the transmitted data correctly. Finally, the data owner can still request the TPA to check the data integrity on Cloud A or Cloud B regularly (Fig. 1).

3 The Construction of Their Scheme

Combining RMHT with the PDP scheme of Yuan [10] and Yu [11], they construct their provable data transfer scheme. The detail constructions are as follows.

G and G_T are two multiplicative cyclic groups with an order of large prime q, g is the generator of G and u is a random value chosen from G. $f_{\vec{a}(x)} = a_0 + a_1x + a_2x^2 + \ldots + a_{s-1}x^{s-1}$ is a polynomial with coefficient vector $\vec{a} = (a_0, a_1, \ldots, a_{s-1})$. $H_1 : \{0,1\}^* \rightarrow Z_q^*$ is a hash function which is used to construct Merkle Hash Tree. Another hash function $H_2 : \{0,1\}^* \rightarrow Z_q^*$ is a cryptographic hash

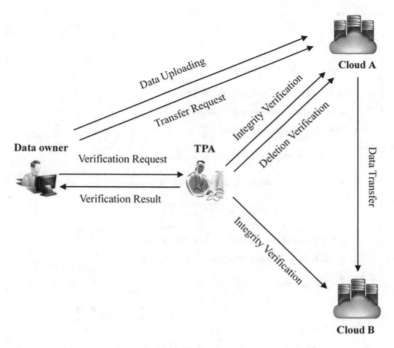

Fig. 1. System model

function. $e : G \times G \to G_T$ is the bilinear map. The public parameters are $(G, G_T, g, q, f_{\vec{a}(x)}, H_1, H_2, e)$.

Their scheme mainly consists of five phases: *KeyGen, Store, Transfer, DeletCheck, IntegCheck*. Because our attack mainly lies in the authentication tag generation process in *Store*, we will describe this phase in detail while other phases briefly. The detail introductions can be found in [1] (Figs. 2, 3, 4, 5 and 6).

4   Our Attack

For each file block m_i, the data owner computes a polynomial-based authentication tag:

$$\sigma_i = (u^{H_1(\mathit{fname})} \cdot \prod_{j=0}^{s-1} g^{m_{ij}\alpha^{j+2}})^{\beta} \tag{1}$$

Let $A = u^{\beta}$, $B_0 = g^{\alpha^2 \beta}$, $B_1 = g^{\alpha^3 \beta}$, $B_2 = g^{\alpha^4 \beta} \ldots B_{s-1} = g^{\alpha^{s+1} \beta}$, the above equation can be converted to:

$$\sigma_i = A^{H_1(\mathit{fname})} \cdot \prod_{j=0}^{s-1} B_j^{m_{ij}} (1 \le i \le n) \tag{2}$$

Data owner	Cloud Server
1. Before outsoucing File F', apply erasure code to generate F;	
2. Divide F into n' blocks and each block includes s sectors;	
3. Add $n - n'$ random sentinel into the n' blocks at random positions, record these positions in a table PF;	
4. Randomly pick a file name $fname \in Z_p^*$ and generate a file $\tau = fname \| n \| C \| Sign(fname \| n \| C)$;	
5. For each block m_i, compute a polynomial-based authentication tag $$\sigma_i = (u^{H_1(fname)} \cdot \prod_{j=0}^{s-1} g^{m_{ij}\alpha^{j+2}})^\beta$$	
6. Construct the RMHT T on the n data blocks, each leaf of RMHT is $h_i = H_1(m_i \| i)$, obtain the root of RMHT R;	
7. Compute $\in = Sign_{ssk}(H_2(R))$,	
	$(\tau, F, \sigma_1, ..., \sigma_n, \in)$ $\longrightarrow$
8. Delete the local file.	

Fig. 2. Store

Data owner
1. Select two random numbers $\beta \in Z_p^*, \alpha \in Z_p^*$ randomly;
2. Compute $\delta = g^\beta, \gamma = g^{\alpha\beta}$;
3. Generate a signed public-private key pair $(spk, ssk) \leftarrow Sign()$;
4. Generate a symmetric key k of AES scheme;
5. Compute $(g^\alpha, g^{\alpha^2}, ..., g^{\alpha^{s+1}})$;
6. Get public key $K = \{\delta, \gamma, spk, g^\alpha, g^{\alpha^2}, ..., g^{\alpha^{s+1}}\}$, private key $S = \{\alpha, \beta, ssk, k\}$;

Fig. 3. KeyGen

By collecting different file block tags equations like above, we can compute the values of $A, B_0, B_1, B_2 . . . B_{s-1}$ through some basic mathematics elimination operations.

To more clearly demonstrate the process of our attack, next we give a simple example:

Assuming that the file F is split into 3 blocks and each block contains 2 sectors, that is to say, $m_{ij}(1 \leq i \leq 3, 0 \leq j < 2)$ can be used to represent the file F. For each file block, the authentication tag $\sigma_i(1 \leq i \leq 3)$ is computed as:

$$\sigma_1 = (u^{H_1(fname)} \cdot g^{m_{10}\alpha^2} \cdot g^{m_{11}\alpha^3})^\beta \tag{3}$$

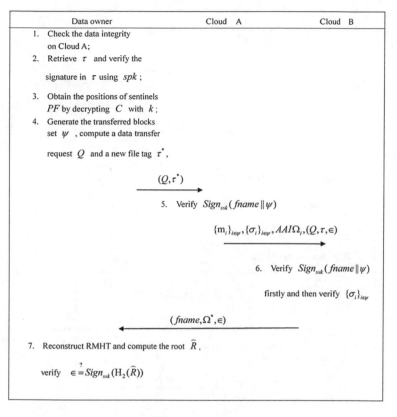

Fig. 4. Transfer

$$\sigma_2 = \left(u^{H_1(fname)} \cdot g^{m_{20}\alpha^2} \cdot g^{m_{21}\alpha^3}\right)^\beta \tag{4}$$

$$\sigma_3 = \left(u^{H_1(fname)} \cdot g^{m_{30}\alpha^2} \cdot g^{m_{31}\alpha^3}\right)^\beta \tag{5}$$

Let $A = u^\beta$, $B_0 = g^{\alpha^2 \beta}$, $B_1 = g^{\alpha^3 \beta}$, then the malicious cloud can get:

$$\sigma_1 = A^{H_1(fname)} \cdot B_0^{m_{10}} \cdot B_1^{m_{11}} \tag{6}$$

$$\sigma_2 = A^{H_1(fname)} \cdot B_0^{m_{20}} \cdot B_1^{m_{21}} \tag{7}$$

$$\sigma_3 = A^{H_1(fname)} \cdot B_0^{m_{30}} \cdot B_1^{m_{31}} \tag{8}$$

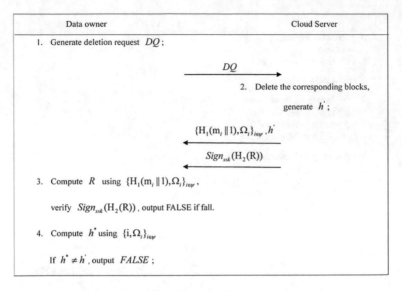

Fig. 5. DeletCheck

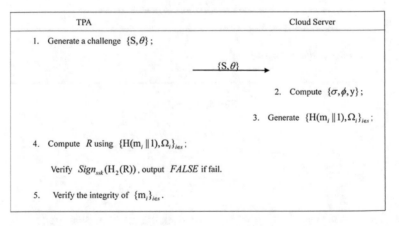

Fig. 6. IntegCheck

Because the values of $\sigma_i(1 \le i \le 3)$, $m_{ij}(1 \le i \le 3, 0 \le j < 2)$ and $H_1(fname)$ are known to the malicious cloud, he can compute A, B_0, B_1 as follows:

$$\frac{\sigma_1}{\sigma_2} = B_0^{m_{10}-m_{20}} \cdot B_1^{m_{11}-m_{21}} \tag{9}$$

$$\frac{\sigma_2}{\sigma_3} = B_0^{m_{20}-m_{30}} \cdot B_1^{m_{21}-m_{31}} \tag{10}$$

Next he computes

$$C = (\frac{\sigma_1}{\sigma_2})^{m_{20}-m_{30}} = B_0^{(m_{10}-m_{20})\cdot(m_{20}-m_{30})} \cdot B_1^{(m_{11}-m_{21})\cdot(m_{20}-m_{30})} \tag{11}$$

$$D = (\frac{\sigma_2}{\sigma_3})^{m_{10}-m_{20}} = B_0^{(m_{20}-m_{30})\cdot(m_{10}-m_{20})} \cdot B_1^{(m_{21}-m_{31})\cdot(m_{10}-m_{20})} \tag{12}$$

Let $x = (m_{11} - m_{21}) \cdot (m_{20} - m_{30})$, $y = (m_{21} - m_{31}) \cdot (m_{10} - m_{20})$, then he can compute

$$E = \frac{C}{D} = B_1^{x-y} \bmod q \tag{13}$$

Thus

$$B_1 = E^{(x-y)^{-1}} \bmod (q - 1) \tag{14}$$

Using the same way, the malicious cloud can compute B_0. After knowing the values of B_0, B_1, A can be computed easily. So a malicious cloud can modify arbitrary block m_i and then forge the corresponding tag which can pass verification successfully.

5 Our Improved Scheme

Based on Liang [1] et al. basic constructions, we give an improved scheme. Let G, G_T be two multiplicative cyclic groups with a composite order $N = pq$ (difficult to factor), where p and q are two different large primes, similar with the RSA scheme, given the value of N, it is difficult to factor it to be p and q. $H_1 : \{0,1\}^* \rightarrow Z_N^*$ is a hash function chosen to construct Merkle Hash Tree. Another hash function $H_2 : \{0,1\}^* \rightarrow Z_N^*$ is a cryptographic hash function. Other constructions and processes are same as their original scheme.

Because N is difficult to be factored, computing $N's$ Euler function $\phi(N) = (p-1)(q-1)$ is also difficult. In our attack, Eq. 14 will be replaced by the following Eq. 15:

$$B_1 = E^{(x-y)^{-1}} \bmod \phi(N) \tag{15}$$

From the above equation, the adversary is difficult to compute the value of B_1. Because knowing the value of N, it is difficult to compute the value of $\phi(N)$. In other words, the polynomial-based authentication tag can't be forged by the adversary. Therefore, our improved scheme enables to resist the attack we have proposed.

6 Conclusion

In this paper, we present a concrete attack on Liang [1] et al.'s Provable Data Transfer from Provable Data Possession and Deletion in Cloud Storage scheme, but we still emphasize that their scheme is of great value for follow-up studies. For the security flaw of their scheme, we propose an improved scheme. The security analysis in Sect. 6 suggests that our improved scheme is more secure and practical.

Acknowledgments. This work is supported by National Cryptography Development Fund of China under grant number MMJJ20170112, National Natural Science Foundation of China (Grant Nos. U1636114, 61772550, 61572521), National Key Research and Development Program of China (Grant No. 2017YFB0802000), Natural Science Basic Research Plan in Shaanxi Province of China (Grant Nos. 2018 JM6028,2016JQ6037).

References

1. Xue, L., Ni, J., Li, Y., et al.: Provable data transfer from provable data possession and deletion in cloud storage. Comput. Stan. Interfaces **54**(P1), 46–54 (2016)
2. Liu, X., Sun, W., Quan, H., Lou, W., Zhang, Y., Li, H.: Publicly verifiable inner product evaluation over outsourced data streams under multiple keys. IEEE Trans. Serv. Comput. https://doi.org/10.1109/tsc.2016.2531665
3. Sun, W., Liu, X., Lou, W., et al.: Catch you if you lie to me: efficient verifiable conjunctive keyword search over large dynamic encrypted cloud data. In: Computer Communications, pp. 2110–2118. IEEE (2015)
4. Wang, C., Ren, K., Lou, W., et al.: Toward publicly auditable secure cloud data storage services. Network IEEE **24**(4), 19–24 (2010)
5. Zhu, Y., Hu, H., Ahn, G.J., et al.: Cooperative provable data possession for integrity verification in multicloud storage. IEEE Trans. Parallel Distrib. Syst. **23**(12), 2231–2244 (2012)
6. Yan, Z.H.U., et al.: Secure collaborative integrity verification for hybrid cloud environments. Int. J. Coop. Inf. Syst. **21**(03), 165–197 (2012)
7. Shacham, H., Waters, B.: Compact proofs of retrievability. In: Advances in Cryptology - ASIACRYPT 2008. Springer, Heidelberg (2008)
8. Chalse, R.R., Katara, A., Selokar, A., et al.: Inter-cloud data transfer security. In: International Conference on Communication Systems & Network Technologies, pp. 654–657. IEEE (2014)
9. Lee, J., Yi, S., Heo, J., et al.: An efficient secure deletion scheme for flash file systems. J. Inf. Sci. Eng. **26**(1), 27–38 (2010)
10. Yuan, J., Yu, S.: Secure and constant cost public cloud storage auditing with deduplication. In: Communications and Network Security, pp. 145–153. IEEE (2013)
11. Yu, Y., et al.: Provable data possession supporting secure data transfer for cloud storage. In: International Conference on Broadband and Wireless Computing, Communication and Applications, pp. 38–42. IEEE (2016)

Chinese POS Tagging Method Based on Bi-GRU+CRF Hybrid Model

Jia-jun Guo, Shu-pei Wang$^{(\boxtimes)}$, Cheng-hai Yu, and Jin-yu Song

School of Information, Zhejiang Sci-Tech University, Hangzhou, China
596061773@qq.com, 1395865929@qq.com,
{ych, songjinyu}@zstu.edu.cn

Abstract. Chinese part-of-speech tagging (POS tagging) is a key part of Chinese Natural Language Processing (CNLP) Research. Using POS tagging can better understand semantics and improve the efficiency of Natural Language Processing. This paper proposes a method of POS tagging based on a bidirectional GRU and CRF hybrid model, which can automatically learn features and reduce operational complexity. Under the same conditions, compared with Bi-LSTM+CRF, CNN+LSTM, LSTM+CRF and HMM, the model obtains the best accuracy.

1 Introduction

Since the Chinese Natural Language Processing (CNLP) problem was put forward, the research in this field has been attracted wide attention. As a hot spot in the field of NLP, the POS tagging is an important prerequisite for information extraction and retrieval. POS tagging provides the correct word-of-speech tagging for every Chinese word in the context, thereby providing support for morphological analysis, syntactic analysis and semantic analysis in the field of CNLP.

In the development of Chinese POS tagging, Hidden Markov model (HMM), Support Vector Machine (SVM) model and Conditional Random Fields (CRF) are widely used. In 2016, Dr. Yang Ronggen conducted research on the Chinese part-of-speech tagging based on the HMM model [4], which verified the validity of HMM Chinese POS tagging and improved the analysis strategy based on the limitations of the HMM model. However, in recent years, with the development of Deep Learning technology, researchers have also put forward a number of effective methods for the part-of-speech tagging based on Deep Neural Networks. At present, the three general methods have been used by academia for POS tagging in different ways.

Combination of POS tagging and syntactic analysis: The researchers found that due to the close correlation between POS tagging and syntactic analysis, the combination of POS tagging and syntactic analysis can significantly improve the accuracy of both tasks.

Heterogeneous data fusion: Scholars consider Multi-Source heterogeneous data to improve accuracy of models, such as methods based on guiding features and methods based on neural network sharing.

© Springer Nature Switzerland AG 2019
F. Xhafa et al. (Eds.): INCoS 2018, LNDECT 23, pp. 453–460, 2019.
https://doi.org/10.1007/978-3-319-98557-2_41

Methods based on Deep Learning: Deep Learning can automatically use the non-linear activation function to accomplish feature extraction. In addition, Deep Learning makes initial word vector input has already described similarity information between words, which is important for POS tagging. In the 2016 ACL academic conference, Xuezhe Ma proposed the combined model of LSTM and CRF for POS tagging in English, which made the tagging performance was significantly improved.

This paper proposes a method based on Bidirectional GRU and CRF model for POS tagging by acquiring context information, generating word representation features as the input of the next layer. The method can make full use of GRU sequence, and combine CRF model with sentences decoding which can extract the corresponding tags from the entire sentence efficiently. Experiments show that Chinese POS tagging performance is significantly improved based on the Bidirectional GRU and CRF hybrid model.

2   Neural Network Structure

The paper designs a neural network model based on bidirectional GRU and CRF. Taking "今天天气非常适合游玩" as an example shown in Fig. 1, the model is divided into Word Vector, Bi-GRU and CRF. In this section, the model describes the components of the neural network architecture, and introduces the neural layers in the neural network one by-one from bottom to top.

2.1   Word Vector

The goal of research on the Chinese POS tagging is the further use of the Word Vector. Therefore, the model firstly makes the words vectorization, digitizes the symbols and map them into the k-dimensional vector space. This paper uses the Google open source project "Word2vec" to translate Chinese words into a corpus into word Vectors.

2.2   Bidirectional GRU Model Structure

2.2.1   GRU

The GRU neural network model is a special RNN. A GRU unit consists of an update gate and a reset gate. It is the special structure that can choose which information is forgotten and which information is remembered.

- The reset gate r_t:r_t is used to control the influence of the previous hidden layer unit h_{t-1} on the current word x_t. If h_{t-1} is not important to x_t, that is to say the new meaning starts from the current word x_t, regardless of previous influences. The r_t switch can be turned on, so that h_{t-1} has no effect on x_t.
- The update gate z_t:z_t is used to decide whether to ignore the current word x_t. Similar to the input gate in LSTM, z_t can determine whether the current word x_t is important for the meaning of the expression. When the z_t switch turns on, the model will ignore the current word x_t and at the same time constitute a "short-circuit connection" from h_{t-1} to h_t, which gradient has effectively propagated back.

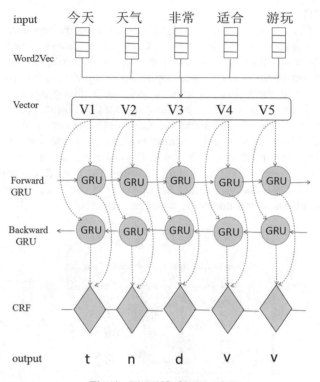

Fig. 1. Bi-GRU+CRF model

At time t, the components of the GRU unit are updated:

$$r_t = \sigma(W_r x_t + U_r h_{t-1} + b_r) \tag{1}$$

$$z_t = \sigma(W_z x_t + U_z h_{t-1} + b_z) \tag{2}$$

$$\bar{h}_t = \tanh(W x_t + r_t U h_{t-1} + b) \tag{3}$$

$$h_t = (1 - z_t)\bar{h}_t + z_t h_{t-1} \tag{4}$$

σ represents the sigmoid activation function, x_t represents the input vector at time t. U_z W_z represents the input matrix and the connection matrix of the hidden layer to the reset gate at the previous time, and W_r U_r represents the input matrix and the connection matrix from the hidden layer to the updated gate at the previous time. W U represents the connection matrix of the input and the hidden layer $\bar{h}_t$ to the optional state at the previous time. The model structure shows that the GRU model is very powerful in the sequence modeling. It can obtain long-term context information and fit non-linearity, which is superior to the traditional models.

2.2.2   Bi-GRU

In the classical Recurrent Neural Network, researchers learn that the transmission of the state is unidirectional. However, in Chinese POS tagging, the current output is not only related to the previous state but also the back one. For example, the word "习惯" can be both a noun and a verb in a sentence, judging it needs previous information and back information. Thus, Bi-GRU is an efficient way to solve such problems.

It is easy to learn the Bi-GRU is combined by two GRU units. From Fig. 2, at each time t, the output is determined by the two GRU units. At time t, the output layer is affected by the hidden layer h_t and the input layer x_t. However, the output layer y_t at time t is independent from other time y_m. If y_t and y_m has a strong dependency relationship between the current time and other time (the behind words of adjective is generally a noun), GRU has less ability to realize these constraints in model. For this reason, the model introduces constraint to solve such constraints well.

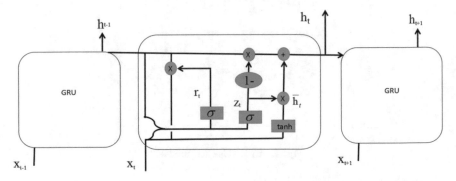

Fig. 2. The structure of gated recurrent unit

2.3   CRF

For the sequence tagging task, researchers generally consider co-decoding because of constraint information. For example, adjectives are generally more likely to follow a noun. The model builds label sequence jointly using CRF instead of decoding each label independently. CRF can avoid the problems because of the independent assumption output in HHM, which leads to its inability to consider the characteristics of the context. It also avoids the problem of marked bias which is created by the possibility of arbitrary selection of features in ME model. That means CRF can obtain the global optimal value.

$x = (x_1, x_2, x_3 \ldots, x_n)$ represents the vector of words, and $y = (y_1, y_2, y_3 \cdots y_n)$ represents the tag sequence corresponding to x. y(x) denotes all possible sets of tag sequences for x. CRF defines a conditional probability $P(y|x)$ for all possible tag sequences given in x as follows:

$$P(y|x) = \frac{\exp(\sum_{i,k} \lambda_k t_k(y_{i-1}, y_i, x, i) + \sum_{i,l} \mu_l \delta_l(y_i, x, i))}{Z(x)} \qquad (5)$$

Where $Z(x)$ is the normalization factor, t_k and δ_1 are the corresponding feature functions, λ_k and μ_1 are weights of the functions. t_k is defined as the edge of the feature function, also called the transfer feature that depends on the current and previous position. δ_1 is a feature function defined on a node and is called a state feature. The value of the feature function sets 1 under the feature condition, otherwise it sets 0. Given a conditional random field $P(y|x)$ and an input sequence x, the conditional probability maximum output sequence y_{max} is found. The prediction algorithm uses the Viterbi algorithm in this model.

3   Experiment

3.1   Data Set

This paper considers the corpus of "People's Daily" (PFR) as experimental data. PFR is often used as data set in academia. In order to ensure the objectivity of the experiment, the corpus is randomly divided into training set, development set and test set. In addition, the number of tags directly affects the tagging accuracy. The more tags, the lower the accuracy. In recent years, academia has various revision extensions for tagging sets. Therefore, in order to ensure uniformity, the tagging sets in this paper are all in the People's Daily Standard Edition.

3.2   Experiments

Experimental machine configuration is CPU:Inter (R) Xeon (R) CPU E3-1230 v5@ 3.4 GHz, GPU:NVIDIA GeForce GTX 1060 3 GB, Memory: 8G;

To get the best experimental results, the setting of model parameters is particularly important. In this paper, we need to set reasonable parameters such as number of hidden layer nodes, learning rate, and word vector dimension. The researchers got the following results by experiments.

Figures 3 and 4 show that the results are best when the hidden layer number is 128 and the word vector dimension is 64. As for the learning rate, we find that it is related to the number of iterations and can obtain the best value during run time.

In order to explore the validity of the model proposed in this paper, researchers set up the experiment to compare the results of different models under the same conditions, i.e. using the same corpora and setting the same parameters to use LSTM+CRF, Bi-LSTM+CRF, HMM, and CNN+LSTM and the models proposed in this paper for Chinese POS tagging. Experimental standards are mainly compared in terms of accuracy and training time.

Experiment 1: Chinese POS tagging accuracy experiment.

Hyperparameter parameters: word vector length 64, number of hidden layer nodes 128, learning rate $\alpha_t = \alpha_0/(\rho(t-1)+1)$, PFR corpus, experimental results are shown in the following table:

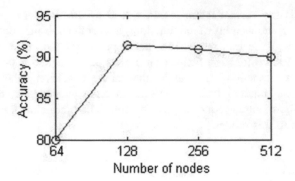

Fig. 3. The accuracy of different nodes

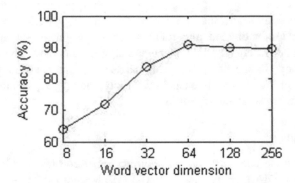

Fig. 4. The accuracy of different word vector

From the experimental results, the five models achieved greater than 90% accuracy in the Chinese POS tagging. The accuracy of the POS tagging methods from high to low are: Bi-GRU+CRF>Bi-LSTM+CRF>CNN+LSTM>LSTM+CRF>HMM. The Bi-GRU+CRF model proposed in this paper is particularly prominent in the comparison of experimental results. It is 2%, 1.27%, 1.23%, and 0.42% higher than the HMM, LSTM +CRF, CNN+LSTM, and Bi-LSTM+CRF models respectively. The Bi-GRU+CRF and Bi-LSTM+CRF results are larger than CNN+LSTM and LSTM+CRF, which fully proves the limitations of unidirectional models in relational context information. It also proves that bidirectional neural network models are used in the field of POS tagging by contextual information has greater advantages than the unidirectional neural network model, and it also highlights the superiority of the model in the Chinese POS tagging. In this experiment, we did not add any prior knowledge to the model for training. In order to further verify the results of the Chinese POS tagging of Bi-GRU+CRF, more comparison experiments were practiced in this paper.

Experiment 2: Chinese POS tagging time comparison experiment.

Hyperparameter parameters: word vector length 64, number of hidden layer nodes 128, learning rate $\alpha_t = \alpha_0/(\rho(t-1)+1)$, take the same 60 KB of data to mark the time test, the experimental results are shown in Table 1 (for easy to compare):

Table 1. Accuracy of different models

Model	Accuracy (%)	Time (ms)
LSTM+CRF	90.76	212
Bi-LSTM+CRF	91.61	240
HMM	90.03	191
CNN+LSTM	90.80	198
Bi-GRU+CRF	92.03	200

Experiments show that the Bi-GRU+CRF proposed in this paper has increased the complexity of the model, thus increasing the computational complexity. However, from the analysis of the overall experimental results, the Bi-GRU+CRF model is still optimal in the Chinese POS tagging task.

In order to improve the accuracy and speed of labeling, we put some constraint information into the model according to the criteria of experiments 1 and 2 for comparison (Table 2):

Table 2. The result of Bi-GRU+CRF+Priori

Model	Accuracy (%)	Time (ms)
Bi-GRU+CRF+Priori	92.32	188

Experiments show that after adding the constraint conditions, the accuracy of Chinese POS tagging and the tagging speed have greatly improved.

Experiment 3: Comparison with other research results.

Table 3 shows the results of the POS tag model of this article comparing with the other four Chinese POS tagging research results. Our model is significantly better than Senna (Collobert et al. 2011). It is a feed-forward neural network model using uppercase and discrete suffix features. In addition, the accuracy of our model is improved by 1.99% compared to the domain-adaptive compound word segmentation tagging method. This is a fusion support vector machine model and conversion learning-based method. It summarizes grammatical features and language phenomena. By comparing with these research results, we can fully prove the effectiveness of Bi-GRU for continuous data modeling and the importance of joint decoding and structured prediction. Our results have reached the most advanced accuracy.

Table 3. Comparison with other research results

Other research	Accuracy (%)
Collobert et al.	89.59
Yang et al.	90.20
Zhang and Zhao	90.33
Xuezhe Ma et al.	91.21
This paper	92.32

4 Conclusion

This paper proposes a Chinese POS tagging model based on Bi-GRU+CRF and adding a priori knowledge and obtains the best experimental results through experimental comparison. However, due to the existence of multiple corpus and annotation set, how to find a new method that takes into account more annotation sets, and further improve the accuracy of Chinese POS tagging is the next step in later research.

Acknowledgments. This work is funded by Public Welfare Technology Application Projects of Zhejiang Province under Grant (2016C31072), Research Project of Educational Reform in Zhejiang (jg20160053), Zhejiang Natural Science Foundation (LQ17E050013), Education Department-Autodesk Inc specialized comprehensive reform Project (No. joz201401), State Scholarship Fund.

References

1. Bach, N.X., Linh, N.D., Tu, M.P.: An empirical study on POS tagging for Vietnamese social media text. Comput. Speech Lang. **50**, 1–15 (2017)
2. Zhang, C.R., Zhao, Q.: Research on domain-adaptive POS tagging for compound words. Appl. Res. Comput. (2018)
3. Collobert, R., Weston, J., Bottou, L., Karlen, M., Kavukcuoglu, K., Kuksa, P.: Natural language processing (almost) from scratch. J. Mach. Learn. Res. **12**, 2493–2537 (2011)
4. Ronggen, Y.: Research on Chinese POS tagging based on HMM. J. Jinling Inst. Sci. Technol. **33**(1), 20–23 (2017)
5. Yi, X.: Chinese POS tagging based on CNN and LSTM hybrid models. J. Wuhan Univ. **63**(3), 246–250 (2017)
6. Guillaume, L., Ballesteros, M. Neural Architectures for named entity recognition. In: California: Association for Computational Linguistics, pp. 260–270 (2016)
7. Xuefen, X., Zhou, G.: Deep learning for natural language processing. J. Autom. **42**(10), 1445–1465 (2016)
8. ADaud, W.K., Che, D.: Urdu language processing: a survey. Artif. Intell. Rev. (47), 1–33 (2017)
9. Liu, B., Du, J., Nie, B., Liu, L., Zhang, X.: The diagnosis of ancient Chinese part of speech tagging based on second-order HMM. Comput. Eng. **43**(7), 211–216 (2017)
10. Tan, Y., Yang, L., Hu, D.: A method of part-of-speech tagging for Chinese students' english articles. J. Beijing Univ. Posts Telecommun. **40**(2), 16–20 (2017)
11. Sun, S., Lin, H., Meng, J., Liu, H.: Use of source domain structure granular migration learning and part-of-speech tagging applications. J. Chin. Inf. Process. **31**(1), 66–74 (2017)

The 4th International Workshop on Theory, Algorithms and Applications of Big Data Science (BDS-2018)

Load Balancing on Cloud Analyst Using First Come First Serve Scheduling Algorithm

Faizan Saeed, Nadeem Javaid[(⊠)], Muhammad Zubair, Muhammad Ismail,
Muhammad Zakria, Muhammad Hassaan Ashraf,
and Muhammad Babar Kamal

COMSATS University, Islamabad 44000, Pakistan
nadeemjavaidqau@gmail.com
http://www.njavaid.com

Abstract. Cloud computing is major component in our daily life; Integration of Cloud with smart grid brings an important role in electricity management. Fog computing concept is also introduced in this paper which helps to minimize the load on cloud. Many techniques are introduced in papers that includes Round Robin (RR), Genetic Algorithm (GA) and Binary Particle Swarm Optimization (BPSO) etc. In this paper authors introduce First Come First Serve (FCFS) load balancing technique with the broker policy of Closest Data Center to allocate resources for Virtual Machines (VM). FCFS algorithm results are compared with existing known algorithms which includes RR and Throttled algorithm. The Response Time (RT) is less in some clusters as compared to RR and Throttled algorithm. The main goal is to optimise the Response Time (RT) on cloud.

1 Introduction

Electricity is very important for surviving in this world. Traditional grids are not much efficient than Smart Grids (SG). SGs is a combination of traditional grid and computer communication, its a two way communication between clients and utilities [1]. SGs provides electricity in a more efficient and effective way that build trust and safety for the consumers. Cloud computing is network based platform that connects heterogeneous system with different types of network that includes public, private and hybrid. All applications or services that are served to user are accessed by internet which means user can request for electricity with the help of web page or with mobile devices. The resources that are including in cloud are processing power, memory management, VM allocations are managed by Cloud Service Provider (CSP) [2]. In Public cloud all user that are connected by internet can access this cloud from anywhere in the world. Public cloud provides access to all over the world [3]. Private cloud is available for only some particular organization and can only be accessed by the members of some organization. This type of cloud can be placed in any organization or company.

© Springer Nature Switzerland AG 2019
F. Xhafa et al. (Eds.): INCoS 2018, LNDECT 23, pp. 463–472, 2019.
https://doi.org/10.1007/978-3-319-98557-2_42

Public users will not able to access this cloud whereas Hybrid cloud is mixture of both public and private cloud. Anyone can access this cloud whether user will be from some organization or in public place [3]. In different papers authors introduced a variety of software and hardware solutions. Implement energy efficienct issues in cloud operations by minimizing the impact of the cloud. Virtualization technology can be used to improve reducing resource isolation and reducing energy consumption through field relocation and reduction Integration. In this paper, Authors considered a scenario of three layered architecture. In the first layer, There are clusters that contains some buildings and in these buildings there are some homes that are connected with smart meters that monitor the user electricity usage. In second layer there are fogs that are connected with homes and centralized cloud. In each fog there is VM installed that contains operating system and storage. Fog Computing is introduced by the Computer Information System Company (CISCO) [1]. Fog is a communicationchannel between users and cloud any request that is generated by user first goes to fog, if fog is able to handle this request than it acknowledge the request of user; otherwise fog will send this request to cloud. In cloud and fog computing there are three services model that is used in any cloud based application and these are:

- Infrastructure as a Service (IaaS),
- Platform as a Service (PaaS),
- Software as a Services (SaaS).

IaaS provides the platform of physical machines, virtual machines and load balancer. PaaS gives us a platform that will be used for development and deployment tools and SaaS provides software service that will be used on web browser or deployed on local servers of the system. A large part of these associations shows the enthusiasm for cloud computing, given that we can acquire assets from the cloud in an adaptable and secure manner with minimal effort. Cloud shares its assets with multiple customers. The cost of assets depends primarily on the design using them. Therefore, the effective management of assets is very positive for cloud providers and cloud users. The implementation of any hypervisor in the cloud depends largely on adaptability. In this paper Cloud computing concept is used to manage the large amount of users and more computational power. Cloud is the distributed computation network rather than local servers. In a cloud handling the large amount of data of users is a difficult job; therefore to overcome these challenges authors extend cloud to fog computing due to which fogs are placed near to users and as a result the RT is also less than cloud.

1.1   Motivation

In traditional electricity there is not any intelligent system that controls the demand of electricity [10], in traditional grid there is lot of electricity wastage without any check and balance. The motivation of this paper is to combine the traditional grid with cloud system to make Smart Grid (SG) system that helps consumers to use the electricity in an efficient way. The smart appliances and smart meter can fulfil their demand of electricity from nearer grid stations. In short integration of cloud-fog based platform helps us to utilize the electricity resources in efficient way with less RT. In order to balance the load on cloud different load balancing techniques are introduced in order to avoid this problem. For load balancing algorithm different authors used different load balancing techniques like Round Robin (RR), Particle Swarm Optimization (PSO), and Genetic Algorithms (GA) etc. [1] In this paper FCFS load balancing technique is introduced in order to load the balance on cloud. Fogs are placed near to buildings due to which low latency and faster response time is happened between cloud server and user [11]. FCFS load balancing algorithm is intrdouced in this paper.

1.2   Contribution

- Reduce the overall response time on cloud servers,
- Fogs are placed at the edge of user as result faster response time than cloud,
- With the help of fogs location awareness of smart devices will be provided,
- Introduced FCFS algorithm with closest data service broker policy that helps to allocate VM's.

Rest part of the paper is presented as: in Sect. 2 related work is presented. Proposed system model is presented in Sect. 3. Simulations and result presented in section 4 and conculusion is presented in Sect. 5.

2   Related Work

Many techniques are implemented by different authors around the world that helps to balance the load of SGs and cloud. Itrat *et al.* in [1] implement the service broker policy for existing load balancing algorithms. Authors take the scenario of six regions. Each region contains two fogs. In the first tier, there is a cluster of buildings. Each cluster contains 50 to 100 buildings that have 50 to 80 apartments. Fog is connected with cloud and users. They compared the results of RR and throttled algorithms with different services broker policies. The results are better with throttled instead of RR load balancing algorithm. Saman *et al.* in [2] implement the Particle Swarm Optimization (PSO) technique to balance the load on cloud. In this paper, two buildings are considered that contains ten numbers of homes, and each home has its own renewable energy resource that is connected to fogs. They compared the PSO algorithm with other load balancing algorithms and conclude that their proposed technique gives better result than other techniques. Xia *et al.* in [6] analyzed that due to increasing number of users, cloud computing faces some challenges that need to be resolved; like high latency, lack of reliability, location awareness and many more. To overcome this problem many researchers working on [7–12]. Xu *et al.* in [8] used the idea of uniform random initialization, the idea of binary search, combined with the idea of Artificial Bee Colony (ABC) and proposed a new immigration policy based on a live heuristic to cloud environment. They also use Bayesian theorem to moreoptimize the ABC-based improvement process to get the final better solution. As a result, the entire method achieves efficient long-term optimization of energy savings. The simulation and experimental results show that the PS-ABC significantly drops the overall power and better protects the existing research performance compared to running and migrating in a virtual machine. Mevada *et al.* in [9], they proposed a virtual machine placement strategy to get better energy efficiency in a cloudto achieve load balancing. The authors proposed an improved version of the power-based virtual machine placement algorithm to reduce power consumption, better load balancing, and optimize virtual machine placement.

3 System Model

In this paper, Authors integrate cloud and fog environment. In [11] fog computing helps to reduce the load on cloud. The placement of VMs will be done efficiently with less RT and latency [13]. In this proposed model they present three layered architecture, The first layer consists of clusters and each cluster contains some

---

Algorithm 1. FCFS

1: initilization
2: *INPUT ← process, burst time*
3: calculate waiting time
4: **for** wait time wt i=1 to process n **do**
5: | wt = burst time + waiting time
6: **end for**
7: process wait
8: $wt[0] = 0$
9: turn around time
10: **for** i=1 to process n **do**
11: | tat = burst time + waiting time
12: **end for**
13: calculate averge time
14: total wait time / no of processes
15: find avg turn around time
16: totat tat / no of processes
17: end

---

buildings that have a controller which manages the electricity of home. In the second Layer fogs are presented in this layer, each fog is connected with the cluster and in third layer clouds are connected with fogs. Consumers are not directly connected with fog and neither cloud directly communicates with the consumer. Only fog should communicated with consumers and cloud. When the consumer will send request to fog for electricity, It will first communicate with closest Micro Gird (MG), If MG is able to fulfil the demand of electricity then fog will send its request back to the consumer otherwise fog will send the request to cloud and cloud will provide nearer MG facility to the consumer. In this paper Authors consider the whole world that is divided into six regions which are also called continents [1]. Each region has two fogs that serves two clusters in the same region. Each fog is nearer to MG that fulfils the demand of electricity. User will request the demand of electricity to fog then fog will communicate with the nearer MG, if MG is not able to fulfil this demand then fog search for another MG nearer to it and so on, if none of the MG not able to fulfil the demand of user then fog send request to cloud and cloud will fulfil the demand of the user. In Fig. 1 proposed system model is introduced and Algorithm 1 is about FCFS.

4 Problem Formulation

In this section, authors will discuss problem formulation of this paper.

4.1 Expected Response Time

In cloud based environment authors need to minimize the RT of cloud [15]. In Eq. 1 The response time of request will be calculated by subtracting the finish time and arrival time then adding with transmission delay. The formal representation of response time is

$$RT = F_t - A_t + T_d \tag{1}$$

F_t is finish time of user request, A_t is arrival time of request and T_d is transmission delay.

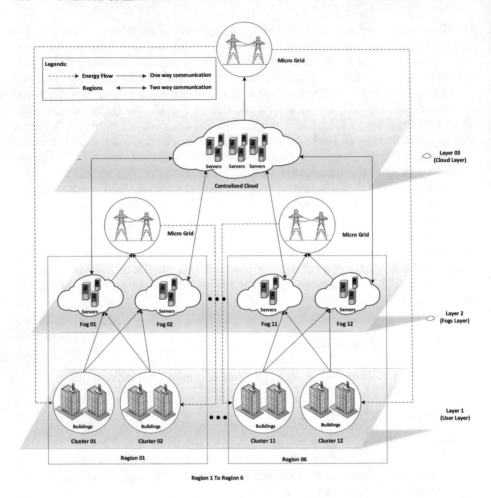

Fig. 1. Proposed system model

4.2   Transmission Delay

In Eq. 2 transmission delay will be calculated by adding transmission latency and tranmission transfer. The equation is:

$$T_D = T_{latency} + T_{transfer} \tag{2}$$

$T_{latency}$ is network latency and $T_{transfer}$ is time taken to tansfer the size of data D.

$$T_{transfer} = \frac{D}{BW_{peruser}} \tag{3}$$

BW is bandwidth of each user which is calculated as:

$$BW_{peruser} = \frac{BW_{total}}{N} \tag{4}$$

BW_{total} is total bandwidth and N is the number of user request.

5   Simulation and Discussion

In this section, authors will discuss simulation setup and conducted results related to this work.

5.1   Setup

In this paper, simulation is done using Cloud Analyst tool that is JAVA based. Simulations are performed on the HP Pavilion 15R with 8 gigabytes of RAM on Windows 10 Operating System. Cloud Analyst tool helps us to compare the load balancing technique with different service broker policy in order to analyze and manage the behaviour of energy demand. For this, they take twelve clusters of buildings that are located in the six different regions of the world. The disribution of regions is in Table 1 taken from [1]. Authors consider two fogs that are connected to cloud. Each fog will contain memory, VM and storage which means that when requests come to fog, it uses service broker policy with load balancing algorithms to assign VM.

Table 1. Distribution of regions

Regions	Id
North America	1
South America	2
Europe	3
Asia	4
Africa	5
Oceania	6

5.2   Discussion

In this paper, Authors use FCFS algorithm with the closest service broker policy. In closest service broker policy, the closest fog is selected on the basis of minimum latency in the network. Homes get quickly response when they are communicating with fog. Following are the minimum, maximum and average response time of the cluster of buildings in Fig. 2. The maximum response time of cluster is almost 70 (ms) in C9 and C10, Minimum response time is in cluster 4 that is about 38 (ms) and average response time is in cluster C7 and C8 that is almost 50 (ms). Fog request servicing time is started when request sends to fog. The time that is taken to process the request is response time of fog. The average, minimum and maximum time of fog can be shown in Fig. 3. In fog performance the maximum RT values starts from 2.5 (ms) to 11 (ms) whereas minimum RT is less than 3 (ms) and average RT is between 7 (ms) and 1 (ms). In Fig. 4, there is comparison of the average results of RR, Throttled and FCFS techniques.

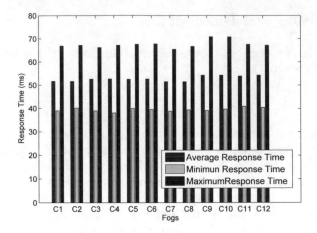

Fig. 2. Performance of clusters

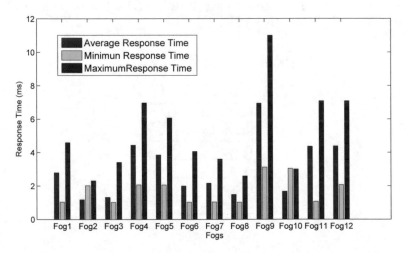

Fig. 3. Performance of fogs

In some clusters, The average response time of fogs is almost same except C4, C8 and C11. In these three clusters, the average response time of FCFS is less than the RR and Throttled. In clusters C4, C8 and C11 the RT is slightly less in proposed technique as compared to RR and Throttled. In Fig. 5, there is cost of each fog with RR, Throttled and with proposed technique. In some cases, cost of fogs is higher then our proposed technique; however, in fog2 and fog6 cost is less than RR and Throttled. Cost of each fog is different according to algorithms; in fog4 the cost of all algorithm is same whereas in other fogs the cost is different for every algorithm.

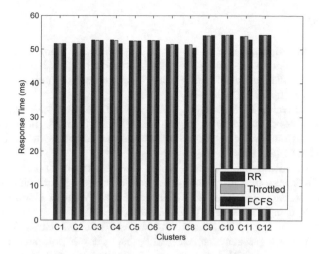

Fig. 4. Comparison of proposed technique

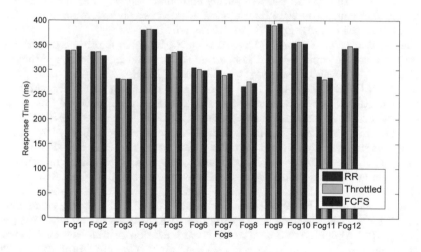

Fig. 5. Comparison of cost

6   Conclusion

In this paper an effective fog-cloud based platform is introduced in order to minimize the RT and latency on cloud. The purpose of this work is to manage the energy requirements of homes and buildings in less RT. Simulations are performed on Cloud Analyst JAVA based platform. FCFS algorithm is proposed and implemented in this paper in order to allocate the VMs with efficient resource management. However, the purposed FCFS technique not able to outperform and beat the other algorithms likes RR and Throttled. In future results should be better if we apply dynamic priority scheduling algorithms.

References

1. Fatima, I., Javaid, N., Iqbal, M.N., Shafi, I., Anjum, A., Memon, U.: Integration of cloud and fog based environment for effective resource distribution in smart buildings. In: 14th IEEE International Wireless Communications and Mobile Computing Conference (IWCMC-2018) (2018)
2. Zahoor, S., Javaid, N., Khan, A., Muhammad, F.J., Zahid, M., Guizani, M.: A cloud-fog-based smart grid model for efficient resource utilization. In: 14th IEEE International Wireless Communications and Mobile Computing Conference (IWCMC-2018) (2018)
3. Yasmeen, A., Javaid, N., Rehman, O.U., Iftikhar, H., Malik, M.F., Muhammad, F.J.: Efficient resource provisioning for smart buildings utilizing fog and cloud based environment. In: 14th IEEE International Wireless Communications and Mobile Computing Conference (IWCMC-2018) (2018)
4. Abbasi, B., Javaid, S., Bibi, S., Khan, M., Malik, M.N., Butt, A.A., Javaid, N.: Demand side management in smart grid by using flower pollination algorithm and genetic algorithm. In: 12th International Conference on P2P, Parallel, Grid, Cloud and Internet Computing (3PGCIC) (2017)
5. Aazam, M., Huh, E.-N.: Fog computing and smart gateway based communication for cloud of things. In: 2014 International Conference on Future Internet of Things and Cloud (Fi Cloud), pp. 464–470. IEEE (2014)
6. Xia, Z., Wang, X., Zhang, L., Qin, Z., Sun, X., Ren, K.: A privacy-preserving and copy-deterrence content-based image retrieval scheme in cloud computing. IEEE Trans. Inf. Forensics Secur. **11**(11), 2594–2608 (2016)
7. Domanal, S.G., Reddy, G.R.M.: Optimal load balancing in cloud computing by efficient utilization of virtual machines. In: 2014 Sixth International Conference on Communication Systems and Networks (COMSNETS), pp. 1–4. IEEE (2014)
8. Xu, G., Ding, Y., Zhao, J., Hu, L., Fu, X.: A novel artificial bee colony approach of live virtual machine migration policy using bayes theorem. Sci. World J (2013)
9. Mevada, A., Patel, H., Patel, N.: Enhanced energy efficient virtual machine placement policy for load balancing in cloud environment. Int. J. Cur. Res. Rev. **9**(6) (2017)
10. Guo, M., Guan, Q., Ke, W.: Optimal scheduling of VMs in queueing cloud computing systems with a heterogeneous workload. IEEE Access. **6**, 15178–15191 (2018)
11. Jena, S.R., Ahmad, Z.: Response time minimization of different load balancing algorithms in cloud computing environment. Int. J. Comput. Appl. **69**(17) (2013)
12. Latiff, M.S., Abd, S.H., Madni, H., Abdullahi, M.: Fault tolerance aware scheduling technique for cloud computing environment using dynamic clustering algorithm. Neural Comput. Appl. **29**(1), 279–293 (2018)
13. Ye, X., Yin, Y., Lan, L.: Energy-efficient many-objective virtual machine placement optimization in a cloud computing environment. IEEE Access. **5**, 16006–16020 (2017)
14. Ibrahim, H., Aburukba, R.O., El-Fakih, K.: An integer linear programming model and adaptive genetic algorithm approach to minimize energy consumption of cloud computing data centers. Comput. Electr, Eng (2018)
15. Hemamalini, M., Srinath, M.V.: Response time minimization task scheduling algorithm. Int. J. Comput. Appl. **145**1 (2016)

On the Need for a Novel Intelligent Big Data Platform: A Proposed Solution

Jeffrey Ray[✉] and Marcello Trovati

Department of Computer Science, Edge Hill University, Ormskirk, UK
{rayj,trovatim}@edgehill.ac.uk

Abstract. This article discusses some existing techniques and methods in data analytics, which aim to identify, extract and assess insights from both unstructured and structured datasets. However, most of them have limitations due to either an over-specialisation or an attempt to provide a general "data panacea". In this work, we will also suggest a potential approach, which aims to overcome such challenges. In fact, specific enhancements to the existing methods are likely to lead to the creation of a tool that is more suited to rapid decision making in business, through the use of clear visualisations supported by novel machine learning methods.

1 Introduction

The creation and availability of large datasets has changed how businesses, scientific and academic fields design their experiments, collect data and ultimately, how usable insights are extracted. As a consequence, over the last decade many data analytics environments and platforms have been proposed to enhance the decision making process. In particular, real-time machine learning combined with detailed and interactive visualisations are likely to enhance the users' ability to process and create actionable outcomes from data.

However, the majority of the proposed solutions have serious limitations, which only allow information extraction and analysis, under specific conditions. Despite the extensive advertising campaigns carried out by businesses, R&D organisations, as well as academic institutions, a comprehensive platform, which utilises a multi-disciplinary approach has not yet been developed. In fact, the way data is gathered, analysed, visualised and adjusted according to the users' needs depends on numerous parameters, which are rooted in various disciplines and approaches.

The purpose of this article is to discuss the processes and rationale behind the need for a new type of data analytics platform, and discuss the initial stages of a proposed solution. The article is structured as follows: Sect. 2 provides an overview of the existing technology and approaches and Sect. 3 describes the proposed solution. Finally, Sect. 4 discusses the next steps and concludes the work.

© Springer Nature Switzerland AG 2019
F. Xhafa et al. (Eds.): INCoS 2018, LNDECT 23, pp. 473–478, 2019.
https://doi.org/10.1007/978-3-319-98557-2_43

2   Discussion of Existing Approaches

As discussed above, there has been a remarkable expansion of data analytics platforms in terms of capabilities, aims and targets.

Alteryx [3] offers a unified machine-learning platform, which facilitates the design of models in a single workflow. With the combination of machine learning algorithms and some visualisation techniques, the Alteryx solution attempts to unify all aspects of data analytics. In particular, it focuses on business users and data scientists, and it attempts to address the skills shortage in this field, by enabling the creation and execution of models in an intuitive manner.

However, Alteryx is regarded as only a data preparation solution vendor. Whilst the software does include some other features for the reporting and visualisation of data, they remain comparatively weak with respect to the intuitiveness of the rest of the platform. This area is a large downfall to the overall solution. A lack of overall support and compatibility with Unix systems creates would limit enterprise roll out to Windows-based organisations.

The Anaconda Suite [4] is an open-source development environment based on the Python and R programming languages. It is a centralised repository for open source libraries to reside and be continually upgraded to ensure the latest version is available to all users. The system enables real-time processing and live data analytics with the use of a Python, R-notebook or live environment (Jupyter). Whilst a large portion of libraries are well maintained and have been carefully developed, it requires users to report fallacious or malicious libraries as the Anaconda developers have limited control over the quality and reliability of the code. The open source nature of Python and R allows developers the autonomy to fully customise and tailor the approach taken in order to maximise the output so that it aligns with the business aims and objectives. In fact, the approach taken by the Anaconda team is to create a platform that can be freely customised. If a developer decides additional functionality is required, it is entirely possible to add such functionality within the Anaconda framework with additional Python code. The machine learning capabilities of Python are far reaching, as it is currently one of the most popular choices for creating machine learning solutions.

However, every component provided by Anaconda must be custom developed. As a consequence, extensive Python or R knowledge is required to developed and implement a solution within the Anaconda system.

IBM [5] offers a wide range of machine learning and data analytics solutions, with over 40 different products available, each focusing on a small analytics or machine learning solutions. The text analysis and data mining solution SPSS and Watson allow the identification of business-ready results, which makes it a popular choice across the business world, with IBM currently possessing the market share of users at around 9%. Their intuitiveness enables all users to harness the machine learning processes with the data preparation and model management aspects.

However, a large number of solutions provided by IBM propose a fragmented set of the machine learning options, which can create confusions as to what exact learning algorithm and output would be most suitable to a specific business application. Furthermore, the majority of the solutions are single-use applications with no ability to tailor the software and its interface (as expected from closed source software).

Microsoft [6] much like IBM, provides a wide range of products suited and aimed at the machine learning and data analytics community. The use of Microsoft Azure cloud services allows a connected system that provides a scalable cloud solution. The two major routes Microsoft offers include Azure Machine learning (Studio) and its onsite solution of SQL server with Machine Learning Services. The overall solution offered is limited in its advanced options but does provide a suitable introduction to Machine Learning for business intelligence. Being one of the most recognisable names in technology Microsoft has substantial market awareness and capital to create a product that meets the consumers' needs. However, the ability to provide a cloud-based solution, enables a smaller business to enjoy the benefits of computationally expensive machine learning at a fraction of the cost.

This solution requires trust in the cloud services to maintain data confidentiality with potentially significant risks. The lack of ability to customise the application to address specific business needs can also cause problems if the domain is highly focused. The machine learning studio offering lacks in some critical areas, including code synthesis, containerisation and external code access. Users are also unable to maintain a stable working version. If Microsoft pushes an update to the services it is impossible for users to retain the current version.

The University of Waikato, New Zealand, created two software solutions: Weka [7] with a primary focus on regression and classification tools, and MOA [8] for streaming data analysis. These packages are both free (GNU licensed) and they contain a collection of visualisation tools and algorithms for data analysis and predictive modelling. Its simple GUI provides access to all its functionality with no coding knowledge or specialist training. The machine learning functions are implemented in Java and simple data pre-processing is also available. The tool originated with the processing of agricultural data, but has since been developed in a full machine learning based analysis platform, capable of running across any system capable of running a Java executable. The Weka and MOA solutions allow for simple analysis and models to be built with no complications, which is the main reason for its popularity. The system is based on a "plug and play" model aimed at entry level machine learning and data analysis. In comparison to the other methods available, Weka is significantly less flexible for statistical analysis and data exploration. Unlike its competitors it tends to be difficult to modify and clean datasets. Furthermore, the package lacks the ability to explore and transform data sets without fully understanding the source code and altering the application and the packages source code requires a high level of expertise.

3   Proposal of Integrated Solution

As discussed above, one of the most recurrent issues found across the current market offerings is the lack of a full integration of visualisation and interactivity

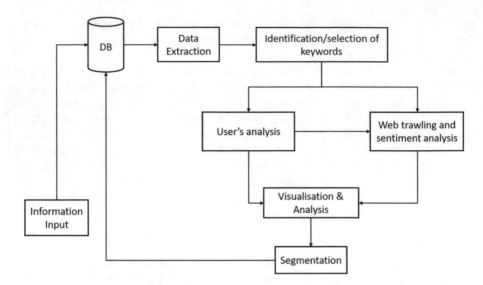

Fig. 1. The main components.

capabilities, as well as exhaustive data analysis. As part of our current research efforts, we are aiming to create a novel data analytics tool, which combines suitable features to address the above challenges. To achieve this, a group of multi-disciplinary researchers have been working on the different components, which define the proposed solution.

In this article, we will briefly describe part of the general architecture, which aims to create a dynamic, responsive and accurate tool to enhance the decision making process. The main components of the proposed approach are outlined in Fig. 1. In particular, this shows how the interactivity between individual components is key to a useful system that provides business advantage over existing solutions. An outline of each component is discussed below:

- Data extraction: the relevant data is mined and assessed to identify the relevant information, which will lead to the identification of actionable insights.
- Identification and selection of keywords: this process focuses on the assessment and definition of specific keywords, if textual input data sources are available. This will be used to refine the searching and trawling component by specifying a context or scenario, which will enable a greater precision by specifying the semantic target.
- User's analysis: during this stage, the user is able to interact with the system via a UI. Specific information will be used to gather information regarding properties and attributes of concepts, topics and products.
- Web trawling and sentiment analysis: general and more specific websites are identified based on pre-defined keywords, as well as other search criteria. Social platforms will also be investigated.
- Visualisation: the results are visualised via an appropriate UI.

- Segmentation and analysis: during his process, the information retrieved and entered into the system will be used to identify and analyse specific segments of the data, which will be utilised to identify specific trends and market needs.

In particular, the ability to gather information across key global impact factors such as social media feeds and web forums or web sites, can provide a vital insight into customers' opinions with regards to a product service or brand. In particular, the ability to monitor such feeds in an automated but reliable manner is likely to create valuable direction for product development. Sentiment analysis has developed significantly over the past 10 years and will continue to become more accurate, while requiring fewer resources as new techniques are introduced [9, 11]. Any data analytics system must be able to quickly embrace these changes to maintain a cutting-edge position as an industry leader in the implementation of machine learning for business intelligence.

In [1], the authors emphasise that an efficient approach must capture the audience's attention and help them to engage with the data so that the overall understanding is increased. Furthermore, when data is accessible, users can better understand how a machine learning algorithm has reached its final decision. The level of trust and the value people perceive from machine derived decisions is expected to increase, especially if the machine learning algorithms will lead to correct decisions and predictions [2].

Another important component is the segmentation and analysis elements. Whilst it is entirely possible for the machine learning to identify and classify what segment of the data fits the purpose without a visualisation, this could lead to a lack of understanding and a subsequent rejection of the analysis carried out. However, requiring continuous and not optimised human interaction will lead to a slower and more costly decision-making process. The segmentation and analysis components are part of a set of tools, which have been developed to integrate novel algorithms based on network theory, topological data analysis, text mining techniques and automated extraction of decision models [10, 12–14]. These are part of an IP process and as a consequence, at the moment they cannot be described in details.

4   Conclusion

The initial work discussed in this article aims to provide a unified solution of interactive visualisations that help shape the automatic decision-making process and give greater insight to the user as to what parameters have been used to influence the overall decision. The ability to create a tool that not only aids novice users but can be adapted by industry experts, is likely to create a unique but intuitive user experience. This aspect will aid with the usability of the advanced machine learning, whilst providing easy access to powerful machine learning tools to enhance the business decision making process.

References

1. Evergreen, S., Metzner, C.: Design principles for data visualization in evaluation. New Dir. Eval. **140**, 5–20 (2013)
2. Lee, M.K.: Understanding perception of algorithmic decisions: fairness, trust, and emotion in response to algorithmic management **5**(1) (2018) https://doi.org/10.1177/2053951718756684
3. Alteryx. https://www.alteryx.com/. Accessed 14 Jun 2018
4. Anaconda. https://www.anaconda.com/what-is-anaconda/. Accessed 14 Jun 2018
5. IBM. https://www.ibm.com/analytics/. Accessed 14 Jun 2018
6. Microsoft. https://azure.microsoft.com/en-gb/services/machine-learning-studio/. Accessed 14 Jun 2018
7. Holmes, G., Donkin, A., Witten, I.H.: Weka: a machine learning workbench (PDF). In: Proceedings of the Second Australia and New Zealand Conference on Intelligent Information Systems, Brisbane, Australia (1994). Accessed 25 Jun 2007
8. Bifet, A., Holmes, G., Kirkby, R., Pfahringer, B.: MOA: massive online analysis. J. Mach. Learn. Res. **99**, 1601–1604 (2010)
9. Liu, B.: Sentiment Analysis and Opinion Mining. Morgan and Claypool Publishers (2012)
10. Trovati, M., Bessis, N., Huber, A., Zelenkauskaite, A., Asimakopoulou, E.: Extraction, identification and ranking of network structures from data sets. In: Proceedings of CISIS, pp. 331–337 (2014)
11. Trovati, M., Bessis, N.: An influence assessment method based on co-occurrence for topologically reduced big datasets. In: Soft Computing, Springer, Heidelberg (2015)
12. Trovati, M., Asimakopoulou, E., Bessis, N.: An analytical tool to map big data to networks with reduced topologies. In: Proceedings of InCoS 2014, pp. 411–414 (2014)
13. Trovati, M.: Reduced topologically real-world networks: a big-data approach. Int. J. Distrib. Syst. Technol. (2015)
14. Trovati, M., Hayes, J., Palmieri, F., Bessis, N.: Automated extraction of fragments of Bayesian networks from textual sources. Appl. Soft Comput. **60**, 508–519 (2017)

Place of Analytics Within Strategic Information Systems: A Conceptual Approach

Martina Halás Vančová[✉]

Faculty of Management, Odbojárov 10, P. O. Box 95,
820 05 Bratislava 25, Slovakia
martina.halas@fm.uniba.sk

Abstract. The paper is focused on the topic of analytics and its connection to strategic information systems. Analytics strategy and IT strategy, however, are not separated areas within an organization – they are mutually connected and moreover, they are connected to the overall business strategy as well as other strategies of functional areas within management. Thus, the paper covers the connection of analytics strategy with IT strategy and the overall business strategies. Based on literature review, there are provided visualizations expressing the placement of analytics strategy to other strategies in management.

1 Introduction

The term "analytics" is nowadays well-known every-day routine for many organizations. However, it is important to highlight that analytics lies in the interface between business and its functional areas and IT and therefore it is necessary to theoretically integrate the field of analytics into theories of management. Thus, we have proposed a research question: What is the connection between strategic information systems and analytics? The research question has been answered and the proposed paper was elaborated based on the available scientific literature with the addition of own conceptual framework, which was created on the basis of the literature research.

2 Company's Strategy and Analytics

The concept of "analytics" was for the first time introduced in 2005 [2]. Since then, there have been many conceptual frameworks and analytical approaches applied in order to explain the phenomenon. The field of analytics is often confused with related concepts such as big data or business intelligence or its derivatives such as business analytics, predictive analytics, prescriptive analytics or even service analytics. Thus, it is necessary to firstly theoretically determine where analytics belong from the perspective of business.

Nowadays, organizations are not and cannot be focused only on actual activities or past success or failure. They must consider future steps and the big picture. Therefore, organizations should create a formal strategic plan based on which all further plans will be created. Then, naturally, strategical planning influences operational as well as tactical

© Springer Nature Switzerland AG 2019
F. Xhafa et al. (Eds.): INCoS 2018, LNDECT 23, pp. 479–485, 2019.
https://doi.org/10.1007/978-3-319-98557-2_44

plans. Logically, a strategy is not used only in business and management; it has been commonly used in wars, sports and even in ordinary life events. It can be defined as "the view from the top", a roadmap or an action plan providing people an overview of how to achieve desired goals. In business, a strategy is a base stone for managing people and processes, basically, without a precise definition of business strategy organizations would be hardly able to achieve competitive advantage [1].

M. Porter defined five forces influencing business strategy. These are:

- the threat of new entrants
- the threat of substitute products and services
- the threat of rivalry among existing competitors
- the threat of bargaining power of suppliers
- the threat of bargaining power of buyer [12].

Strategies of organizations are holistically influenced by these threats; they arise from the internal as well as the external environment and since they influence a strategy, they influence all the processes, including information systems and the selection of strategic information systems [8].

Generally, information systems in an organization can be divided according to various criteria. From the perspective of planning, they are usually categorized into strategic (IS for top management), tactical (IS for support, decision-making, and management) and operational (transactional systems). In the past, information systems fulfilled purely the support function in organizations. Nowadays, they are tools that help to gain a competitive advantage. Naturally, many organizations may not benefit from strategic information systems in the form of gaining a competitive advantage, but still, they are important tools to improve market dominance [13]. However, as mentioned above, the main goal of strategic information systems is to be aligned with business in order to help to achieve goals of the business strategy, support innovative approaches, help to create new products and product ranges, attract new clients as well as decrease costs [6]. There are several types of business strategies to which strategic IS should be aligned – e.g. a strategy of cost leadership, a strategy of differentiation, a focus strategy, a blue ocean strategy, etc. [11].

It is undeniable that information technology is nowadays considered as an enabler for strategic management. If a company wants to be successful in delivery of services and products, this alignment is a core factor influencing its position in the market as it supports all business processes, not only the delivery [18]. Alignment of IT with business strategy usually brings:

- Innovative applications enable an organization to improve business processes and do more in less time with lower costs
- Improved business processes bring an organization closer to its goal of competitive advantage, they enable to carry our necessary business steps easier, with fewer employees and, as mentioned above, with lower costs
- Simpler connection with suppliers and customers (e.g. through emails, chats, instant messaging, video calls, dashboards with the actual state of supplies, information system shows a customer a current state of goods in stock, etc.)

- Relationship with suppliers and customers is linked to the previous point. IT enables simple omnipresent connection with suppliers and customers and continuous daily communication has the positive impact on the relationship with them.
- And last, but not least competitive intelligence [7].

Additionally, it can be noted that nowadays business companies are dependent on fast and reliable flow of data and information and moreover, information services must be complex and flexible [19].

Researchers have taken several approaches on how to define competitive intelligence. According to Kahaner, "competitive intelligence is a systematic program for gathering and analysing information about competitors' activities and general business trends to further your own company's goals" [5]. Society of Competitive Intelligence defines competitive intelligence as "the process of ethically collecting, analysing and disseminating accurate, relevant, specific, timely, foresighted and actionable intelligence regarding the implications of the business environment, competitors and the organization itself" [17].

Competitive intelligence can be achieved in many ways; however, information technology and the Internet are important tools which can make an organization "intelligent" [7]. Referring to Porter's five forces that influence a strategy, an organization must have enough data about itself, customers, suppliers as well as actual and potential competitors [12]. However, pure data are not equal to intelligence. Data firstly need to be understood as information, which requires present as well as past data. In addition to this, information must be connected with human expertise in order to create knowledge. After an organization obtained sufficient knowledge, it can lead to intelligence provided that an organization is able to make decisions and control own actions and processes based on it [10]. Competitive intelligence is focused on information that helps an organization to make better decisions, as mentioned above, it is the process of data collecting, seeing opportunities and beating competitors. Thus, competitive intelligence could be understood as a theoretical approach to intelligence, but on the other hand, there is business intelligence (BI), which is represented by physical applications, software, and technology aimed at actual gathering, storing and analysing data [16]. Thus, in order to carry out data analysis, an organization needs to implement tools for analytics. Competitive intelligence is often connected with business intelligence; however, business intelligence is a single piece of a whole "analytics" field. There are three types of approaches to analytics: descriptive, predictive and prescriptive and in relation to them it is necessary to distinguish analytics approaches: business intelligence or business analytics. While business intelligence looks into current and past data in order to find actual patterns, information or intelligence for decision-making. However, analytics go deeper – into predictive and prescriptive approaches. When analytics is used within an organization we consider it as business analytics [15]. Nowadays, business analytics, as a form of advanced analytics, is considered as a source of competitive advantage [2, 3]. The phenomenon of a competitive advantage has become very important for organizations and they are trying to improve and innovate all the business areas and processes they consider as the most influential to overcome competitors – higher quality of customer service, more innovations, higher quality of end products or services, etc. All of the mentioned aspects and many others on top of that are definitely a part of a strategy, but their effect may be long-term, and they may not be reflected in profit immediately. However, using effective and coordinated analytics may help organizations to reach strategic goals easier and assure higher organizational effectiveness [9].

Proposed Framework. Analytics can cover many areas of business – marketing, sales, finances, stock, etc. However, it is, and it will always be connected with information technology and it will be relying on it since the analytics tools cannot be used without proper information systems.

We can say that analytics lies in the interface between information technologies and business and is dependent on their mutual cooperation and alignment [14]. Thus, we may derive the relationship that analytics strategy is strongly connected to the overall business strategy as well as IT strategy. Within the business strategy, it may be connected to other sub-strategies such as marketing strategy, financial strategy, etc.

The alignment of IT strategy with business strategy is not a new phenomenon. In 1993, Henderson and Venkatraman introduced the Strategic Alignment Model for Information Technology [4]. The model shows the strategic fit and linkage between business strategy and IT strategy. Based on the model, business strategy and IT strategy mutually influence each other, i.e. business strategy influences the definition of IT strategy and vice-versa, IT strategy has a continuous impact on business strategy. Moreover, the mutual influence is continuous. Based on the proposed model, we have derived a similar conceptual framework that contains also analytics strategy as a phenomenon lying in between of business strategy and IT strategy.

Looking at the individual parts of the model shown in Fig. 1, the business scope has an impact on business governance and distinctive competencies. In addition to this, business strategy influences IT strategy and its scope, governance, and competencies. This is the

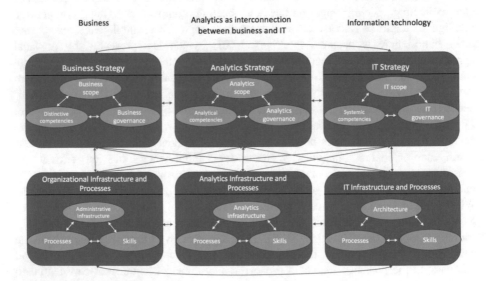

Fig. 1. The strategic alignment model for information technology by Henderson and Venkatraman [4] extended by analytics strategy

same for analytics – the analytics scope is influenced by and at the same time influences the two other strategies. Within the Analytics Strategy, there is the scope which has to be determined and it influences and is influenced by analytical competencies and governance. In line with this logic, analytics infrastructure, processes and skills should be aligned with the same aspects within IT strategy and the overall business strategy.

Figure 2 shows a visualization of the relationship between analytics strategy, overall business strategy, and strategies of 5 functional areas within the management. In traditional management theories, there are several functional areas the within management of an organization, e.g. marketing, HR, finance, office management and production and IT. Some of them are considered as support functions, however, their strategies have to be aligned with each other as well as to the overall business strategy. Analytics strategy should be aligned as derived from all of them, as well as the overall business strategy. For example, HR strategy should evaluate the benefits of analytics – how analysis of HR data can contribute to the HR area but also to the whole business and competitive advantage. But the relationship is mutual – what is required from the HR area if an organization wants to achieve perfection in analytics? The relationship is similar for all other functional areas within management, however, in case of IT strategy it is necessary to highlight that IT is an enabler for analytics – without information systems, it would (nowadays) be impossible to carry out analytics that would be able to bring competitive advantage.

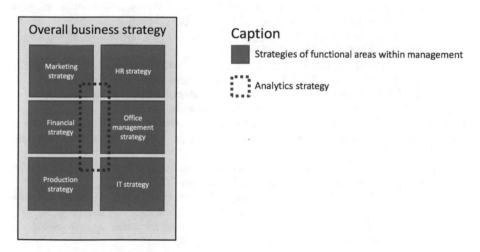

Fig. 2. Relationship between analytics strategy, overall business strategy, and other substrategies

3 Conclusion

The proposed paper covers the topic of analytics and analytics strategy and its placement within management, or more specifically its connection to strategic information systems and the overall business strategy. The paper suggests that analytics strategy should be considered from the very beginning of the formation of the overall

484   M. H. Vančová

business strategy and sub-strategies, i.e. strategies of individual functional areas within management. It has been mentioned that strategic information systems help an organization to achieve competitive advantage. The logic is the same in case of analytics. The effective and coordinated use of analytics in combination with strategic information systems can even leverage the opportunity of achieving a competitive advantage. However, in order to increase chances of having a competitive advantage, organizations should take a closer look on their strategies – the overall business strategy and sub-strategies must be aligned to each other as well as to analytics strategy. In addition to this, analytics strategy is dependent on strategic information systems and IT strategy, because nowadays, information systems are an enabler of analytics. The paper covers these relationships, and they are visualized in conceptual frameworks expressing the connection of analytics strategy to the overall business strategy, IT strategy and other sub-strategies in management.

References

1. Altaf, M., Khalil, M.: Strategic information system: a source of competitive advantage. J. Inf. Knowl. Manage. **6**(9), 24–34 (2016)
2. Daventport, T.: Competing on analytics. Harv. Bus. Rev. (2005). https://hbr.org/2006/01/competing-on-analytics
3. Fromm, H., Habryn, F., Satzger, G.: Service analytics: leveraging data across enterprise boundaries for competitive advantage. In: Bäumer, U., Kreutter, P., Messner, W. (eds.) Globalization of Professional Services, pp. 139–149. Springer, Berlin, Heidelberg (2012). https://doi.org/10.1007/978-3-642-29181-4_13
4. Henderson, J., Venkatraman, N.: Strategic alignment: leveraging information technology for transforming organizations. IBM Syst. J. (1993). https://doi.org/10.1147/sj.382.0472
5. Kahaner, L.: Competitive Intelligence: How to Gather Analyze and Use Information to Move Your Business to the Top. Simon and Schuster (1997)
6. Kamariotou, M., Kitsios, F.: Strategic information systems planning. In: Encyclopedia of Information Science and Technology, Fourth Edition, pp. 912–922. IGI Global Publishing (2018)
7. Koutsoukis, N.-S., Mitra, G.: Decision Modelling and Information Systems: The Information Value Chain. Springer Science & Business Media, New York (2012)
8. Lederer, A.L., Hannu, S.: Toward a theory of strategic information systems planning. J. Strateg. Inf. Syst. **5**, 237–253 (1996). https://doi.org/10.1016/S0963-8687(96)80005-9
9. Levenson, A.: Strategic Analytics: Advancing Strategy Execution and Organizational Effectiveness. Berrett-Koehler Publishers, Oakland (2015)
10. Liew, A.: DIKIW: Data, Information, Knowledge, Intelligence, Wisdom and their Interrelationships. Bus. Manage. Dyn. **2**, 49–62 (2013)
11. Papula, J., Papulová, Z.: Stratégia a strategický manažment ako nástroje, ktoré umožňujú súperenie i spolužitie Dávida s Goliášom., 1st edn. Bratislava: Iura Edition (2012)
12. Porter, M.: Competitive Strategy: Techniques for Analyzing Industries and Competitors, 1st edn. The Free Press, New York (1980)
13. Rackoff, N., Wiseman, C., Ullrich, W.A.: Information systems for competitive advantage: implementation of a planning process. MIS Q. **9**, 285–294 (1985). https://doi.org/10.2307/249229

14. Saxena, R., Srinivasan, A.: Business Analytics: A Practitioner's Guide. Springer Science & Business Media, New York (2012)
15. Schniederjans, M., Schniederjans, D., Starkey, C.: Business Analytics Principles, Concepts, and Applications. Pearson Education Inc., Pearson (2014)
16. Tarver, E.: What is the difference between business intelligence and competitive intelligence? (2018)
17. Vriens, D.J.: Information and Communication Technology for Competitive Intelligence. Idea Group Inc (IGI) (2004)
18. Molnár, E., Kryvinska, N., Greguš, M.: Customer driven big-data analytics for the companies' servitization. In: Baines, T., Clegg, B., Harrison, D. (eds.) Conference: The Spring Servitization Conference, 12–14 May 2014, Aston Business School, Aston University, UK, Aston Business School, Aston University, UK (2014)
19. Engelhardt-Nowitzki, C., Kryvinska, N., Strauss, C.: Strategic demands on information services in uncertain businesses: a layer-based framework from a value network perspective. In: Internation Conference on Emerging Intelligent Data and Web Technologies (2011)

Big Data Inconsistencies: A Literature Review

Olayinka Johnny[1] and Marcello Trovati[2(✉)]

[1] Department of Computing and Maths, University of Derby, Derby, UK
fabyinka@yahoo.com
[2] Department of Computer Science, Edge Hill University, Ormskirk, UK
Marcello.Trovati@edgehill.ac.uk

Abstract. This article presents an overview of data inconsistencies and a review of approaches to resolve various levels of data inconsistencies. It provides a discussion of the approaches, which motivates a Bayesian Network approach in inconsistency resolution.

1 Introduction

An important research area in big data analysis and integration is the extraction of meaningful, accurate, and relevant information [16]. By undertaking such analysis, concepts can be linked and classified accordingly [4]. Such classification will lead to a better understanding and prediction of the properties of the systems that are modelled by specific networks [17].

Datasets from different sources consist of expressions that may contain inconsistencies. In particular, relationships among data are typically inferred at an abstract level, rather than at a data granularity level, where a high proportion of inconsistencies are present. Therefore, in order to extract meaningful and relevant information from data, it is important that the inconsistencies present are investigated and addressed. This article focuses on a review of inconsistencies present in data and different data format, and it is structured as follows: in Sect. 2 we introduce the motivation behind our work. Sections 3 and 4 provide an overview of data inconsistencies and some approaches to address them. Section 5 discusses such approaches, with particular emphasis on their limitations, and Sect. 6 concludes the article.

2 Motivation

This study is motivated by the research in one of the 5 Vs (volume, veracity, value, velocity and variety) of big data. We are particularly interested in variety and veracity properties of big data as elements of inconsistencies are identified at these properties. Variety refers to different forms of structured and unstructured data sets that are collected for use. Veracity focuses on the quality of the data, which is regarded as good, bad, or undefined due to data inconsistency or incompleteness [9]. Research in the variety of big data requires dealing with data in multiple formats, both structured and unstructured, such as numeric, textual, video, audio, and other media and sources. Data Inconsistencies can typically occur when data from the different sources are analyzed

© Springer Nature Switzerland AG 2019
F. Xhafa et al. (Eds.): INCoS 2018, LNDECT 23, pp. 486–492, 2019.
https://doi.org/10.1007/978-3-319-98557-2_45

and integrated to create a business context. These sources can potentially represent the same information in diverse ways. This create a risk of inconsistencies and challenge with respect to the overall business context.

3 Data Set Inconsistencies

As stated in [19], inconsistencies exist at schema level, data representation level and data value level, which is regarded as a data analysis and integration problem. At the data value level, inconsistencies may exist in the facts that are presented from several sources with respect to values that describe the same objects [1].

Consider the integration of the two tables, which have been adapted from [5]. They contain different information about missing persons and admitted to a hospital, respectively (Table 1).

Table 1. Information about missing persons as given by the police and admitted to a hospital, respectively.

Relation *Police*

	Name	Birthdate	Sex	Address
1	Miller	7/7/1959	m	234 Main St.
2	Miller	⊥	⊥	234 Main St.
3	Peters	1/19/1953	m	43 First St.
8	Smith	8/9/1970	m	Mass Ave.

Relation *Hospital*

	Name	Birthdate	Sex	Blood
4	Peters	1/19/1953	⊥	AB
5	Peters	1/19/1953	m	⊥
6	Miller	⊥	f	B
7	Miller	7/7/1959	m	o
9	Smith	9/8/1970	m	A

A possible result of the integration of the two tables is presented in Table 2, which is the result of an outer union relational algebra operation [6]. Succinctly put, relational algebras contain operands, which are relations or variables representing relations. The standard set of operations includes selection, projections, join and union. This result in Table 2 includes all the tuples from both tables; therefore, one could argue that it is a complete integration of the relation. However, the result is not concise, as there is still more than one representation for one person. For example, there are four representations for Miller. The goal of integration is to increase the completeness, conciseness and correctness of data [8].

Another possible result of integration based on the *Name* and *Sex* attributes is

Table 2. Outer Union result of the two tables

Outer Union of *Police* and *Hospital*

	Name	Birthdate	Sex	Address	Blood
1	Miller	7/7/1959	m	234 Main St.	⊥
2	Miller	⊥	⊥	234 Main St.	⊥
3	Peters	1/19/1953	m	43 First St.	⊥
4	Peters	1/19/1953	⊥	⊥	AB
5	Peters	1/19/1953	m	⊥	⊥
6	Miller	⊥	f	⊥	B
7	Miller	7/7/1959	m	⊥	o
8	Smith	8/9/1970	m	Mass Ave.	⊥
9	Smith	9/8/1970	m	⊥	A

presented in Table 3, which is the result of a join relational algebra operation based on equality of the predicates. The result shows that Miller and Peter have consistent result. However, in the case of Smith, there is still a contradiction of the data value on the *Birthday* attribute, which is an example of data value inconsistency.

Big data can be categorised into three broad levels of data, information and

Table 3. Join result of the two example tables based on Name and Sex attributes

Join of *Police* and *Hospital*

	Name	Sex	P.Birthdate	H.Birthday	P.Address	H.Blood
1⋈7	Miller	m	7/7/1959	7/7/1959	234 Main St.	o
3⋈5	Peters	m	1/19/1953	1/19/1953	43 First St.	⊥
8⋈9	Smith	m	8/9/1970	9/8/1970	mass Ave.	A

knowledge. Data is the first level which is the value that is drawn from the source. The data values are mostly drawn from structured or unstructured sources. The second level is information, which is partially structured and it associated with the meaning attached to the various data values. The third level is knowledge where there are well structured and more enhanced representations of the information level. This is the domain that informs decisions making. In [20], the author define these levels as layers of knowledge content and states that inconsistencies can occur at these various granularities of knowledge content from the levels of data, information, and knowledge and by extension meta-knowledge and expertise knowledge. For example, at the data level, inconsistencies involve conflicting values from two instances in the same data set that has different labels, while inconsistencies at information level manifest in terms of how information are functionally dependent on the other [20]. This example of inconsistencies at the data level is equivalent to the data value level inconsistencies described in [1]. However, the difference is that while [1] took the multiple data sources perspective, [20] captured the data level inconsistencies from a single large data set perspective. This study takes the perspective outlined in [20].

3.1   Types of Inconsistencies

Various types of inconsistencies based on the type of big data can be found. This largely depends on the sources of data. To this extent we recognize text, functional dependency, temporal and spatial types of inconsistencies for the different kinds of big data sets.

Text is usually defined as unstructured datasets and it is a rich source of information. For example, causal relations among concepts can be extracted, even though there are a number of issues to consider [10]. In particular, text encodes information in a form that is difficult to decipher automatically. A more relevant problem is that of inconsistencies. In general, text inconsistencies arise from semantics and syntax issues, which characterise natural language in textual data. In [14], the authors use intention and dependency analysis approach to implement TAKMI (Text Analysis and

Knowledge MIning). First, it extracts concepts from the content in textual data by creating semantic dictionary using natural language processing. Next, they apply statistical analysis functions to the extracted concepts to detect associations. The approach is based on the assumption that concepts in the same semantic category have a similar distribution of associations in comparison to other concepts. In [15], an overview of semantic issues is introduced, which discusses unstructured datasets. Furthermore, a survey of the text representation approaches and categorization of text data is presented, with particular emphasis to inconsistencies issues.

Other types of data inconsistencies include functional inconsistencies, which refer to a violation of integrity constraints and spatial-temporal data inconsistencies, which are characterised by location based or time-series data conflicts.

4   Approaches to Resolve Inconsistencies in Big Data

In [2], a framework based on the concept of repair and consistent query answer is presented. The former leaved the database as it is, while the issue of inconsistency is handled at query time, during which the database is consistent with respect to given integrity constraints.

In [11], a system called EQUIP is presented, which focuses on resolving inconsistencies in relational database. This uses Binary Integer Programming (BIP) to repeatedly searches for repairs to eliminate candidate consistent answers until no further such candidates can be eliminated. This approach is in the direction of resolving inconsistency involving functional dependencies.

In [19], a data inconsistency approach utilising data source quality criteria to resolve data inconsistencies is proposed. In particular, it extends an approximate object-oriented data model with defined data source quality criteria with quantitative and qualitative values. It subsequently applies fuzzy multi-attribute decision making approach to select the most appropriate data source as the data inconsistency solution.

In [18], the authors apply probabilistic partial values to address inconsistencies, which are based on relational operators to manipulate such relations. This provides users with the ability to compare tuples of queries whilst having a better understanding of query results in order to decide. This approached is expanded in [12] by utilizing probabilities as well as extended algebraic operations to cope with this additional data.

A possible drawback of these probabilistic approaches is that probabilistic information must be provided for every data item in the scope of the conflicting data [13]. Also, if such probabilities are aggregated, extra constraints from probability consistency checking are added.

In other words, these approaches integrate probabilities, rather than conflicting data. Moreover, such techniques give a probabilistic value as output. Hence, the issue of traceability to the actual conflicting data value item associated with the probabilistic value becomes a challenge. Nevertheless, they further argue that probabilistic approaches are elegant because the values are more general and the type of output of their resolution process is the same as the types of its input that is probabilistic values. This study subscribes to this argument.

A rule-based approach to identify conflicting relationships within an ontology is discussed in [3]. In this approach, experts define a group of conflicting rules, which are then validated against a populated ontology. The approach will then list any cases were these rules are violated. Users use the conflicting statements to improve quality of the ontology. This approach gives the leverage to users to identify properties that are considered to be in conflict. An important challenge with rule-based approach is how to address the ambiguity of the syntax and semantics of natural language. This makes the development of rule-based approaches very challenging to address even in very limited domains of text [17].

5 Discussion

Most of the approaches address inconsistencies at the schema level, otherwise referred to as the structural or schematic inconsistencies. These are the preliminary steps of duplicate detection and schema mapping. Duplicate detection techniques contribute to identify different representations of same real-world objects, and schema mappings are used to identify common representations if data originates in different sources. More specifically, schematic and representation inconsistencies are commonly addressed by relational database approaches [7, 11]. Most of these approaches are based on an implicit assumption that the contents of all the information sources are mutually consistent [13]. In other words, there are no data conflicts within the datasets, hence these approaches lack semantic richness and therefore cannot be used to identify contextual and hidden inconsistencies. The focus is on identifying inconsistencies within big data (structured and unstructured datasets) at the data value level. Such identification is only possible when both schema inconsistencies and data representation inconsistencies have been resolved. The identification of inconsistencies at the data value level would enable the identification of "hidden" inconsistencies, which might not be "universal" but rather contextual [16]. In particular, when large datasets are analyzed, the probability of generating inconsistencies, such as cycles or different probability evaluations representing the same real-world entity, increases almost exponentially.

One strategy to resolve inconsistencies is to first use Natural Language Processing (NLP) and Bayesian Network (BN) approaches to understand and resolve inconsistency. The aim is to extract BNs from text mining/NLP techniques, and in doing so, it is important to identify any inconsistency created. More specifically, NLP can extract the textual information, and subsequently use a BN approach to model and generate the graph about the textual information. The BNs will enable the understanding of the intersection of inconsistencies in the dataset, which could arise from the aggregation of information from textual data sources. In particular, when large datasets are analysed, the probability of generating inconsistencies increases almost exponentially. We note however, that datasets that contain different concepts will not have issues if analysed independently, but could exhibit the inconsistency issues when analyzed and aggregated with other datasets in the set of interconnected statements. Using a BN allows us to reason about probabilistic inconsistencies in unstructured dataset. Its main characteristic is the ability of capturing the probabilistic relationship among variables [17].

6   Conclusion

In this article, we have presented data inconsistencies, whilst providing a discussion on various approaches used in resolving data inconsistencies. Some of the drawbacks of the various approaches have also been investigated, which motivates the proposal of a Bayesian Network approach in inconsistency resolution. This is part of our current research efforts.

References

1. Anokhin, P., Motro, A.: Data integration: inconsistency detection and resolution based on source properties. In: Proceedings of the International Workshop on Foundations of Models for Information Integration (FMII 2001) (2001)
2. Arenas, M., Bertossi, L., Chomicki, J.: Consistent query answers in inconsistent databases. In: Proceedings of the Eighteenth ACM SIGMOD-SIGACT-SIGART Symposium on Principles of Database Systems, pp. 68–79. ACM, May 1999
3. Arpinar, I.B., Giriloganathan, K., Aleman-Meza, B.: Ontology quality by detection of conflicts in metadata. In: Proceedings of the 4th International EON Workshop, May 2006
4. Bansal, S.K., Kagemann, S.: Integrating big data: a semantic extract-transform-load framework. Computer 3, 42–50 (2015)
5. Bleiholder, J.: Data fusion and conflict resolution in integrated information systems (Doctoral dissertation), University of Potsdam (2010). http://www.hpi.unipotsdam.de/fileadmin/hpi/Forschung/Publikationen/Dissertationen/Diss_Bleiholder.pdf
6. Bratbergsengen, K.: Relational algebra operations. In: Parallel Database Systems, pp. 24–43. Springer, Heidelberg (1991)
7. Chomicki, J., Marcinkowski, J., Staworko, S.: Computing consistent query answers using conflict hypergraphs. In: Proceedings of the Thirteenth ACM International Conference on Information and Knowledge Management, pp. 417–426. ACM, November 2004
8. Dong, X.L., Naumann, F.: Data fusion: resolving data conflicts for integration. Proc. VLDB Endow. 2(2), 1654–1655 (2009)
9. Elgendy, N., Elragal, A.: Big data analytics: a literature review paper. In: Industrial Conference on Data Mining, pp. 214–227. Springer, Cham, July 2004
10. Hearst, M.A.: Untangling text data mining. In: Proceedings of the 37th Annual Meeting of the Association for Computational Linguistics on Computational Linguistics, pp. 3–10. Association for Computational Linguistics, June 1999
11. Kolaitis, P.G., Pema, E., Tan, W.C.: Efficient querying of inconsistent databases with binary integer programming. Proc. VLDB Endow. 6(6), 397–408 (2013)
12. Lim, E.P., Srivastava, J., Shekhar, S.: Resolving attribute incompatibility in database integration: an evidential reasoning approach. In: 10th International Conference on Data Engineering, Proceedings, pp. 154–163. IEEE, February 1994
13. Motro, A., Anokhin, P.: Fusionplex: resolution of data inconsistencies in the integration of heterogeneous information sources. Inf. Fusion 7(2), 176–196 (2006)
14. Nasukawa, T., Nagano, T.: Text analysis and knowledge mining system. IBM Syst. J. 40(4), 967–984 (2001)
15. Stavrianou, A., Andritsos, P., Nicoloyannis, N.: Overview and semantic issues of text mining. ACM Sigmod Rec. 36(3), 23–34 (2007)
16. Trovati, M., Bessis, N.: An influence assessment method based on co-occurrence for topologically reduced big data sets. Soft Comput. 20, 1–10

17. Trovati, M., Bagdasar, O.: Influence discovery in semantic networks: an initial approach. In: Proceedings of UKSim (2014)
18. Tseng, F.S.C., Chen, A.L., Yang, W.P.: A probabilistic approach to query processing in heterogeneous database systems. In: Second International Workshop on Research Issues on Data Engineering, Transaction and Query Processing, pp. 176–183. IEEE, February 1992
19. Wang, X., Huang, L., Xu, X., Zhang, Y., Chen, J.Q.: A solution for data inconsistency in data integration. J. Inf. Sci. Eng. 27(2), 681–695 (2011)
20. Zhang, D.: Granularities and inconsistencies in big data analysis. Int. J. Softw. Eng. Knowl. Eng. 23(06), 887–893 (2013)

The 1st International International Workshop Machine Learning in Intelligent and Collaborative Systems (MaLICS-2018)

Matching a Model to a User - Application of Meta-Learning to LPG Consumption Prediction

Michał Kozielski[1]([✉]) and Zbigniew Łaskarzewski[2]

[1] Institute of Informatics, Silesian University of Technology,
Akademicka 16, 44-100 Gliwice, Poland
michal.kozielski@polsl.pl
[2] AIUT Ltd., Wyczółkowskiego 113, 44-109 Gliwice, Poland
Zbigniew.Laskarzewski@aiut.com

Abstract. When predicting consumption of Liquefied Petroleum Gas (LPG) it is profitable to know the consumer type. Different consumers, like a single-family house, a bakery or a primary school, require different predictive models to produce qualitative results. Application of meta-learning makes possible an automated approach to LPG consumption prediction through model selection. Additionally, such solution improves the overall prediction quality.

1 Introduction

Liquefied petroleum gas (LPG) is a propane and butane gas mixture. It has many applications such as: car fuel, fuel for heating buildings, fuel for gas cookers, fuel used in industrial processes. In this work application as a car fuel is omitted and petrol station supply is not taken into consideration. Typically, LPG is stored in tanks located at the client sites and the gas in the tank must be topped up whenever it reaches 20% of the tank's volume. Of course, refuelling can be carried out earlier.

The task undertaken in this work is to predict LPG consumption in a short term horizon (1–7 days). The analysis is performed on the tanks that are equipped with the meters, that can communicate with the supplier IT infrastructure sending measurements of a gas level. Having regular sensor readings reporting the LPG level it is easy to correctly react and refuel the tank when the gas is scarce at a given location. However, from the economical point of view the most profitable is sending a transport that refuels the tank as rare as possible. It means that sending a tanker truck to a client location where there is still a significant fuel supply can be classified as mismanagement in most cases. In order to safely balance on the edge of fuel availability, the thorough predictions are required.

Another case is a situation when the number of transports necessary on a given day exceeds the capacity available for a supplier. It is required then to

© Springer Nature Switzerland AG 2019
F. Xhafa et al. (Eds.): INCoS 2018, LNDECT 23, pp. 495–503, 2019.
https://doi.org/10.1007/978-3-319-98557-2_46

minimise the risk of delayed delivery to the chosen clients. Again, evaluation of such a risk should be based on the thorough predictions.

To make an accurate prediction, it is necessary to select the appropriate predictive model. Such model must be based on data that properly describes the analysed client. Due to the fact that there are different LPG consumers (like e.g., a single-family house, a bakery or a primary school), having different characteristics, it is required to utilize different predictive models to produce qualitative results.

The goal of this work is an automatic match of a predictive model type and a customer with specific consumption characteristics. The contribution of this work consists of a proposed solution of the above issue and of the analysis of the application domain. The proposed solution is based on a meta-learning approach called stacking that is applied to the automatic selection of one of the specialised models or their combination.

The structure of the paper is as follows. Section 2 presents an overview of previous research related to the presented topics. The task of LPG prediction is discussed in Sect. 3. Section 4 outlines an introduced solution. Experiments that were performed and the obtained results are presented in Sect. 5. Final conclusions are presented in Sect. 6.

2   Related Work

According to no free lunch theorem [12] in terms of average quality of results for all classes of problems, there is no best method. Therefore, it is required to learn which learning method is most suitable for a given issue. Such "learning about learning" approach is called meta-learning. Meta-learning in general was surveyed in several papers, e.g. [4,5,9,10] and the recent advances in the field are presented in [2]. A typical approach to meta-learning is based on analysis of the data set characteristics (meta-features) and method performances [1,9].

One of the meta-learning applications is to indicate the best method from the set of available ones. Such issue is called algorithm selection and it was discussed, among others, in [8]. Using this approach, it is possible to choose from different classes of machine learning algorithms, e.g. Support Vector Machines, Neural Networks, Random Forests, Decision Trees, and Logistic Regression the one that is predicted to give the best results on a given data set [7].

Another meta-learning based approach is named *stacked generalisation* (stacking) [11]. Within this method a meta-model (typically a linear model) calculates the predicted value on the basis of base-models predictions. Therefore, it does not select the best method explicitly, but it generates a model, where the weights represent method selection or combination.

A very interesting work in the domain of the given work was presented in [3]. It presents the analysis of LPG usage for different types of buildings. Gas consumption depends on the building destiny, therefore there were three classes identified in this work: housing, housing with high thermal capacity, and indus-try. Depending on the building class and its characteristics different models should be applied in order to predict gas consumption. The models utilised in the research were based on auto-correlations of gas usage time series and different cross-correlations between gas consumption and temperature time series.

3   LPG Consumption Prediction

LPG consumers can be of different types and can include e.g. a single-family house, a bakery, a primary school, a farm or a factory. Their gas consumption can be, therefore, temperature dependent, constant, seasonal or of other type. Figure 1 presents two examples of consumption plot in time, where month of a year is marked as an axis label. Temperature dependent consumption (Fig. 1b) is characterised by higher values during winter months and lower values during summer. In case of the other plot (Fig. 1a) no dependency on temperature or other seasonal changes can be recognised.

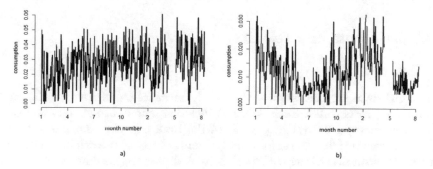

Fig. 1. Consumption values within months - (a) non-temperature dependent characteristic, (b) temperature dependent characteristic

Therefore, looking for the possibly best prediction solution for the given task it is required not so much to choose between the algorithms, such as e.g. linear regression, regression tree or kNN in order to generate the prediction model of the best quality. The basic task is rather to find a model or combination of models matching the given client characteristics. Such models can be generated on the basis of utterly different features characterising clients and which features have to be derived from data.

4   Proposed Solution

The proposed solution is based on a meta-learning approach, what means that it consists of the base-models and a meta-model. The base-models in this approach correspond to the classes of clients utilising LPG. They can be labelled by a type of the location, e.g., a single-family house, or by a type of consumption characteristic, e.g., temperature dependent. However, it is assumed that there can be many (unknown number) classes of clients and their characteristics can be a composition of several factors. As a consequence, a single client can be characterised by a combination of the predefined models described by labels.

Therefore, the idea is to use a meta-model that can learn what should be the association between the models for a given client and combine their predictions taking under consideration the model weights. Such approach promotes application of stacked generalisation (stacking) to the presented solution.

Stacking has been generally presented in Sect. 2. When applied to the issue undertaken in this work, it generates a meta-model predicting the LPG consumption value on the bases of base-models predictions. The meta-model is a linear combination of base-predictions and therefore, it has the following form: $y = a_1 * x_1 + ... + a_k * x_k + b$, where k is a number of base-models.

Stacking has two phases of model generation. At first, base-models have to be created utilising the first part of training data. Next, the meta-model is created utilising the predictions of base-models applied to the second part of training data. In this way an ensemble of base-models and a meta-model are trained.

5   Experiments and Results

5.1   Data Set

The original data set consisted of the hourly measurements collected for over three hundred LPG tanks. Each data row included among others date and time of a measurement, gas level in percent of the tank capacity and outside temperature. It was decided to perform daily aggregations of the original data and the resulting data describing each tank consisted of: date, median gas level and median outside temperature.

Gas consumption was calculated as a difference in gas level for a given day and a day before. Consumption data may contain artifacts that require further processing. Tank refuelling results in a large negative consumption value that should be set to zero. A gauge low accuracy can result in the small negative consumption values that should be also set to zero. In such case, an initial gas level should be retained in order to be used for further calculation of consumption in a next day. There is a number of measurements that are missing in a data set. It was decided that a given day is represented by a missing value if there are 12 or more missing hourly values per day.

It was decided to reduce a number of the analysed tanks in order to make it more manageable and easier to illustrate. Therefore, the analysis was performed on a data set containing data of 23 LPG tanks. Each tank data consisted of at least 495 days of measurements (maximal number of measurements was 613).

5.2   Experimental Setup

Two simple models were designed as base-models predicting LPG consumption calculated as a tank percent. The models are intuitive and correspond with the conclusions resulting from the Fig. 1. Additionally, the models are based on utterly different features.

The first approach is based on an assumption that the consumption values predicted in a short term should not be much different from the recent ones. The attributes utilised by this model consist of:

- a mean consumption values calculated of the recent 3, 5, and 7 days,
- differences of mean consumption in recent 3 and 5 days, and 3 and 7 days.

The second approach is based on an observation that there is a number of clients whose gas consumption depends on the outside temperature (see Fig. 1b). In this case short term weather forecast can be utilised to calculate temperature-based consumption. In case of the presented work the collected outside temperature measurements were utilised as forecasts.

The last approach is based on stacking, where the two base-models presented above create an ensemble. Final consumption value is calculated as a linear combination of the values predicted by base-models.

The solution that is presented is applicable to each tank, treating it as a separate source of measurements (data set). The three prediction methods are evaluated on each data set (tank). Prediction horizon was set to 1–7 days. Therefore, seven prediction scenarios were generated for each model type - for each scenario the consumption in day h starting from the current day is predicted, where $h \in \{1, ..., 7\}$.

Each data set was divided into training and testing data. When a temperature-based model is used, it is required to generate it with data that cover the whole range of possible temperatures. The measurements in the analysed data sets start in winter, therefore, the minimum period of time that can satisfy the above condition is half a year. The base-models were generated on initial 180 examples. Next, the meta-model had to be trained on the basis of base-models results, what required further 180 examples. The rest of each data set was labelled as test data.

The undertaken issue of LPG consumption prediction is a regression task. Therefore, it was decided that every model (both base- and meta-model) would be generated by means of the M5 algorithm [6] that induces a regression tree.

5.3   Results

In order to evaluate and compare the developed models, prediction error was calculated. Figures 2 and 3 visualise Mean Absolute Error (MAE) values for 23 data sets (tanks), for prediction horizon equal 1 an 7 days respectively. The figures have a form of linear plot to make the visualisation more transparent. For the same reason the tanks were sorted according to the error values of meta-model.

Comparing the base-models it can be noticed that when a very short-term prediction is taken into account (1 day) the temperature based model has significantly worse quality in comparison to recent consumption based model

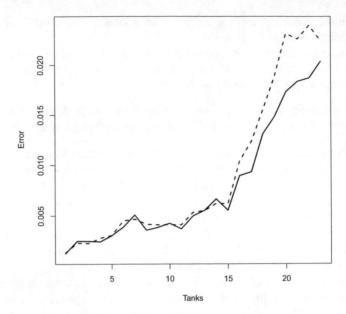

Fig. 2. Prediction Error (MAE) of the compared models for 23 tanks, for the one day prediction horizon (solid line - recent consumption based model, dashed line - temperature based model, dotted line - meta-model)

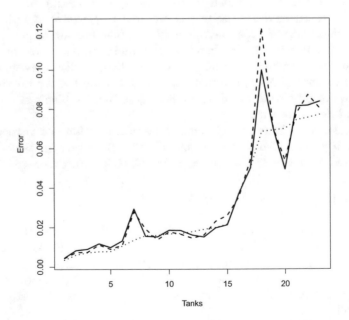

Fig. 3. Prediction Error (MAE) of the compared models for 23 tanks, for the one day prediction horizon (solid line - recent consumption based model, dashed line - temperature based model, dotted line - meta-model)

(see Fig. 2). The difference decreases with the increase of the prediction horizon and is not that clear in case of 7 day horizon (see Fig. 3).

The created meta-model (marked with a dotted line on the figures) is characterised with the lowest MAE in most cases. In order to make this comparison unequivocal a ranking of methods was calculated. The results are presented in Table 1, where a number of times a given model was ranked as the best one is showed. The conclusions drawn from Table 1 are analogous as above.

Table 1. Number of times a given model was ranked as the best one taking into consideration the analysis for the given prediction horizon (M1 - recent consumption based model, M2 - temperature based model, M3 - meta-model)

Prediction horizon	M1	M2	M3
1	6	2	15
2	5	1	17
3	7	2	14
4	6	3	14
5	7	3	13
6	5	5	13
7	6	4	13

In addition to the numerical evaluation presented above it is also interesting to verify if the application of the proposed solution utilizing the meta-model was justified. If there was only one scheme of base models combination used throughout the analysed tanks, then there would be no point in application of the automatic generation of such model. It could be set statically and matching a model to a user would not be required.

Having two base-models, a meta-model being a linear combination of these models results can have one of the following forms:

1. $y = a_1 * x_1 + a_2 * x_2 + b$
2. $y = a * x + b$
3. $y = b$

The first model occurs when the LPG client can be characterised in the best way by a combination of both: recent consumption and temperature dependency. The second model shows strict dependency only on one of the approaches. The last model means that gas consumption is constant and no dependencies are required to model it.

Analysis of the meta-models generated in the experiments showed that all the possible linear meta-model types were generated, what justifies application of a meta-model.

6   Conclusions

The presented research was aimed at proposing and evaluating the approach to automatic matching of a learning model to the characteristics of a client. Such an approach should allow for better quality of LPG consumption prediction.

The proposed solution can be assessed as successful. Application of the meta-model based stacking approach to the presented task resulted in the increased prediction quality in comparison to the base-models. Additionally, the analysis of the results revealed how the relative quality of the chosen base-models change with a prediction horizon. Finally, the proposed approach does not require additional domain experience in the field of matching a predictive model to the characteristics of the LPG consumer. It is the meta-model that decides whether to predict consumption as static, based on outside temperature or recent consumption or based on a combination of these approaches.

Presented work is an initial approach to the analysed issue. Future work that is considered may concern two directions. The first one is related to other learning approaches, especially dedicated to time series analysis. The second direction stems from the fact that in this work it is not taken into account that the data characteristics can evolve in time. Such changes in data named as concept drift may be caused by various factors - change in the number of household members, building insulation, change in the nature of production in a factory. Application of the methods able to adapt to such concept drift may allow further improvement of the results and can make the solution more robust.

Acknowledgements. This research was realised in co-operation of Silesian University of Technology and AIUT Ltd. company and co-funded by the Polish agency National Center for Research and Development within grant POIR.01.01.01-00-0104/17 (System for decision support and management of operational and process knowledge for the LPG gas distributors market). The work was also carried out within the statutory research project of the Institute of Informatics, Silesian University of Technology: BK213/RAU2/2018.

References

1. Blachnik, M.: Instance selection for classifier performance estimation in meta learning. Entropy **19**(11) (2017)
2. Brazdil, P., Giraud-Carrier, C.: Metalearning and algorithm selection: progress, state of the art and introduction to the 2018 special issue. Mach. Learn. **107**(1), 1–14 (2018). https://doi.org/10.1007/s10994-017-5692-y
3. Domino, K., Głomb, P., Łaskarzewski, Z.: Classification of LPG clients using the hurst exponent and the correlation coeficient. Theor. Appl. Inf. **27** (2015)
4. Jankowski, N., Grąbczewski, K.: Universal meta-learning architecture and algorithms. In: Jankowski, N., Duch, W., Grabczewski, K. (eds.) Meta-Learning in Computational Intelligence, Studies in Computational Intelligence, vol. 358, pp. 1–76. Springer, Heidelberg (2011). https://doi.org/10.1007/978-3-642-20980-2-1
5. Lemke, C., Budka, M., Gabrys, B.: Metalearning: a survey of trends and technologies. Artif. Intell. Rev. **44**(1), 117–130 (2015). https://doi.org/10.1007/s10462-013-9406-y

6. Quinlan, J.R., et al.: Learning with continuous classes. In: 5th Australian Joint Conference on Artificial Intelligence, vol. 92, pp. 343–348. Singapore (1992)
7. Reif, M., Shafait, F., Goldstein, M., Breuel, T., Dengel, A.: Automatic classifier selection for non-experts. Pattern Anal. Appl. **17**(1), 83–96 (2014). https://doi.org/10.1007/s10044-012-0280-z
8. Smith-Miles, K.A.: Cross-disciplinary perspectives on meta-learning for algorithm selection. ACM Comput. Surv. **41**(1), 6:1–6:25 (2009). https://doi.org/10.1145/1456650.1456656
9. Vanschoren, J.: Understanding machine learning performance with experiment databases. Lirias. Kuleuven. be, no, May 2010
10. Vilalta, R., Giraud-Carrier, C., Brazdil, P.: Meta-learning - concepts and techniques. In: Maimon, O., Rokach, L. (eds.) Data Mining and Knowledge Discovery Handbook, pp. 717–731. Springer (2010). https://doi.org/10.1007/978-0-387-09823-4-36
11. Wolpert, D.H.: Stacked generalization. Neural Netw. **5**(2), 241–259 (1992). http://www.sciencedirect.com/science/article/pii/S0893608005800231
12. Wolpert, D.H., Macready, W.G.: No free lunch theorems for optimization. IEEE Trans. Evol. Comput. **1**(1), 67–82 (1997). https://doi.org/10.1109/4235.585893

Automated Optimization of Non-linear Support Vector Machines for Binary Classification

Wojciech Dudzik, Jakub Nalepa, and Michal Kawulok$^{(\boxtimes)}$

Future Processing and Silesian University of Technology, Gliwice, Poland
{wojciech.dudzik,jakub.nalepa,michal.kawulok}@polsl.pl

Abstract. Support vector machine (SVM) is a popular classifier that has been used to solve a broad range of problems. Unfortunately, its applications are limited by computational complexity of training which is $O(t^3)$, where t is the number of vectors in the training set. This limitation makes it difficult to find a proper model, especially for non-linear SVMs, where optimization of hyperparameters is needed. Nowadays, when datasets are getting bigger in terms of their size and the number of features, this issue is becoming a relevant limitation. Furthermore, with a growing number of features, there is often a problem that a lot of them may be redundant and noisy which brings down the performance of a classifier. In this paper, we address both of these issues by combining a recursive feature elimination algorithm with our evolutionary method for model and training set selection. With all of these steps, we reduce both the training and classification times of a trained classifier. We also show that the model obtained using this procedure has similar performance to that determined with other algorithms, including grid search. The results are presented over a set of well-known benchmark sets.

Keywords: Support vector machine · Evolutionary algorithms
Model selection · Training set selection · Feature set selection

1 Introduction

Nowadays, the datasets are becoming fairly large much faster than the available computing power. As a result, machine learning systems have to scale up greatly with this growth. This has led to Support Vector Machines (SVMs) being unable to be applied to huge datasets. The main problem is the SVM training time (hence, elaborating a trained model $\mathcal{M}$). It has computational complexity of $O(t^3)$ and memory complexity of $O(t^2)$, where t is the cardinality of a traning set ($\boldsymbol{T}$). It may seem that improvement of hardware made this problem irrelevant, but recent works in the field of implementing SVMs on modern GPGPU units [21,23] denies that. Besides the problem of training time, there is also a need for proper tuning of the SVM hyperparameters. For SVM, there is a slack penalty

© Springer Nature Switzerland AG 2019
F. Xhafa et al. (Eds.): INCoS 2018, LNDECT 23, pp. 504–513, 2019.
https://doi.org/10.1007/978-3-319-98557-2_47

coefficient (C) and hyperparameters of the kernel function in the case of non-linear SVM. Without this process, the resulting $\mathcal{M}$ may have poor performance. This concerns both the generalization error and time needed for classification of new examples.

When it comes to a problem of expensive training, it can be tackled with proper selection of T, as for most cases only small part of a dataset is chosen as Support Vectors (S). The challenging part is that selection of $\mathcal{M}$ and selection of refined training set $(T'$, where $T' \subset T)$ are mutually dependent. Unfortunately, T' cannot be selected *a priori*, because $\mathcal{M}$ has to be trained and evaluated.

Another problem is that there can be a lot of features that are redundant or noisy. These features may bring down the classifier performance, thereby process of feature selection is needed (selecting refined feature set (F') from full feature set (F), where $F' \subset F)$. All of that makes it hard to use SVM classifiers for non-expert users as extensive knowledge in the field of machine learning is required. We propose a new technique that addresses all of the mentioned problems, coupling evolutionary algorithms (EA) with feature selection process.

1.1 Contribution

In this paper, we introduce a schema for automated selection of T', F' and $\mathcal{M}$ for the SVM optimization. We show that with this approach we can decrease both the training and classification times for the obtained classifier. We accomplish that by applying Recursive Feature Elimination (RFE) algorithm, whose results are taken and incorporated into evolutionary optimization of model and training set. Main goal of selecting F' (refined feature set) is to remove redundant and noisy features from the original dataset. This will help to produce better models in a shorter time as the number of features also affects the training time. Evolutionary methods for selecting $\mathcal{M}$ and T' provide easy to use and reliable process to obtain good model. As a result of this, it is easier to train SVM classifier without having expert knowledge about selection process of $\mathcal{M}$, T' and F'. The key aspects of our contribution are as follows: (i) we make it possible to run a single optimization process to select all $\mathcal{M}$, T' and F' at once, (ii) we establish a new scheme, in which every step is performed by other algorithm, (iii) our process is hands-free, therefore it does not require expert knowledge for executing all of these steps. However, it is possible to exploit additional expert knowledge in selecting $\mathcal{M}$, T', or F'.

1.2 Paper Structure

In Sect. 2 we analyze related literature. In Sect. 3, we present our approach and algorithms with focus on our evolutionary methods. Section 4 shows the results obtained over the benchmark datasets alongside with the discussion on the performance and behavior of algorithm. Section 5 concludes the paper.

2   Related Literature

To the best of our knowledge, there are no methods for selecting $\mathcal{M}$, T' and F' in an automated manner, however there are many approaches to solve these problems independently from each other, or by combining only two of those processes. The baseline method for model selection of SVM classifier is Grid Search (GS) as most often there are only two hyperparameters (C and one hyperparameter of kernel function). Although there is a lot of work put in other methods, evolutionary algorithms are often choice for this as PSO with local refinements [1,24]. In another recent work [3], the SVM hyperparameters are optimized by fast messy GA. One more evolutionary strategy propose a bat algorithm [20], which can outperform the ones mentioned in [10,14]. Other methods of model selection couple it with features selection by special coding of individuals in evolutionary algorithms. Genetic algortithms [10] and PSO [14] are used in this case. When it comes to selection of T' recently published review on that topic summarizes the current advances in the field [16]. These methods can be divided into two groups: (i) methods whose complexity is dependent on the cardinality of T, and (ii) those that are independent. The algorithms from the first group exploit the information about the layout of T, including clustering method [19,22], algorithms benefitting from the statistical properties of the T vectors [8], and neighborhood analysis methods [9]. The second group includes among others the induction trees [2] and evolutionary algorithms [15,17]. Decreasing the size of T makes SVM training feasible for large datasets, but it also allows for reducing S, which accelerates the classification (which is linearly dependent on the number of Support Vectors). Beyond that, there is also an interesting finding on the topic in [12], where the authors check how hyperparameters can be obtained on subsets of full training set. As a result of this analysis, they claim that there is very little difference in hyperparameters between full and reduced training sets. Some of the mentioned work already had solution to feature selection. Besides that, other work in this area include wrapper method of feature selection with a lot of work put into using evolutionary algorithms [6,7]. There exist approaches which focus on using and modifying l_2 norm [18] that is used in the SVM training process to gain information about features that should be used.

3   Proposed Algorithm

In this work, we present FSALMA—**F**eature **S**election followed by **AL**ternating **Me**metic **A**lgorithm. This is an extended version of our previous algorithm (ALGA) [11]. Here, we focus on non-linear SVM with a kernel using radial basis function (RBF): $\mathcal{K}(\boldsymbol{u}, \boldsymbol{v}) = \exp\left(-\gamma\|\boldsymbol{u} - \boldsymbol{v}\|^2\right)$, where $\boldsymbol{u}$ and $\boldsymbol{v}$ are the input vectors, and γ is the kernel width. A general schema of this algorithm is presented on Fig. 1. First, our workflow starts with feature selection described in more detail in Sect. 3.2. Then, a reduced set of features (F') is applied to dataset which is used by EA.

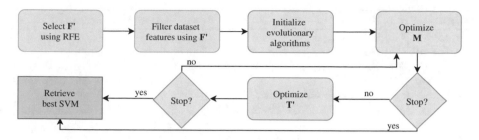

Fig. 1. Flowchart of FSALMA

We initialize training set population by randomly drawing N sets and evaluate them with the default value of $\gamma = 1$, $C = 1$. The best individual of this evaluation is then taken as a training set for the $\mathcal{M}$ phase. The model selection process is presented in Sect. 3.4. This process lasts till the there is no improvement in the average fitness (Sect. 3.1) of population, which is the local stop condition. When this condition is fulfilled, then the algorithm switches to optimize T', described in Sect. 3.3. This switching process lasts as long as one of the phases improves. If no improvement can be made (global stop condition), we take the best model from the last phase. In the next sections details of each process will be presented.

3.1 Fitness Metric

For measuring fitness in feature selection we used accuracy as a metric to compare different models within the cross-validation. Individuals in EA use area under the receiver operating characteristic curve (AUC) as metric [4]. As SVM does not provide probability needed to calculate AUC, we used algorithm described in [13], to convert distance from hyperplane into probability value.

3.2 Feature Selection

For selection of feature set, we exploit Recursive Feature Elimination with cross-validation (RFECV). It works by training some classifier (R) on a full feature set, and then pruning the least important features in an iterative manner. This process is presented in Algorithm 1. We utilize a step (Z) of 10% of features (line 1) and in each of those steps we perform 5-fold cross-validation (lines 3–7). In each fold we train classifier $(R_{T'})$ and score it (lines 5–6). Before getting to next smaller set of features we average score from CV (line 8). After the training process (line 10), we select Z_{best} that yields the best score. Finally, we train R one more time for the entire T (line 11) and return F' pruning the least important features from R_{best} to obtain Z_{best} number of features (line 12). Here, we use an extremely randomized tree classifier [5]—the motivation for using this method is its short training time, its non-linearity and discarding of features based on importance that is natural for this classifier. It can happen that no features are discarded by this process.

Algorithm 1. Recursive Feature Elimination with Cross-Validation (RFECV).

1: $\{Z_i\} = \{1.0, 0.9, 0.8, ..., 0.1\}$;   ▷ Set of steps
2: **for all** $\{Z_i\}$ **do**
3:      T', Ψ ← Split T with $F' = F \times Z_i$ for 5-fold CV.
4:      **for all** T' **do**   ▷ cross-validation for T
5:          $R_{T'}$ ← Train classifier on T'
6:          $\{Score_{T'}\}$ ← EVALUATE($R_{T'}$, Ψ)
7:      **end for**
8:      $\{Score_{T'}\}$ ← AVERAGE($\{Score_{T'}\}$)
9: **end for**
10: Z_{best} ← Take Step for MAX($\{Score_{T'}\}$)
11: R_{best} ← Train classifier for T
12: **return** Take F' from R_{best} with Z_{best}

3.3   Training Set Selection

For training set selection, we use adaptive memetic algorithm (MASVM) presented in [15]. In this algorithm, each of individuals consists of training set created from equal number of examples from each class (K) taken from T. It is ensured that these examples (vectors) are unique. The population is processed with crossover operator in which, two of individuals are summed. Afterwards, new individual is created by randomly selecting no more than K examples for each class. Next, each individual is subject to mutation with the probability $P_m = 0.3$ (as proposed in [15]). This consists in replacing $f_m = 0.3$ of examples to random ones taken from T. The crucial part is local education where we exploit information about previous individuals and adaptive growing of K. By using this technique we can discard the importance of hyperparameter K. We put a improvement on stop conditions of growth of K. We introduced parameter which will stop its size at desired rate in comparison to original training set (W). Its role is to prevent algorithm from over-fitting to data in case where validation data is very similar to training data. The default value is $W = 0.8$, it is based on 5-fold cross-validation split. At the end, the $N = 10$ individuals with the highest fitness survive to maintain the constant population size.

3.4   Model Selection

The process of optimizing the hyperparameters utilize a genetic algorithm. Each individual consist of two things, set of hyperparamaters and kernel type. The whole population (Q) share the same T provided by the training set selection. At the begging, the population is initialized deterministically to cover a large range of the values with a logarithmic step. The hyperparameters are set to $\gamma \in \{0.01, 0.1, 1, 10, 100\}$ and $C \in \{0.1, 1, 10, 100\}$, hence the size of population is $|Q| = 20$. In each iteration, a new population of size Q is created by the crossover process of two individuals x_a and x_b. The child parameter values x_{a+b} become $x_{a+b} = x_a + \alpha \cdot (x_a - x_b)$, where α is the crossover weight randomly drawn from the interval $[-0.5, 1.5]$ to diversify the search (and $x_{a+b} > 0$). Next, each individual with the probability of h is mutated. This consists in modifying a value x within a range $x \in [x - u \cdot x, x + u \cdot x]$, where x is γ or C. To keep population at constant size, only $|Q|$ fittest individuals are taken to the next generation.

4   Experimental Validation

All of the experiment where executed on a virtual machine equipped with Intel Xeon E5-2698 v3 2.30 GHz with 16 cores, 64 GB DDR4 RAM. For training and prediction, we used SVM implemented OpenCV 3.3.1 library. Most of the proposed algorithms were implemented in C++ and complied with Microsoft Visual C++ 2017 Compiler. Only RFE algorithm was implemented in Python 3.5 with Sci-kit learn library. We used Anaconda3 environment for Python.

For all of presented results, we used 5-fold cross-validation procedure. We split all of the data into the training and test sets. In case of Grid Search algorithm internal 5-fold cross-validation on T was performed. Grid Search is looped 3 times as its steps decrease logarithmically with each iteration.

To compare our algorithms with the state of the art, we utilized five popular benchmark datasets from the UCI Repository (Table 1). It is important to note that *Gisette* and *Madelon* datasets originate from the NIPS 2003 Feature Selection challenge. Because of that feature selection is vital here. The sizes in brackets for *Gisette* and *Madelon* are original sizes but we could not get access to test set. We used only training and validation parts which were combined.

Table 1. UCI Repository benchmark dataset

Name	Size	Number of features	Number of classes
German	1000	24	2
Ionosphere	351	34	2
Gisette	7000 (13500)	5000	2
Madelon	2600 (4400)	500	2
Wisconsin	699	9	2

4.1   Results and Discussion

Each evolutionary algorithm was run 50 times (10 times for each fold) and Grid Search (GS) was run 5 times (once per fold). The Table 2 shows average of all these experiments together with the standard deviation of results. Times listed in Table 2 is only for the part of M and T' selection. Feature selection was not counted in as it time remain constant independently of the algorithm used. Average times in seconds were respectively: 2.9 for *German*, 3.8 for *Ionosphere*, 4.7 for *Madelon*, 3.9 for *Wisconsin* and 16.5 for *Gisette*. The process of selection K values for presented datasets is as follows. We take $F/4$ and logarithmically increase it with the step of 4 until $K < 512$. In case of the *Gisette* dataset, values were chosen arbitrary to 250 and 1000 as with that many features this heuristic would provide bad results with long training time and poor performance. ALGA is faster than FSALMA for small datasets and have similar fitness score. Although it is better for certain K values it is not easy to determine this value so algorithm has to be run many times in order to check which K values are appropriate. ALMA provided higher variances in both fitness and the training time because of its adaptive manner. One of the reasons is that, in around 5% of cases the condition for growing initial K was not fulfilled and the algorithm encountered stop condition. It is not a problem as re-running algorithm several times is still faster than running a single GS optimization process. The strong point of FSALMA is its good generalization capability. As it can be seen over all tested datasets there are no cases that clearly identify overfitting. This is not so obvious in the case of ALGA as for higher K values on *German* and *Wisconsin* the fitness of V stays high, but the fitness for Ψ starts to drop down which may indicate overfitting of the resulting $\mathcal{M}$. As it can be seen over the *Madelon* dataset, feature selection is important, as without it the classifier performance is notably worse, both in fitness values and classification time. It also shows (see the results of GS) that feature selection can make training process much faster. The differences can be seen on *Gisette* dataset where GSFS was almost 2.7 times faster than GS. For FSALMA these differences are much greater, as for *Gisette* it is 170 times faster than GS and 64 times faster than GSFS. On all other datasets presented FSALMA was on average 3 times faster then GSFS. We believe that with using larger datasets those differences might be even bigger. It is hard to measure the speed-up versus ALGA as proper K is really crucial in here. In most cases, FSALMA was faster and achieved comparable results. We executed statistical tests to verify the importance of the retrieved results in AUC scores using one-tailed Wilcoxon test. The difference in scores was significant at p-value of 0.05 for all except FSALMA vs. GSFS for *German, Madelon* and *Wisconsin*. These values are the result of using feature selection with Grid Search which yield similar results.

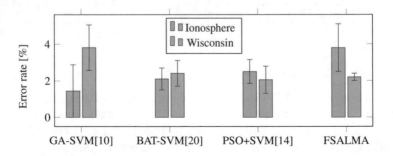

Fig. 2. Comparison of error rate [%] with others [10, 14, 20]

Table 2. Results obtained on UCI datasets. Symbols are as follows: S denote number of Support Vector for $\mathcal{M}$, Fit. V is AUC score for validation set, Fit. Ψ is AUC score on test set, and GSFS means Grid Search with Feature Selection run beforehand.

Set	Algorithm	K	Training time [s]	S	Fit. V	Fit. Ψ	Classification Time V [ms]
German	ALGA	6	0.24 ± 0.05	11.2 ± 1.1	.792 ± .008	.775 ± .030	1.33 ± 2.8
	ALGA	24	0.39 ± 0.10	41.8 ± 5.0	.812 ± .010	.774 ± .029	2.00 ± 2.9
	ALGA	96	0.93 ± 0.29	185.7 ± 4.6	.875 ± .008	.700 ± .037	6.69 ± 6.1
	FSALMA		0.65 ± 0.85	70.6 ± 90.2	.812 ± .043	.766 ± .032	2.20 ± 3.6
Ionosphere	GSFS		2.14 ± 0.89	463.2 ± 13.3	.818 ± .026	.790 ± .028	12.06 ± 9.6
	GS		2.50 ± 1.05	458.6 ± 22.9	.836 ± .027	.793 ± .031	10.56 ± 0.4
	ALGA	8	0.14 ± 0.03	15.7 ± 0.7	.984 ± .006	.975 ± .018	0.42 ± 0.0
	ALGA	32	0.19 ± 0.05	56.3 ± 8.2	.995 ± .003	.976 ± .018	0.87 ± 0.1
	FSALMA		0.17 ± 0.07	34.7 ± 23.8	.987 ± .007	.973 ± .019	0.89 ± 2.9
Madelon	GSFS		0.27 ± 0.02	179.6 ± 37.4	.998 ± .003	.975 ± .016	1.74 ± 0.5
	GS		0.32 ± 0.01	167.2 ± 16.8	.997 ± .006	.984 ± .012	1.77 ± 0.3
	ALGA	125	38.64 ± 7.48	250.0 ± 0.1	.811 ± .026	.599 ± .018	58.10 ± 10.6
	ALGA	500	149.77 ± 48.35	1000.0 ± 0.0	.717 ± .023	.622 ± .026	217.12 ± 11.4
	FSALMA		9.66 ± 6.90	710.9 ± 364.9	.836 ± .061	.801 ± .065	19.33 ± 12.3
Wisconsin	GSFS		13.96 ± 1.44	1551.8 ± 22.3	.833 ± .044	.853 ± .018	32.74 ± 0.4
	GS		105.45 ± 0.37	1600.0 ± 0.0	.649 ± .042	.640 ± .010	297.89 ± 9.5
	ALGA	8	0.06 ± 0.01	8.9 ± 3.9	.996 ± .001	.995 ± .005	0.53 ± 0.0
	ALGA	32	0.08 ± 0.01	21.1 ± 15.5	.996 ± .001	.993 ± .007	0.63 ± 0.1
	ALGA	128	0.22 ± 0.03	141.5 ± 31.6	.999 ± .000	.983 ± .012	1.61 ± 0.4
	FSALMA		0.06 ± 0.01	6.5 ± 1.7	.996 ± .001	.995 ± .004	0.49 ± 0.1
Gisette	GSFS		0.37 ± 0.08	164.4 ± 108.9	.996 ± .002	.991 ± .012	2.07 ± 1.7
	GS		0.34 ± 0.05	158.4 ± 102.5	.996 ± .001	.995 ± .005	1.55 ± 0.8
	ALGA	250	237.33 ± 64.69	329.4 ± 19.0	.991 ± .002	.990 ± .001	2735.24 ± 277.8
	ALGA	1000	1631.34 ± 391.72	793.5 ± 53.8	.995 ± .001	.994 ± .001	6735.90 ± 810.1
	FSALMA		45.58 ± 100.76	147.9 ± 169.1	.985 ± .009	.982 ± .010	593.75 ± 1202.4
	GSFS		2914.52 ± 2917.72	1513.6 ± 267.5	.997 ± .001	.996 ± .000	3155.69 ± 3571.8
	GS		7828.45 ± 749.40	1847.4 ± 179.1	.996 ± .001	.996 ± .000	9959.75 ± 1115.2

As shown in the Fig. 2 we are achieving results comparable to state-of-the-art. Furthermore, we were able to achieve that in a fully-automated manner. What is vital, our model is very fast to train. However, due to different hardware specifications and different implementations of SVM, it is hard to compare times in easy manner. Also most of the papers do not show number of S so it is impossible to compare these results.

5   Conclusions and Outlook

In this work, we presented algorithm, which is the first to provide hand-free optimization of all aspects of SVM training process (optimization of T', F' and $\mathcal{M}$). As results show our approach achieves results comparable to state of the art. Besides that, it does not show tendencies to overfit to data and is quick to train. What is crucial, we manage to keep number of S at much smaller value compared to other algorithms which allow fast classification process.

Our future work will be directed into improving model and feature selection part. We also consider further feature selection as another part of an alternating manner algorithm, possibly with use of evolutionary algorithms. Other interesting topic might be co-evolution of T' and $\mathcal{M}$ populations with exchange of information between them instead of using alternating manner.

Acknowledgement. This work was supported by the National Science Centre under Grant DEC-2017/25/B/ST6/00474, and by the Silesian University of Technology, Poland, funds no. BKM-509/RAu2/2017.

References

1. Bao, Y., Hu, Z., Xiong, T.: A PSO and pattern search based memetic algorithm for SVMS parameters optimization. Neurocomputing **117**, 98–106 (2013)
2. Cervantes, J., Lamont, F.G., López-Chau, A., Mazahua, L.R., Ruíz, J.S.: Data selection based on decision tree for SVM classification on large data sets. Appl. Soft Comput. **37**, 787–798 (2015)
3. Chou, J.S., Cheng, M.Y., Wu, Y.W., Pham, A.D.: Optimizing parameters of support vector machine using fast messy genetic algorithm for dispute classification. Expert Syst. Appl. **41**(8), 3955–3964 (2014)
4. Fawcett, T.: An introduction to ROC analysis. Pattern Recogn. Lett. **27**(8), 861–874 (2006)
5. Geurts, P., Ernst, D., Wehenkel, L.: Extremely randomized trees. Mach. Learn. **63**(1), 3–42 (2006)
6. Ghamisi, P., Benediktsson, J.A.: Feature selection based on hybridization of genetic algorithm and particle SWARM optimization. IEEE Geosci. Remote Sens. Lett. **12**(2), 309–313 (2015)
7. Ghamisi, P., Couceiro, M.S., Benediktsson, J.A.: A novel feature selection approach based on FODPSO and SVM. IEEE Trans. Geosci. Remote Sens. **53**(5), 2935–2947 (2015)
8. Guo, L., Boukir, S.: Fast data selection for SVM training using ensemble margin. Pattern Recogn. Lett. **51**, 112–119 (2015)
9. He, Q., Xie, Z., Hu, Q., Wu, C.: Neighborhood based sample and feature selection for SVM classification learning. Neurocomputing **74**(10), 1585–1594 (2011)
10. Huang, C.L., Wang, C.J.: A GA-based feature selection and parameters optimizationfor support vector machines. Expert Syst. Appl. **31**(2), 231–240 (2006)
11. Kawulok, M., Nalepa, J., Dudzik, W.: An alternating genetic algorithm for selecting SVM model and training set. In: Mexican Conference on Pattern Recognition, pp. 94–104. Springer (2017)
12. Klein, A., Falkner, S., Bartels, S., Hennig, P., Hutter, F.: Fast bayesian optimization of machine learning hyperparameters on large datasets. CoRR abs/1605.07079 (2016)
13. Lin, H.T., Lin, C.J., Weng, R.C.: A note on platt's probabilistic outputs for support vector machines. Mach. Learn. **68**(3), 267–276 (2007)

14. Lin, S.W., Ying, K.C., Chen, S.C., Lee, Z.J.: Particle swarm optimization for parameter determination and feature selection of support vector machines. Expert Syst. Appl. **35**(4), 1817–1824 (2008)
15. Nalepa, J., Kawulok, M.: A memetic algorithm to select training data for support vector machines. In: Proceedings of the 2014 Annual Conference on Genetic and Evolutionary Computation, GECCO 2014, pp. 573–580. ACM, New York (2014)
16. Nalepa, J., Kawulok, M.: Selecting training sets for support vector machines: a review. Artif. Intell. Rev. 1–44 (2018)
17. Nalepa, J., Siminski, K., Kawulok, M.: Towards parameter-less support vector machines. In: Proceedings of the ACPR, pp. 211–215 (2015)
18. Neumann, J., Schnörr, C., Steidl, G.: Combined SVM-based feature selection and classification. Mach. Learn. **61**(1), 129–150 (2005)
19. Shen, X.J., Mu, L., Li, Z., Wu, H.X., Gou, J.P., Chen, X.: Large-scale support vector machine classification with redundant data reduction. Neurocomputing **172**, 189–197 (2016)
20. Tharwat, A., Hassanien, A.E., Elnaghi, B.E.: A BA-based algorithm for parameter optimization of support vector machine. Pattern Recogn. Lett. **93**, 13–22 (2017). Pattern Recognition Techniques in Data Mining
21. Vanek, J., Michalek, J., Psutka, J.: A GPU-architecture optimized hierarchical decomposition algorithm for support vector machine training. IEEE Trans. Parallel Distrib. Syst. **28**(12), 3330–3343 (2017)
22. Wang, D., Shi, L.: Selecting valuable training samples for SVMs via data structure analysis. Neurocomputing **71**, 2772–2781 (2008)
23. Wen, Z., Shi, J., He, B., Li, Q., Chen, J.: ThunderSVM: A fast SVM library on GPUs and CPUs. To appear in arxiv (2018)
24. Zhang, X., Qiu, D., Chen, F.: Support vector machine with parameter optimization by a novel hybrid method and its application to fault diagnosis. Neurocomputing **149**, 641–651 (2015)

Cloud and Fog Based Smart Grid Environment for Efficient Energy Management

Maria Naeem, Nadeem Javaid[✉], Maheen Zahid, Amna Abbas,
Sadia Rasheed, and Saniah Rehman

COMSATS University, Islamabad 44000, Pakistan
nadeemjavaidqau@gmail.com
http://www.njavaid.com

Abstract. Cloud is a pool of virtualized resources. Integrating cloud in a smart grid environment helps to efficiently utilize the energy resources while fulfilling the energy demands of residential users. However, when number of users increase it is difficult to efficiently utilize the cloud resources to handle so many user requests. Fog reduce the latency, processing and response time of user requests. In this paper, cloud-fog based environment for efficient energy management is proposed. The objective of achieving maximum performance is also formulated mathematically in this paper. Simulations in CloudAnalyst are performed to compare and analyze the performance of load balancing algorithms: Round Robin (RR), Throttled, and Weighted Round Robin (WRR) and service broker policies: Service Proximity Policy, Optimize Response Time, Dynamically Reconfigure with Load, and New Dynamic Service Proximity. Simulation results showed that Throttled load balancing algorithm give better response time than RR and WRR.

1 Introduction

According to US Department of Energy report, electricity demand has increased 2.5% annually over last 20 years [1]. Traditional grids are unable to handle this large energy demand, therefore alternative energy resources is only solution. Consumers in traditional grid cannot monitor and control their excessive energy consumption. Thus, there is a need for energy distribution grid with integrated alternative energy resources and ability to allow users to control and monitor their energy resources [2].

To address all these issues, Smart Grids (SGs), the next generation of energy distribution grid, are introduced [3]. SGs allow two way communication of information and electricity. And also enables the service providers and consumers to control and monitor the energy consumption, production and pricing. This helps the service providers to efficiently manage the energy load, and consumers to minimize their electricity bills.

© Springer Nature Switzerland AG 2019
F. Xhafa et al. (Eds.): INCoS 2018, LNDECT 23, pp. 514–525, 2019.
https://doi.org/10.1007/978-3-319-98557-2_48

Smart meters are used in SG to collect and transmit the customers personal and energy consumption information, to cloud [4]. This large amount of data is stored on cloud and used for efficient energy management. Cloud has large number of resources like hardwares, softwares, processors, and storage. But when number of users increase, efficient utilization of these cloud resources become difficult. Further, SG and cloud should ensure privacy, flexibility, security, reliability and scalability.

Fog is integrated between cloud and SG to provide efficient utilization of cloud resources. Fog is deployed on network edge to decrease latency, response, and processing time.

In this paper, a cloud-fog based environment is proposed for efficient utilization of energy resources. The objective function of achieving maximum performance is also formulated mathematically in this paper. Through simulations in CloudAnalyst, different load balancing algorithms: Round Robin (RR), Throttled, and Weighted Round Robin (WRR) and service broker policies: service proximity, optimized response time, dynamically reconfigure with load, and new dynamic service proximity are compared and analyzed.

Rest of the paper is divided into six sections as follows. Related work is discussed in Sect. 2. Proposed model is explained in Sect. 4. Load balancing algorithms and service broker policies are listed in Sects. 5 and 6, respectively. Section 7 contains the simulation results and discussion. Finally conclusion of work is discussed in Sect. 8.

1.1 Motivation

Authors in [5,6] have used the cloud services in smart grid environments to efficiently manage the resources. It is observed that by incorporating cloud and fog in smart grid resources can be utilized efficiently to fulfill the requirements of energy consumers. Load balancing plays important role in efficient utilization of cloud and fog resources, when number of users increase. Many load balancing algorithms are proposed for cloud computing, like in [7,8]. This motivates to propose the cloud-fog based system model for SG and compare the performance of different proposed load balancing algorithms.

2 Related Work

Cloud and fog is integrated in SG to efficiently manage the energy resources. Efforts by research community are done to improve the efficiency of SG and provide better services to energy consumers.

In [9], authors have proposed a cloud-fog based environment for SG and compared performance of load balancing algorithms: RR and Throttled. They have also proposed a new service broker policy. However, they have not considered Virtual Machines (VMs) with different capabilities. A cloud-fog based framework is proposed in [10] to efficiently manage the information in SG. Authors have used RR, Ant Colony Optimization (ACO), Particle Swarm Optimization

(PSO), Throttled, and Artificial Bee Colony (ABC) load balancing algorithms. Authors in [10] have proposed hybrid of ACO and ABC, HABACO load balancing algorithm. However, they have not considered different service broker policies.

Authors in [11] have proposed new Cloud to Fog to Consumer based framework (C2F2C) for efficient management of energy resources in residential buildings. They have also proposed a new algorithm Shortest Job First and show that their proposed technique out perform other techniques. There is a trade off between processing time of algorithms and cost of VMs. To reduce the load of consumers in SG, authors in [12] have proposed a three layer model based on cloud and fog and load balancing algorithm Particle Swarm Optimization with Simulated Annealing (PSOSA). By simulations, they show that proposed PSOSA is better than Round Robin, Throttled, and Cuckoo Search.

In [13] authors have considered charging and discharging of Electric Vehicles (EVs). They have proposed communication architecture for cloud and SG, and priority based scheduling algorithms for EVs to manage energy resources efficiently. This helps to manage energy demand of EVs in peak hours and improve grid stability. A new model is proposed in [14] to allocate the cloud computing resources for demand side management in such a way that cost is optimized. They have developed two algorithms: Simulated Annealing (SA) and modified priority list (MPL) for cost optimization, where MPL outperforms SA. They claim that proposed model has high potential in SG.

A fog computing based SG model is proposed in [3] and its advantages over cloud are analyzed. It is shown that efficiency of cloud based SG can be increased by fog computing. In [15], a new dynamic approach is proposed for load balancing in cloud. Authors have shown that new proposed paradigm Cloud Load Balancing (CLB) is able to balance the load on servers by considering their processing power and current load. Energy management is implemented as service on fog in [16], which provides data privacy, flexibility, and interoperability. For demonstration, authors have implemented and experimented two prototypes of micro grid level energy management and home energy management.

In [17], stochastic model is used in SG by energy consumers for shifting load from peak hours in order to reduce cost. Proposed model is intended to create small energy hubs for users. Authors in [18] has proposed a fog computing paradigm for SG. They have used Bee Swarm algorithm for optimized task scheduling. In [19], authors have proposed a multitenant cloud based nanogrid in SG for efficient energy management. The proposed model reduces the execution time and improves the performance of parameters observed.

3   Problem Formulation

The motivation of this paper is load balancing of VMs in fog. The purpose of load balancing is to improve the system performance by efficient resource utilization by achieving minimum response time, cost and resource wastage. By efficient resource utilization, we mean that load is balanced on all servers such that no server is underutilized or overloaded.

Fog consists of m Physical Machines (PMs), the set of PMs is PM $= \{PM_1, PM_2, \ldots, PM_m\}$. And n number of VMs, the set of VMs is VM $= \{VM_1, VM_2, \ldots, VM_n\}$. k number of incoming user requests are represented as R $= \{r_1, r_2, \ldots, r_k\}$.

3.1 Objective Function

Our main objective is to achieve maximum system performance by minimizing the response time, cost, and resource wastage. The objective function is

$$MaximizeP = \frac{1}{w_1RT + w_2C + w_3RW} \tag{1}$$

Where w_1, w_2, and w_3 are weights, such that, $w_1 + w_2 + w_3 = 1$. P is performance of systems to be maximized, response time, cost, and resource wastage are denoted by RT, C, and RW, respectively. Performance is inversely proportional to response time, cost, and resource wastage. If response time, cost, and resource wastage are minimum than performance is maximum and vice versa.

Cost is total cost of all resources like storage, CPU, memory etc.

Response time RT is defined as,

$$RT = \sum_{i=1}^{k}(FT - AT + L) \tag{2}$$

Where AT, FT, and L are request arrival time, finish time and latency, respectively.

Resource wastage mean that all resources are efficiently utilized and no resource is underutilized. It is defined as,

$$RW = Res_{Total} - Res_{Consumed} \tag{3}$$

4 Proposed System Model

A cloud-fog based system model is proposed to efficiently manage the energy demands of the residential users. The proposed system model, represented in Fig. 1, comprised of three layers:

1. End user layer: The end user layer consists of N number of buildings. Each building has M number of homes and is connected with one fog. Further, each home has multiple electrical appliances and a micro grid. Micro grid include renewable energy resources with storage devices to fulfill the energy demand.
2. Fog layer: Fog layer is used to reduce the latency, processing time and response time of user requests. Each fog in fog layer consists of hardware resources like storage, processor, and bandwidth, which are virtualized. V number of virtual machines are installed on each fog to process user requests. Data center controller on fog route the user requests to virtual machines using VM load balancer.

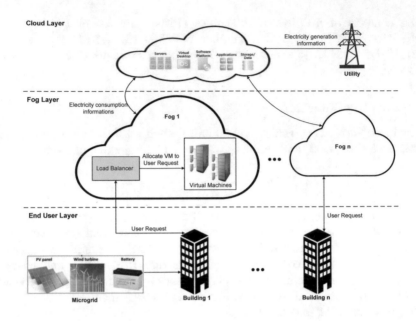

Fig. 1. Proposed system model

3. Cloud layer: Cloud layer contains data centers. Cloud stores the information related to energy consumption of all users and energy generation by utility.

   User requests from end user layer first goes to load balancer in fog. Load balancer allocate VM to user request in such a way that requests load is distributed on all VMs and no VM is underutilized or overloaded. Then fog send the summarized user information to cloud and cloud also have the information about utility. At the end request response is send back to the users to fulfill their requirements. In this way fog is used to efficiently manage the energy resources in SG. And load balancer is used in fog for efficient utilization of fog resources.

5   Load Balancing Algorithms

Load balancing is used to efficiently utilize the cloud and fog resources. Fog has many resources like hardware, storage, and bandwidth. VMs are installed on fog to handle user requests. Data Centers on fog are responsible for managing VMs and routing user requests to VMs. Data Centers run load balancing algorithms to route user request to VMs in such a way that load on all VMs is equal. In his paper, following load balancing algorithms are compared and analyzed in terms of their processing and response time.

5.1   Round Robin

RR algorithm distribute the user requests equally in a circular fashion among VMs. First VM is chosen randomly and then VMs are allocated in a line.

5.2 Throttled

This load balancing algorithm assign the user request to VM which is able to handle the request load.

5.3 Weighted Round Robin

RR distribute user requests equally among VMs, considering current load of VM. But it is possible that VMs have different capabilities (like processor, bandwidth, and storage) to handle requests. In this paper, WRR load balancing algorithm is implemented to deal with this scenario. WRR assign more requests to VM with more weight than the VM with less weight, to get better performance than RR. The weight of VM represent how much requests it can handle. The steps of WRR are represented in Algorithm 1.

Algorithm 1. Weighted Round Robin

1: Input: W
2: $i \leftarrow -1$
3: $currentWeight \leftarrow 0$
4: **while** TRUE **do**
5:    $i = (i + 1)$ mod n
6:    **if** $i == 0$ **then**
7:        $currentWeight = currentWeight - gcd(W)$
8:        **if** $currentWeight < 0$ **then**
9:            $currentWeight = max(W)$
10:            **if** $currentWeight == 0$ **then**
11:                return -1
12:            **end if**
13:        **end if**
14:    **end if**
15:    **if** $W_i \geq currentWeight$ **then**
16:        return i
17:    **end if**
18: **end while**

---

6   Service Broker Policies

The job of service broker policy is to select the fog for user requests. The service broker policy route the user request to fog which can provide best services to users. Following service broker policies are considered in this paper:

6.1 Service Proximity Policy

Service proximity policy maintains the table of all fogs. When a request is received, it route the request to closest fog with minimum latency.

6.2 Optimized Response Time

Optimized response time policy select the fog with best response time. The request is then routed to selected fog.

6.3 Dynamically Reconfigure with Load

Dynamically reconfigure policy is hybrid of service proximity and optimized response time policies. This policy selects the fog with minimum latency and minimum response time.

6.4 New Dynamic Service Proximity

This policy consider the latency and current load on fog to select the fog. Then the user request is routed to selected fog.

7 Simulations and Discussions

This section discusses the simulations and their results.

7.1 Setup

CloudAnalyst tool is used for the simulations. The world is divided into six regions, to map with real world six continents. Each region has 2 or 3 clusters with 10–40 buildings in each cluster. Each cluster is connected to two fogs. Further, three load balancing algorithms are used for experiments, Round Robin, Throttled, and Weighted Round Robin. Performance of these load balancing algorithms is compared with respect to each service broker policy: service proximity, optimized response time, dynamically reconfigure with load, and new dynamic service proximity. Response time of algorithms is analyzed.

7.2 Simulations

Simulations and their results are discussed in this section.

7.2.1 Service Proximity Policy

In service proximity policy, the closest fog is selected to handle user request. Once the closest fog is selected, the user request is routed to that fog. Response Time (RT) is defined as the time from when request is sent to fog and when response is received. The average RT for each cluster of buildings is shown in Fig. 2.

Comparison of Round Robin, Throttled and Weighted Round Robin in terms of response time is shown in Table 1. Results show that in service proximity policy, Throttled give best average response time than RR and WRR.

Table 1. Response time comparison in service proximity policy

Algorithms	Avg (ms)	Min (ms)	Max (ms)
Round robin	51.86	38.36	66.17
Weighted round robin	51.84	38.05	66.21
Throttled	51.79	38.36	66.17

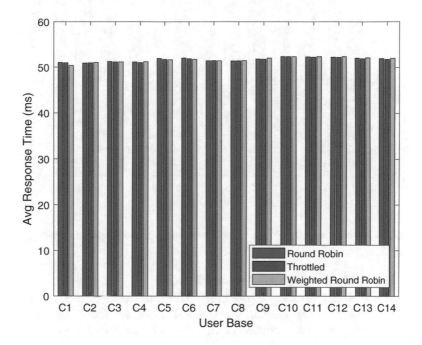

Fig. 2. Average response time comparison in service proximity policy

7.2.2   Optimize Response Time

In this policy, the fog with best response time is select. Once the fog is selected, the user request is routed to that fog. The average RT for each cluster of buildings is shown in Fig. 3.

Comparison of response time of load balancing algorithms Round Robin, Throttled and Weighted Round Robin is shown in Table 2. The results show that Throttled outperforms RR and WRR in terms of average response time.

Table 2. Response time comparison in optimize response time policy

Algorithms	Avg (ms)	Min (ms)	Max (ms)
Round robin	51.80	36.82	67.60
Weighted round robin	51.77	38.59	66.91
Throttled	51.71	36.82	67.60

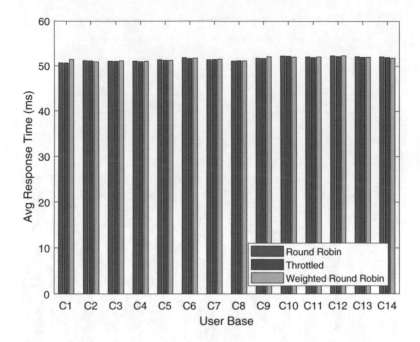

Fig. 3. Average response time comparison in optimize response time policy

7.2.3   Dynamically Reconfigure with Load

This service is similar to service proximity policy. In this service the fog selected based on allocated virtual machines and processing time. Once the closest fog is selected, the user request is routed to that fog. The average RT for each cluster of buildings is shown in Fig. 4.

Comparison of response time of load balancing algorithms Round Robin, Throttled and Weighted Round Robin is shown in Table 3. Results show that in service proximity policy, Throttled give best average response time than RR and WRR.

Table 3. Response time comparison dynamically reconfigure with load policy

Algorithms	Avg (ms)	Min (ms)	Max (ms)
Round robin	52.06	37.96	77.90
Weighted round robin	52.70	45.61	62.20
Throttled	51.95	38.36	66.511

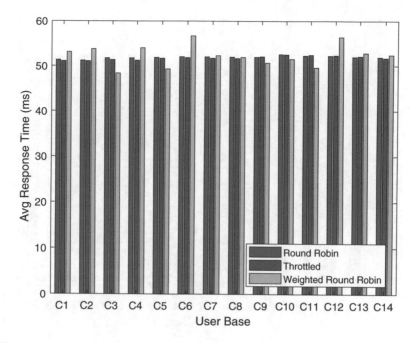

Fig. 4. Average response time comparison in dynamically reconfigure with load policy

7.2.4   New Dynamic Service Proximity

In this service policy the fog is selected on the basis of minimum network delay and allocated virtual machines. Once the fog is selected, the user request is routed to that fog. The average RT for each cluster of buildings is shown in Fig. 5.

Comparison of Round Robin, Throttled and Weighted Round Robin in terms of response time is shown in Table 4. The results in table show that, in optimize response time policy Throttled outperforms RR and WRR in terms of average response time.

Table 4. Response time comparison in new dynamic service proximity policy

Algorithms	Avg (ms)	Min (ms)	Max (ms)
Round robin	52.09	38.36	82.66
Weighted round robin	52.77	45.61	63.65
Throttled RT	51.94	37.96	66.51

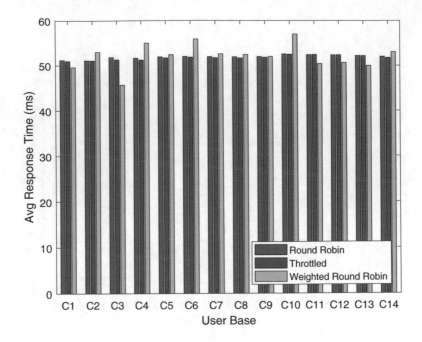

Fig. 5. Average response time comparison in new dynamic service proximity policy

7.3   Results Discussion

Simulations in CloudAnalyst are performed to compare and analyze the performance of load balancing algorithms: Round Robin (RR), Throttled, and Weighted Round Robin (WRR) and service broker policies: Service Proximity Policy, Optimize Response Time, Dynamically Reconfigure with Load, and New Dynamic Service Proximity. Simulation results showed that Throttled load balancing algorithm give better response time than RR and WRR.

8   Conclusion

In this paper, a cloud-fog based model is proposed in Smart Grid environment for efficient utilization of resource and energy management to fulfill the energy demands of residential users. The objective of achieving maximum performance is also formulated mathematically in this paper. CloudAnalyst tool is used for simulations. In simulation, performance of load balancing algorithms: RR, Throttled, and WRR and service broker policies: service proximity, optimized response time, dynamically reconfigure with load, and new dynamic proximity is analyzed. Simulation results showed that Throttled load balancing algorithm give better response time than RR and WRR.

References

1. Vehbi, C.G., Sahin, D., Kocak, T., Ergut, S., Buccella, C., Cecati, C., Hancke, G.P.: Smart grid technologies: communication technologies and standards. IEEE Trans. Ind. Inf. **7**(4), 529–539 (2011)
2. Farhangi, H.: The path of the smart grid. IEEE Power Energy Mag. **8**(1) (2010)
3. Okay, F.Y., Ozdemir, S.: A fog computing based smart grid model. In: 2016 International Symposium on Networks, Computers and Communications (ISNCC), pp. 1–6. IEEE (2016)
4. McKenna, E., Richardson, I., Thomson, M.: Smart meter data: balancing consumer privacy concerns with legitimate applications. Energy Policy **41**, 807–814 (2012)
5. Luo, F., Zhao, J., Dong, Z.Y., Chen, Y., Xu, Y., Zhang, X., Wong, K.P.: Cloud-based information infrastructure for next-generation power grid: conception, architecture, and applications. IEEE Trans. Smart Grid **7**(4), 1896–1912 (2016)
6. Xing, H., Minyue, F., Lin, Z., Mou, Y.: Decentralized optimal scheduling for charging and discharging of plug-in electric vehicles in smart grids. IEEE Trans. Power Syst. **31**(5), 4118–4127 (2016)
7. Dam, S., Mandal, G., Dasgupta, K., Dutta, P.: An ant-colony-based meta-heuristic approach for load balancing in cloud computing. In: Applied Computational Intelligence and Soft Computing in Engineering, pp. 204–232. IGI Global (2018)
8. Chen, S.-L., Chen, Y.-Y., Kuo, S.-H.: Clb: a novel load balancing architecture and algorithm for cloud services. Comput. Elect. Eng. **58**, 154–160 (2017a)
9. Fatima, I., Javaid, N., Iqbal, M.N., Shafi, I., Anjum, A., Memon, U.: Integration of cloud and fog based environment for effective resource distribution in smart buildings. In: 14th IEEE International Wireless Communications and Mobile Computing Conference (IWCMC-2018), pp. 2–6 (2018)
10. Zahoor, S., Javaid, N.: A cloud-fog based smart grid model for effective information management (2018)
11. Aslam, S., Munir, K., Javaid, S., Javaid, N., Alam, M.: A cloud to fog to consumer based framework for intelligent resource allocation in smart buildings (2018)
12. Yasmeen, A., Javaid, N.: Exploiting load balancing algorithms for resource allocation in cloud and fog based infrastructures (2018)
13. Chekired, D.A., Khoukhi, L.: Smart grid solution for charging and discharging services based on cloud computing scheduling. IEEE Trans. Ind. Inf. **13**(6), 3312–3321 (2017)
14. Wan, C., Song, Y., Zhang, Y., Zijian, C., Lin, J., Wang, X.: Optimal cloud computing resources allocation for demand side management in smart grid. IEEE Trans. Smart Grid **8**(4), 1943–1954 (2017)
15. Chen, S.-L., Chen, Y.Y., Kuo, S.-H.: Clb: a novel load balancing architecture and algorithm for cloud services. Comput. Elect. Eng. **58**, 154–160 (2017b)
16. Faruque, M.A.A., Vatanparvar, K.: Energy management-as-a-service over fog computing platform. IEEE Internet Things J. **3**(2), 161–169 (2016)
17. Hossein, M., Moghaddam, Y., Leon-Garcia, A., Moghaddassian, M.: On the performance of distributed and cloud-based demand response in smart grid. IEEE Trans. Smart Grid (2017)
18. Bitam, S., Zeadally, S., Mellouk, A.: Fog computing job scheduling optimization based on bees swarm. Enterp. Inf. Syst. **12**(4), 373–397 (2018)
19. Fang, B., Yin, X., Tan, Y., Li, C., Gao, Y., Cao, Y., Li, J.: The contributions of cloud technologies to smart grid. Renew. Sustain. Energy Rev. **59**, 1326–1331 (2016)

Multi-scale Voting Classifiers
for Breast-Cancer Histology Images

Jakub Nalepa$^{(\boxtimes)}$, Szymon Piechaczek, Michal Myller,
and Krzysztof Hrynczenko

Future Processing and Silesian University of Technology, Gliwice, Poland
jakub.nalepa@polsl.pl,
{szympie318,michmyl139,krzyhry690}@student.polsl.pl

Abstract. Breast cancer is the most pervasive form of cancers in
women, therefore automated algorithms for cancer detection, and analysis of hematoxylin and eosin stained breast-cancer histopathology images
are being actively developed worldwide. In this paper, we propose
multi-scale voting classifiers which operate on clinically-relevant features
extracted from such images, and apply them to classify real-life breast-
cancer data. The extensive experiments (encompassing cross-validation
scenarios backed up with statistical tests) showed that our models deliver
high-quality classification, can be learned quickly, and offer instant
operation.

Keywords: Multi-scale analysis · Classification · Ensemble · Voting

1 Introduction

Breast cancer is still one the main causes of death in women worldwide. Since the
assessment of hematoxylin and eosin (H&E) stained breast-cancer histopathology images is a challenging and very user-dependent task, designing robust
automated approaches for detecting cancers from such images is an important
research stream in medical image processing and analysis. Although conventional image-analysis and machine-learning techniques are still being actively
developed [5], deep learning-powered methods are getting attention [1,4] (however, interpreting the internals of deep neural nets is still problematic, and may
be an obstacle in applying such approaches in clinical practice). In this paper,
we follow the former research pathway, and propose new multi-scale voting classifiers for H&E images, and apply them to the real-life image data within the
BACH breast-cancer challenge.

In the proposed classification engine, various clinically-relevant features are
extracted from (original or color-normalized) *patches*, being the (sub-)images.
Since the size of patches may vary, and our classifiers can operate on various
(and mixed) sized patches, we employ the weighted voting scheme to elab-
orate the final class label for an input image (we tackle four-class classification of histological images, with *normal, benign, in-situ carcinoma,* and *invasive*

© Springer Nature Switzerland AG 2019
F. Xhafa et al. (Eds.): INCoS 2018, LNDECT 23, pp. 526–534, 2019.
https://doi.org/10.1007/978-3-319-98557-2_49

carcinoma examples). An extensive experimental study which involved 10-fold cross-validation, and was coupled with Wilcoxon statistical tests (executed to check the significance of the retrieved results) showed that our approaches offer very high-quality classification, can be trained in short time, and perform instant classification of incoming (unseen) images. Additionally, we analyzed the impact of the pivotal components of our method on its capabilities and behavior.

Our multi-scale voting algorithm is presented in Sect. 2. The experiments are thoroughly discussed in Sect. 3. Section 4 concludes the paper.

2   Proposed Algorithm

In this paper, we propose a voting classifier for breast-cancer histology images. Such microscopy images are to be assigned to one (out of four) class: *normal*, *benign*, *in-situ carcinoma*, or *invasive carcinoma*, according to the cancer type present in the corresponding frame. Our approach consists in extracting multi-scale features to build a training set T, and learning a *voting classifier* (which operates on *patches*, being the parts of an input image). For each patch, its label is elaborated, and the final label for an input image becomes the result of voting of all classifiers. Such a technique leaves us with an opportunity to focus on the analysis of whole images, as well as on the details that can be manifested in smaller patches (therefore, a multi-scale analysis is performed). In the following sections, we present the pivotal components of the introduced method.

2.1   Feature Extraction

Given an input breast-histology image, we build feature vectors by extracting different types of image characteristics, including *shape*, *color*, *wavelet* and *texture* features (importantly, most of the extracted features can be easily clinically interpreted by a reader [5]). Additionally, we exploit the histogram of oriented gradients (HOG) feature extractor (with 9 orientations) [2]. To quantify texture, we utilize two Tamura's features—coarseness and contrast, where the former relates to distances of spatial variations of gray levels (it may be casted to the size of a texel), and the latter measures the variability of gray levels [8]. For shape features, we segment nuclei from an input image in the first step (Fig. 1). An image is converted into the LAB color space (since this color space allows for splitting an image into independent layers of brightness and color information, and can be effectively used for correcting uneven lighting in microscopical images [6]). Afterwards, the AB space is clustered into three clusters (using k-means), and—based on the cluster centroids—we determine which cluster exhibits pixels of nuclei. We perform Otsu's thresholding to discard false positives. Once the nuclei are segmented, the following shape features are extracted:

- Number of all nuclei and the total area of all nuclei,
- Sum of the nucleus areas relative to the size of the patch,

- Min., avg., max., and the standard dev. of the nucleus areas, perimeters, eccentricity $\epsilon = M/m$, where M is the length of the major axis of a nucleus, and m is the length of its minor axis, and solidity $\mathcal{S} = \mathcal{A}_N/\mathcal{A}_{CH}$, where $\mathcal{A}_N$ denotes the nucleus area, and $\mathcal{A}_{CH}$ is the area of its convex hull.

Input image   Conversion to LAB k-means

Fig. 1. Consecutive steps of an example segmentation of nuclei (from left to right).

To extract color features, we convert the input image (RGB) into the HSV space. Then, we append the average and standard dev. of each HSV component of the given patch to the feature vector. An additional color-normalization step may be utilized here [7] (Fig. 2). For wavelet features, an image is converted into grayscale, and the 2D discrete Daubechies wavelet transform is executed. We extract the min., avg., skewness (s), and kurtosis (k) from a filtered image:

$$s = \frac{n}{(n-1)(n-2)} \frac{\sum_{x,y}(v(x,y) - \bar{v})^3}{sd^3}, \tag{1}$$

$$k = \frac{n(n+1)}{(n-1)(n-2)(n-3)} \frac{\sum_{x,y}(v(x,y) - \bar{v})^4}{sd^4} - 3\frac{(n-1)^2}{(n-2)(n-3)}, \tag{2}$$

where $\bar{v}$ and n are the average pixel values and total number of pixels within the region of interest, and sd denotes the standard deviation of pixel values.

Normal   Benign   In-situ   Invasive

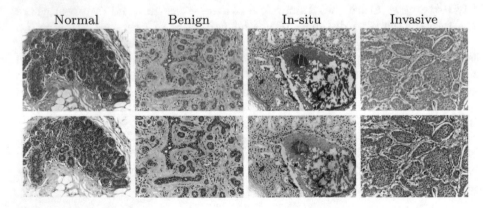

Fig. 2. Example normalization: original (first row) and color-normalized images.

2.2   (Weighted) Voting Classifiers

In this work, we exploit multi-scale classification—for each *patch* in a training input image (i.e., belonging to the training set T), we extract features (as discussed in Sect. 2.1). The training set is fed into a learning procedure of a corresponding classifier, which is later used for labeling unseen frames constituting the validation set V. A process of classifying an incoming image into a class is visualized in Fig. 3 (an optional color-normalization step is visualized in light orange). Here, we present an example of three scales at which features may be extracted (the granularity of patches can be easily increased or decreased, if necessary)—the entire input image (which can be represented as an 1×1 grid of patches), four (2×2 grid), and sixteen (4×4 grid) patches (sub-images). Then, a class label is assigned by a trained model for *each* patch, and the final label c_i (assigned to the i-th input image, and being one out of the following labels: N—normal, B—benign, IS—in-situ, or IN—invasive) is a result of majority (possibly *weighted*) voting, and it becomes:

$$c_i = \arg \max \{\alpha_j \, |c_j|\}, \tag{3}$$

where $\alpha_j \in \{\alpha_N, \alpha_B, \alpha_{IS}, \alpha_{IN}\}$ are the pre-defined weights, and $|c_j|$ is the number of patches assigned with a label c_j, where $c_j \in \{N, B, IS, IN\}$. If all α weights are equal, then we perform a standard majority-voting classification. On the other hand, the *importance* of selected classes may be manifested in their increased weight values. This approach can help minimize the number of false negatives, especially in the context of in-situ and invasive carcinoma in the breast-cancer image classification (the number of false positives for these classes may slightly increase, however it is much more affordable for medical applications).

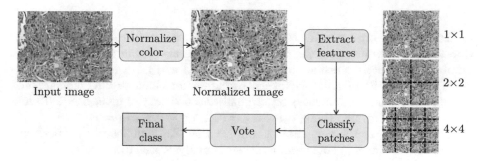

Fig. 3. The process of classifying an unseen (validation) input image.

3   Experiments

The proposed feature extractors and multi-scale classification engines were implemented in Python. The experiments were executed on a computer equipped with Intel Core i3 3130M CPU, 4 GB RAM and 1 TB HDD. In this work, we focused on the BACH dataset[1] composed of 100 normal, 100 benign, 100 in-situ carcinoma, and 100 invasive carcinoma images. To verify the generalization capabilities of our classifiers, we followed 10-fold cross-validation—the entire training set T of 400 images has been divided into a reduced training set T', and a validation set V ten times without overlaps, where $T' \cap V = \emptyset$, and $|V| = 0.1\,|T|$ (the sets are balanced). Since our classification engine is fairly flexible (any base classifiers can be applied), we investigated random forests (RFs), artificial neural nets (ANNs) with two hidden layers (150 and 20 neurons), support vector machines (SVMs) with radial-basis kernels ($\gamma = 1/\,|F|$ and $C = 1$, where F is the number of features), and k-nearest neighbors ($k = 15$) during the experiments.

Feature extraction constitutes an initial step in our breast-cancer image analysis system. In Table 1, we gather the feature extraction times τ_e (divided into all feature types), averaged across all training images (from all classes). The results show that τ_e remains fairly stable across different classes, with the nuclei segmentation being the most-time consuming procedure (due to k-means clustering executed over large, 2048×1536 pixel images). It can be easily accelerated, e.g., by using a more efficient clustering algorithm [3].

Table 1. Average feature extraction times (in seconds) for all image classes.

Class	Nuclei segment	Shape features	Color normal	Color features	Wavelet features	Tamura's features	HOG features	Total time
Normal	35.05	.53	3.01	1.12	.55	6.64	.49	47.33
Benign	35.65	.74	3.13	1.20	.56	7.88	.42	49.59
In-situ	36.14	.54	3.06	1.12	.45	6.69	.49	48.50
Invasive	38.78	.62	2.91	1.15	.47	6.73	.46	51.11

To verify the impact of the color-normalization procedure alongside the selection of base classifiers on the classification accuracy (denoted as η), i.e., the number of correctly classified images, we trained and assessed different classifiers (RF, ANN, SVM, and k-NN; their prime versions indicate classifiers learned using color-normalized images) on original and normalized images in a 10-fold cross-validation fashion (here, $\alpha_N = \alpha_B = 0.4$, $\alpha_{IS} = 0.8$, $\alpha_{IN} = 1$).

The results reported in Table 2 confirm that RF significantly outperforms other base classifiers in almost all multi-scale analysis scenarios (the 8×8 grid is the only exception). Interestingly, color normalization does seem to worsen the classification performance—we executed the Wilcoxon test to double-check if the differences between classifiers learned using features extracted from original and normalized images differ in a statistically significant way, and the null hypothesis saying that "classifiers trained using original and normalized images deliver the same-quality classification" can be rejected (at $p < .0001$). On the contrary, the number of patches in an input image has a big impact on the accuracy (only the

---

[1] https://iciar2018-challenge.grand-challenge.org/dataset/.

Table 2. Classification accuracy of the investigated classifiers.

Grid	RF	ANN	SVM	k-NN	RF′	ANN′	SVM′	k-NN′
10-fold cross-validation								
1×1	**.50 ± .07**	.32 ± .12	.31 ± .07	.33 ± .09	**.50 ± .07**	.32 ± .11	.46 ± .13	.36 ± .08
2×2	**.67 ± .08**	.53 ± .10	.45 ± .09	.42 ± .09	.62 ± .10	.52 ± .07	.38 ± .04	.40 ± .06
4×4	**.67 ± .10**	.61 ± .12	.45 ± .11	.43 ± .10	.64 ± .11	.57 ± .12	.40 ± .07	.42 ± .08
8×8	.67 ± .10	**.68 ± .09**	.40 ± .06	.37 ± .05	.65 ± .09	.64 ± .09	.40 ± .05	.37 ± .07
Training on and classifying the entire training set T								
1×1	**1.00**	.43	**1.00**	.54	**1.00**	.51	.64	.65
2×2	**1.00**	.57	.50	.66	**1.00**	.67	.40	.67
4×4	**1.00**	.76	.52	.77	**1.00**	.79	.42	.80
8×8	**1.00**	.93	.43	.70	**1.00**	.95	.42	.76

ensembles for 4×4 and 8×8 grids do not notably differ, $p = .61$). We also trained our classifiers using T, and all RF-based classifiers delivered perfect operation.

In Fig. 4, we render the training and classification times of all investigated classification engines. We can appreciate that our approach offers almost instant classification for smaller number of patches (up to 4×4). If the number of patches increase, the base classifiers need to be selected with care to deliver

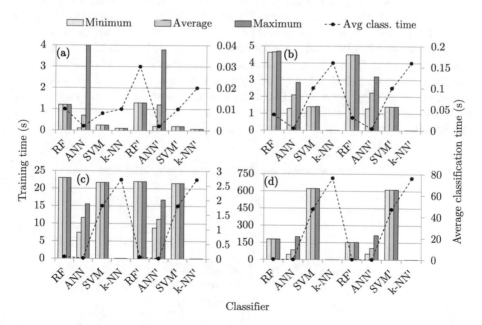

Fig. 4. Training (minimum, average and maximum), and average classification times (in seconds) for the classifiers over the (a) 1×1, (b) 2×2, (c) 4×4, and (d) 8×8 grids.

Table 3. Classification accuracy of the voting classifiers: (a) 10-fold cross-validation, and (b) RF, (c) ANN, (d) SVM, and (e) k-NN trained on and classifying T.

	A	B	C	D	A$'$	B$'$	C$'$	D$'$
(a)	.66 ± .06	**.72 ± .07**	.70 ± .07	.70 ± .06	.61 ± .07	.67 ± .07	.66 ± .07	.66 ± .06
(b)	**1.00**	**1.00**	**1.00**	**1.00**	**1.00**	**1.00**	**1.00**	**1.00**
(c)	.48	.38	—	.38	.40	**.54**	—	.49
(d)	**1.00**	**1.00**	—	**1.00**	**1.00**	**1.00**	—	**1.00**
(e)	.70	.79	—	**.87**	.57	.68	—	.75

not only high-quality classification, but also to be trainable in affordable time (see e.g., SVMs in Fig. 4d, whose training is the most time-consuming—they also perform relatively slow classification, most likely due to a large number of elaborated support vectors). On the other hand, k-NN do not require training, but their classification time easily explodes and becomes infeasible.

To benefit from multi-scale analysis, we integrated patches of several sizes into weighted voting classifiers. Here, we also analyzed the impact of the alpha weights, and our classifiers were as follows (similarly, the prime counterparts were trained on color-normalized images) **A**: (1×1 and 2×2 grids, $\alpha_N = \alpha_B = 0.4$, $\alpha_{IS} = 0.8$, $\alpha_{IN} = 1$), **B**: (1×1, 2×2, and 4×4 grids, $\alpha_N = \alpha_B = 0.4$, $\alpha_{IS} = 0.8$, $\alpha_{IN} = 1$), **C**: (1×1, 2×2, and 4×4 grids, $\alpha_N = \alpha_B = \alpha_{IS} = \alpha_{IN} = 1$), and **D**: ($1 \times 1$, 2×2, 4×4, and 8×8 grids, $\alpha_N = \alpha_B = 0.4$, $\alpha_{IS} = 0.8$, $\alpha_{IN} = 1$). As previously, we performed both 10-fold cross-validation and classification of the entire T (Table 3). The **B** ensemble delivered the best cross-validated η values averaged across all folds, and it performed very well for the full T. Interestingly, classifiers learned from non color-normalized images offers better classification accuracy (at $p = .0002$). Finally, our grid-searched α values helped boost the η values (see e.g., **B** and **C**). The **B** model has been submitted as our final classifier for the BACH breast-cancer challenge. It does not only gives best-quality classification, but can be trained quickly and infers very fast (Fig. 5).

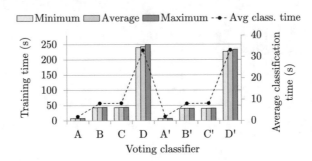

Fig. 5. Training (minimum, average and maximum) and average classification time (in seconds) for weighted voting classifiers in the 10-fold cross-validation scenario.

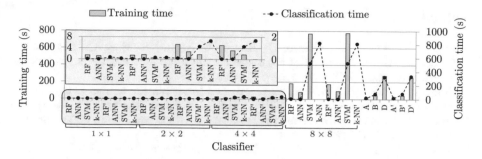

Fig. 6. Training and classification times (in seconds) of all classifiers over the entire training set T. A part of this plot has been zoomed in for clarity.

In Fig. 6, we present the training and operation times of our models over the full training set T. It can be seen that a majority of the voting classifiers (for different granularities of patches) can be trained quickly, and operate fast. One notable exception are the classifiers with SVMs as the base models. This problem can be alleviated by careful selection of refined training sets for SVMs [9] (large T's affect training and classification times of this learner).

4   Conclusions

In this paper, we proposed a weighted voting classifiers which operate on clinically-interpretable features extracted from breast-cancer histopathology images. We applied our technique for classification of a real-life four-class H&E set within the BACH challenge framework. Our experiments, backed up with statistical tests, revealed that the introduced classifiers offer high-quality classification, can be quickly trained and infer really fast. We verified how different aspects of our algorithm affects its performance and abilities.

Our current work focuses on comparing the proposed ensembles with deep neural networks, and on evolving deep architectures and hyperparameters for medical image analysis. It is a pivotal step in building a deep learning-powered engine, since improperly selected architectures can not only deteriorate the classification accuracy but can also be very prone to overfitting (especially for small ground-truth datasets). Finally, we work on applying features extracted using deep nets in our evolved ensembles.

Acknowledgments. This research was supported by the National Centre for Research and Development (POIR.01.02.00-00-0030/15). Calculations were carried out on the cluster (http://www.ziemowit.hpc.polsl.pl) funded by the Silesian BIO-FARMA project (POIG.02.01.00-00-166/08).

References

1. Araujo, T., Aresta, G., Castro, E., Rouco, J., Aguiar, P., Eloy, C., Polonia, A., Campilho, A.: Classification of breast cancer histology images using convolutional neural networks. PLOS ONE **12**(6), 1–14 (2017)
2. Dalal, N., Triggs, B.: Histograms of oriented gradients for human detection. In: Proceedings of the CVPR, pp. 886–893. IEEE Computer Society, USA (2005)
3. Frey, B.J., Dueck, D.: Clustering by passing messages between data points. Science **315**(5814), 972–976 (2007)
4. Han, Z., Wei, B., Zheng, Y., Yin, Y., Li, K., Li, S.: Breast cancer multi-classification from histopathological images with structured deep learning model. Sci. Rep. **4172**, 1–10 (2018)
5. Kumar, R., Srivastava, R., Srivastava, S.: Detection and classification of cancer from microscopic biopsy images using clinically significant and biologically interpretable features. J. Med. Eng. **2015**, 1–15 (2015)
6. Kuru, K.: Optimization and enhancement of H&E stained microscopical images by applying bilinear interpolation method on lab color mode. Theor. Biol. Med. Model. **11**(1), 9 (2014)
7. Macenko, M., Niethammer, M., Marron, J.S., Borland, D., Woosley, J.T., Guan, X., Schmitt, C., Thomas, N.E.: A method for normalizing histology slides for quantitative analysis. In: Proceedings of the IEEE ISBI, pp. 1107–1110, June 2009
8. Majtner, T., Svoboda, D.: Extension of tamura texture features for 3D fluorescence microscopy. In: Proceedings of the IEEE 3DIM, pp. 301–307 (2012)
9. Nalepa, J., Kawulok, M.: Adaptive memetic algorithm enhanced with data geometry analysis to select training data for SVMs. Neurocomputing **185**, 113–132 (2016)

Correction to: Two-Factor Blockchain for Traceability Cacao Supply Chain

Andi Arniaty Arsyad, Sajjad Dadkhah, and Mario Köppen

Correction to:
Chapter "Two-Factor Blockchain for Traceability
Cacao Supply Chain" in: F. Xhafa et al. (Eds.):
Advances in Intelligent Networking and Collaborative Systems,
LNDECT 23, https://doi.org/10.1007/978-3-319-98557-2_30

In the original version of the book, typographical error in second author name "Sajjad Dhadkah" should be corrected to read as "Sajjad Dadkhah" in chapter "Two-Factor Blockchain for Traceability Cacao Supply Chain". The correction chapter and the book have been now updated with the change.

The updated online version of this chapter can be found at
https://doi.org/10.1007/978-3-319-98557-2_30

© Springer Nature Switzerland AG 2019
F. Xhafa et al. (Eds.): INCoS 2018, LNDECT 23, p. E1, 2019.
https://doi.org/10.1007/978-3-319-98557-2_50

Author Index

Printed in the United States
By Bookmasters